Medicinal Plant Biotechnology

The Gayatri Mantra

ॐ भूर्भुवः स्वः तत्सवितुर्वरेण्यं ।
भर्गो देवस्य धीमहि, धीयो यो नः प्रचोदयात् ।।

Aum Bhur Bhuvah Swah, Tat Savitur Varenyam
Bhargo Devasya Dhimahi, Dhiyo Yo Nah Prachodayat

****

Aum: Symbol of the Para Brahman; The primeval sound
Bhur: Physical plane
Bhuvah: Astral plane
Swah: Celestial plane, spiritual world
Tat: Transcedent Parmatman
Savitur: The Sun, Creator, Preserver
Varenyam: Fit to be worshipped; most adorable, enchanting
Bhargo: Remover of sins and ignorance; Glory Effulgence
Devasya: Resplendent; shining
Dheemahi: We meditate upon
Dhiyo: The Intellect; Understanding
Yo: May this light
Nah: Our
Prachodyat: Enlighten; Guide; Impel, Inspire

****

"We meditate on the Divine Mother's glory, Who is the Protector, the basis of all life, Who is self-existent, Who is free from all pains and Whose very contact frees the soul from all troubles, Who pervades the Universe and sustains all, the Creator and Energizer of the whole Universe, the Giver of happiness, Who is the embodiment of Knowledge and Light, Who is Pure and the Purifier of all, let us embrace the Divine Mother, so that She may enlighten our intellects."

****

This book is dedicated with love to Her children
Tanmay and *Geetansh*

Medicinal Plant Biotechnology

Editor

Rajesh Arora, PhD

Scientist
Defence Research and Development Service
Institute of Nuclear Medicine and Allied Sciences
(Defence Research and Development Organization)
Delhi, India

www.cabi.org

CABI is a trading name of CAB International

CABI Head Office
Nosworthy Way
Wallingford
Oxfordshire OX10 8DE
UK

Tel: +44 (0)1491 832111
Fax: +44 (0)1491 833508
E-mail: cabi@cabi.org
Website: www.cabi.org

CABI North American Office
875 Massachusetts Avenue
7th Floor
Cambridge, MA 02139
USA

Tel: +1 617 395 4056
Fax: +1 617 354 6875
E-mail: cabi-nao@cabi.org

A catalogue record for this book is available from the British Library, London, UK.

Library of Congress Cataloging-in-Publication Data

Medicinal plant biotechnology / editor: Rajesh Arora.
p. cm.
Includes bibliographical references and index.
ISBN 978-1-84593-678-5 (alk. paper)
1. Medicinal plants--Biotechnology. I. Arora, Rajesh. II. C.A.B. International. III. Title.

TP248.27.P55M436 2011
615'.32--dc22

2010016962

ISBN-13: 978 1 84593 678 5
CABI South Asia Edition: 978 1 84593 855 0

Commissioning editor: Rachel Cutts
Production editor: Fiona Chippendale

Printed and bound in the UK from copy supplied by the authors by CPI Antony Rowe, Chippenha

Contents

SECTION VII
Bioinformatics: Interfacing Biotechnology and Bioinformatics

SECTION VIII
Molecular Characterization of Medicinal Plants

SECTION IX
Molecular Farming

SECTION X
Leads from Chemistry, Nanotechnology and Pharmacology for Future Biotechnological Interventions

Contributors

Paramvir Singh Ahuja, Institute of Himalayan Bioresource Technology (CSIR), Palampur, Himachal Pradesh, 176061, India

Kenji Akiyama, RIKEN Plant Science Center, Japan

Rajesh Arora, Institute of Nuclear Medicine and Allied Sciences, Brig. S.K. Mazumdar Road, Timarpur, Delhi-110009, India

Amita Bhattacharya, Institute of Himalayan Bioresource Technology (CSIR), Palampur, Himachal Pradesh, 176061, India

Thomas Bley, Institute of Food Technology and Bioprocess Engineering, Dresden University of Technology, Bergstrasse 120, D-01069 Dresden, Germany

Usha Bughani, Laboratory of Drug Discovery, Design and Research, Department of Cell Biology, 455 Whitehead Biomedical Research Building, Emory University School of Medicine, 615 Michael Street, Atlanta, GA-30322-3030, USA

Suman Chandra, National Center for Natural Products Research, School of Pharmacy, The University of Mississippi, MS 38677, USA

Abha Chaudhary, Natural Plant Products Division, Institute of Himalayan Bioresource Technology, Council of Scientific and Industrial Research, Palampur-176061, India

Mahmoud A. ElSohly, National Center for Natural Products Research, School of Pharmacy, The University of Mississippi, MS 38677, USA and Department of Pharmaceutics, School of Pharmacy, The University of Mississippi, MS 38677, USA.

Milen I. Georgiev, Department of Microbial Biosynthesis and Biotechnologies–Laboratory in Plovdiv, Institute of Microbiology, Bulgarian Academy of Sciences, 26 Maritza Blvd., 4002, Plovdiv, Bulgaria

C.M. Govil, Department of Botany, Chaudhary Charan Singh University, Meerut-250005, India

Atul Grover, University School of Biotechnology, Guru Gobind Singh Indraprastha University, Kashmere Gate, Delhi-110403, India

Suvi Häkkinen, Plant Biotechnology, VTT Technical Research Centre of Finland, PO Box 1000, FI-02044 VTT, Finland

Iliana Ionkova, Biotechnology Section, Faculty of Pharmacy, Department of Pharmacognosy, 2 Dunav Str., 1000 Sofia, Medical University of Sofia, Bulgaria

Ankit Jain, University School of Biotechnology, Guru Gobind Singh Indraprastha University, Kashmere Gate, Delhi-110403, India

Harish C. Joshi, Laboratory of Drug Discovery, Design and Research, Department of Cell Biology, 455 Whitehead Biomedical Research Building, Emory University School of Medicine, 615 Michael Street, Atlanta, GA-30322-3030, USA

Deepshikha Pande Katare, Amity Institute of Pharmacy (AIP), Amity University Uttar Pradesh (AUUP), Sector 44, Noida-201303, India

Ikhlas A. Khan, National Center for Natural Products Research, School of Pharmacy, The University of Mississippi, MS 38677, USA and Department of Pharmacognosy, School of Pharmacy, The University of Mississippi, MS 38677, USA

Feroz Khan, Metabolic and Structural Biology Division, Central Institute of Medicinal and Aromatic Plants (Council of Scientific and Industrial Research), Lucknow 226015, India

Amit C. Kharkwal, Amity Institute of Microbial Technology (AIMT), Amity University Uttar Pradesh (AUUP), Block 'E-3', IVth Floor, Sector 125, Noida-201303, India

J. Calvin Kouokam, James Graham Brown Cancer Center, University of Louisville, Owensboro Cancer Research Program, 1020 Breckenridge Street, Suite 201, Owensboro, Kentucky 42303, USA

Hemant Lata, National Center for Natural Products Research, School of Pharmacy, The University of Mississippi, MS 38677, USA

Ajay K. Mathur, Division of Plant Tissue Culture, Central Institute of Medicinal and Aromatic Plants, Council of Scientific and Industrial Research, PO CIMAP, Kukrail Picnic Spot Road, Lucknow 226015, India

Archana Mathur, Division of Plant Tissue Culture, Central Institute of Medicinal and Aromatic Plants, Council of Scientific and Industrial Research, PO CIMAP, Kukrail Picnic Spot Road, Lucknow 226015, India

Abha Meena, Metabolic and Structural Biology Division, Central Institute of Medicinal and Aromatic Plants (Council of Scientific and Industrial Research), Lucknow 226015, India

Jutta Ludwig-Müller, Institute of Botany, Dresden University of Technology, Zellescher Weg 20b, D-01062, Dresden, Germany

Noor Hasima Nagoor, Faculty of Science, University of Malaya, Kuala Lumpur, Malaysia

Pradeep K. Naik, Department of Bioinformatics and Biotechnology, Jaypee University of Information Technology, Waknaghat, Solan-173215, Himachal Pradesh, India

Rofina Yasmin Othman, Head, Centre for Research in Biotechnology for Agriculture (CEBAR) ISB, Faculty of Science, University of Malaya, 50603, Kuala Lumpur, Malaysia

Kenneth E. Palmer, James Graham Brown Cancer Center, University of Louisville, Owensboro Cancer Research Program, 1020 Breckenridge Street, Suite 201, Owensboro, Kentucky 42303, USA

John R. Porter, Department of Biological Sciences, University of the Sciences in Philadelphia, 600 S. 43rd Street, Philadelphia PA 19104, USA

Om Prakash, Division of Biotechnology, Institute of Himalayan Bioresource Technology, PO Box No. 06, Palampur-176019, Himachal Pradesh, India

Simone Reichwein, Technical University Darmstadt, Institute for Botany, Schnittspahnstrae 3-5, 64287 Darmstadt, Germany

Anneli Ritala, Plant Biotechnology, VTT Technical Research Centre of Finland, PO Box 1000, FI-02044 VTT, Finland

Kazuki Saito, RIKEN Plant Science Center, Japan

Tetsuya Sakurai, Unit Leader, Integrated Genome Informatics Research Unit, Metabolomics Research Group, RIKEN Plant Science Center, Japan

Asmita Salil, Laboratory of Drug Discovery, Design and Research, Department of Cell Biology, 455 Whitehead Biomedical Research Building, Emory University School of Medicine, 615 Michael Street, Atlanta, GA-30322-3030, USA

Ashok Sharma, Biotechnology Division, Central Institute of Medicinal and Aromatic Plants (Council of Scientific and Industrial Research), PO CIMAP, Kukrail Picnic Spot Road, Lucknow 226015, India

P.C. Sharma, University School of Biotechnology, Guru Gobind Singh Indraprastha University, Kashmere Gate, Delhi-110403, India

Shri Ram, Department of Bioinformatics and Biotechnology, Jaypee University of Information Technology, Waknaghat, Solan-173215, Himachal Pradesh, India

Bikram Singh, Natural Plant Products Division, Institute of Himalayan Bioresource Technology, Council of Scientific and Industrial Research, Palampur-176061, India

Dharam Singh, Lab No. 419, Institute of Molecular Biology, Academia Sinica, Nankang, Taipei-115 29, Taiwan

Harvinder Singh, Department of Bioinformatics and Biotechnology, Jaypee University of Information Technology, Waknaghat, Solan-173215, Himachal Pradesh, India

Ajit Varma, Amity Institute of Microbial Technology (AIMT), Amity University Uttar Pradesh (AUUP), Block 'E-3', IVth Floor, Sector 125, Noida-201303, India

Rajeshwar Verma, Department of Chemistry, Pomona College, 645 N. College Avenue, Claremont, CA 91711, USA

Heribert Warzecha, Technical University Darmstadt, Institute for Botany, Schnittspahnstrae 3-5, 64287 Darmstadt, Germany

Christoph Wawrosch, Department of Pharmacognosy, Plant Tissue Culture, PharmaCenter, Vienna, Althanstr.14, A-1090, Vienna, Austria

Preface

The ancient physicians of India said and proved that '*there is no plant in this world which is not a medicine*'. Scientists too have begun to realize that if we intend to provide affordable and accessible healthcare globally, we should sustainably utilize the available medicinal plant biodiversity. Plant-derived drugs can be a panacea, provided we utilize them on a scientific basis. Medicinal and aromatic plants (MAPs) are nature's gift and have been used for allaying disease and improving the quality of life for ages. The area of medicinal plants is once again witnessing burgeoning interest in view of their colossal potential to prevent and cure virtually any disease. MAPs continue to and will remain invaluable sources of new chemical entities in future.

In recent years compounds derived from MAPs have been commercially exploited by the pharma, food, flavour, fragrance, dyeing and pesticide industry. The disparate relationship between demand and supply has necessitated conventional and modern research in the field. A paradigm shift towards biotechnological applications is evident since modern biotech tools offer novel opportunities not only for improved production of useful compounds, but also development of designer plants, a feat impossible to accomplish by conventional approaches. The future appears bright, but the proof of the pudding lies in the fact that the designer crops should find commercial application and a long way would have to be traversed before the lab to land to market concept becomes a reality.

The present book has been conceived with the intention of providing the state-of-the-art in the subject and discusses the various facets in the light of contemporary developments. The book is organized into ten major sections.

Section I highlights the emerging trends in medicinal plant biotechnology, emphasizing the current state of expertise available at the hands of researchers for production of a plethora of medicinal compounds in plants cell factories.

Indiscriminate population growth, coupled with urbanization and overharvesting has led to erosion of precious genetic resources. In addition, extensive human intervention has resulted in global climate change, which is taking a heavy toll on natural biodiversity. Further climate change might usher in more vagaries of environment and several medicinal plant species may become endangered or extinct in coming times. The need of the hour is to conserve the MAPs and step up efforts to develop improved diverse environment-tolerant strains. Section II discusses the tools available for conservation of genetic resources, and the potential and opportunities of microsatellite markers in genetic diversity screening.

Section III of this book provides a glimpse of the role micropropagation can play in mass multiplication of elite genotypes of MAPs. The finer nuances of *in vitro* secondary metabolite production form the basis of Section IV.

Hairy roots offer immense potential for the production of secondary metabolites as they are amenable to upscaling in bioreactors. Hairy roots can be utilized for phytoremediation, and via metabolic engineering it is feasible to overexpress genes and produce recombinant proteins. Section V deliberates on the role of hairy roots in secondary metabolite production and discusses the biotechnological applications of plant–microbe interactions.

Genomic and proteomic approaches have been utilized for understanding the intricate mechanisms involved in secondary metabolite synthesis. Section VI discusses important analytical platforms and databases ranging from plant transcriptomics to metabolomics.

A new trend is fast emerging with the hitherto inconceivable amalgamation of computers and biology/biotechnology. Section VII focuses on the bioinformatics tools for virtual screening of anticancer drugs. Though still in a stage of infancy, bioinformatic approaches are going to become the mainstay of medicinal plant biotechnology in future.

Section VIII discusses the potential of molecular tools for characterization of phytodiversity, while Section IX exclusively focuses on molecular farming for production of biopharmaceuticals. Transgenic medicinal plants that produce edible HIV, hepatitis, cholera vaccines and the like will be in great demand in coming years.

Section X deliberates on the use of a multipronged approach in the discipline of medicinal plant biotechnology so that leads from diverse disciplines like chemistry, pharmacology and nanotechnology can be integrated.

The hope and hype associated with developments in the field will have to be confronted head on with a greater degree of maturity in the future.

I must humbly admit that it is impossible to cover all aspects of medicinal plant biotechnology in a small book like this; nonetheless, an attempt has been made to focus on some important recent developments and upcoming trends. It is hoped that the readers will be benefited from this endeavour.

20 October, 2010

Rajesh Arora
(rajesharoradr@rediffmail.com)

The views expressed here and elsewhere in the book are my own and not those of the Government of India.

Foreword

Medicinal plants have been man's best friend for a long time. Early civilizations relied exclusively on the medicinal plants for treating their various ailments and slowly what started as primitive herbal medicine around the world evolved into full-fledged systems like Ayurveda, Unani, Chinese, European, Japanese and Korean systems of medicine, primarily focusing on holistic healing encompassing prophylaxis and therapeutics. The utilization of medicinal plants for a plethora of applications has resulted in renewed interest for they offer practical and safe medical solutions for use by the masses.

Though earth has been bestowed with a rich biodiversity, overexploitation of medicinal plants has led many of them to the verge of extinction. On the other hand, medicinal plants often have to tolerate extreme milieu; they are vulnerable to bacterial/fungal/viral infections themselves and in order to survive continuously evolve and develop mechanisms to synthesize secondary metabolites. However, the yield of secondary metabolites in plants grown in the field is often low at times to be able to meet the burgeoning demand.

It has been realized over the years that plant improvement is essential, especially for obtaining higher yields of high-value, low-volume medicinally useful compounds. Conventional genetic approaches have helped in myriads of ways by improving plant strains and productivity of secondary metabolites. None the less, the need has been felt for improved procedures and protocols for medicinal plant improvement and it is here that plant biotechnology has carved a niche for itself.

Medicinal plant biotechnology continues to be an emerging area and besides effecting improved secondary metabolite production, it has positively impacted conservation of elite strains. Biotechnological interventions have helped in selection, multiplication, improvement and better molecular analysis of elite medicinal plants. Genetic transformation has served as a powerful tool for enhancing the productivity of secondary metabolites and production of novel medicinal plants. Several new strains/varieties of medicinal plants have been developed that are highly resistant to diseases, vagaries of environment, and produce entirely different novel compounds that are not found in nature's repertoire, utilizing rDNA technology and taking leads from advances in genomics and proteomics. Up-scaling of secondary metabolite production in bioreactors has led to reduction of cost to a great extent, while DNA microarrays are increasingly being utilized as high throughput screening tools for the simultaneous analysis of multiple genes and gene expression to gain information on the intricate regulatory molecular pathways. The newly emerging field of bioinformatics of medicinal plants is another area where there is immense scope for research. Edible vaccines have already been produced in plants and will shortly hit the market. In coming years, we might witness several more exclusive designer medicinal plants with multiple traits which could serve not only as a panacea, but also beat climate change.

I am extremely happy to learn that Dr Rajesh Arora has envisaged a book on this important emerging cross-cutting area, covering the latest aspects in the field of Medicinal Plant Biotechnology, by bringing together eminent scientists from across the globe. I am sure the book will be useful to researchers, academicians and students alike.

I congratulate Dr Arora on his endeavour and wish the book all success.

Dr W. Selvamurthy
Distinguished Scientist and Chief Controller Research and Development
Defence Research and Development Organization (Government of India), New Delhi, India

Acknowledgements

I would like to express my deepest sense of gratitude to the following for providing inspiration and support in manifold ways:

Dr V.K. Saraswat
Scientific Adviser to Defence Minister
Secretary and Director General Research and Development (DRDO)
Ministry of Defence, Government of India
New Delhi

Dr W. Selvamurthy
Distinguished Scientist and Chief Controller Research and Development (LS)
DRDO Headquarters
New Delhi, India

Dr R.P. Tripathi
Outstanding Scientist and Director, INMAS, Delhi, India

My family has always been a source of inspiration! Mrs Shanti Arora and Mr Gopal Kumar Arora, my parents, kindled hope in me greatly influencing my thoughts. My wife, Mrs Preeti Arora, has been a pillar of strength and kept pushing me to complete my targets whenever I was indolent. Without her support, this book would not have seen the light of the day. No doubt, I had to ignore my family during this project, but they never minded my deliberate neglect nor expressed it, rather supported me. I can't thank them enough! My bigger family has in no less way contributed to my endeavours. I am thankful to all my other family members and friends for moral support. My teachers were a constant source of inspiration and persuaded me to put in my best.

I deem it my duty to sincerely thank all the contributors, who cooperated wholeheartedly to make this book a success. While I neglected my immediate family, I had another family and obviously what else could I ask for! It was a wonderful experience working with this family too!

Another part of my family has been CABI Publishing. They not only set the deadlines, but also helped me meet it. My sincere thanks go to Ms Rachel Cutts, Associate Editor, CABI Publishing, Wallingford, Oxon, UK for persuading me to compile this book and constantly encouraging me. I am extremely grateful to Dr Nigel Farrar, Book Publisher, CABI, who whole-heartedly supported the publication of this book and became a dear friend, who not only discussed the book but spiritual issues above the human plane. It has been my pleasure knowing him in person. I thank Ms Fiona Chippendale nee Harrison, Production Editor, CABI, for taking care of the finer nuances of publication.

Last, but not the least, I am grateful to Almighty for giving me the requisite strength and enlightening my intellect to compile this book! Like Mother Teresa used to say, I must admit that I have only been a pencil in the hands of a writing God!

October, 2010 Rajesh Arora

The Editor

Dr Rajesh Arora is a senior scientist with India's Defence Research and Development Organization (DRDO). His contributions in the area of novel drug design and development, particularly radiation countermeasure agents of natural origin and augmentation of medicinally useful secondary metabolites using biotechnological interventions have received wide acclaim. Dr Arora's name figures in the *Who's Who of the World*, USA, *Who's Who in Science and Engineering*, USA and *International Biography*, UK. He is a recipient of several prestigious awards and fellowships. Dr Arora has been a Visiting Scientist in the European Union during 2009–10. Dr Arora has published extensively in peer reviewed international journals and has 12 patents and 6 books to his credit. He is a life member of several professional societies and is on the editorial board of numerous journals. Several of his students occupy responsible positions in the government/academia/industry. Dr Arora received his PhD in Biotechnology jointly from Central Institute of Medicinal and Aromatic Plants, Lucknow, and University of Meerut. He also holds post-graduate diplomas in Human Resource Development and Marketing Management and an MBA degree in Human Resource Management. Dr Arora was professionally trained at the Central Drug Research Institute, Lucknow, Sanjay Gandhi Post-Graduate Institute of Medical Sciences, Lucknow, Defence Institute of Advanced Technology, Pune, Defence Institute of Physiology and Allied Sciences, Delhi, and Institute of Technology Management, Mussoorie. Dr Arora also coordinates Technical and Human Resource Development activities for leveraging human capital in cutting edge areas of S&T.

He can be contacted via e-mail: rajesharoradr@rediffmail.com

Chapter 1

Emerging Trends in Medicinal Plant Biotechnology

Rajesh Arora, Archana Mathur and Ajay K. Mathur

Introduction

The consumption of herbal medicines is increasing world over, primarily due to the benefits it offers. As per World Health Organization estimates, nearly 80% of the world's population in developing countries relies primarily on herbal medicine for meeting their healthcare needs (Vines, 2004; Sharma and Arora, 2006; Arora, 2010). Normally, medicinal plants are harvested from the wild, which causes habitat destruction leading to depletion of the irreplaceable genetic diversity. Systematic cultivation of medicinal plants has been proposed and adopted quite often as a substitute to collection from the wild and has the following advantages: (i) optimization of yield and high quality product; (ii) no problem of mistaken identity; (iii) reduction of genetic and phenotypic variability; (iv) the issue of contaminants is taken care of; (v) the problem of variability in extracts and their instability is reduced; and (vi) availability in large quantities without disturbing the natural environment. However, sometimes the spectra of compounds produced by medicinal and aromatic plants (MAPs) under cultivation are quite different from those of wild populations. In order to obtain elite germplasm with enhanced qualities conventional plant-breeding methods have been extensively employed to improve agronomical as well as medicinal traits and molecular marker assisted breeding has been used with substantial returns on investment. None the less, it has been felt that significant improvement of MAPs for meeting the ever increasing demand of the industry in future is possible only by employing biotechnological interventions.

Phytomedicines Drive the Biotech Industry

MAPs have been utilized for the prevention and treatment of diseases for a long time. In recent years, a plethora of drugs have been derived from medicinal plants e.g., reserpine, an antihypertensive alkaloid from *Rauvolfia serpentina*, vinblastine, an antitumour alkaloid from *Catharanthus roseus*, podophyllotoxin from *Podophyllum hexandrum* and *P.*

peltatum, morphine, a narcotic derived from *Papaver somniferum*, cocaine, a local anaesthetic and a potent CNS stimulant from *Erythroxylon coca*, strychnine, a nerve stimulant from *Strychnos nux vomica*, caffeine from tea, coffee beans and cocoa, quinine from *Cinchona ledgeriana* and *C. succirubra*, artemisinin, an antimalarial drug from *Artemisia annua*, atropine, a parasympatholytic agent from *Atropa belladonna*, camptothecin, the chemotherapy drug from *Camptotheca acuminata*, taxol, the anticancer drug from *Taxus brevifolia* and *Taxus baccata*, papaverine from *Papaver somniferum*, silybin from *Silybum marianum*, hyperoside from *Crateaegus laevigata*, anticancer genistein and diadzein from *Glycine max*, hypericin from *Hypericum perforatum* etc. (Sharma and Arora, 2006; Arora, 2010). MAPs are not only important as drugs, but are a repository of future drugs that can solve emerging diseases e.g., AIDS, novel H1N1 flu (Arora *et al.*, 2011) and emerging problems of modernization like chemical, biological, radiological and nuclear (CBRN) threats (Arora, 2008).

Natural plant products have served as important sources of raw materials for food, pharma, cosmetics and nutraceutical industries (Phillipson, 2003; Oksman-Caldentey and Inze, 2004). The medicinal importance of MAPs has necessitated research and this in turn is the driving force for the biotech industry.

Conservation of Medicinal Plants: the Crucial Role of Biotechnology

Natural selection over the course of years has led to the evolution of MAPs with diverse traits. This genetic diversity has been harvested for human use *per se* and also forms the basis for selection of new and improved varieties of MAPs, which researchers utilize for development of new chemical entities with diverse medical applications. There has been a rapid erosion of the important natural resource of medicinal and aromatic plants due to a variety of reasons including: (i) overharvesting from the wild for commercial use since it is cheaper; (ii) rapid and extensive population growth leading to irreparable environmental damage; (iii) extensive industrialization; (iv) deforestation; (v) substitution of local genotypes with improved varieties and hybrids; (vi) forest fires; (vii) natural disasters; (viii) climate change; (ix) development of land for agriculture; (x) changes in agricultural practices; and (xi) abuse of pesticides, weedicides and other agrochemicals etc. Consequently, a rapid and profound erosion of invaluable bioresource has occurred, often leading to loss of several species whose true potential had hardly been explored.

The herbal industry has paid scant attention so far to the efficient utilization and conservation of natural resource in the environment. It is, therefore, imperative to conserve this precious bioresource. Both *in situ* and *ex situ* methods of conservation have been utilized in recent years and are essential. However, it has been unanimously felt that biotechnological interventions are essential for selection, maintenance and multiplication of the elite genotypes of MAPs.

Tissue culture has been widely employed for conservation of MAPs. However, tissue culture-based approaches aimed at *in vitro* conservation of MAPs should be able to efficiently and rapidly multiply and preserve genetic stability of the plant material. Conservation of germplasm using tissue culture techniques can be envisaged either as short- and medium-term conservation or as long-term conservation/cryopreservation. Molecular techniques are increasingly being employed for germplasm characterization. However, a lot needs to be done in this area. The coming years are likely to witness more efforts in the direction of conservation of medicinal and aromatic plants since climate change is taking a heavy toll and several useful plant species are on the verge of extinction.

Chapters 3 and 4 of this book discuss some of the issues relating to *ex situ* and *in situ* conservation of *Podophyllum hexandrum*, an important medicinal plant species, of the Himalayan region.

Micropropagation (Quality Assurance and Quality Control)

Plant tissue culture techniques offer a viable solution for the production of standardized quality phytopharmaceuticals through mass-production of consistent plant material for physiological characterization and analysis of active ingredients. Micropropagation protocols for cloning a variety of MAPs have been developed over the years (Rout *et al.*, 2000; Nalawade and Tsay, 2004; Rathore *et al.*, 2010). Integrated approaches of micropropagation are needed to provide a basis for the development of novel, safe, effective, and high-quality commercial products.

Micropropagation can be utilized for rapid multiplication of elite clones. The advantages offered by micropropagation of medicinal plants are many and are discussed in Chapter 6. Although micropropagation is a multi-billion dollar industry, the share of MAPs as of now is miniscule, though efficient protocols for several MAPs are available. It is anticipated that with the realization of the importance of quality medicinal plants for the pharmaceutical industry and rapid depletion of natural populations, more attention will be paid to this area. Image-guided robotics for automation of micropropagation will help reduce costs, while assuring quality assurance and quality control.

Micropropagation by conventional techniques is a labour-intensive method. Automation of micropropagation in bioreactors has been suggested as an economically viable solution (Cazzulino *et al.*, 1991). Photoautotrophic micropropagation systems (entirely dependent on photosynthesis of plants *in vitro*) usually result in improved growth, lower loss due to contamination, better quality, elevated percentage survival *ex vitro* and lower production costs and have been advocated for commercial production of micropropagated plantlets of medicinal plant species.

Cell and Tissue Cultures: Do They Represent Factories of the Future?

Secondary metabolites are often produced in miniscule quantities in the plant and the ever increasing demands of the industry can hardly be met through this approach. Therefore, attention has been paid to development of biotechnological methods. Plant cell and tissue cultures represent a viable renewable resource of industrially important plant natural products, required for food, drug, fragrance, flavour and dye industries (Phillipson, 2003; Mulabagal and Tsay, 2004; Oksman-Caldentey and Inze, 2004). Some of the exclusive advantages that cell and tissue culture-based production technology provide include the following: (i) control of product supply independent of the geographical availability of plant; (ii) no constraint of seasonal fluctuations; (iii) uniform growth under defined environment; (iv) shorter life cycle free of seasonal or batch-to-batch variation; (v) bio-transformation through feeding and elicitation is possible; (vi) simpler organization provides minimum structural constraints and neighbouring cell interference; (vii) ease with which cultivation conditions can be altered and optimized to up or down regulate the pathway expression; (viii) complications related to long distant transport, permeability and segregation of metabolic pools are minimized; (ix) sterility ensures that biogenesis of target molecule is not mediated through associated microflora prevalent *in vivo*; (x) easy dissection of metabolic steps at interphase of primary and secondary metabolism; (xi) biosynthetic steps expressed at low level or for very limited time in intact plants can be prolonged; (xii) no use of harmful pesticides and herbicides that are generally used during field; (xiii) cultivation can be completely avoided and better quality can be assured; (xiv) industrial upscaling in bioreactor is possible; (xv) less complex downstream extraction processes; and (xvi) reliable and apredictable production under a defined set of conditions.

The experimental strategies that have been followed in attempts to produce increased levels of a desired plant metabolite in cell culture system include: (i) optmization of growth medium for biomass and/or metabolite synthesis and accumulation; (ii) selection of high yielding cell/tissue lines; (iii) designing of a suitable cell or tissue type through genetic transformation with the aim to amplify a pathway gene to augment the carbon flux towards the synthesis of a desired molecule or to block the diversion of a key branch point precursor /intermediate towards a competitive pathway; (iv) defining elicitation factors that can regulate the expression of master regulators (transcription factors) of an intended pathway for global activation of several pathway genes via signal transduction mechanisms; (v) minimization of catabolism (post-biogenetic conversions) of desired metabolite etc.

The commercial successes have not been proportionate to the level of expectations that were generated initially. However, a few notable exceptions are the production of ginsenosides by Nitto Denko Corporation; berberine from *Coptis japonica* and *Thalictrum minus* by Mitsui Petrochemical Industries and paclitaxel from Taxus by ESCA Genetics and PhytonBiotech GmbH, Germany (Dörnenberg, 2008). Besides these, shikonin from *Lithospermum erythrorhizon* (Fujita, 1988) and rosmarinic acid from *Coleus blumei* (Ulbrich *et al*., 1985) have also been reported to be produced at large scale in cell cultures. There can be several reasons for this slow pace of advancement such as poor understanding of the biogenetic pathway of the target molecule(s), low expression of key pathway genes/enzymes in undifferentiated cultures, genetic instability of cells *in vitro*, tissue or organ specific synthesis and/or accumulation of desired product, and too much emphasis on employing a wild type cell or tissue for the hyperexpression of a native pathway under a set of defined culture conditions rather than designing or tailoring a cell or tissue to perform a definite biochemical task (Alfermann and Peterson, 1995; Yeoman and Yeoman, 1996; Wu and Zhong, 1999; Bourgaud *et al*., 2001; Julsing *et al*., 2007). Bioreactors have been utilized for production of secondary metabolites (Wang *et al*., 2010).

Metabolomics and Metabolic Engineering

Recent advances have led to generation of new knowledge and better understanding of the intricate genetic and biochemical mechanisms involved in biosynthesis of plant secondary metabolites. A plethora of genes involved in secondary metabolite synthesis have been isolated and a cascade of steps of metabolic pathways have been delineated. There is general consensus now that to achieve improved production of secondary metabolites, it is imperative to understand the biogenetic pathways in greater detail at gene/enzyme level. The rate limiting steps of targeted pathway can only be overcome if we understand their intrinsic regulation, both *in vitro* and *in planta* (Saito *et al.*, 1992; Verpoorte and Alfermann, 2000; Liao, 2004; Koffas and Cardayre, 2005). These holistic considerations are collectively studied under a new emerging area of research called metabolic engineering (ME). Metabolic engineering takes into account the identification of major blocks or control points of a pathway at molecular level, followed by removal of these limitations with the aid of biotechnological tools (Kutchan, 2005; Hall, 2006; Gomez-Galera *et al.*, 2007; Chang *et al.*, 2007; Petersen, 2007; Schwender, 2008; Allen *et al.*, 2009). A significant increment in the yield of a desired metabolite can be brought about by: (i) optimal induction of genes/enzymes of an intended pathway at the right time and right place; (ii) assured supply of starting precursor(s) and/or limiting intermediates, particularly those that lie at the border line of primary and secondary metabolisms; and (iii) sufficient sink capacity to store the synthesized product (Mathur *et al.*, 2006). Metabolic engineering, therefore, is a sum of all the optimization efforts culminating in the upregulation of required biochemical and genetical expression of a metabolic flux (Tyo *et al.*, 2007).

Metabolic engineering of cultured cells and tissues to provide efficient production technology for phytomolecules allows the search for rate controlling steps of a metabolic pathway that can be engineered for better expression (Verpoorte *et al.*, 2002). The majority of pathway upregulation attempts generally aim to 'push' the metabolic flux of a pathway towards a target end-product by overexpressing the enzymatic steps located upstream in the biogenetic route (Charlwood and Pletsch, 2002; Julsing *et al.*, 2007). However, it has been observed that such manipulations do not normally result in channelization of the flux in the expected direction as native regulatory mechanism of a cell always tries to neutralize the effect of a hyper-expressed gene. An alternative approach advocated is to 'pull' the flux towards a targeted product by upregulating a terminal step of the pathway located downstream in the sequence of biochemical events. This has been elegantly demonstrated by Zhang *et al.* (2004) in hairy roots of *Hyoscymus niger*. Parallel evidences for several other metabolic pathways resulting in better productivity of anthocyanins, flavonoids and terpenes in cell cultures are accumulating (Zhou and Wu, 2006).

Renewed momentum in the area of metabolic engineering has been due to rapid advancements made in the fields of pathway elucidation at enzyme and gene level, '-omic' science interventions to up- or down-regulate the expression of genes, and the ease with which cells can now be engineered with foreign genes for homo-to-hetero-logous expression of a native or novel metabolic pathway (Oksman-Caldentey and Saito, 2005; Terryn *et al.*, 2006; Julsing *et al.*, 2007).

Hairy Root Culture Technology

A plethora of MAPs have been genetically tranformed to yield hairy roots. Hairy roots of several medicinal plant species have been grown in bioreactors not only for improved productivity, but also for *de novo* synthesis of secondary metabolites. The other main applications of transformed root cultures include production of high-value metabolites and biotransformation, phytoremediation, regeneration of whole plants and production of artificial seeds (Giri *et al.*, 2001; Guillon *et al.*, 2006; Georgiev *et al.*, 2007; also see Chapters 9–11 and 13 of this book).

Hairy roots have a number of advantages over cell cultures as a production platform for secondary metabolites. In recent years, several advances have taken place in areas like elicitation, precursor feeding, cellular permeability, trapping of secondary metabolites released in the milieu etc. However, the true potential of hairy roots has so far not been fully harnessed. Some areas that need more attention in the future include studies on culture conditions, metabolic engineering, productivity, secretion, growth in bioreactors, upscaling and harvesting of secondary metabolites. Hairy roots can be employed as a novel system for molecular farming. Hairy roots have been shown to produce a variety of compounds including functional antibodies, fusion proteins, poly (3-hydroxybutyrate), etc. There is a need to develop better upscaling systems for hairy roots so that their true potential can be realized. The production of antimalarial drug Artemisinin has been shown in bioreactors from hairy root cultures of *Artemisia annua*, as have the anticholinergic drugs Atropine and hysocamine from hairy root cultures of *Atropa belladonna* and *Datura stramonium*.

Environmental pollution is on the rise and extensively affects human populations. Physicochemical technologies of environmental cleanup are expensive and, therefore, plants are being viewed as potential agents for phytoremediation. Hairy roots are amenable to genetic transformation and transgenic approaches have been used to study candidate genes that effect pollutant removal. Hairy roots have been used for bioremediation of toxic wastes, including removal of important water and soil pollutants such as metals, explosives, radionuclides, uranium, insecticides and antibiotics (Eapen *et al.*, 2003; Flocco and Giulietti, 2007; Suza *et al.*, 2008; Abhilash *et al.*, 2009). The advantage offered by hairy roots for use in phytoremediation is due to their ability to grow rapidly in microbe-free conditions, providing a greater surface area of contact between contaminant and tissue, and their genetic and metabolic stablity (Gujarathi *et al.*, 2005).

Molecular Farming

The term molecular farming was initially used in the 1980s to refer to the production of high-value, low-volume compounds in transgenic animals, but is now mainly used to refer to the production of pharmaceutically important and commercially valuable proteins in plants (Fisher and Emans, 2000). Pharmaceuticals have been produced in mammalian or bacterial cells. Though the bacterial production system has several advantages, i.e. easy handling, simple fermentation requirements, known genetic make up, short generation time, and the potential to accumulate transgenic proteins to high titres, a major drawback with bacterial systems is their inability to perform post-translational modifications like *N*- and *O*-glycosylation, fatty acylation, phosphorylation and disulphide bond formation, which are often required for proper assembly and functionality of a protein. Plant cells possess: (i) full post-translational modification potential; (ii) have simple growth requirements; and (iii) unrestrained scalability of whole plants in the field. In addition, plants allow targeting of the recombinant proteins produced to different organs or subcellular compartment, thereby allowing improved protection against proteolysis (Benchabane *et al.*, 2008). Another advantage that plants possess is that they do not harbour human or zoonotic pathogens and this makes them a safe model for the production of pharmaceutically important compunds. Over the years, a large number of compounds have been produced in plants. The first therapeutically important recombinant protein to be expressed in plant system was the human growth hormone. Subsequently, more complex proteins were produced e.g, IgG1 antibody in tobacco (Hiatt *et al.*, 1989), human serum albumin in tobacco and potato (Sijmons *et al.*, 1990), human alpha-interferon in rice (Zhu *et al.*, 1994), glucocerbrosidase and human protein C serum protease in tobacco (Cramer *et al.*, 1996), IL-12 in tomato (Gutiérrez-Ortga *et al.*, 2005), recombinant camelid antibody with specificity for cancer associated mucin in *Nicotiana tabacum* (Ismaili *et al.*, 2007), a model full-length humanized monoclonal antibody (Law *et al.*, 2006) andan HIV-neutralizing antibody

(Ramessar *et al.*, 2008) in maize. A variety of edible plants have been utilized to express vaccines and therapeutic antibodies e.g., potato, carrot, spinach, lettuce, tomato, banana (Karg and Kallio, 2009; Sharma and Sharma, 2009).

The use of plants as bioreactors is an emerging area where significant advances have been made. Plant cells are ideal bioreactors for the production and oral delivery of vaccines and biopharmaceuticals. The advantages include low production cost, product safety and easy scale up. Besides, the use of plants obviates the need for purification, cold storage, transportation and sterile delivery. Though plant vaccines have been developed for the past two decades, none has advanced beyond Phase I. Only some plant-made biopharmaceuticals have advanced to Phase II and Phase II human clinical trials (Daniell *et al.*, 2009).

More research is needed in the following areas so that plant-based vaccines can become a reality, *viz.*, development of better functional assays, large-scale production, purification, functional characterization, oral delivery, transgene silencing in nuclear transgenic plants, transgene containment, preclinical and clinical evaluation etc.

The Biotechnology–Bioinformatics Interface

Developments in the area of biotech have been interspersed with advances in computers leading to a new discipline, bioinformatics. Bioinformatics holds immense potential for drug development. Drug discovery from natural products is a complex, time consuming process that involves huge investments. The time taken for development of a candidate drug up to its commercial availability can vary from about 5–30 years and usually involves investments to the tune of several million dollars. Computational biology and bioinformatics approaches have the potential of not only speeding up the drug discovery process, thereby reducing the costs, but also of changing the way drugs are designed. In view of this, a lot of attention is being focused in this area. Bioinformatics is a new and fast evolving field that is being driven by developments and practical application of the various '-omics technologies'. Applied bioinformatic approaches are being used employing different crops and model plants.

Bioinformatics is going to play a key role in the development of new drugs in the future since the amount of data genereated is growing exponentially and there is a parallel growth in the demand for tools and methods in data management, visualization, integration, analysis, modeling and prediction, particulaly relating to sequence, gene expression, protein, metabolite and phenotypic data. Novel databases are being generated for specific species or experimental sets and these databases and tools may greatly assist in the application of medicinal plant biotechnology in future.

Since most medicinal plant experts are unfamiliar with the available bioinformatics methods, tools and databases, there is a need to train the future generation in this emerging area in coming years if the fruits of this technology have to harvested holistically. Some fundamental issues related to biological sequence analyses, transcriptome analyses, computational proteomics, computational metabolomics, bio-ontologies and biological databases need to be explored to fully exploit this emerging technology (Edwards, 2006). Chapter 15 discusses virtual screening of anticancer drugs.

With rapid advances in the development of advanced automated DNA sequencers and allied technology, immense data on gene sequences, proteins, gene and protein expression have been generated over the years. The integration of information technology providers and biologists has resulted in an approach to analyse the data on sequence information stored in databases and this approach is referred to as computational biology, which mainly aims to: (i) identify genes of interest in stored DNA sequences; (ii) develop novel approaches to predict the structure or function of newly discovered protein(s); (iii) develop new protein models; iv) categorize protein sequences into families of related sequences; (v) generate phylogenetic trees by alignment of similar proteins; and (vi) examine evolutionary relationships between proteins.

It is anticipated that in the future bioinformatic tools will be applied to more MAPs since better software will become available that would be more user friendly and less expensive; *in silico* testing will become easy and less complicated, computing speeds, storage capabiities and data sharing will improve, the accuracy and the ability of DNA and protein microarray technologies to detect small differences will improve substantially. This would definitely result in a better understanding of the intricate mechanisms and lead to designing better drugs from medicinal plants.

The Emergence of Nanotechnology

Advances in the field of nanotechnology have begun to impact plant biotechnology. A honeycomb mesoporous silica nanoparticle (MSN) system with 3-nm pores was used to transport DNA and chemicals into isolated plant cells and intact leaves (Torney *et al.*, 2007). Further developments such as pore enlargement and multifunctionalization of these MSNs may offer new possibilities in target-specific delivery of proteins, nucleotides and chemicals in plant biotechnology. Quantum dots (QDs) have been introduced in recent years as a promising new tool in life sciences in view of their unique optical properties (Michalet *et al.*, 2005; Müller *et al.*, 2006). They are highly stable during excitation and have characteristic absorption and emission spectra. These nanoparticles would have widespread applications in plant biotechnology in coming years.

Nanotechnology will impact medicinal plant biotechnology in a major way. With rapid advances in the field of nanotechnology, it is anticipated that micro- and nanoreactors with high surface to volume ratio will find application in the field of medicinal plant biotechnology. Micro- and nanoreactors could be initially used for high-value, low-volume compounds, e.g. pharmaceuticals, nutritional compounds, flavours, fragrances and dyes etc. and later on for bulky production processes. The acceptance of pharmaceuticals produced by rDNA technology and nanotechnology in the future will be an issue where it would be necessary to gain the trust of the consumers based on risk/benefit evaluation.

Conclusion

New trends in biotechnology reveal that although advances in the area of medicinal plant biotechnology have been rapid, there is a need for a multipronged strategy and the amalgamation of contemporary biotechnological tools and high-throughput and ultra-high-throughput genome analysis, transcriptomics, proteomics, metabolomics, nanotechnology and bioinformatics with conventional tools in medicinal plant biotechnology, coupled with leads from diverse areas such as chemical engineering, bioprocess engineering, robotics etc. is likely to yield fruits in coming times. The ability to express genes from biosynthetic pathways in heterologous organisms has paved the way to combinatorial biosynthesis. In future, more advances are likely to be made in the area of plant combinatorial biosynthesis, particularly for new drug development and this would have far reaching consequences in the area of pharmaceuticals due to the ease of production and economic viability. It is also expected that transgenic plants expressing bacterial or mammalian genes involved in xenobiotic metabolism will improve the efficiency and safety of phytoremediation, leading to wider applications. More attention should be paid towards the development of functional genomics resources (EST databases and microarrays) in MAPs, offering new opportunities for improvement of genotypes using different markers and via genetic transformation techniques. Medicinal plant biotechnology is also likely to witness increased application of microarrays for the analysis of gene expression. A number of medicinal compounds have been produced in bioreactors, e.g. rosmarinic acid from *Coleus blumei*, berberine from *Coptis japonica*, ginsenosides from *Panax ginseng*, podophyllotoxin from *Podophyllum hexandrum* and *Linum album*, sanguinarine from *Papaver somniferum* and antitumour drug taxol from *Taxus baccata*. In future, more such compounds may be produced in macro-, micro- or nano-bioreactors. Potential applications of nanotechnology in medicinal plant biotechnology will be in the areas of high-throughput DNA sequencing and nano-fabricated gel-free systems, DNA and protein microarrays and expression profiling, speedy and accurate disease diagnosis, bionanostructures for inserting functional molecules into plant cells, nanobiosensors for manipulation of culture conditions in bioreactors, micro- and nanobioreactors and simulated nanocells for secondary metabolite production. Though too early to predict, it is anticipated that nanoparticles could be produced in fields of genetically engineered crops, possibly MAPs also, leading to 'particle farming'. A cautious monitoring of the developments in the field of medicinal plant biotechnology is also essential to prevent the use of highly toxic medicinal compounds for abetting terrorism. Medicinal plant biotechnology does have a bright future, none the less; for this, scientists and researchers from diverse disciplines, including chemistry and pharmacology, nanotechnology, molecular biology, engineering, physics etc. will have to innovate and synergize their efforts in coming years.

References

Abhilash, P.C., Jamil, S. and Singh, N. (2009) Transgenic plants for enhanced biodegradation and phytoremediation of organic xenobiotics. *Biotechnology Advances* 27, 474–488.

Alfermann, A.W. and Petersen, M. (1995) Natural product formation by plant cell biotechnology. Results and perspectives. *Plant Cell Tissue and Organ Culture* 43, 199–205.

Allen, D.K., Libourel, I.G.L. and Shachar-Hill, Y. (2009) Metabolic flux analysis in plants: coping with complexity. *Plant, Cell and Environment* 32, 1241–1257.

Arora, R. (2008) *Herbal Radiomodulators: Applications in Medicine, Homeland Defence and Space*. CABI Publishing, Wallingford, Oxon, UK, 332 pp.

Arora, R. (2010) *Herbal Medicine: A Cancer Chemopreventive and Therapeutic Perspective*. Jaypee Brothers Medical Publishers (P) Ltd, Delhi, India, 634 pp.

Arora, R., Chawla, R., Marwah, R., Arora, P., Sharma, R.K., Kaushik, V., Goel, R., Kaur, A., Silambarasan, M., Tripathi, R.P. and Bhardwaj, J.R. (2011) Potential of complementary and alternative medicine in preventive management of novel H1N1 Flu (swine flu) pandemic: Thwarting potential disasters in the bud. *Evidence-based Complementary and Alternative Medicine* (doi:10.1155/2011/586506, p.15).

Benchabane, M., Goulet, C., Rivard, D., Faye, I., Gomord, V. and Michaud, D. (2008) Preventing unintended proteolysis in plant protein biofactories. *Plant Biotechnology Journal* 6, 633–648.
Bourgaud, F., Gravot, A., Milesi, S. and Gontier, E. (2001) Production of plant secondary metabolites: A historical perspective. *Plant Science* 161, 839–851.
Cazzulino, D., Pedersen, H. and Chin, C.K. (1991) Bioreactors and image analysis for scale-up and plant propagation. In: Vasil, I.K. (ed.) Scale-up and Automation in Plant Propagation. *Cell Culture and Somatic Cell Genetics of Plants*, Vol. 8. Academic Press, Inc., San Diego, USA, pp. 147–177.
Chang, M.C.Y., Eachus, R.A., Trieu, W., Ro, D.K. and Keasling, J.D. (2007) Engineering *E. coli* for the production of functionalized terpenoids using plant P450s. *Nature Chemical Biology* 3(5), 274–277.
Charlwood, B.V. and Pletsch, M. (2002) Manipulation of natural product accumulation in plants through genetic engineering. *Journal of Herbs, Spices & Medicinal Plants* 9, 139–151.
Cramer, C.L., Wissenborn, D.L., Oishi, K.K., Grabau, E.A., Bennett, S. and Ponce, E. (1996) Bioproduction of human enzymes in transgenic tobacco. *Annals of New York Academy of Sciences* 792, 62–71.
Daniell, H., Singh, N.D., Mason, H. and Streatfield, S.J. (2009) Plant-made vaccine antigens and biopharmaceuticals. *Trends in Plant Science* 14(12), 669–679.
Dörnenberg, H. (2008) Plant cell culture technology – harnessing a biological approach for competitive cyclotides production. *Biotechnology Letters* 30, 1311–1321.
Eapen, S., Suseelan, K.N., Tivarekar, S., Kotwal, S.A. and Mitra, R. (2003) Potential for rhizofiltration of uranium using hairy root cultures of *Brassica juncea* and *Chenopodium amaranticolor*. *Environmental Research* 91(2), 127–133.
Edwards, D. (ed.) (2006) *Plant Bioinformatics: Methods and Protocols*. Human Press, USA, 522 pp.
Fisher, R. and Emans, N. (2000) Molecular farming of pharmaceutical proteins. *Transgenic Research* 9, 279–299.
Flocco, C.G. and Giulietti, A.M. (2007) *In vitro* hairy root cultures as a tool for phytoremediation research. In: Willey, N. (ed.) *Methods in Biotechnology*, Vol. 23; Phytoremediation: Methods and Reviews, Humana Press Inc., Iotowa, NJ, USA.
Fujita, Y. (1988) Industrial production of shikonin and berberine. In: Bock, G. and Hansh, J. (eds) *Application of Plant Cell and Tissue Culture*, Vol. 137, John Wiley and Sons, Chichester, UK, CIBA Foundation Symposium, pp. 228–235.
Georgiev, M.I., Pavlov, A.I. and Bley, T. (2007) Hairy root type plant *in vitro* systems as sources of bioactive substances. *Applied Microbiology and Biotechnology* 74, 1175–1185.
Giri, A., Dhingra, V., Giri, C.C., Singh, A., Ward, O.P. and Narasu, M.L. (2001) Biotransformations using plant cells, organ cultures and enzyme systems: current trends and future prospects. *Biotechnology Advances* 19, 175–199.
Gomez-Galera, S., Pelacho, A.M., Gene, A., Capell, T. and Christon, P. (2007) The genetic manipulation of medicinal and aromatic plants. *Plant Cell Reports* 26, 1689–1715.
Guillon, S., Tremouillaux-Guiller, J., Pati, P.K., Rideau, M. and Gantet, P. (2006) Hairy root research: recent scenario and exciting prospects. *Current Opinion in Plant Biotechnology* 9, 341–346.
Gujarathi, N.P., Haney, B.J., Park, H.J., Wickramasinghe, S.R. and Linden, J.C. (2005) Hairy roots of *Helianthus annuus*: a model system to study phytoremediation of tetracycline and oxytetracycline. *Biotechnology Progress* 21, 775–780.
Gutiérrez-Ortga, A., Sabdivak-Montes, C., de Olivera-Flores, T.J., Santos-Argumendo, L. and Gómez-Lim, M.A. (2005) Expression of functional interleukin-12 from mouse in transgenic tomato plants. *Transgenic Research* 14, 877–885.
Hall, R.D. (2006) Plant metabolomics: From holistic hope to hype to hot topic. *New Phytologist* 169, 453–468.
Hiatt, A., Cafferkey, R. and Bowdish, K. (1989) Production of antibodies in transgenic plant. *Nature* 342, 76–78.

Ismaili, A., Jalali-Javaran, M., Rasaee, M.J., Rahbarizadeh, F., Forouzandeh-Moghadam, M. and Memari, H.R. (2007) Production and characterization of anti-(mucin MUCI) single-domain antibody in tobacco (*Nicotiana tabacum* cultivar Xanthi). *Biotechnology and Applied Biochemistry* 47, 11–19.

Julsing, M.K., Quax, W.J. and Kayser, O. (2007) The engineering of medicinal plants: Prospects and limitations of medicinal plant biotechnology. In: Kayser, O. and Quax, W.J. (eds) *Medicinal Plant Biotechnology. From Basic Research to Industrial Applications*. Wiley-VCH Verlag GmbH & Co.KGaA, Weinheim, pp. 3–8.

Karg, S.R. and Kallio, P.T. (2009) The production of biopharmaceuticals in plant systems. *Biotechnology Advances* 27, 879–894.

Koffas, M. and Cardayre, S.D. (2005) Evolutionary metabolic engineering. *Metabolic Engineering* 7, 1–3.

Kutchan, T.M. (2005) Predictive metabolic engineering in plants: still full of surprises. *Trends in Biotechnology* 23, 381–383.

Law, R.D., Russel, D.A., Thompson, L.C., Schroeder, S.C., Middle, C.M. and Tremainie, M.T. *et al.*, (2006) Biochemical limitations to high-level expression of humanized monoclonal antibodies in transgenic maize seed endosperm. *BBA-General Subjects* 1760, 1434–1444.

Liao, J.C. (2004) Custom design of metabolism. *Nature Biotechnology* 22, 29–36.

Mathur, A.K., Mathur, A., Seth, R., Verma, P. and Vyas, D. (2006) Biotechnological interventions in designing speciality medicinal herbs for twenty first century: Some emerging trends in pathway modulation through metabolic engineering. In: Sharma, R.K. and Arora, R. (eds) *Herbal Drugs: A Twenty First Century Perspective*. Jaypee Brothers Medical Publishers, New Delhi, India, pp 83–94.

Michalet, X., Pinaud, F.F., Bentolila, L.A., Tsay, J.M., Doose, S., Li, J.J., Sundaresan, G., Wu, A.M., Gambhir, S.S., Weiss, S. (2005) Quantum dots for live cells, *in vivo* imaging, and diagnostics. *Science* 307, 538–544.

Müller, F., Houben, A., Barker, P.E., Xiao, Y., Käs, J.A. and Melzer, M. (2006) Quantum dots – a versatile tool in plant science? *Journal of Nanobiotechnology* 4:5 doi:10.1186/1477-3155-4-5.

Mulabagal, V. and Tsay, H.S. (2004) Plant cell cultures– an alternative and efficient source for the production of biologically important secondary metabolites. *International Journal of Applied Sciences and Engineering* 2, 29–48.

Nalawade, S.M. and Tsay, H.S. (2004) *In vitro* propagation of some important Chinese medicinal plants and their sustainable usage. *In Vitro Cellular and Developmental Biology - Plant* 40(2), 143–154.

Oksman-Caldentey, K.M. and Inze, D. (2004) Plant cell factories in post-genomic era: New ways to produce designer secondary metabolites. *Trends in Plant Sciences* 9, 433–440.

Oksman-Caldentey, K.M. and Saito, S. (2005) Integrating genomics and metabolomics for engineering metabolic pathways. *Current Opinion in Biotechnology* 16, 174–179.

Petersen, M. (2007) Current status of metabolic phytochemistry. *Phytochemistry* 68, 2847–2860.

Phillipson, J.D. (2003) Fifty years of medicinal plant research: every progress in methodology is a progress in science. *Planta Medica* 69(6), 491–493.

Ramessar, K., Rademacher, T., Sack, M., Stadlmann, J., Platis, D., Stiegler, G. *et al.*, (2008) Cost-effective production of a vaginal protein microbicide to prevent HIV transmission. *Proceedings of the National Academy of Sciences* USA 105, 3727–3732.

Rathore, M.S., Gehlot, H.S. and Shekhawat, N.S. (2010) Biotechnological approaches for propagation and prospecting of important medicinal plants from Indian Thar Desert. *International Journal of Plant Production* 4(1), 67–72.

Rout, G.R., Samantaray, S. and Das, P. (2000) *In vitro* manipulation and propagation of medicinal plants. *Biotechnology Advances* 18, 91–120.

Saito, K., Yamazaki, M. and Murakoshi, I. (1992) Transgenic medicinal plants: Agrobacterium-mediated foreign gene transfer and production of secondary metabolites. *Journal of Natural Products* 55, 149–162.

Schwender, J. (2008) Metabolic flux analysis as a tool in metabolic engineering of plants. *Current Opinion in Biotechnology* 19, 131–137.

Sharma, A.K. and Sharma, M.K. (2009) Plants as bioreactors: recent developments and emerging opportunities. *Biotechnology Advances* doi. 10.1016/j.biotechnadv.2009.06.004.

Sharma, R.K. and Arora, R. (eds) (2006) *Herbal Drugs: A Twenty First Century Perspective.* Jaypee Brothers Medical Publishers, New Delhi, India, 688 pp.

Sijmons, P.C., Dekker, B.M., Schrammeijer, B., Verwoerd, T.C., van den Elzen, P.J. and Hoekema, A. (1990) Production of correctly processed human serum albumin in transgenic plants. *Biotechnology* 8, 217–221.

Suza, W., Harris, S.R. and Lorence, A. (2008) Hairy roots: from high-value metabolite production to phytoremediation. *Electronic Journal of Integrative Biosciences* 3(1), 56–57.

Terryn, N., Montagu,V., Inze, D. and Goossens, A. (2006) Functional genomic approaches to study and engineer secondary metabolism in plant cell cultures. In: Bogers, R.J., Craker, L.E. and Lange, D. (eds). *Medicinal and Aromatic Plants.* Springer Verlag, Netherlands, pp. 291–300.

Torney, F., Trewyn, B.G., Liu, V.S.-Y. and Wang, K. (2007) Microporous silica nanoparticles deliver DNA and chemicals into plants. *Nature Nanotechnology* 2, 295–300.

Tyo, E.T., Alper, H.S. and Stephanopoulos, G. (2007) Expanding the metabolic engineering toolbox: more option to engineer cells. *Trends in Biotechnology* 25, 132–137.

Ulbrich, B., Wiesner, W. and Arens, H. (1985) Large-scale production of rosmarinic acid from plant cell cultures of *Coleus blumei* Benth. In: Newmann, K.H., Barz, W. and Reinhard, E. (eds.) *Primary and Secondary Metabolism of Plant Cell Cultures.* Springer- Verlag, Berlin, Heidelberg, pp. 293–303.

Verpoorte, R. and Alfermann, A.W. (2000) *Metabolic Engineering of Plant Secondary Metabolism.* Kluwer Academic Publishers, Dordrecht, The Netherlands.

Verpoorte, R., Contin, A. and Memelink, J. (2002) Biotechnology for the production of plant secondary metabolites. *Phytochemistry Reviews* 1, 13–25.

Vines, G. (2004) Herbal harvests with a future: towards sustainable sources for medicinal plants, Plantlife International; www.plantlife.org.uk.

Wang, G.R., Qi, N.M., and Wang, Z.M. (2010) Application of a stir-tank bioreactor for perfusion culture and continous harvest of *Glycyrrhiza inflata* suspension cells. *African Journal of Biotechnology* 3, 347–351.

Wu, J. and Zhong, J.J. (1999) Production of ginseng and its bioactive components in plant cell culture: current technological and applied aspects. *Journal of Biotechnology* 68, 89–99.

Yeoman, M.M. and Yeoman, C.L. (1996) Manipulating secondary metabolism in cultured plant cells. *New Phytologist* 134, 553–569.

Zhang, L., Ding, R., Chai, Y., Bonfill, M., Moyano, E., Caldentey, K.M.O., Xu, T., Pi, Y., Wang, Z., Zhang, Y., Kai, G., Liao, Z., Sun, X. and Tang, K. (2004) Engineering tropane alkaloid biosynthetic pathway in *Hyoscyamus niger* hairy root cultures. *Proceedings of the National Academy of Sciences USA* 101(17), 6786–6791.

Zhou, L.G. and Wu, J.Y. (2006) Development and application of medicinal plant tissue cultures for production of herbal medicines in China. *Natural Product Reports* 23, 789–810.

Zhu, Z., Hughes, K.W., Huang, L., Sun, B.L., Liu, C.M. and Li, Y. (1994) Expression of human alpha-interferon cDNA in transgenic rice plants. *Plant Cell Tissue and Organ Culture* 36, 197–204.

Chapter 2

Medicinal Compounds Produced in Plant Cell Factories

Suvi T. Häkkinen and Anneli Ritala

Introduction

The importance of plants as a source of medicinal compounds is widely acknowledged nowadays, with approximately 80% of the world's population depending on herbal medicine as a primary health care, according to the World Health Organization (Vines 2004). This causes considerable pressure on many species, since the majority of the material is collected from wild plants. Globally, approximately two thirds of the total of 50,000–70,000 plant species used for medicinal purposes are collected from the wild (Schippmann *et al*., 2006), and even in Europe only 10% of the medicinal species used commercially are cultivated (Vines, 2004). Unfortunately, the limited quantity of active metabolites in the plant, slow growth rates and destruction of natural supplies are problems encountered when exploiting plants for medical needs. Thus, alternative sustainable and renewable production systems are urgently needed to protect and preserve plant diversity.

During the 25-year period 1981–2006, a total of 1184 NCEs (new chemical entities) were approved as drugs (Newman and Cragg, 2007). Of these, 52% have a natural product connection and 30% are purely synthetic. It is interesting to note that 52% of all anti-inflammatory drugs and 51% of all anticancer drugs hitherto approved are directly or indirectly derived from natural sources (Newman and Cragg, 2007). A total of 24 unique natural products discovered in the last 35 years were launched on the market, consisting of molecules isolated from soil microorganisms (79%) and plants (21%) (Ganesan, 2008).

Plants have been utilized as medicines for thousands of years. Morphine, produced by the opium poppy, was the first active component isolated from plants in the early 19th century (Samuelsson, 2004). Since then many important drugs, such as artemisinin, atropine, camptothecin, cocaine, codeine, digoxin, papaverine, pilocarpine and podophyllotoxin, have been discovered from plants. Drug discovery from medicinal plants has played an important role especially in fighting against cancer, with most clinical applications of plant secondary metabolites being targeted to cancer treatment (Butler, 2004). Information provided by traditional medicine has been of great value for the discovery of many new drugs and hundreds of pharmacologically active leads for synthetic modifications. Currently, a large number of natural products are produced solely from massive quantities of whole plant parts. The production of pharmaceutically important plant metabolites has been a target for practical application of plant cell cultures for several decades. However, only a few compounds have reached the commercial production scale, including shikonin and paclitaxel.

The first recombinant protein, human serum albumin, produced in plants was reported nearly 20 years ago (Sijmons *et al.*, 1990). Since then, a number of recombinant proteins have been produced in plant cell cultures. The first technical plant-produced recombinant proteins to be marketed were avidin, trypsin and β-glucuronidase. Proof-of-concept has also been established for the plant-based production of many therapeutic proteins including antibodies, blood products, cytokines, growth factors, hormones, recombinant enzymes and vaccines. Currently, several plant-produced pharmaceuticals are in clinical trials and approaching commercial release within the next few years, insulin and glucocerebrosidase being perhaps the most advanced examples.

Selection of the Production System

Choice of the production system is the first and the most important decision to be made and highly depends on the compound to be produced. Taking into consideration the long and expensive path from the initial pharmaceutical/medicinal invention to production on the market, through clinical stages and regulatory approval, the decision concerning the manufacturing system must be made at a very early stage (Fig. 2.1). The manufacturing system carries a long-lasting impact since it affects the viability of the whole chain. The chosen manufacturing process cannot be changed later without substantial economical loss.

Different production hosts offer different pros and cons, and these factors must be evaluated against the properties of the final product and against total production costs. Depending on whether the final product is e.g. a plant-based small molecule or a therapeutic protein, the method chosen has different requirements. Commercial production of plant natural compounds using cell and tissue culture systems is calculated to become viable when a product has high value, with a price exceeding $500–1000/kg (Sajc *et al.*, 2000). Figure 2.2 presents the advantages displayed by various production hosts to be considered when choosing a system for production of biomedical compounds.

Various plant secondary metabolites are used as medicinal compounds, due to their physiological actions. Thus, many efforts have been made to produce them in plant cell cultures under controlled environment in bioreactors, since these compounds often accumulate in very low amounts (less than 1% dry weight) in whole plants. Moreover, their production and accumulation is commonly organ-specific or they are produced only in specific developmental stages. Many plant-derived pharmaceutical compounds are family- or even species specific, as for example alkaloids and terpenoids, resulting in highly versatile designing of production platforms for each species and compound. On the other hand, the applications of recombinant protein production have mainly been performed using tobacco plants as expression hosts. However, along the development of plant cell culture systems, tobacco BY-2 (Nagata *et al.*, 2004) and NT-1 (Nagata and Kumagai, 1999) cell cultures have served as production hosts in a great majority of cases.

In addition to plant cell cultures, alternative production systems for both secondary metabolites and recombinant proteins are harvest and extraction of the desired compounds from whole plants, or the use of microbial cells as production hosts.

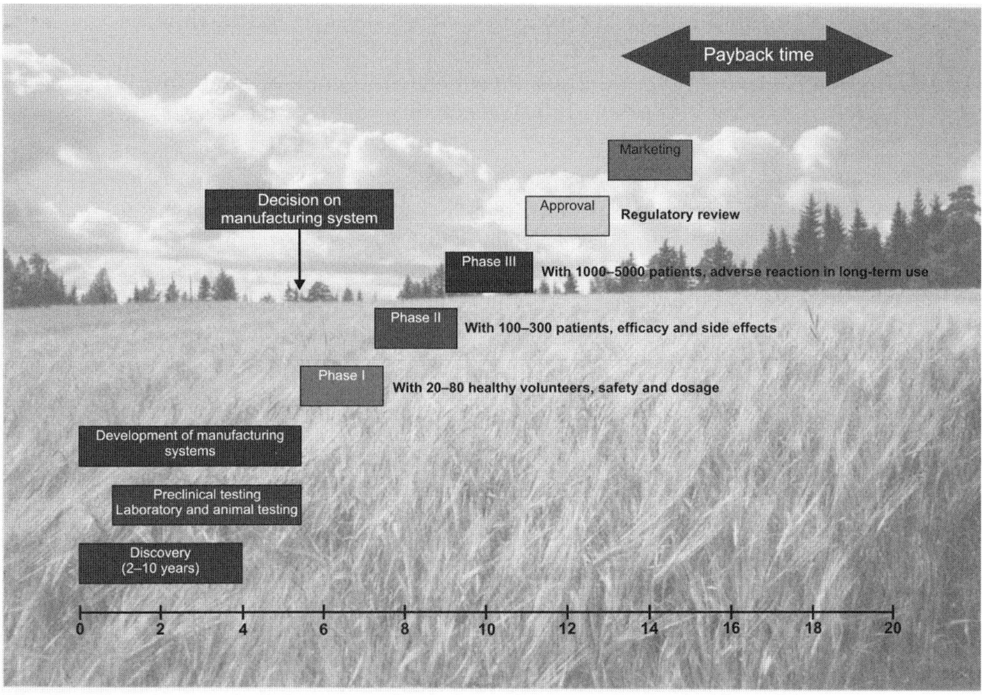

Fig. 2.1. Development of pharmaceuticals from discovery to market launching.

Animal cells have hitherto been the most applied host for recombinant pharmaceutical protein production. On the other hand, secondary metabolites can also be synthesized chemically or via semisynthesis. Each method offers distinctive advantages and disadvantages, depending on the compound of interest, and this will be further discussed in the following sections.

Production in microbial hosts

If possible, the transfer of biochemical plant pathways to microbial cells offers an attractive alternative as an environmetally friendly process. Due to their smaller genome size, the degree of complexity in microorganisms is significantly lower than in plant cells, and this has generated considerable interest in their exploitation for the synthesis of high-value plant metabolites. Microorganisms have fewer intracellular organelles compared to plant cells, and thus metabolite and enzyme compartmentalization can be negligible (Leonard *et al.*, 2009). In addition, the exploitation of microbial cells offers the possibility to use inexpensive carbon sources in production processes, allowing the use of simple, synthetic media. When it comes to large-scale cultivation, microbial cells are less sensitive to shear stress than plant cells, allowing a wider selection of bioreactors. Moreover, the shorter doubling time of microbial cells offers faster production rates.

Microbial hosts also have benefits in downstream processing, since typically products are secreted out of the cells to the surrounding medium, whereas secondary metabolites produced by plant cells are generally accumulated intracellularly. However, a bottleneck in the exploitation of microbial cells in the production of plant-based pharmaceuticals is lack of knowledge of the biosynthetic enzymes and intermediates of the target compounds. For example, reconstruction of the plant biosynthetic pathway of morphine in microbes, starting from tyrosine, would require the functional expression of more than 17 enzymes (Leonard *et al.*, 2009).

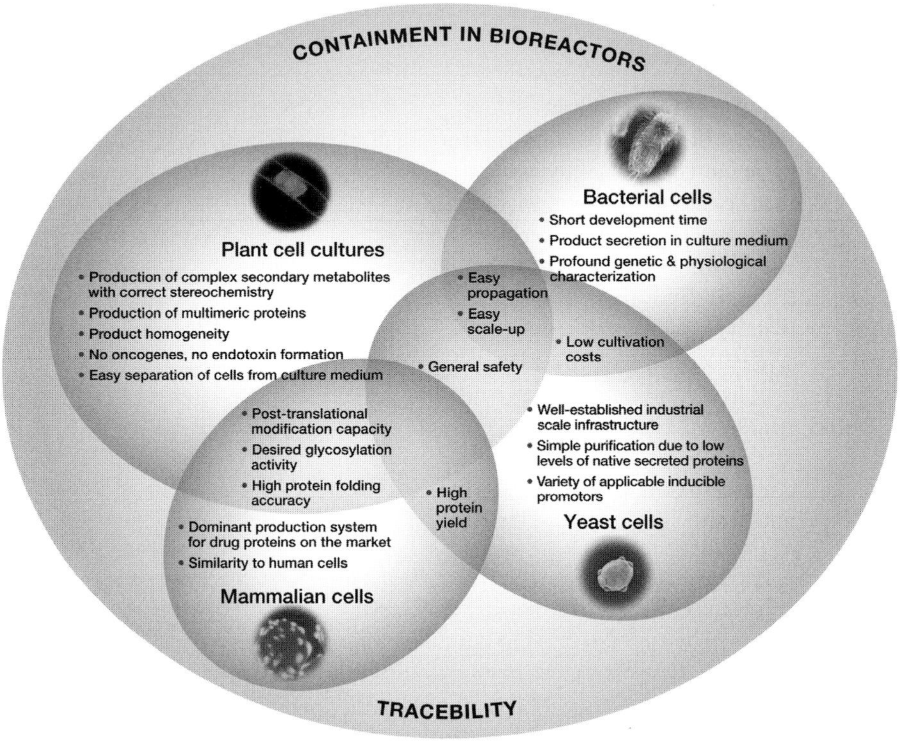

Fig. 2.2. Production of biomedical compounds – advantages offered by different production hosts.

In addition, various enzymatic steps in plant secondary metabolite pathways involve cytochrome P450, for example the biosynthesis of terpenoids, and the expression of redox-partners for P450 enzymes in microbes has hitherto been rather challenging (Khosla and Keasling, 2003). Furthermore, the lack of intracellular S-adenosyl-L-methionine, required for various methylation reactions typical for alkaloid biosynthesis, might become a limiting factor in microbial production systems. Efficient production of a plant-derived high-value drug has been achieved with artemisinin, which is used for the treatment of malaria. Its worldwide demand has led to severe shortages of natural artemisinin sources. Artemisinin has not yet been successfully produced in plant cell or tissue cultures with economical yields. The technology for production of artemisinin and its derivatives in a microbial system combined with chemical synthesis strategies has been established by expressing bacterial, yeast and plant genes in *Escherichia coli* (Chang *et al.*, 2007; Hale *et al.*, 2007) and yeast (Ro *et al.*, 2006). To illustrate the importance of this kind of technology development, the Bill and Melinda Gates Foundation granted $42.6 million funds in 2004 to this synthetic biology approach, aiming to get semi-synthetic artemisinin into poor countries in Asia and Africa by 2012.

Due to its extensive genetic and physiological characterization, short generation time and established cultivation know-how, *E. coli* is the most widely applied prokaryotic host for recombinant protein production. Various pharmaceuticals are being commercially produced with *E. coli*, including somatotropines, insulin and interferon gamma (Schmidt, 2004). However, when it comes to recombinant protein production, being prokaryotes, bacterial cells lack the machinery of plant and animal cells to carry out many desired post-translational modifications, which limits their use for production of a number of medicinal recombinant proteins. In addition, the secretion system in *E. coli* is still not efficient enough to be exploited in the production of many drugs with high volume demands. Like *E. coli*, yeasts can be cultivated rapidly and easily and moreover, their secretion capacity is considerably higher. Besides *Saccharomyces cerevisiae*, which is the best characterized and most extensively used yeast for recombinant protein production, applications using *Pichia* as production host have increased rapidly during the past two decades, with more than 400 proteins reported in 2002 (Lin Cereghino *et al.*, 2002). Examples of commercial pharmaceutical proteins produced by *S. cerevisiae* include hepatitis B vaccine and platelet-derived growth factor (Schmidt, 2004).

Production in animal cells

Animal cell cultures possess the highest similarity to human cells with respect to their ability to carry out post-translational modifications. Today, approximately 60–70% of all pharmaceutical recombinant proteins are produced by mammalian cells (Wurm, 2004). The most frequently applied production host is immortalized Chinese hamster ovary (CHO) cells. Other cell lines, such as mouse myeloma (NS0), baby hamster kidney (BHK), human embryo kidney (HEK-293) and human retinal cells have received regulatory approval for recombinant protein production. Several advances in expression strategies used in mammalian systems have been made, including improvements in vector constructions and selectable marker systems, as well as in gene targeting and high-throughput screening (Andersen and Krummen, 2002). Controllable expression using e.g. tetracycline- or streptogramin-based gene regulation has been demonstrated to be a useful tool for multicomponent control strategies and especially for the expression of products which possess cytotoxic properties (Fussenegger *et al.*, 2000). In addition, recent advances in glycosylation control (Weikert *et al.*, 1999) and reduced lactate accumulation strategies (Chen *et al.*, 2001) have led to improved process performance and product quality. However, the main drawbacks with mammalian production systems are still the relatively expensive media used in culturing, and overall high production costs. Mammalian cells are used for commercial production of e.g. blood coagulation factors, erythropoietin and follitropin (Schmidt, 2004). In addition to mammalian cells, insect cells such as *Trichoplusia* have been efficiently used for production of human collagenase IV. Due to their higher stress resistance and easier handling, insect systems might even be more desirable options for high-throughput protein expression than mammalian cells (Schmidt, 2004).

Chemical synthesis

During the 25-year period 1981–2006, a total of 30% of the new chemical entities approved as drugs were synthetic and 23% were made by semisynthesis (Newman and

Cragg, 2007). Plant-based medicinal compounds can be produced by chemical synthesis if the stereochemistry or chirality of the compound in question does not limit the use of the method. However, medicinal plant compounds are often structurally complex, which makes their chemical synthesis difficult or very expensive. In addition, in many cases synthesis would require the use of harsh solvents, which may ultimately lead to low product yields. Further, problems often encountered in the synthetic substitution of natural compounds are the lack of specificity and efficacy compared to natural compounds. Many natural compounds have important roles as lead molecules which serve as a backbone for synthetic derivatives, such as production of irinotecan (an anticancer compound) from camptothecin. Galanthamine, a drug used in the treatment of Alzheimer's disease, is an alkaloid that was initially isolated from the snowdrop (*Galanthus woronowii* Losinsk.) in the early 1950s, and it has since also been found from other plants in the Amaryllidaceae family (Howes *et al.*, 2003). However, due to the limited availability of plants producing this compound, galanthamine is now produced by total synthesis.

Protein-based drugs are used to diagnose, prevent and treat diseases. The traditional way of manufacturing these proteins was extraction from animal and plant sources. Today, after the recombinant technology revolution, safe and cost-efficient large-scale biotechnological production processes have replaced the old systems. Intense research is continuing to create synthetic proteins with improved functions and stabilities. The long term goal is to develop new tailor-made protein therapeutics that could be administered orally. For this purpose proteins composed of β-amino acids are considered to be one possibility (Petersson and Schepartz, 2008).

Production in plant cell cultures

Use of plant cell cultures offers an attractive alternative for the production of plant-based pharmaceuticals. Compared to field-grown plants, cell cultures can be cultivated in a controlled and contained environment. Production in cell cultures also offers possibilities for production optimization, independently of climatic or environmental effects (Fig. 2.3). Problems often associated with extracts from whole plants include lack of reproducibility in the bioactivities and variation in the biochemical profiles depending on the cultivation time and location. In addition, valuable compounds in aerial parts of the plants are often present in very low amounts and their detection can be even more difficult due to the presence of pigments and polyphenols, which in many cases may interfere in the drug screens (Poulev *et al.*, 2003).

In many cases the production levels of secondary metabolites in undifferentiated cell cultures are very low and often secondary metabolism is strongly linked to cellular differentiation. Moreover, the stability of the production rates in undifferentiated cell cultures is poor. This is partly related to the somaclonal variation and selection of cells during long subculturing. As an example, production of the terpenoid indole alkaloids vincristine and vinblastine has not hitherto been successful in undifferentiated cells and it is recognized that final dimerization of vindoline and catharantine requires cellular differentiation (Samanani and Facchini, 2006).

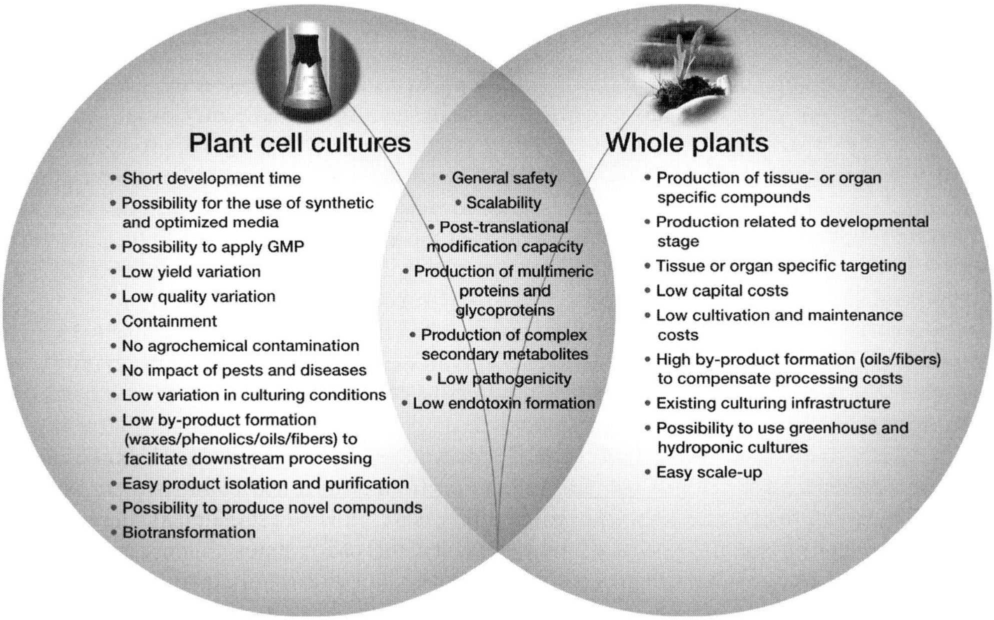

Fig. 2.3. Production of medicinal compounds by plant cell cultures and whole plants.

Hairy root cultures, generated as a result of *Agrobacterium rhizogenes* infection, have proved to be good alternatives as production hosts, producing the same compounds and even in higher quantities than the mother plants (Sevón and Oksman-Caldentey, 2002). In addition, hairy roots gain biomass rather quickly and have simple cultivation medium requirements, being able to grow without phytohormones. They commonly display higher genetic stability and more stable metabolite production than that of undifferentiated cell cultures. Hundreds of plant species have been transformed with *A. rhizogenes* for induction of hairy roots. The majority of cases has involved the use of alkaloid-producing species (Oksman-Caldentey, 2007).

Medicinal compounds produced by plant cell cultures in commercial or semi-commercial scale are presented in Table 2.1. In addition, even though not commercially produced, examples of cell culture systems in which higher production has been reported than in intact plants include diosgenin by *Dioscorea* (Sahai and Knuth, 1985) and ubiquinone-10 by *Nicotiana tabacum* (Fontanel and Tabata, 1987).

Compared to microorganisms, plant cells are larger in size and have a rigid cell wall, which makes them more sensitive to shear stress during bioreactor cultivation and mixing. Another major challenge in exploitation of plant cell cultures as production hosts for secondary metabolites is the high variability often encountered in production rates. Cells often tend to form aggregates composed of several individual cells. The individual cells within aggregates are exposed to differential microenvironmental conditions, which increases the risks for labile production (Roberts, 2007).

In addition, the slow multiplication rate of plant cells limits the rate of product accumulation. Methods to increase the yield of plant secondary metabolites and recombinant proteins obtained by cell culture systems include screening and selection of high-producing cell lines, precursor feeding, optimization of the nutrient medium and cultivation parameters, elicitation and immobilization (Rao and Ravishankar, 2002). As an example, a two-phase cultivation system, in which cells are first cultivated in the medium optimal or preferable for cell growth, and then changed into production medium, can be exploited. This strategy has successfully been exploited in the production of shikonin (Tabata and Fujita, 1985), anthocyanins (Hirasuna *et al.*, 1991), indole alkaloids (Asada and Shuler, 1989) and paclitaxel (Tabata, 2004).

Optimization of Production in Plant Cell Cultures

Although plant cell cultures have been extensively studied for their possibilities to produce medicinal high-value compounds, so far very few applications have reached the stage of commercial production. The problem most frequently encountered is the low production rate and yield, and many efforts have been made to increase production levels in plant cell cultures (Table 2.2). These include optimization of the cultivation medium, especially levels of sugar, nitrate and phosphate, as well as other culturing parameters such as light, temperature, hormones and aeration (see e.g. Fujita and Tabata, 1987; Rao and Ravishankar, 2002). Immobilization of producing cells has been achieved by using calcium alginate beads, stainless steel and foam particles (Hall *et al.*, 1988). Such platforms are still rather expensive and require product accumulation in the culture medium, which is often challenging in plant cell culture systems, since in most cases the products formed in plant cells are stored inside the cells. Nevertheless, the main advantage in immobilized systems is that the producing cells can be separated from product allowing continuous or semi-continuous production. Transient permeabilization of cells has been achieved using e.g. organic solvents or polysaccharides such as chitosan to enhance the release of intracellular product. In addition, metabolic engineering can be exploited for active transportation of products outside the cells, by engineering the specific transporter proteins functioning in secondary metabolism (Yazaki, 2006).

Secondary metabolite production in plants is often highly inducible, and this can be exploited by eliciting the cultures with elicitor compounds, such as jasmonates and their derivatives, salicylic acid, chitosan, heavy metals or with biotic elicitors, such as bacterial or fungal extracts. Jasmonates, octadecanoid-derived signalling molecules, are capable of inducing the biosynthesis of a wide range of secondary compounds, such as alkaloids, terpenoids, flavonoids and anthraquinones (Gundlach *et al.*, 1992). In addition to the production enhancement, elicitor induction has been extensively used for modern functional genomics-based strategies to generate differential transcriptional expression of genes involved in the biochemical pathway of the product (Yamazaki and Saito, 2002; Goossens *et al.*, 2003; Achnine *et al.*, 2005; Rischer *et al.*, 2006; Saito *et al.*, 2007). These studies have led to many advances in mapping of plant secondary metabolite pathways, including biosynthesis of anthocyanins, terpenoids and alkaloids.

Increased product formation has also been achieved by feeding cultures with putative product precursors or intermediates. As an example, phenylalanine addition caused increased production of rosmarinic acid in cell cultures of *Salvia officinalis* and *Coleus blumei* (Ellis and Towers, 1970; Ibrahim, 1987), and of paclitaxel in *Taxus cuspidata* cultures (Fett-Neto *et al.*, 1994).

However, feeding putative precursors, e.g. amino acids or even more proximate precursors to cultures does not always result in elevated levels of the end products (Lockwood and Essa, 1984; Hamill *et al.*, 1990). This may be due to the complex metabolite regulation and feedback systems in plant cells, or sometimes to the high toxicity of the administered precursor (Robins *et al.*, 1987; Chintapakorn and Hamill, 2003).

Table 2.1. Commercially or semi-commercially cell culture-produced plant medicinal compounds (adapted from Eibl and Eibl, 2002, 2008; Basaran and Rodríguez-Cerezo, 2008).

Secondary compounds	Compound	Species	Manufacturer	Reference
	Catharanthine	*Catharanthus roseus*	Mitsui Chemicals (Japan)	Misawa, 1994
	Echinacea polysaccharides	*Echinacea purpurea, Echinacea angustifolia*	Diversa (Germany)	Ritterhaus *et al.*, 1990
	Geraniol	*Geranium* sp.	Mitsui Petrochemicals Ind. (Japan)	Misawa, 1994
	Paclitaxel	*Taxus brevifolia, Taxus chinensis*	ESCAgenetics (USA), Phyton Biotech(USA/ Germany), Nippon Oil (Japan)	Smith, 1995; Tabata, 2004
	Podophyllotoxin	*Podophyllum* sp.	Nippon Oil (Japan)	Misawa, 1994
	Protoberberines	*Coptis japonica*	Mitsui Petrochemical Ind. (Japan)	Sato and Yamada, 1984
	Rosmarinic acid	*Coleus blumei*	Nattermann (Germany)	Ulbrich *et al.*, 1985
	Scopolamine	*Duboisia* sp.	Sumitomo Chemical (Japan)	Misawa, 1994
	Shikonin	*Lithospermum erythrorhizon*	Mitsui Petrochemicals Ind. (Japan)	Fujita and Tabata, 1987; Payne *et al.*, 1991
Recombinant proteins	Glucocerebrosidase	*Daucus carota*	Protalix Biotherapeutics (Israel)	Shaaltiel *et al.*, 2007
	ADDC[1]-mediated antibodies	*Physcomitrella patens*	Greenovation (Germany)	Nechansky *et al.*, 2007; Schuster *et al.*, 2007

[1] Antibody-dependent direct cytotoxicity

The other method exploiting the cell's own metabolic capacity to convert available substrates into more valuable end-products is called biotransformation (Hamada *et al.*, 1998; Giri *et al.*, 2001). Plant cells have an excellent capacity to catalyze stereospecific reactions, resulting in optically pure compounds. In addition, they can carry out regiospecific modifications that are not easily achieved by chemical synthesis or by e.g. microorganisms. As an example, β-methyldigitoxin has been converted into much more valuable cardiac glycoside β-methyldigoxin by *Digitalis lanata* cell cultures (Alfermann *et al.*, 1980), and coniferyl alcohol into podophyllotoxin by *Podophyllum hexandrum* cell cultures (Van Uden *et al.*, 1995).

Table 2.2. Strategies to enhance the production of pharmaceuticals in plant cell cultures.

Secondary metabolites	Recombinant proteins
Selection of high-producing cell lines	Codon optimization of foreign gene sequences
Growth medium and environment optimization	Protein fusions to improve stability
Immobilization of cells	Protein targeting
Use of elicitors	Use of inducible promoters
Precursor feeding	Growth medium and environment optimization
Permeabilization of cells	Use of protein antidegradation agents
Metabolite adsorption	Use of protein synthesis stimulants
Metabolic engineering	Use of secretion inhibitors (brefeldin A)

Recombinant proteins have been produced by plant cell cultures with levels up to 4% of total soluble protein, although the productivities can vary greatly (Hellwig *et al.*, 2004). Whether the product is produced inside the cells or secreted to the culture medium is an important factor affecting the downstream processing. In some cases, downstream processing can cause the majority of the overall process expenses. However, unlike in the case of secondary compounds, for which product secretion in the culture medium is generally a considerable benefit, secretion may seriously alter the stability of the produced recombinant protein due to protein degradation in the culture medium (Tsoi and Doran, 2002). Hence, sometimes protein targeting into the endoplastic reticulum may result in higher levels than in the case of secretion to the medium (Schouten *et al.*, 1996; Streatfield, 2007). On the other hand, recovery of the product from plant cell suspensions is commonly considered to be easier than from microbial suspensions, due to less interfering proteins in the supernatant. Certain stabilizing agents including dimethylsulfoxide and polyethylene glycol or secretion inhibitors such as brefeldin A added to the culture medium may improve the product recovery (see Hellwig *et al.*, 2004 and references therein).

Plant Metabolic Engineering

The major challenge in efficient production of plant secondary metabolites is the discovery of the biosynthetic steps leading to the final product formation. Hitherto, only a few pathways have been well characterized, including flavonoid and terpenoid indole alkaloid (e.g. vincristine) as well as isoquinoline alkaloid (e.g. berberine, morphine) pathways (Grothe *et al.*, 2001; Winkel-Shirley, 2001; Samanani and Facchini, 2002; Hashimoto and Yamada, 2003). These achievements have been reached as a result of very long and laborious research work. Although metabolic engineering has been very extensively applied in microbes, the properties of higher eukaryotic cells such as plant cells set unique challenges, which have complicated the discovery of plant metabolic pathways. In plants, the metabolic pathways are often long and coordinately regulated by several enzymes. Furthermore, the enzymes often have high specifities for their substrates, which can be difficult to obtain, and they have low abundance in plant cells and are sometimes very labile for use in research purposes. In order to successfully perform metabolic engineering in plant cells, the regulation of multiple steps in parallel or engineering of the regulatory genes, such as transcription factors, controlling the complete metabolic pathways is needed.

Metabolic profiling has become an integral part of plant functional genomics (Fiehn *et al.*, 2000; Oksman-Caldentey and Saito, 2005). However, until recently, almost all plant metabolomics studies have been applied to primary metabolites. Compared to primary compounds, profiling of secondary compounds is far more challenging, due to their highly divergent chemical structures and sensitivities in extraction conditions. The huge variety of different chemical structures, possessing a range of physical and chemical properties, sets great challenges for analytical tools when profiling multiple metabolites in parallel (Oksman-Caldentey *et al.*, 2004). Currently, no single analytical technique provides the ability to profile the complete metabolome and this obstacle has been addressed by using selective extraction and combination of analytical platforms. The key for understanding pathway regulation is to define intermediates and to measure flux through the pathway. By fluxomics approaches integrated with transcriptomic data, a novel means for mapping pathways at the systems level from gene to metabolite can be created.

In the case of medicinal plants, the first breakthrough example of metabolic engineering was achieved by Yun *et al.* (1992). They cloned the gene *H6H* (hyoscyamine-6β-hydroxylase) from *Hyoscyamus niger* and transferred it into *Atropa belladonna*, a well-known hyoscyamine-producing species. As a result, the majority of the hyoscyamine was converted into scopolamine. Later, this finding resulted in the engineering of this final step in the tropane alkaloid pathway in various other Solanaceae species (Jouhikainen *et al.*, 1999; Zhang *et al.*, 2004; Oksman-Caldentey, 2007). Recent development of functional genomics tools as well as the sequencing of the full genome of model plants such as *Arabidopsis* and *Medigaco*, and crops such as rice, maize, soybean and many others in the near future, can ameliorate the pathway elucidation by a systems biology approach. However, for many medicinal plants, there is only very limited or no existing data of ESTs or entire genome sequences. For these purposes, differential display methods, such as cDNA-AFLP (Yamazaki and Saito, 2002; Goossens *et al.*, 2003; Rischer *et al.*, 2006) have proved to be excellent tools. The advantages for genome-wide expression analysis methods such as cDNA-AFLP include the possibility for quantitative transcript profiling and, moreover, it is applicable to any organism without the need for prior gene sequence information. It also allows the discovery of completely novel genes, which is very important when discovering unknown steps in the biosynthesis. Recent advances in high-throughput genome sequencing methods, such as pyrosequencing (Ronaghi *et al.*, 1998; Vera *et al.*, 2008), will allow novel means for gene discovery, giving close to one million 400 bp sequences per run with the possibility to analyse multiple samples in parallel.

Case: Functional genomics to discover alkaloid biosynthesis in two Solanaceae species

In our studies, we aimed to improve understanding of the regulation of alkaloid biosynthesis in two Solanaceae species, *Nicotiana tabacum* and *Hyoscyamus muticus*, by establishing a functional genomics-based technology. We applied cDNA-AFLP transcript profiling in combination with targeted metabolite analysis in order to map the biosynthetic genes involved in the tobacco alkaloid pathway. In these studies, *Nicotiana tabacum* BY-2 cell suspension culture was used and the alkaloid production was elicited in these cultures with methyl jasmonate (Goossens *et al.*, 2003). Altogether 459 unique genes were found to be differentially regulated by methyl jasmonate, and homology searches performed with these genes revealed that 58% of the genes displayed similarity with already known genes, whereas no homology was found for 26% of the genes. Alkaloid accumulation took place 12 h after methyl jasmonate application with varying kinetics, and in addition to common tobacco alkaloids such as nicotine, anabasine and anatabine, anatalline was also shown to accumulate in tobacco cell culture after elicitation (Häkkinen *et al.*, 2004). Interestingly, nicotine was only a minor alkaloid accumulating in these cultures, whereas anatabine and anatalline formed the main alkaloid pool. However, the limiting step in nicotine biosynthesis was not observed to be very early in the pathway, since *N*-methylputrescine was produced in high amounts. This result suggested that the limiting step in nicotine biosynthesis occurred after *N*-methylputrescine. These results were supported by the recent findings of Shoji and Hashimoto (2008), who showed that BY-2 cell cultures, contrary to the case in tobacco hairy roots, exhibit very low activity for MPO (*N*-methylputrescine oxidase), the next step in the pathway resposible for *N*-methylputrescine conversion.

Functional testing of the selected genes found from tobacco BY-2 was performed by overexpressing these genes both in tobacco and in the related species *Hyoscyamus muticus*. Two novel genes were found to activate tobacco alkaloid metabolism either in undifferentiated cultures or in differentiated hairy root cultures (Häkkinen *et al.*, 2007). Interestingly, one of the genes also caused a marked increment in tropane alkaloid derivative production in *H. muticus* hairy roots. This result suggests that this gene might have a broader function in alkaloid metabolism in Solanaceae. On the other hand, when the tropane alkaloid gene *H6H* was overexpressed in *Nicotiana tabacum* hairy roots, 45% of the hyoscyamine supplied to the culture was converted into scopolamine (Häkkinen *et al.*, 2005). Most interestingly, up to 85% of the produced scopolamine was secreted out of the cells. This is noteworthy, since tropane alkaloids are commonly retained in the root tissue in hairy roots of endogenous tropane alkaloid producers (Hashimoto *et al.*, 1993; Jouhikainen *et al.*, 1999). The transport mechanism of scopolamine in *Hyoscyamus* and *Nicotiana* is not known, although it is likely that different transporter proteins are involved in the two species because scopolamine is an endogenous product in the former but a foreign metabolite in the latter. However, recently the set of tobacco BY-2 genes, which we discovered earlier by the cDNA-AFLP analysis, revealed the presence of a MATE-type transporter involved in alkaloid transport and it was co-regulated by methyl jasmonate in a similar manner as other tobacco alkaloid biosynthetic genes (Morita *et al.*, 2009). Earlier, it has been shown that the alkaloid berberine is transported by an MDR-type transporter in *Coptis japonica* (Shitan *et al.*, 2003), for which berberine is an endogenous metabolite, whereas in *Arabidopsis* another type of transporter MATE is involved (Li *et al.*, 2002). Thus, due to its ability to take up and convert exogenously applied hyoscyamine, *Nicotiana* could be of interest when designing a candidate production host for tropane alkaloids or possibly for other secondary compounds which are currently still being extracted from whole plants for commercial production.

Case: Production of recombinant collagen and gelatin-related proteins with predetermined composition and structure

Collagen and its denaturated form gelatin are used extensively in various applications in the pharmaceutical industry. Annually over 50 thousand metric tonnes of gelatin are processed to capsules and tablets, plasma expanders, stabilizers for biologics, and to structural components for tissue engineering purposes. Traditionally gelatin is extracted from animal tissues resulting in inhomogeneous products consisting of molecules having different molecular weights and gelling properties. Particularly in the case of medical applications it is important to have collagen-related products that are homogeneous and have a defined composition; thus there is a need to apply recombinant technology for their manufacturing.

The production of recombinant collagen and collagen fragments with designed and predetermined characteristics has been demonstrated in mammalian cells, insect cell cultures, yeast, milk of transgenic animals and in tobacco plants (e.g. Geddis and Prockop, 1993; Vuorela *et al.*, 1997; Nokelainen *et al.*, 1998; John *et al.*, 1999; Ruggiero *et al.*, 2000; Olsen *et al.*, 2003). In our studies, we aimed to produce a recombinant full-length collagen type I α1 (rCIa1) and a related 45-kDa rCIa1 fragment in barley (Ritala *et al.*, 2008; Eskelin *et al.*, 2009). We selected barley to be the production host because barley is an important crop plant with well established agronomic, harvesting, transport, storage and processing practices. Barley has been adapted to a wider variety of climates than any other cereal, from sub-Arctic to sub-tropical areas. Barley is a self-pollinating plant and our earlier study has verified that gene flow through pollen dispersal is extremely low (Ritala *et al.*, 2002). In addition, barley stores large amounts of protein in its seed endosperm and desiccated mature seeds can be stored for several years without substantial protein degradation or loss of activity.

Our first study (Ritala *et al.*, 2008) was designed to demonstrate the feasibility of barley cell culture as a rapid and efficient means of prototype characterization. Both transient and stable expression modes were applied to predict whether barley is a suitable host for production of recombinant collagen and gelatin. The rCIa1 accumulation was targeted to the endoplasmic reticulum (ER) by fusing the CIa1 gene to an ER-directing signal peptide sequence and an ER retention signal HDEL. The gene delivery was carried out by particle bombardment into immature barley half embryos or barley cell suspension. Preliminary evaluation of the suitability of barley as a production host for rCIa1 was obtained by using a transient expression system. Less than 1 week was required to determine expression, avoiding the time-consuming and labour-intensive production of stably transformed suspension cell lines and genetically modified plants. After the preliminary transient testing, stable transgenic barley cell lines were produced and grown as suspension cultures in flasks and in a Wave bioreactor. The best producing cell line synthesized rCIa1 similar to that purified from the yeast *Pichia pastoris* based on Western blotting, pepsin resistance and mass spectroscopy analyses. However, the barley cell culture-derived CIa1 intracellular accumulation levels were low, ranging from 2 to 9 μg/l. The other published barley cell culture production system reached a level of 0.5 mg/l secreted heat-stable endo-β-glucanse (Aspegren *et al.*, 1995). It is difficult to compare productivities of different systems since e.g. the nature of the product itself, factors regulating the gene expression, integration sites and copy numbers all determine the overall outcome. Nevertheless, the obtained production levels illustrated the need for further process improvement in order to use this technology to supply material for product development activities.

The promising results obtained using barley cell culture as a production host for rCIa1 motivated us to carry on to produce a related 45-kDa rCIa1 fragment in barley seeds (Eskelin *et al.*, 2009). The 45-kDa rCIa1 fragment expression, also targeted to the ER, was controlled by three different promoters (a constitutive maize ubiquitin, seed endosperm-specific rice glutelin and germination-specific barley alpha-amylase fusion) in order to compare their effects on 45-kDa rCIa1 fragment accumulation. Highest accumulation of the 45-kDa rCIa1 was obtained with the glutelin promoter (140 mg/kg seed), whereas the lowest accumulation was obtained with the alpha-amylase promoter (2 mg/kg seed). The study of Qu and Takaiwa (2004) showed that the glutelin promoter GluB1 also used in our experiments was among the weakest. Thus it is possible to increase the expression level by selecting a stronger promoter carrying high transcriptional activity. Mass spectroscopy and amino acid composition analysis of the purified 45-kDa rCIa1 fragment revealed that a small percentage of prolines were hydroxylated, with no additional detectable post-translational modifications. In conclusion, we successfully expressed recombinant collagen and gelatin in barley. With the obtained accumulation levels we calculated that by dedicating 10% of the land area under barley cultivation in Finland we could fulfill the estimated global annual need of specialized collagens.

Discussion

From the point of view of the pharma-industry, plants are and will remain an important source of new drugs, and also have a significant role as a source of lead compounds for the design and development of novel pharmaceuticals for research and development (Wang, 2008). The plant kingdom possesses a remarkable collection of efficient and highly active medicinal compounds, the majority of them yet uncharacterized. More than 200,000 secondary metabolite structures have hitherto been discovered from plants (Pichersky and Gang, 2000). The majority of plant-derived compounds currently undergoing clinical trials are directed to the oncological area, including new analogs of known anticancer drugs based on the camptothecin-, taxane-, podophyllotoxin-, or vinblastine-type skeletons (Butler, 2005; 2008; Arora *et al.*, 2008; Arora, 2010). It remains to be seen whether the current global economic crisis will slow down the breakthrough of biopharmaceuticals, but it has been forecast that their total value will be \$70 billion in 2010 (Walsh, 2006). However, biopharmaceuticals account for 32% of all new medical entities. Approximately 165 biopharmaceuticals have been approved and about half of them belong to the five major target categories of cancer, diabetes, growth disturbances, haemophilia and hepatitis. Erythropoietin has paved the way as one of the flagships of biopharmaceuticals, with sales exceeding \$10 billion in 2005 (Walsh, 2006). The drug discovery process has made great advances as a result of modern automated high-throughput technology and data handling, which allows thousands of plant extracts to be screened in a few days. Plant cell cultures have been shown to be good alternatives for production of high-value compounds, which are produced in whole plants often in very minute amounts. Paclitaxel (Taxol®), an important anticancer drug, is a good example of the importance of plant-based compounds, as well as of their complexity and the necessity of finding alternative sources for their production. Theoretically, in order to produce 2.5 kg of taxol, 27,000 t of *Taxus brevifolia* bark is required and altogether 12,000 trees must be cut down (Rates, 2001). Paclitaxel is currently produced by plant cell cultures (Tabata, 2004).

Major advantages in plant cell-based production platforms include low production costs, easy scale-up and particularly the general safety attributed to these methods. Due to these factors, plant cell-based systems might one day outweigh other production systems, especially those using animal cells as hosts (Ma *et al.*, 2003). Furthermore, plant cells are capable of carrying out many desirable post-translational modifications for recombinant protein production, which cannot be obtained by e.g. microbial hosts to the same extent. However, the limitations faced by plant cell cultivation technology still include the generally low product yields and variable production rates. The production of plant-derived compounds sometimes requires the use of two-phase cultivation, since the biosynthesis of secondary metabolites is often linked to growth stage. Advances in bioreactor technology have led to a wide variety of different types of reactors suitable for plant cell and tissue cultures (Eibl and Eibl, 2008), and nowadays the cultivation of fragile and demanding plant cells, traditionally considered rather challenging, can no longer be considered to be the major limiting factor in the development of production processes. However, difficulties with downstream processing and the presence of non-authentic glycan residues in recombinant human proteins have caused complications in regulatory and approval issues concerning production of plant-based biopharmaceuticals (Miele, 1997). Coping with these factors is essential for the overall production process, since for example downstream processing may account for even 80% of all production costs (Hellwig *et al.*, 2004). In the case of downstream processing, plant cell cultures offer a good choice for easier and economically more viable production, since the formation of by-products, such as waxes and phenolics, is typically lower in cell cultures than in differentiated cultures or whole plants. In plant cell culture-based platforms, an essential prerequisite for a viable system is the banking of the production cell line (Hellwig *et al.*, 2004). In the case of plant cell cultures this means that cryopreservation standard operation protocols must be drawn up and a storage system with back-ups must be established. The cryopreservation operates in two basic modes, either by slow-freezing or via rapid cooling of the cell line by vitrification or encapsulation-dehydration. Both modes require empirical studies, optimization and the use of cryoprotectants. However, rapid cooling can be considered favourable, if successful, due the simplicity of the process. Furthermore, this system does not require expensive or highly specialized equipment (e.g. Winkelmann *et al.*, 2004; Grout, 2007). As decisions concerning the manufacturing process are taken at an early stage, considerable emphasis must be placed on development of the storage of the production lines in question. This is a stage that cannot be neglected and should be included in calculations when making profitability estimations and overall calculations of product costs.

The regulatory process concerning plant-made pharmaceuticals (PMPs) still requires development. The main problem is that the existing guidelines of the Food and Drug Administration (FDA, 2002) and the European Agency for the Evaluation of Medicinal Products (CPMP, 2002) are focused on production in whole plants, making the regulatory issues concerning plant cell cultures rather obscure. Contained production using plant cell cultures will fall under different regulations than those for field-grown products. Thus it is expected that the regulation can be less stringent (Spök *et al.*, 2008). From the regulatory point of view, the whole production chain must comply both with the guidelines concerning GM crops and with the requirements set for the production of pharmaceuticals

(Sparrow *et al.*, 2007; Spök *et al.*, 2008). Currently, the roles of regulatory authorities are overlapping both in the USA and in the EU. Of the applicant point of view, a clarified guidance would be beneficial to support the application process and speed up the launching of plant-derived pharmaceuticals.

With regard to general public approval, in March 2006 the EU comission published a eurobarometer tracking European opinions in biotechnology. On the basis of this query biotechnology related both to medicine and to industrial use, e.g. bioplastics and bioproduction of pharmaceuticals, are widely supported in EU countries. On the other hand, opinions towards biotechnology related to agriculture are still sceptical, with 58% of the respondents having a negative attitude towards agricultural exploitation of biotechnology. Directed to the public, and especially to patients worldwide, in 2005 the International Alliance of Patients Organizations (IAPO) launched a Briefing Paper on the topic of plant-made pharmaceuticals, providing accurate and independent information based on evidence-based research. It is clear that when developing modern biotechnological applications, especially those exploiting genetic engineering, it is of great importance to consider the ethical issues relating to them. Particularly in the case of the agricultural and food sector, environment and healthcare, biotechnological applications are connected to new processes and ethical questions, which should be discussed in open forums. In addition, it should be considered that prevailing values are dependent on the cultural background and may vary considerably between countries.

The hopefully soon-to-be success stories of insulin and glucocerebrosidase will clear the way for other plant-made biopharmaceuticals. Safflower-produced insulin was shown to be bioequivalent to already marketed recombinant human insulin (Eli-Lilly's Humulin) in Phase III clinical trials in the UK in 2009 (www.sembiosys.com). The demand for insulin is expected to grow exponentially in coming years as incidences of diabetes are expected to reach 380 million cases in 2025. In addition, the development of new nasal and inhaled administrations requiring high doses and frequent dosing will further increase the consumption of insulin (Nykiforuk *et al.*, 2006). Fortunately, it is estimated that the safflower-based production system is so efficient that two middle-sized Canadian farms could provide sufficient insulin for global markets (Moloney, 2007). The new glucocerebrosidase (GCD) production system expected to be approved in the near future utilizes carrot cells as manufacturing host (Shaaltiel *et al.*, 2007). Recently, in clinical Phase I studies six healthy volunteers were treated with plant cell-produced recombinant GCD (prGCD) by intravenous infusion. The prGCD was well tolerated, non-immunogenic and even showed a prolonged half-life when compared to a commercial enzyme manufactured in mammalian cells. Phase III clinical trials are currently in progress (Aviezer *et al.*, 2009). These examples are very welcome, especially as they will cause the regulatory approval process to take shape. These products will pave the way for plant- and plant cell culture-manufactured biopharmaceuticals. The global demand for new drugs is constantly increasing as the world's population continues to grow. This global burden makes sustainable and renewable drug production processes even more valuable, even essential. Although not yet fully exploited, plant cell culture-based production offers a huge potential to overcome diseases and to contribute to human well-being.

Acknowledgements

We thank Kirsi-Marja Oksman-Caldentey for critically reviewing the manuscript and Michael Bailey for revising the language of this chapter.

References

Achnine, L., Huhman, D.V., Farag, M.A., Sumner, L.W., Blount, J.W. and Dixon, R.A. (2005) Genomics-based selection and functional characterization of triterpene glycosyltransferases from the model legume *Medicago truncatula*. *Plant Journal* 41, 875–887.

Alfermann, A.W., Schuller, I. and Reinhard, E. (1980) Biotransfomation of cardiac glycosides by immobilized cell cultures of *Digitalis lanata* L. *Planta Medica* 40, 218–223.

Arora, R., Singh, S., Puri, S.C. and Sharma, R.K. (2008) Himalayan Mayapple (*Podophyllum hexandrum* Royle): Traditional Uses, Clinical Indications and Future Prospects. In: Watson, R. and Preedy, V.R. (eds) *Botanical Medicine in Clinical Practice*. CABI Publishing, Wallingford, Oxon, United Kingdom, pp. 71–84.

Arora, R. (2010) *Herbal Medicine: A Cancer Chemopreventive and Therapeutic Perspective*. Jaypee Brothers Medical Publishers (P) Ltd, Delhi, India, 634 pp.

Andersen, D.C. and Krummen, L. (2002) Recombinant protein expression for therapeutic applications. *Current Opinion in Biotechnology* 13, 117–123.

Asada, M. and Shuler, M.L. (1989) Stimulation of ajmalicine production and excretion from *Catharanthus roseus*: effects of adsorption *in situ*, elicitors and alginate immobilization. *Applied Microbiology and Biotechnology* 30, 475–481.

Aspegren, K., Mannonen, L., Ritala, A., Puupponen-Pimiä R., Kurtén, U., Salmenkallio-Marttila, M., Kauppinen, V. and Teeri, T.H. (1995) Secretion of a heat-stable fungal beta-glucanase from transgenic, suspension-cultured barley cells. *Molecular Breeding* 1, 91–99.

Aviezer, D., Brill-Almon, E., Shaaltiel, Y., Hashmueli, S., Bartfeld, D., Mizrachi, S., Liberman, Y., Freeman, A., Zimran, A. and Galun, E. (2009) A plant-derived recombinant human glucocerebrosidase enzyme – A preclinical and phase I investigation. *Public Library of Science ONE* 4, e4792.

Basaran, P. and Rodríguez-Cerezo, E. (2008) Plant Molecular Farming: Opportunities and Challenges. *Critical Reviews in Biotechnology* 28, 153–172.

Butler, M.S. (2004) The role of natural product chemistry in drug discovery. *Journal of Natural Products* 67, 2141–2153.

Butler, M.S. (2005) Natural products to drugs: natural product derived compounds in clinical trials. *Natural Product Reports* 22, 162–195.

Butler, M.S. (2008) Natural products to drugs: natural product derived compounds in clinical trials. *Natural Product Reports* 25, 475–516.

Chang, M.C., Eachus, R.A., Trieu, W., Ro, D.K. and Kiesling, J.D. (2007) Engineering *Escherichia coli* for production of functionalized terpenoids using plant P450s. *Nature Chemical Biology* 3, 274–277.

Chen, K., Liu, Q., Xie, L., Sharp, P.A. and Wang, D.I.C. (2001) Engineering of a mammalian cell line for reduction of lactate formation and high monoclonal antibody production. *Biotechnology and Bioengineering* 72, 55–61.

Chintapakorn, Y. and Hamill, J.D. (2003) Antisense-mediated down-regulation of putrescine N-methyltransferase activity in transgenic *Nicotiana tabacum* L. can lead to elevated levels of anatabine at the expense of nicotine. *Plant Molecular Biology* 53, 87–105.

CPMP (2002) Point to consider on quality aspects of medicinal products containing active substances produced by stable transgene expression in higher plants (CPMP/BWP/764/02). The European Agency for the Evaluation of Medicinal Products, London, United Kingdom.

Eibl, R. and Eibl, D (2002) Bioreactors for plant cell and tissue cultures. In: Oksman-Caldentey, K.M. and Barz, W. (eds) *Plant Biotechnology and Transgenic Plants*. Marcel Decker, Inc., New York, USA, pp.163–199.

Eibl, R. and Eibl, D. (2008) Design of bioreactors suitable for plant cell and tissue cultures. *Phytochemistry Reviews* 7, 593–598.

Ellis, B.E. and Towers, G.H.N. (1970) Biogenesis of rosmarinic acid in *Mentha*. *Journal of Biochemistry* 118, 297–297.

Eskelin, K., Ritala, A., Suntio, T., Blumer, S., Holkeri, H., Wahlström, E.H., Baez, J., Mäkinen, K. and Nuutila, A.M. (2009) Production of a recombinant full length collagen type i $\alpha 1$ and of a 45-kda collagen type i $\alpha 1$ fragment in barley seeds. *Plant Biotechnology Journal* 7, 657–672.

FDA. Guidance for industry (2002) Drugs, biologics, and medical devices derived from bioengineered plants for the use in human and animals. Food and Drug Administration, Rockville, Maryland, USA.

Fett-Neto, A.G., Melanson, S.J., Nicholson, S.A., Pennington, J.J. and DiCosmo, F. (1994) Improved taxol yield by aromatic carboxylic and amino acid feeding to cell cultures of *Taxus cuspidata*. *Biotechnology and Bioengineering* 44, 967–971.

Fiehn, O., Kopka, J., Dörmann, P., Altmann, T., Trethewey, R.N. and Willmitzer, L. (2000) Metabolite profiling for plant functional genomics. *Nature Biotechnology* 18, 1157–1161.

Fontanel, A. and Tabata, M. (1987) Production of secondary metabolites by plant tissue and cell cultures. Present aspects and prospects. *Nestlé Research News* 1986/87, 92–103.

Fujita, Y. and Tabata, M. (1987) Secondary metabolites from plant cells– pharmaceutical applications and progress in commercial production. *Plant Physiology and Biochemistry* 3, 169–185.

Fussenegger, M., Morris, R.P., Fux, C., Rimann, M., von Stockar, B., Thompson, C.J. and Bailey, J.E. (2000) Streptogramin-based gene regulation systems for mammalian cells. *Nature Biotechnology* 18, 1203–1208.

Ganesan, A. (2008) The impact of natural products upon modern drug discovery. *Current Opinion in Chemical Biology* 12, 306–317.

Geddis, A.E. and Prockop, D.J. (1993) Expression of human COL1A1 gene in stably transfected HT1080 cells: the production of a thermostable homotrimer of type I collagen in a recombinant system. *Matrix* 13, 399–405.

Giri, A., Dhingra, V., Giri, C.C., Singh, A., Ward, O.P. and Narasu, M.L. (2001) Biotransformation using plant cells, organ cultures and enzyme systems: Current trends and future prospects. *Biotechnology Advances* 19, 175–199.

Goossens, A., Häkkinen, S.T., Laakso, I., Seppänen–Laakso, T., Biondi, S., De Sutter, V., Lammertyn, F., Nuutila, A.M., Söderlund, H., Zabeau, M., Inzé, D. and Oksman-Caldentey, K.-M. (2003) A functional genomics approach toward the understanding of secondary metabolism in plant cells. *Proceedings of the National Academy of Sciences, USA* 100, 8595–8600.

Grothe, T., Lenz, R. and Kutchan, T.M. (2001) Molecular characterization of the salutaridinol 7-O-acetyltransferase involved in morphine biosynthesis in opium poppy *Papaver somniferum*. *Journal of Biological Chemistry* 276, 30717–30723.

Grout B.W. (2007) Cryopreservation of plant cell suspensions. *Methods in Molecular Biology* 368, 153–161.

Gundlach, H., Müller, M.J., Kutchan, T.M. and Zenk, M.H. (1992) Jasmonic acid is a signal transducer in elicitor-induced plant cell cultures. *Proceedings of the National Academy of Scences, USA* 89, 2389–2393.

Häkkinen, S.T., Rischer, H., Laakso, I, Maaheimo, H., Seppänen-Laakso, T. and Oksman-Caldentey, K.-M. (2004) Anatalline and other methyl jasmonate inducible nicotine alkaloids from *Nicotiana tabacum* cv. BY–2 cell cultures. *Planta Medica* 70, 936–941.

Häkkinen, S.T., Moyano, E., Cusidó R.M., Palazón, J., Piñol M.T. and Oksman-Caldentey, K.-M. (2005) Enhanced secretion of tropane alkaloids in *Nicotiana tabacum* hairy roots expressing heterologous hyoscyamine-6β-hydroxylase. *Journal of Experimental Botany* 56, 2611–2618.

Häkkinen, S.T., Tilleman, S., Świątek, A., De Sutter, V., Rischer, H., Vanhoutte, I., Van Onckelen, H., Hilson, P., Inzé, D., Oksman-Caldentey, K.-M. and Goossens, A. (2007) Functional characterisation of genes involved in pyridine alkaloid biosynthesis in tobacco. *Phytochemistry* 68, 2773–2785.

Hale, V., Keasling, J.D., Renninger, N. and Diagana, T.T. (2007) Microbially derived artemisinin: a biotechnology solution to the global problem of access to affordable antimalarial drugs. *American Journal of Tropical Medicine and Hygiene* 77, 198–202.

Hall, R.D., Holden, M.A. and Yeoman, M.M. (1988) *The Immobilization of Higher Plant Cells*. In: Bajaj, Y.P.S. (ed.) *Immobilization of Plant Cells*. Springer Verlag, New York, USA, pp. 136–156.

Hamada, H., Miyamoto, Y., Nakajima, N. and Furuya, T. (1998) Highly selective transformation by plant catalysis. *Journal of Molecular Catalysis B: Enzymatic* 5, 187–189.

Hamill, J.D., Robins, R.J., Parr, A.J., Evans, D.M., Furze, J.M. and Rhodes, M.J.C. (1990) Overexpression a yeast ornithine decarbozylase gene in transgenic roots of *Nicotiana rustica* can lead to enhanced nicotine accumulation. *Plant Molecular Biology* 15, 27–38.

Hashimoto, T. and Yamada, Y. (2003) New genes in alkaloid metabolism and transport. *Current Opinion in Biotechnology* 14, 163–168.

Hashimoto, T., Yun, D.-J. and Yamada, Y. (1993) Production of tropane alkaloids in genetically engineered root cultures. *Phytochemistry* 32, 713–718.

Hellwig, S., Drossard, J., Twyman, R.M. and Fischer, R. (2004) Plant cell cultures for the production of recombinant proteins. *Nature Biotechnology* 22, 1415–1422.

Hirasuna, T.J., Shuler, M.L., Lackney, V.K. and Spanswick, R.M. (1991) Enhanced anthocyanin production in grape cell cultures. *Plant Science* 78, 107–120.

Howes, M.J.R., Perry, N.S.L. and Houghton, P.J. (2003) Plants with traditional uses and activities relevant to the management of Alzheimer's disease and other cognitive disorders. *Phytotherapy Research* 17, 1–8.

Ibrahim, R.K. (1987) *Regulation of Synthesis of Phenolics*. In: Constabel, F. and Vasil, I.K. (eds.) *Cell Cultures and Somatic Cell Genetics of Plants*, Vol. 4. Academic Press, San Diego, USA, pp. 77–95.

John, D.C.A., Watson, R., Kind, A.J., Scott, A.R., Kadler, K.E. and Bulleid, N.J. (1999) Expression of an engineered form of recombinant procollagen in mouse milk. *Nature Biotechnology* 17, 385–389.

Jouhikainen, K., Lindgren, L., Jokelainen, T., Hiltunen, R., Teeri, T., and Oksman-Caldentey, K.-M. (1999) Enhancement of scopolamine production in *Hyoscyamus muticus* L. hairy root cultures by genetic engineering. *Planta* 208, 545–551.

Khosla, C. and Keasling, J.D. (2003) Metabolic engineering for drug discovery and development. *Nature Reviews in Drug Discovery* 2, 1019–1025.

Leonard, E., Runguphan, W., O'Connor, S. and Prather, K.J. (2009) Opportunities in metabolic engineering to facilitate scalable alkaloid production. *Nature Chemical Biology* 5, 292–300.

Li, L., He, Z., Pandey, G.K., Tsuchiya, T. and Luan, S. (2002) Functional cloning and characterization of a plant efflux carrier for multidrug and heavy metal detoxification. *Journal of Biological Chemistry* 277, 5360–5368.

Lin Cereghino, G.P., Lin Cereghino, J., Ilgen, C. and Cregg, J.M. (2002) Production of recombinant proteins in fermenter cultures of the yeast *Pichia pastoris*. *Current Opinion in Biotechnology* 13, 329–332.

Lockwood, G.B. and Essa, A.K. (1984) The effect of varying hormonal and precursor supplementations on levels of nicotine and related alkaloids in cell cultures of *Nicotiana tabacum*. *Plant Cell Reports* 3, 109–111.

Ma, J.K.C., Drake, P.M.W. and Christou, P. (2003) The production of recombinant pharmaceutical proteins in plants. *Nature Genetics Reviews* 4, 794–805.

Miele, L. (1997) Plants as bioreactors for biopharmaceuticals: regulatory considerations. *Trends in Biotechnology* 15, 45–50.

Misawa, M. (1994) Plant Tissue Culture: An alternative for production of useful metabolites. *FAO Agricultural Services Bulletin*, Rome, 1–87.

Moloney, M. (2007) Production, recovery and bioequivalence of human insulin obtained from transgenic oilseeds. *PSE Congress, Plants for Human Health in the Post-Genome Era, Helsinki*, Finland, 29.8.2007, oral presentation.

Morita, M., Shitan, N., Sawada, K., Van Montagu, M.C.E., Inzé, D., Rischer, H., Goossens, A., Oksman-Caldentey, K.-M., Moriyama, Y. and Yazaki, K. (2009) Vacuolar transport of nicotine is mediated by a multidrug and toxic compound extrusion (MATE) transporter in *Nicotiana tabacum*. *Proceedings of the National Academy of Sciences, USA* 106, 2447–2452.

Nagata, T., Hasezawa, S. and Inzé, D. (2004) Tobacco BY-2 Cells. *Biotechnology in Agriculture and Forestry*. Vol. 53. Springer Verlag, Heidelberg, Germany, 347 p.

Nagata, T. and Kumagai, F. (1999) Plant cell biology through the window of the highly synchronized tobacco BY-2 cell line. *Methods Cell Science* 21, 123–127.

Nechansky, A., Schuster, M., Jost, W., Siegl, P., Wiederklum, S., Gorr, R. and Kircheis, R. (2007) Compensation of endogenous IgG mediated inhibition of antibody-dependent cellular cytotoxicity by glycol-engineering of therapeutic antibodies. *Molecular Immunology* 44, 1826–1828.

Newman, D.J. and Cragg, G.M. (2007) Natural Products as Sources of New Drugs over the Last 25 Years. *Journal of Natural Products* 70, 461–477.

Nokelainen, M., Helaakoski, T., Myllyharju, J., Notbohm, H., Pihlajaniemi, T., Fietzek, P.P. and Kivirikko, K.I. (1998) Expression and characterization of recombinant human type II collagens with low and high contents of hydrolysine and its glycosylated forms. *Matrix Biology* 16, 329–338.

Nykiforuk C.L., Boothe, J.G., Murray, E.W., Keon, R.G., Goren, H.J., Markley N.A. and Moloney, M.M. (2006) Transgenic expression and recovery of biologically active recombinant human insulin from *Arabidopsis thaliana* seeds. *Plant Biotechnology Journal* 4, 77–85.

Oksman-Caldentey, K.-M. (2007) Tropane and nicotine alkaloid biosynthesis – novel approaches towards biotechnological production of plant-derived pharmaceuticals. *Current Pharmacological Biotechnology* 8, 203–210.

Oksman-Caldentey, K.-M. and Saito, K. (2005) Integrating genomics and metabolomics for engineering plant metabolite pathways. *Current Opinion in Biotechnology* 16, 174–179.

Oksman-Caldentey, K.-M., Inzé, D. and Oresic, M. (2004) Connecting genes to metabolites by a systems biology approach. *Proceedings of the National Academy of Sciences USA* 101, 9949–9950.

Olsen, D., Yang, C., Bodo, M., Chang, R., Leigh, S., Baez, J., Carmichael, D., Perälä, M., Hämäläinen, E-R., Jarvinen, M. and Polarek, J. (2003) Recombinant collagen and gelatine for drug delivery. *Advances in Drug Delivery Reviews* 55, 1547–1567.

Payne, G.F., Bringi, V., Prince, C. and Shuler, M.L. (1991) Plant cell and tissue culture in liquid systems. *Biotechnology Techniques* 6, 287.

Petersson, E.J. and Schepartz, A. (2008) Toward β-amino acid proteins: Design, synthesis and characterization of a fifteen kilodalton β-peptide tetramer. *Journal of American Chemical Society* 130, 821–823.

Pichersky, E. and Gang, D. (2000) Genetics and biochemistry of secondary metabolites: an evolutionary perspective. *Trends in Plant Science* 5, 439–445.

Poulev, A., O'Neal, J.M., Logendra, S., Pouleva, R.B., Timeva, V., Garvey, A.S., Gleba, D., Jenkins, I.S., Halpern, B.T., Kneer, R., Cragg, G.M. and Raskin, I. (2003) Elicitation, a new window into plant chemodiversity and phytochemical drug discovery. *Journal of Medicinal Chemistry* 46, 2542–2547.

Qu, L.Q. and Takaiwa, F. (2004) Evaluation of tissue specificity and expression strength of rice seed component gene promoters in transgenic rice. *Plant Biotechnology Journal* 2, 113–125.

Rao, S.R. and Ravishankar, G.A. (2002) Plant cell cultures: Chemical factories of secondary metabolites. *Biotechnology Advances* 20, 101–153.

Rates, S.M.K. (2001) Plant as source of drugs. *Toxicon* 39, 603–613.

Rischer, H., Orešič, M., Seppänen-Laakso, T., Katajamaa, M., Lammertyn, F., Ardiles–Diaz, W., Van Montagu, M.C.E., Inzé, D., Oksman-Caldentey, K.-M. and Goossens, A. (2006) Gene-to-metabolite networks for terpenoid indole alkaloid biosynthesis in *Catharanthus roseus* cells. *Proceedings of the National Academy of Sciences, USA* 103, 5614–5619.

Ritala, A., Nuutila, A.M., Aikasalo, R., Kauppinen, V. and Tammisola, J. (2002) Measuring gene flow in the cultivation of transgenic barley. *Crop Science* 42, 278–285.

Ritala, A., Wahlström, E., Holkeri, H., Hafren, A., Mäkeläinen, K., Baez, J., Mäkinen, K. and Nuutila, A.-M. (2008) Production of a recombinant industrial protein using barley cell cultures. *Protein Expression and Purification* 59, 274–281.

Ritterhaus, E., Ulrich, J. and Westphal, K. (1990) Large-scale production of plant cell cultures. *International Association of Plant Tissue Culture Newsletter* 61, 2–10.

Ro, D.K., Paradise, E.M., Ouellet, M., Fisher, K.J., Newman, K.L., Ndungu, J.M., Ho, K.A., Eachus, R.A., Ham, T.S., Kirby, J., Chang, M.C.Y., Withers, S.T., Shiba, Y., Sarpong, R. and Keasling, J.D. (2006) Production of the antimalarial drug precursor artemisinic acid in engineered yeast. *Nature* 440, 940–943.

Roberts, S.C. (2007) Production and engineering of terpenoids in plant cell culture. *Nature Chemical Biology* 3, 387–395.

Robins, R.J., Hamill, J.D., Parr, A.J., Smith, K., Walton, N.J. and Rhodes, M.J. (1987) Potential for use of nicotinic acid as a selective agent for isolation of high–nicotine producing lines of *Nicotiana rustica* hairy root cultures. *Plant Cell Reports* 6, 122–126.

Ronaghi, M., Uhlén, M. and Nyrén, P. (1998) A sequencing method based on real-time pyrophosphate. *Science* 17, 363–365.

Ruggiero, F., Exposito, J.Y., Bournat, P., Gruber, V., Perret, S., Comte, J., Olagnier, B., Garrone, R. and Theisen, M. (2000) Triple helix assembly and processing of human collagen produced in transgenic tobacco plants. *FEBS Letters* 469, 132–136.

Sahai, O. and Knuth, M. (1985) Commercializing plant tissue culture processes: Economics, problem and prospects. *Biotechnology Progress* 1, 1–9.

Saito, K., Hirai, M.Y. and Yonekura–Sakakibara, K. (2007) Decoding genes with coexpression networks and metabolomics – 'majority report by precogs'. *Trends in Plant Science* 13, 36–43.

Sajc, L., Grubisic, D. and Vunjak-Novakovic, G. (2000) Bioreactors for plant engineering: and outlook for further research. *Biochemical Engineering Journal* 4, 89–99.

Samanani, N. and Facchini, P.J. (2002) Purification and characterization of norcoclaurine synthase. The first committed enzyme in bezyl-isoquinoline alkaloid biosynthesis in plants. *Journal of Biological Chemistry* 277, 33878–33883.

Samanani, N. and Facchini, P.J. (2006) Compartmentalization of plant secondary metabolism. In: Romeo, T.J. (ed.) *Recent Advances in Phytochemistry* Vol. 40, Elsevier, New York, pp. 53–83.

Samuelsson, G. (2004) *Drugs of Natural Origin. A Textbook of Pharmacognosy,* 5th revised edition, Swedish Pharmaceutical Press, Stockholm, Sweden, pp. 473–575.

Sato, F. and Yamada, Y. (1984) High berberine producing cultures of *Coptis japonica* cells. *Phytochemistry* 23, 281–285.

Schippmann, U., Leaman, U. and Cunningham, A.B. (2006) Plants as source of medicines: new perspectives. In: Bogers, R.J., Craker, L.E. and Lange, D. (eds.) *Medicinal and Aromatic Plants – Agricultural, Commercial, Ecological, Legal, Pharmacological and Social Aspects.* Springer, Dordrecht, The Netherlands, p. 75.

Schmidt, F.R. (2004) Recombinant expression systems in the pharmaceutical industry. *Applied Microbiology and Biotechnology* 65, 363–372.

Schouten, A., Roosien, J., van Engelen, F.A., de Jong, G.A.M., Borst-Vrenssen, A.W.M., Zilverentant, J.F., Bosch, D., Stiekema, W.J., Gommers, F.J., Schots, A. and Bakker, J. (1996) The C-terminal KDEL sequence increases the expression level of a single-chain antibody designed to be targeted to both the cytosol and the secretory pathway in transgenic tobacco. *Plant Molecular Biology* 30, 781–793.

Schuster, M., Jost, W., Mudde, G.C., Wiederklum, S., Schwager, C., Janzek, E., Altmann, F., Stdlmann, J., Stemmer, C. and Gorr, G. (2007) *In vivo* glyco-engineered antibody with improved lytic potential produced by an innovative non-mammalian expression system. *Biotechnology Journal* 2, 700–708.

Sevón, N. and Oksman-Caldentey, K.-M. (2002) *Agrobacterium rhizogenes*-mediated transformation: root cultures as a source of alkaloids. *Planta Medica* 68, 859–868.
Shaaltiel, Y., Bartfeld, D., Hashmueli, S., Baum, G., Brill-Almon, E., Galili, G., Dym, O., Boldin-Adamsky, S.A., Silman, I., Sussman, J.L., Futerman, A.H. and Aviezer, D. (2007) Production of glucocerebrosidase with terminal mannose glycans for enzyme replacement therapy of Gaucher's disease using a plant cell system. *Plant Biotechnology Journal* 5, 579–590.
Shitan, N., Bazin, I., Dan, K., Obata, K., Kigawa, K., Ueda, K., Sato, F., Forestier, C. and Yazaki, K. (2003) Involvement of CjMDR1, a plant multidrug-resistance-type ATP-binding cassette protein, in alkaloid transport in *Coptis japonica. Proceedings of the National Academy of Scences, USA* 100, 751–756.
Shoji, T. and Hashimoto, T. (2008) Why does anatabine, but not nicotine, accumulate in jasmonate-elicited cultured tobacco BY-2 cells? *Plant Cell Physiology* 49, 1209–1216.
Sijmons, P.C., Dekker, B.M.M., Schrammeijer, B., Verwoerd, T.C., van den Elzen, P.J.M. and Hoekema, A. (1990) Production of correctly processed human serum albumin in transgenic plants. *Nature Biotechnology* 8, 217–221.
Smith, M.A.L. (1995) Large Scale Production of Secondary Metabolites. In: Terzi, M. (ed) *Current Issues in Plant Molecular and Cell Biology*. Kluwer Academic, New York, pp. 669–674.
Sparrow, P.A.C., Irwin, J.A., Dale, P.J., Twyman, R.M. and Ma, J.K.C. (2007) Pharma-planta: Road testing the developing regulatory guidelines for plant-made pharmaceuticals. *Transgenic Research* 16, 147–161.
Spök, A., Twyman, R.M., Fischer, R., Ma, J.K.C. and Sparrow, P.A.C. (2008) Evolution of regulatory framework for pharmaceuticals derived from genetically modified plants. *Trends in Biotechnology* 26, 506–517.
Streatfield, S.J. (2007) Approaches to achieve high-level heterologous protein production in plants. *Plant Biotechnology Journal* 5, 2–15.
Tabata, H. (2004) Paclitaxel production by plant-cell-culture technology. *Advances in Biochemical Engineering/Biotechnology 7,* 1–23.
Tabata, M. and Fujita, Y. (1985) *Production of Shikonin by Plant Cell Cultures*. In: Zaitling, M., Day, P. and Hollaender, A. (eds) *Biotechnology in Plant Science*, Academic Press, Orlando, Florida, USA, p. 207.
Tsoi, B.M. and Doran, P.M. (2002) Effect of medium properties and additives on antibody stability and accumulation in suspended plant cell cultures. *Biotechnology and Applied Biochemistry* 35, 171–180.
Ulbrich, B., Wiesner, W. and Arens, H. (1985) Large-scale Production of Rosmarinic Acid from Plant Cell Cultures of *Coleus blumeii* Benth. In: Neumann, K.H., Barz, W. and Reinhard, E. (eds) *Primary and Secondary Metabolism of Plant Cell Cultures*. Springer, Berlin, 293–303.
Van Uden, W., Bouma, A.S., Waker, J.F.S., Middel, O., Wichers, H.J., Waard, P., Woerdenbag, H., Kellog, R.M. and Pras, N. (1995) The production of podophyllotoxin and its 5-methoxy derivatives through bioconversion of cyclodextrin complexed desoxypodophyllotoxin by plant cell cultures. *Plant Cell Tissue and Organ Culture* 42, 73–79.
Vera, J.C., Wheat, C.W., Fescemyer, H.W., Frilander, M.J., Crawford, D.L., Hanski, I. and Marden, J.H. (2008) Rapid transcriptome characterization for a nonmodel organism using 454 pyrosequencing. *Molecular Ecology* 17, 1636–1647.
Vines, G. (2004) Herbal harvest with a future. Towards sustainable sources for medicinal plants. *PlantLife International*. www.plantlife.ogr.uk
Vuorela, A., Myllyharju, J., Nissi, R., Pihlajaniemi, T. and Kivirikko, K.I. (1997) Assembly of human prolyl 4-hydroxylase and type III collagen in the yeast *Pichia pastoris*: formation of a stable enzyme tetramer requires coexpression with collagen and assembly of a stable collagen requires coexpression with prolyl 4-hydroxylase. *EMBO Journal* 16, 6702–6712.
Walsh, G. (2006) Biopharmaceutical benchmarks. *Nature Biotechnology* 24, 769–776.
Weikert, S., Papac, D., Briggs, J., Cowfer, D., Tom, S., Gawlitzek, M., Lofgren, J., Mehta, S., Chisholm, V., Modi, N., Eppler, S., Carroll, K., Chamow, S., Peers, D., Berman, P.

and Krummen, L. (1999) Engineering Chinese hamster ovary cells to maximize sialic acid content of recombinant glycoproteins. *Nature Biotechnology* 17, 1116–1121.

Winkelmann, T., Mussmann, V. and Serek, M. (2004) Cryopreservation of embryogenic suspension cultures of *Cyclamen persicum* Mill. *Plant Cell Reports* 23, 1–8.

Winkel-Shirley, B. (2001) Flavonoid biosynthesis. A colorful model for genetics, biochemistry, cell biology, and biotechnology. *Plant Physiology* 126, 485–493.

Wurm, F.M. (2004) Production of recombinant protein therapeutics in cultivated mammalian cells. *Nature Biotechnology* 22, 1393–1398.

Yamazaki, M. and Saito, K. (2002) Differential display analysis of gene expression in plants. *Cellular and Molecular Life Sciences* 59, 1246–1255.

Yazaki, K. (2006) ABC transporters involved in the transport of plant secondary metabolites. *FEBS Letters* 580, 1183–1191.

Yun, D.-J., Hashimoto, T. and Yamada, Y. (1992) Metabolic engineering of medicinal plants: transgenic *Atropa belladonna* with an improved alkaloid composition. *Proceedings of the National Academy of Sciences, USA* 89, 11799–11803.

Zhang, L., Ding, R., Chai, Y., Bonfill, M., Moyano, E., Oksman-Caldentey, K.-M., Xu, T., Pi, Y., Wang, Z., Zhang, H., Kai, G., Liao, Z., Sun, X. and Tang, K. (2004) Engineering tropane biosynthetic pathway in *Hyoscyamus niger* hairy root cultures. *Proceedings of the National Academy of Sciences, USA* 101, 6786–6791.

Wang, Y. (2008) Needs for new plant-derived pharmaceuticals in the post-genome era: an industrial view in drug research and development. *Phytochemistry Reviews* 7, 395–406.

Chapter 3

Biotechnological Characterization of Different Populations of an Endangered Medicinal Herb – *Podophyllum hexandrum* Royle.

Hemant Lata and Suman Chandra

Introduction

In the past few decades, due to the sharp decline in the number of plant species, as a result of their ruthless exploitation, many species of medicinal interest are facing the danger of extinction in their natural habitat. Our recent surveys of the Garhwal Himalayan region indicate that the frequency of medicinally important plant *Podophyllum hexandrum* Royle is decreasing considerably in its natural habitat.

P. hexandrum is a herbaceous and perennial species belonging to the family Podophyllaceae (previously Berberidaceae). It is distributed in the interior ranges of the Himalayas at an altitude ranging from 2000–4000m. The mature rhizome of an individual plant bears the aerial reproductive shoot (mostly two leaved) and one to five vegetative shoots. The individual flowering shoot bears a solitary flower attached to the upper leaf petiole. The fruit is a scarlet coloured berry. Roots and rhizomes of *P. hexandrum* produce compounds such as podophyllic acid, podophyllin, podophyllotoxin and quercetin. Use of Indian Podophyllum was used by ancient Hindu physicians for treating a variety of ailments. The inhabitants of Punjab used the rhizome as a bile expellant (Chatterjee, 1952; Arora e*t al.*, 2008). Roots and rhizomes of the plant yield a drug known as podophyllin, which is a resinous alcoholic (water insoluble) extract of the rhizome.

Podophyllin consists of podophyllotoxin, podophyllic acid, podophylloresin and quercetin. Podophyllin was included in the first US pharmacopoeia, dating from 1820, as a catharatic and chalogogue (Chatterjee, 1952). In the same year however, it was reported that venereal warts (Condyloma acuminatum) could be selectively destroyed by the topical application of podophyllin resin. Nowadays, this resin is still the drug of choice in the treatment of this disease (Ayres and Loike, 1990). Podophyllin is cholagogue, alternative, emetic, bitter tonic and a drastic purgative. Purgative property of the drug is due to podophylloresin.

The interest in *Podophyllum* was developed when Kaplan in 1942 demonstrated the curative effect of podophyllin in tumour growth, followed by the finding of King and Sullivan (1946) that the anti-proliferative effect of podophyllin was similar to that of colchicine at the cellular level. However, it has been shown to be twice as effective as colchicine in arresting cell division (MacRae and Towers, 1984). Later it was found that the antitumorous activity of podophyllin was due to the active principle, podophyllotoxin, which is the major constituent of podophyllin (Hartwell and Schrechker, 1958). Podophyllotoxin belongs to the class of natural substances called lignans.

Lignans are defined as dimerization products of two phenylpropane units which are linked by the β-carbon atoms of their side chains (MacRae and Towers, 1984; Pelter, 1986; Dewick, 1989). Lignans are widely distributed in the plant kingdom and fifty-five families of vascular plants were included as lignin producers (MacRae and Towers, 1984). Podophyllotoxin, however, is restricted to only a few plant species. Podophyllotoxin occurs mainly in the families Berberidaceae and Cupressaceae. The highest podophyllotoxin contents are found in the needles of *Callitris drummondii*. However, the podophyllotoxin yields of *P. hexandrum* vary strongly with the location of cultivation and season of collection. It is maximal when the plant has reached the flowering stage (Chatterjee, 1952). From Kumaun Himalayas, Nadeem *et al.* (2007) have reported a positive correlation between podophyllotoxin content and increase in altitude, whereas Pandey *et al.* (2007) have reported an increase in podophyllotoxin contents with increasing plant age.

Of all the lignans, podophyllotoxin is very important and valuable because it is used for the preparation of semisynthetic anticancer drugs such as Proresid oral®, Proresid intravenous®, teniposide and etoposide. Etoposide (VP-16-213) and teniposide (VM-26) are well known anti-tumour agents. Their action is based on the inhibition of topoisomerase II (Beers *et al.*, 1988; Holthuis, 1988) and they are used for the treatment of small lung cell cancer, testicular cancer neuroblastoma and hepatoma (Holthuis, 1988). In addition, more than 550 derivatives of podophyllotoxin were prepared and clinically tested. The constituents of *Podophyllum* plants, chemistry (Hartwell and Shrechker, 1958) and biosynthesis of lignans (Dewick, 1989) and their biological effects (MacRae and Towers, 1984), modification of podophyllotoxin to use as anticancer drugs (Stahelin and Von Wartburg, 1989) have been extensively reviewed. Biological effects of podophyllotoxin include antiviral activity, cytotoxicity, hepatoxic protection, cathartic effects and allergenicity, hypotensive effects, influence on nucleic acids and enzymes, inhibition of respiratory enzymes, insecticidal and piscicidal activity, germination inhibition, antimicrobial and fungistatic activity (MacRae and Towers, 1984; Pelter, 1986). In the recent reports, Arora *et al.* (2005), Chawla *et al.* (2006) and Kumar *et al.* (2009) have reported radioprotective and antioxidant properties of *P. hexandrum* extract.

Owing to ruthless collection from the wild and the absence of organized cultivation practices the population of *P. hexandrum* has dwindled to such a critical level that a ban on its collection has been recommended (Sundaresh, 1982; Gupta and Sethi, 1983).

Considering its endangered status, very small population size and continuous exploitation, this species is of special interest and importance to conservation biologists. It is also important to study its population biology and genetic diversity so that conservation strategies can be successfully developed. This chapter summarizes the work undertaken to study the distribution pattern, population diversity based on morphological, physiological and biochemical characteristics from Garhwal Himalaya, and apply the *in vitro* methods of propagation and conservation strategies for this important species.

Distribution and Population Size

Field surveys were conducted in different alpine and subalpine ranges of the Garhwal Himalaya to study the distribution, frequency, density, natural regeneration and population diversity of *P. hexandrum*. Based on the occurrence and frequency of this species, the subalpine and high altitude areas in Kedarnath, Tungnath, Chopta, Ghangaria and The Valley of Flowers (Chamoli district) and Dayara range (Uttarkashi district) were identified as target areas for detailed studies. Some surveys were also conducted in Rudranath area (Chamoli District).

The areas mentioned above were surveyed at monthly intervals to monitor the frequency, density and natural regeneration of *P. hexandrum*. In the subalpine forests, *P. hexandrum* was found as an under-canopy plant, whereas in high altitudes it was found near boulders and sometimes in open meadows. Compared with several other endangered or threatened species from this region, e.g. *Nardostachys jatamansi*, *Picrorhiza kurrooa* and *Aconitum atrox* the natural population of *P. hexandrum* were distributed in very restricted and small pockets.

This species was found distributed in the subalpine forests, timberline zone and high altitude areas from 2600–4000 m altitude. Among the surveyed areas, highest (16.4 plants/m^2) and lowest (1.8 individuals/m^2) density was recorded in Kedarnath and Chopta areas respectively (Table 3.1). Our regular field observations over the past few years indicate that the population size of this species is continuously decreasing in nature and several populations observed during 1980–1985 have nearly completely disappeared. In Kedarnath range, although the population size of *P. hexandrum* was small (only 200 individual plants), the density was highest. This was due to the fact that plants were found in a restricted pocket. The populations of Chopta and Tungnath were also small in size. Although the population size in Dayara and Ghangaria ranges was found to be much larger (700 and 400 individual plants, respectively) as compared to other sites, the individuals were widespread in a larger area and thus, the density of plants was lower as compared to the Kedarnath and Tungnath areas (Table 3.1).

Population Diversity

Podophyllum hexandrum was found to have a fairly high level of population diversity in plant height, size, shape, lobing and number of leaves, size and colour of seeds, germination pattern and isoenzyme polymorphism. Our regular observations of these areas revealed that the plants growing in different locations in Garhwal Himalaya show considerable morphological variations in plant height, leaf characteristics, fruit weight, seed weight and colour and other traits (Bhadula *et al.*, 1996a; Purohit *et al.*, 1998). Therefore, the plants of each locality were considered to represent a different population. Within a single population, however, morphological variations in plant height as well as in the number, shape, size, lobing and pigmentation of leaves were observed. Thus, because of intra-population variations, the germplasm of individual morphological variants of each population was collected and analysed separately. Plants of a population were classified into different groups based on the shape, size, lobing and number of leaves (three, two or one leaf in fruiting stage) and on the basis of seed colour. Thus, six morphological variants in the Kedarnath population, five each in the Chopta and Ghangaria populations, four each in The Valley of Flowers and Dayara and two in the population from Tungnath were identified (Table 3.1). Within each population, individual plants differing in leaf characteristics as well as in seed colour ranging from black, brown to pink were observed (Bhadula *et al.*, 1996a).

Isozyme and Polypeptide Patterns

In an attempt the find the markers for different populations of seeds of *P. hexandrum*, various isozymes, AAT, ADH, MDH, ACP, and peroxidases were analysed. It was found that these isozymes did not appear to be good markers of species and of populations in dry seeds. In fact, many of the isozyme systems did not seem to be expressed in dry seeds. However, esterases gave excellent patterns of isozymes and it was possible to distinguish species as well as populations of the same species on the basis of the number of isozymes as well as the relative intensity of individual bands. For different populations of *P. hexandrum*, the inter- and intra-population morphological variations and the existence of population polymorphism was further supported by the pattern of esterase isoenzymes (Bhadula *et al.*, 1996b).

The SDS-PAGE (Sodium dodecyl sulphate polyacrylamide gel electrophoresis) patterns of polypeptides extracted from seeds were also analysed in the above mentioned populations of *Podophyllum* using 10% SDS-gels. The qualitative as well as quantitative differences were observed in polypeptides in various populations indicating that there are genetic differences among different seed types of a given population (Bhadula *et al.*, 1996a).

Podophylloresin Content

An analysis of the active ingredients in different populations and the intra-population varaince is a necessary requirement so that the better clones can be selected from natural populations and multiplied on a large scale. The application of biotechnological tools is helpful for further improvement of desirable characters using suitable germplasm.

For different populations of *P. hexandrum*, resin content was analysed from dried rhizomes. The material was extracted with ethanol in a Soxhlet apparatus. The ethanolic extract was distilled *in vacuo* to remove the solvent and weighed for total extract. The residue thus obtained was dissolved in least amount of ethanol and precipitated by acidulated water, 2% HCl. The resin so obtained was weighed for total resin content. Percentage podophylloresin content in different populations of *P. hexandrum* is shown in Table 3.2. Podophylloresin content was found to be maximum in roots obtained from Kedarnath (10.63%) population, whereas it was found minimal in the population from Ghagaria (3.91 %).

In vitro Conservation of *P. hexandrum*

The use of biotechnological tools such as tissue culture techniques in the conservation of endangered species is becoming increasingly important, mainly due to the fact that in many cases, *in situ* conservation is not effective due to partial or complete loss of natural habitats.

Table 3.1. Distribution pattern of various populations of *Podophyllum hexandrum* in Garhwal Himalaya.

Locality/ population[a]	**No. of variants**[b]	**Altitude**[c] **and climatic zone**	**Frequency**[d] **(%)**	**Density**[e]	**Dominant associated vegetation**
Kedarnath (Kd)	6	3700 Alpine	80	16.4 **(200)**	*Danthonia* sp., *Polygonum alpinum*, *Cynanthus* sp.
Tungnath (Tn)	2	3300 Alpine	80	4.6 **(70)**	*Trillium govanianum*, *Anemone* sp., *Rhododendron campanulatum*
Valley of Flowers (Vf)	4	3000 Timberline	60	8.0 **(40)**	*Abies spectabilis*, *Prunus cornuta*, *Doronicum* sp., *Viburnum foetens*, *Inula grandiflora*
Chopta (Ch)	5	2800 Subalpine	100	1.8 **(60)**	*Taxus baccata*, *Acer caesium*, *Abies pindrow*, *Quercus semicarpifolia*, *Rhododendron barbatum*
Ghangaria (Gh)	5	3000 Subalpine	100	2.8 **(400)**	*T. baccata*, *A. caesium*, *A. pindrow*, *Polygonum*, *Rhododendron arboreum*, *Rumex nepalensis*, *Q. semicarpifolia*
Dayara (Dr)	4	2300 Subalpine	60	3.8 **(700)**	*Taxus baccata*, *A. caesium*, *A. pindrow*, *Rhododendron arboreum*, *Q. semicarpifolia*, *Fragaria*, *R. nepalensis*

[a] A population is given an abbreviated name (in parenthesis) after the locality. [b] Number of variants indicates the number of morphologically different types of individuals within one population, e.g. with three, two or one leaf at flowering stage and differing in seed weight color. [c] Metres above sea level. [d] Frequency is defined as the number of quadrates of occurrence/total number of quadrates x 100. [e] Density is defined as total number of individuals/m^2. Figures in parentheses **(bold)** indicate population size (total number of plants in one population).

Table 3.2. Percent podophylloresin content (on dry weight basis) in different populations of *P. hexandrum*.

Populations	Weight of the sample (g)	Weight of the ethanol extract (g)	Extract (%)	Weight of resin (g)	Resin (%)
C1	2.59	0.90	34.75	0.16	6.18
C3	4.39	1.67	38.04	0.39	8.88
D1	18.70	3.85	20.59	1.15	6.15
G1	4.05	0.98	24.19	0.20	4.94
G2	2.90	0.70	27.24	0.12	4.14
G3	5.62	1.50	26.69	0.22	3.91
G5	5.29	1.56	29.49	0.24	4.53
K1	2.37	0.83	26.58	0.13	5.48
K2	6.49	2.52	38.83	0.69	10.63
K3	4.72	1.70	36.01	0.50	10.59
K4	2.88	0.86	29.86	0.15	5.21
K6	2.76	0.93	33.69	0.16	5.79
K5	3.02	1.09	36.09	0.17	5.63
V2	1.32	0.25	18.94	0.07	5.30

C-Chopta, D-Dayara, G-Ghagaria, K-Kedarnath and V-Valley of flowers.

For species with small populations or those suffering from genetic and population bottlenecks, tissue culture techniques can be used to preserve the germplasm.

As reported earlier, the populations of this species are polymorphic with inter- and intra- population variation in vegetative and reproductive characteristics (Bhadula *et al.*, 1996b). Thus it was not known if all the populations would respond to culture medium in a similar way, or if their requirements for nutrients or growth regulators would be different. Thus, it was considered reasonable that if some elite populations with desired characteristics (e.g. high toxin content) can be identified and used for further multiplication and cultivation, the commercial demand can be fulfilled from the cultivated material. This would reduce the over-exploitation of the natural populations and the chances of their survival in the natural habitat would increase. Although reports exist on the *in vitro* propagation of *P. hexandrum* for multiple shoot multiplication (Arumugam and Bhojwani, 1990; Nadeem *et al.*, 2000; Sultan *et al.*, 2006), the present study undertaken is the first report for the biotechnological characterization of different populations from Garhwal Himalayas.

Thus the objectives of this study was two fold: (i) to find out if various polymorphic populations of *P. hexandrum* can be induced to form viable cultures; and (ii) to develop suitable conditions for the multiplication of selected natural populations with desirable characteristics (e.g. high podophyllotoxin content).

Germplasm of various populations and intra-population variants within a population was collected from various localities in Garhwal Himalaya, i.e. the subalpine and high altitude areas in Kedarnath, Tungnath, Chopta, Ghangaria and The Valley of Flowers (Chamoli district) and Dayara range (Uttarkashi district). Propagules were planted in earthern pots under nethouse conditions at Srinagar (550 m) and Tungnath (3600 m). Following the surface sterilization with 0.1% solution of mercuric chloride and 0.5% solution of Tween– 20 for 3–4 min and washed thrice in sterile distilled water under aseptic conditions, 80–100% of the explants yielded aseptic cultures. In the preliminary screening for suitable explants to initiate morphogenetic cultures, pieces of root, rhizome and leaf, shoot buds, whole seeds, endosperm and excised embryos were cultured on MS (Murashige and Skoog, 1962) and MS supplemented with various concentrations of 2, 4-dichlorophenoxyacetic acid (2,4-D), napthalene acetic acid (NAA), indole acetic acid (IAA), and benzyl amino purine (BAP) and thidiazuron (TDZ), individually and in several combinations. Of all the explants cultured, leaf and embryos responded best.

The leaf explants and the zygotic embryos of all the populations gave viable cultures. Since the alpine populations of Kedarnath gave the highest podophylloresin content (Table 3.2), greater efforts were made to propagate this population. In addition to high resin content, the population of Kedarnath also responded much better than other populations in culture medium, followed by the populations of Ghangaria. The plant material that originated from the populations growing in The Valley of Flowers and Dayara range showed poor response. Therefore, the results obtained using the explants that originated from the populations of Kedarnath are presented.

Callus Induction from Leaf Sections

For all the populations viable callus cultures using leaf explants can be produced on MS medium containing 0.5 μM each of NAA and TDZ (Table 3.3). The callus cultures showed vigorous growth on the same medium after subculturing. Also these cultures could be induced to form globular structures and produced multiple shoots. Thidiazuron is a synthetic phenylurea and considered to be one of the most active cytokinins for shoot induction in plant tissue culture (Huetteman and Preece, 1993). Reports suggest that TDZ induces shoot regeneration better than other cytokinins (Thomas, 2003; Thomas and Puthur, 2004; Husain *et al.* 2007). Malik and Saxena (1992) demonstrated the importance of TDZ in stimulating shoot regeneration in seed cultures of dry bean, a recalcitrant legume. Magioli *et al.*, (1998) have reported the superiority of a lower concentration of TDZ on *Solanum melongena* using a similar explant. As this species is endangered and fruit and seed setting is poor, leaf explants are easily available and thus, starting with 2-4 mm explants, excellent callus cultures can be obtained for all the populations. Such callus cultures can be used in future for the production of podophyllotoxin *in vitro*.

Table 3.3. Effect of NAA and TDZ on callus induction from leaf cultures of *P. hexandrum*.

Medium	Growth Regulators (μM)			
	NAA	TDZ	Callus formation	Expansion of leaf
MS	0.0	0.0	NR	NR
MS	0.5	0.5	+ + +	NR
MS	1.0	0.5	+ +	+ +
MS	2.5	0.5	+	+ +
MS	5.0	0.5	+	+
MS	0.5	1.0	+ +	+ + +
MS	0.5	2.5	+	+ + +
MS	0.5	5.0	NR	+ + + +

NR- no response, + Low, + + medium, + + + high and + + + + highest

Zygotic Embryo Culture

In this study, zygotic embryos were also cultured on MS supplemented with various growth regulators, individually or in various combinations. On MS medium containing various concentrations of NAA and BAP, the zygotic embryos showed swelling of the cotyledonary tubes and callus initiation. Excellent callus cultures were produced in the presence of IAA and NAA. In the presence of BAP (2 μM) callus cultures developed vigorously followed by production of multiple shoots. These shoots, however, could not be rooted.

When cultured under dark conditions on MS medium supplemented with 2,4-D, the zygotic embryos of *P. hexandrum* showed massive callus growth and development of shiny globular embryos. Although it was possible to multiply these calli and embryos using subculture, the embryos did not mature in the same or any other medium. It has been reported that somatic embryogenesis can occur under a variety of light-dark regimes (Ammirato, 1983). For example, whereas in *Nicotiana tabacum*, somatic embryogenesis required high light (Haccius and Lakshmanan 1965), in carrot and Carum (Ammirato, 1974) the maturation of embryos was better in total darkness. A significant role of 2, 4-D in the initiation and development of somatic embryos has been reported in *Panax ginseng* (Lee *et al*., 1990). It has also been reported that although 2,4-D is required to induce embryogenesis in many plants the differentiation does not proceed beyond globular stage. Further differentiation only occurs when 2,4-D is reduced or omitted (Bapat and Rao, 1988). Arumugam and Bhojwani (1990) reported a similar effect of 2,4-D on callus development and formation of globular embryos in *P. hexandrum*. However, they found that when transferred to NAA containing medium, dicotyledenous embryos developed. A different response of zygotic embryos of *P. hexandrum* in this study may be due to the fact that there are inter- as well as intra-population variations in morphophysiological and biochemical characteristics of the seeds of this species. Therefore, such differences may be due to the variations in physiological stage of the seeds or due to genetic diversity among and within the populations.

A combination of IAA and BAP in MS medium proved to be the best condition for the *in vitro* plant regenartion of *P. hexandrum* (Fig. 3.1). The zygotic embryos cultured on MS medium supplemented with 0.5 μM IAA and 2.0 μM BAP showed the development of multiple shoots and roots. The frequency of rooting was increased in the presence of higher IAA and lower BAP concentrations. Plants raised using this method were acclimated and hardened in pots (Fig. 3.2) and later transferred to the natural growth environment in a field nursery, located at 2000 m altitude in Garhwal Himalaya for its further cultivation.

Conclusions

Based on the variations observed at the inter- and intra-population level, *P. hexandrum* is certainly considered a complex species. The results indicate that the natural populations of *P. hexandrum* in Garhwal Himalaya are polymorphic. The differences observed in the response of various populations in the culture medium also reflect the population polymorphism of this endangered species. However, as evident from this study, the existing population size of *P. hexandrum* is small and if its exploitation continues at the present rate, it may eventually face the threats of genetic drift, population bottlenecks and inbreeding depression. Therefore, among these populations, screening and selection of podophyllotoxin

Fig. 3.1. Effect of various concentrations of IAA and BAP on zygotic embryo cultures of *P. hexandrum*. A and B: two and six week old cultures, respectively (MS + 0.5 µM IAA + 2.0 µM BAP); C and D: three and four week old cultures, respectively (MS + 1.0 µM IAA + 0.5 µM BAP); E and F: four and three weeks old cultures, respectively (MS + 1.0 µM IAA + 2.0 µM BAP). R = roots.

rich elite clones and their mass propagation using micropropagation protocols can be a helpful tool for the conservation of this species. Once such clones are selected from naturally growing populations and multiplied using micropropagation, a cultivation package can be developed so that the farmers in the upper Himalayan mountain ranges can start cultivating it as a cash crop which will reduce the pressure of extraction on this species from its natural habitats.

Fig. 3.2. Potted *P. hexandrum* plants after hardening and transfer to soil.

Acknowledgements

The authors are highly grateful to the Director of the High Altitude Plant Physiology Research Center, Srinagar- Garhwal, U.P., India for providing research facilities.

References

Ammirato, P.V. (1974) The effect of abscisic acid on the development of embryos from cell of caraway (*Carum carvi* L.). *Botanical Gazette* 135, 328–337.

Ammirato, P.V. (1983) Embryogenesis. In: *Handbook of Plant Cell Culture.* Vol I.1. *Techniques for Propagation and Breeding.* Macmillan Publishing Co., New York, pp. 82–123.

Arora, R., Chawla, R., Puri, S.C., Sagar, R., Singh, S., Kumar, R., Sharma, A.K., Prasad, J., Singh, S., Kaur, G., Chaudhary, P., Qazi, G.N. and Sharma, R.K. (2005) Radioprotective and antioxidant properties of low-altitude *Podophyllum hexandrum* (LAPH). *Journal of Environmental Pathology Toxicology and Oncology* 24, 299–314.

Arora, R., Singh, S., Puri, S.C. and Sharma, R.K. (2008) Himalayan Mayapple (*Podophyllum hexandrum* Royle): Traditional Uses, Clinical Indications and Future Prospects. In: Watson, R. and Preedy, V.R. (eds) *Botanical Medicine in Clinical Practice.* CABI Publishing, Wallingford, Oxon, United Kingdom, pp. 71–84.

Arumugam, N. and Bhojwani, S.S. (1990) Somatic embryogenesis in tissue cultures of *Podophyllum hexandrum. Canadian Journal of Botany* 68, 487–491.

Ayres, D.C. and Loike, J.D. (1990) *Chemistry and Pharmacology of Natural Products. Lignans. Chemical, Biological and Clinical Properties.* University Press, Cambridge, pp. 285–383.

Bapat, V.A. and Rao, P.S. (1988) Sandalwood plantlets from 'Synthetic seeds'. *Plant Cell Reports* 7, 734–736.

Beers, S.A., Imakura, Y., Dai, H.J., Li, D.H., Cheng, Y.C. and Lee, K.H. (1988) Antitumor agents. 99. Synthetic ring C aromatized podophyllotoxin analogues as potential inhibitors of human DNA topoisomerase II. *Journal of Natural Products* 51, 901–905.

Bhadula, S.K., Singh, A., Lata, H., Kuniyal, C.P. and Purohit, A.N. (1996a) Distribution Pattern, Population Diversity and Propagation of Some High Altitude Medicinal Herbs from Garhwal Himalaya: Problems and Prospects for Conservation. In: Pangtey, Y.P.S. and Rawal, R.S. (eds) *High Altitudes of Himalaya.* Vol II, Gyannodaya Prakashan, Nainital, pp. 389–413.

Bhadula, S.K., Singh, A., Lata, H., Kuniyal, C.P. and Purohit, A.N. (1996b) Genetic resources of *Podophyllum hexandrum* Royle, an endangered medicinal species from Garhwal Himalaya, India. *International Plant Genetic Resources Newsletter* 106, 26–29.

Chatterjee, R. (1952) Indian Podophyllum. *Economic Botany* 6, 342–354.

Chawla, R., Arora, R., Singh, S., Sagar, R.K., Sharma, R.K., Kumar, R., Sharma, A., Tripathi, R.P., Puri, S.C., Khan, H.A., Shawl, A.S., Sultan, P., Krishnan, T. and Qazi, G.N. (2006) *Podophyllum hexandrum* offers radioprotection by modulating free radical flux: Role of aryl-tetralin lignans. *Evidence Based Complementary and Alternative Medicine* 3(4), 503–511.

Dewick, P.M. (1989) Biosynthesis of Lignans. In: Atta-ur-Rahman (ed.) *Studies in Natural Product Chemistry. Structure Elucidation* 5 (Part-B). Elsevier Science Publishers, Amsterdam, The Netherlands, pp. 459–503.

Gupta, R. and Sethi, K.L. (1983) Conservation of medicinal plant resources in the Himalayan region. In: Jain, S.K. and Mehra, K.L. (eds) *Conservation of Tropical Plant Resources.* Botanical Survey of India, Howrah, pp. 101–107.

Haccius, B. and Lakshmanan, K.K. (1965) Adventive embryogenesis in *Nicotiana tobaccum* callus. *Planta* 65, 102–104.

Hartwell, J.L. and Schrechker, A.W. (1958) The chemistry of *Podophyllum. Journal of Natural Products* 15, 83–166.

Holthuis, J.J.M. (1988) Etoposide and teniposide: Bioanalysis, metabolism and clinical pharmokinetics. *Pharm Weekle (Sci.)* 10, 101–116.

Huetteman, C.A. and Preece, J.E. (1993) Thidiazuron: a potent cytokinin for woody plant tissue culture. *Plant Cell Tissue and Organ Culture* 33, 105–119.

Husain, M.K., Anis, M. and Shahzad, A. (2007) *In vitro* propagation of Indian Kino (*Pterocarpus marsupium* Roxb.) using Thidiazuron. *In vitro Cell. Dev. Biol. Plant* 43, 59–64.

Kaplan (1942) *New Orleans Med. Surg.* 94, 388.

King, L.S. and Sullivan, M.M. (1946) The similarity of the effect of podophyllin and colchicine and their use in the treatment of Condylomata acuminata. *Science* 104, 244–245.

Kumar, R., Singh, P.K., Arora, R., Chawla, R. and Sharma, R.K. (2009) Radioprotective activities of *Podophyllum hexandrum*: Current Knowledge of the Molecular Mechanisms. *Trees for Life* 4, 1–9.

Lee, H.S., Liu, J.R., Yang, S.G. and Lee, Y.H. (1990). *In vitro* flowering of plantlets regenerated from zygotic embryo derived somatic embryos of ginseng. *Hortscience* 25, 1652–1654.

MacRae, W.D. and Towers, G.H.N. (1984) Biological activities of lignans. *Phytochemistry* 23, 1207–1220.

Magioli, C., Rocha, A.P.M., de Oliveira, D.E. and Mansur, E. (1998) Efficient shoot organogenesis of eggplant (*Solanum melongena* L.) induced by Thidiazuron. *Plant Cell Reports* 17, 661–663.

Malik, K.A. and Saxena, P.K. (1992) Regeneration of *Phaseolus vulgaris* L.: High frequency induction of direct shoot formation in intact seedlings by N^6- benzylaminopurine and thidiazuron. *Planta* 186, 384–389.

Murashige, T. and Skoog, F. (1962) A revised medium for rapid growth and bioassays with tobacco tissue culture. *Physiologia Plantarum* 15, 473–497.

Nadeem, M., Palni, L.M.S., Purohit, A.N., Pandey, H. and Nandi, S.K. (2000) Propagation and conservation of *Podophyllum hexandrum* Royle: an important medicinal herb. *Biological Conservation* 92, 121–129.

Nadeem, M., Palni, L.M.S., Kumar, A. and Nandi, S.K. (2007) Podophyllotoxin content, above and belowground biomass in relation to altitude in *Podophyllum hexandrum* populations from Kumaun region of the Indian Central Himalaya. *Planta Medica* 73, 388–391.

Pandey, H., Nandi, S.K., Kumar, A., Palni, U. and Palni, L.M.S. (2007) Podophyllotoxin content in *Podophyllum hexandrum* Royle plants of known age of seed origin and grown at a lower altitude. *Acta Physiologiae Plantarum* 29, 121–126.

Pelter, A. (1986) Lignans: Some properties and synthesis. *Recent Advances in Phytochemistry* 20, 201–241.

Purohit, A.N., Lata, H., Nautiyal, S.C. and Purohit, M.C. (1998) Some characteristics of four morphological variants of *Podophyllum hexandrum* Royle. *International Plant Genetic Resources Newsletter* 114, 51–52.

Stahelin, H.F. and Von Warburg, A. (1989) From podophyllotoxin glucoside to etoposide. *Progress in Drug Research* 33, 169–266.

Sultan, P., Shawl, A.S., Ramteke, P.W., Jan, A., Chisti, N., Jabeen, N. and Shabir, S. (2006) *In vitro* propagation for mass multiplication of *Podophyllum hexandrum*: a high value medicinal herb. *Asian Journal of Plant Sciences* 5, 179–184.

Sundaresh, I. (1982) Export potential of medicinal plants and their derivatives from India. In: Atal, C.K. and Kapur, B.K. (eds) *Cultivation and Utilization of Medicinal Plants*. Regional Research Laboratory, CSIR, Jammu-Tawi, India, pp. 800–823.

Thomas, T.D. (2003) Thidiazuron induced multiple shoot induction and plant regeneration from cotyledonary explants of mulberry. *Biologia Plantarum (Prague)* 46, 529–533.

Thomas, T.D. and Puthur, J.T. (2004) Thidiazuron induced high frequency shoot organogenesis in callus from *Kigelia pinnata* L. *Botanical Bulletin of Academia Sinica (Taipei)* 45, 307–313.

Chapter 4

Traditional and Biotechnological Strategies for Conservation of *Podophyllum hexandrum* Royle – A Case Study

Amit C. Kharkwal, Dharam Singh, Om Prakash, Deepshikha Pande Katare, Ajit Varma, Amita Bhattacharaya and Paramvir Singh Ahuja

Introduction

Podophyllum hexandrum Royle (English: common name Indian Podophyllum and Himalayan Mayapple) is a small herbaceous monotypic genus of the family Berberidaceae with a diploid chromosome number 2n=12; predominantly occurring in the North Temperate Zone and discontinuously distributed in eastern North America, insular East Asia and Hindu Kush-Himalayas (Good, 1974; Xiang *et al.*, 2000; Arora *et al.*, 2008).

The roots and rhizomes of Podophyllum yield a drug known as podophyllin. Podophyllin consists of podophyllotoxin, podophyllic acid, podophylloresin and quercitin. It is a cholagogue, alterative, emetic, and a strong purgative (Singh *et al.*, 1999). The podophyllotoxins are lignans, which have anti-cancer, anti-mitotic, and immunostimulatory activities (Kaplan, 1942; Loike *et al.*, 1978; Pugh *et al.*, 2001) and whose semi-synthetic derivatives etoposide (VP-16-213), teniposide (VM-26) and etopophos are approved drugs for the treatment of testicular and small cell lung cancer (Stahelin and Warburg, 1991).

Ever increasing demand for this drug in modern medicine, coupled with its existing use in traditional systems has resulted in ruthless uprooting of the underground parts leading to intense collection coupled with lack of organized cultivation. Presently, the rate of its overexploitation far exceeds its natural regeneration potential. Consequently *P. hexandrum* has been declared an endangered species (Airi *et al.*, 1997).

In the present study, survey, standardization of propagation techniques, seed biology study, DNA polymorphism of plants using RAPD and quantification of podophyllotoxin using HPLC was done in the samples procured from five geographically distinct populations in the state of Himachal Pradesh (HP) India.

Survey and distribution

A survey was made for collection of proximal *P. hexandrum* populations of HP from five locations, namely Great Himalayan National Park (GHNP) district Kullu, Koksar and Kukum-Seri of the District of Lahul and Spiti, Pangi and Chamba regions of district Chamba. GPS (Magnell Inc., USA) data were obtained for altitude, latitude, longitude and population size (Table 4.1) from these five populations.

Table 4.1. *P. hexandrum* population size, location and geographic coordinates.

Geographical location of populations	Total no. of individuals	Altitude (msl)	Longitude	Latitude
Chamba	600	2200	32°02'.786N	76°01'.052E
Pangi	150	3600	33°05'.950N	76°26'.046E
Koksar	250	3200	32°24'.868N	77°14'.080E
Kukum-Seri	800	2800	32°42'.335N	76°40'.557E
Great Himalayan National Park (GHNP)	40	2400	31°45'.389N	77°22'.510E

Collection of plant materials

Full-grown rhizomes collected in the months of May and September from five different populations of Himachal Pradesh were used for podophyllotoxin estimation and vegetative propagation respectively. Mature and immature pods were also collected from five different regions.

Characterization

Chemical characterization

Variation in total percentage of resin content and podophyllotoxin was recorded between the five accessions (Table 4.2). No correlation was found between the amount of resin and podophyllotoxin content of the rhizomes.

Table 4.2. Total resin and podophyllotoxin content of five different populations.

Region (Himachal Pradesh)	Extract (%)	Resin (%)	Podophyllotoxin (%)
Chamba	25.3	4.29	1.29
Pangi	39.4	11.8	1.10
Koksar	28	9.4	1.96
Kukumseri	23.4	10.2	5.22
GHNP	25.7	8.57	0.70

Cytological characterization

Cytological studies on five populations revealed no variation in chromosome number and structure among the populations. The representative karyotype is shown in Fig. 4.1. Across the five populations it was observed that the Kukum-Seri population exhibited the breakage of chromosome (Fig. 4.2) and Pangi showed doubling of chromosome followed by cytomixes (Fig. 4.3 a–d).

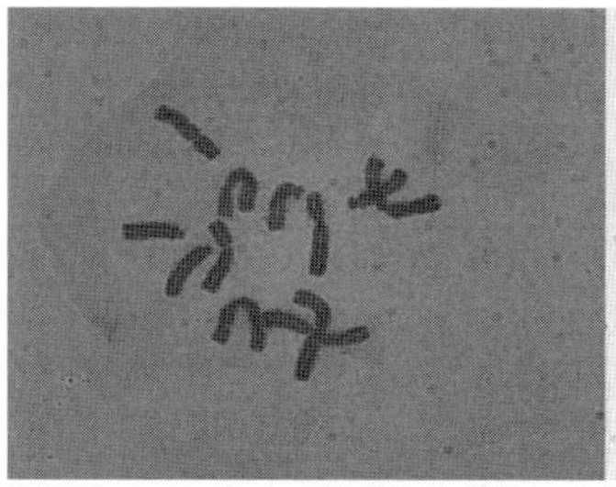

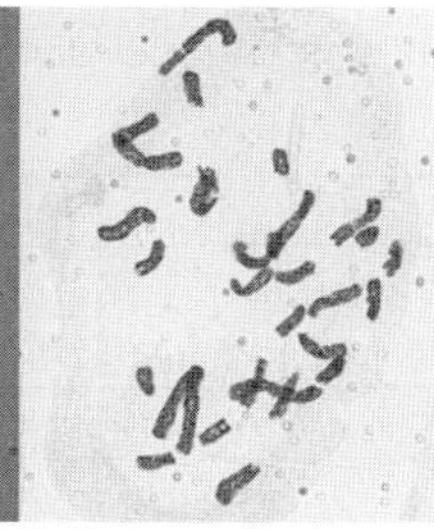

Fig. 4.1. Karyotype of *P. hexandrum* (100X); **Fig. 4.2.** Kukum-Seri population showing breakage of chromosome (100X).

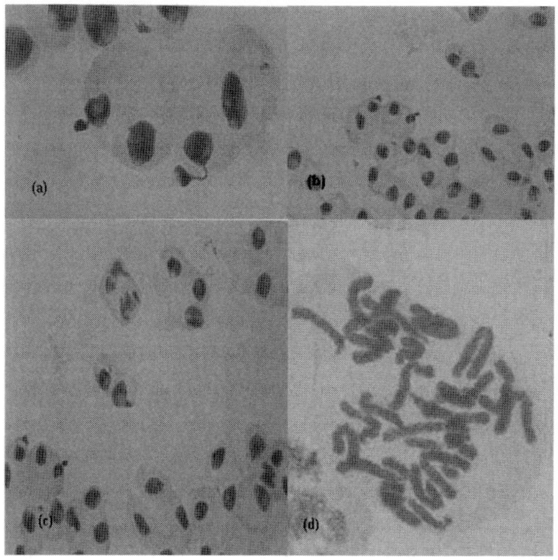

Fig. 4.3. a–d. Pangi population showing cytomixis and chromosome doubling (100X).

Genetic diversity characterization by using randomly amplified polymorphic DNA (RAPD)

Eighty random 10-mer primers were screened for the five populations (9–10 plants/population). Only 11 primers, which gave three or more polymorphic bands, were finally selected for producing consistent, scorable and reproducible results and therefore taken for further analysis. Shannon's index of phenotypic diversity was used to partition the diversity into intra- and inter-population components (Table 4.3). The genetic diversity present within populations, H_{pop}/H_{sp}, compared with that between populations, H_{sp}-H_{pop}/H_{pop}, revealed an average diversity of 62% within populations. However, the distribution of diversity did vary between and within populations with different primers.

Table 4.3. Partitioning of genetic diversity into within and between populations for 11 primers.

Primer	H_{sp}	H_{pop}	H_{pop}/H_{sp}	H_{sp}-H_{pop}/H_{pop}
OPA-05	2.87	1.61	0.56	0.43
OPA-07	1.83	1.22	0.66	0.33
OPA-08	3.14	1.64	0.52	0.47
OPA-14	1.87	1.05	0.56	0.43
OPA-15	1.26	0.87	0.69	0.31
OPA-16	2.6	1.8	0.69	0.31
OPA-18	2.35	1.51	0.64	0.35
OPE-02	2.29	1.45	0.63	0.36
OPG-11	1.94	1.33	0.68	0.31
OPM-04	2.32	1.43	0.61	0.38
OPM-15	1.72	1.07	0.62	0.37
Average	2.19	1.36	0.62	0.36

Phenotypic frequencies detected with 11 primers were calculated and these data were used in estimating the diversity (H_o) within the five populations of *P. hexandrum*. The Kukum-Seri population exhibited lowest intra-population variability as compared to Chamba population, which exhibited highest variability followed by GHNP, Koksar and Pangi populations (Table 4.4).

The principal co-ordinate analysis displayed the relationship of the individuals within the populations. Within each population, most individuals clustered in a single group (Fig. 4.4).

Exceptionally, the GHNP population did not show clear population structure. The narrow genetic base of Kukum-Seri and Pangi populations were displayed in a PC plot where they form a narrow distinct group. Though individuals of Chamba population form a distinct group even then they are distributed throughout the axis due to a broad genetic base. Some of the individuals of Koksar population were mingled with other populations. But the individuals of GHNP populations were distributed randomly throughout the axis suggesting a broad genetic base as well as relationship with other populations. The evaluation of genetic variability distribution within and among populations was measured by AMOVA (analysis of molecular variance). AMOVA values obtained by RAPD data show that 74% of total variation is attributed to the individuals within a population while 26% is due to differences among populations (Table 4.5).

Based on the RAPD data the genetic distances were calculated among populations. The GHNP and Chamba populations showed more genetic similarity with all other populations. The maximum genetic similarity was observed between Pangi and Kukumseri populations, followed by GHNP and Chamba. Though overall genetic distances among populations were below 21%, they form a distinct entity except few individuals. This result can also be attributed with allele similarity in the adjacent population. The Nm value displays the gene flow existing between the populations. The maximum Nm value (Table 4.6) was found between GHNP and Chamba (1.510) followed by Koksar with Chamba (0.913) and Koksar with GHNP (0.903). Based on binary RAPD data, the Nei genetic distances were calculated between the populations. These values were used to construct a dendrogram by the UPGMA method to display the relationship between the populations. There were two distinct clusters formed in the dendrogram (Fig. 4.5); one of them with GHNP, Kukumseri and Pangi populations and another with Koksar and Chamba. Though GHNP is associated with the former population it is distantly placed in that cluster.

Table 4.4. Estimates of genetic diversity (H_o) within five populations of *P. hexandrum.*

Primer	GHNP	Koksar	Kukum-Seri	Pangi	Chamba
OPA-05	1.29	1.71	1.69	1.84	1.51
OPA-07	1.04	0.83	1.08	1.49	1.64
OPA-08	1.86	2.01	1.72	0.32	2.3
OPA-14	0.92	1.13	0.32	1.94	0.93
OPA-15	0.98	0.52	0.85	0.85	1.18
OPA-16	1.65	1.86	1.16	2.16	2.16
OPA-18	2.01	1.71	0.94	1.08	1.83
OPE-02	1.56	1.5	1.35	1.74	1.08
OPG-11	1.52	1.86	0.78	1.08	1.41
OPM-04	2.01	1.25	1.46	0.93	1.49
OPM-15	1.28	0.99	0.32	0.93	1.83
Average	1.46	1.39	1.06	1.30	1.57

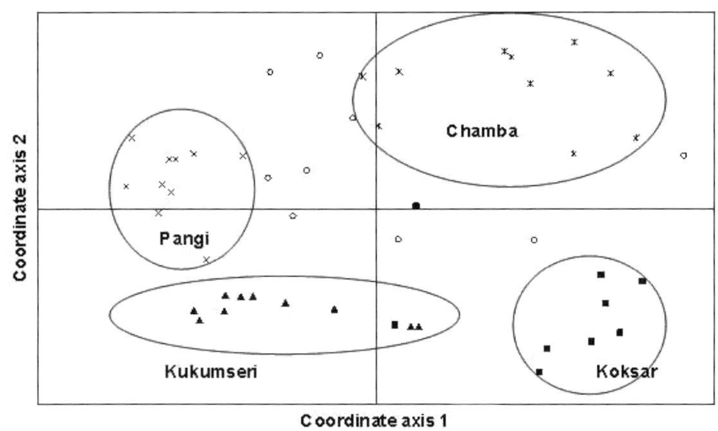

Fig. 4.4. Principal co-ordinate analysis of *P. hexandrum* based on 11 random primers.

Table 4.5. AMOVA in 48 individuals among five populations of *P. hexandrum.*

Source of variation	d.f.	SS	MS	Est. Variance	% of variation	P
Among population	4	145.767	36.442	2.930	26%	0.001
Within population	43	358.233	8.331	8.331	74%	0.001

d.f. – degree of freedom, MS – mean squared deviation, SS. sum of squares, P– significance of the variance components after 999 random permutations

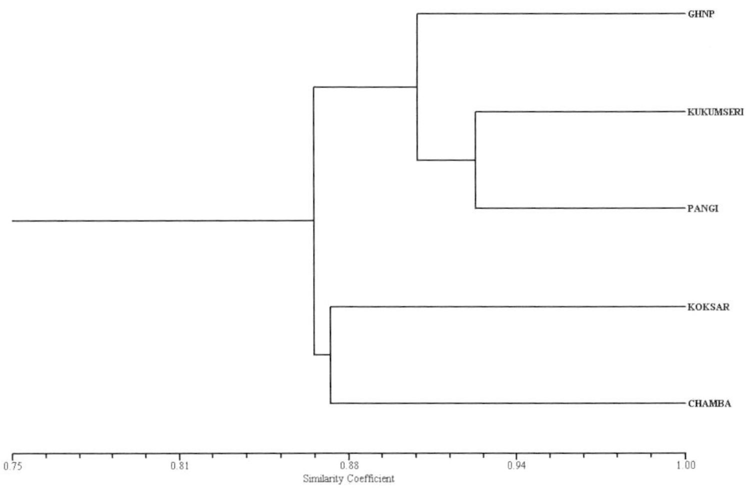

Fig. 4.5. Cluster analysis of *P. hexandrum* based on 11 random primers.

Table 4.6. Estimates of genetic distances (Nei) and gene flow (Nm) between the populations of *P. hexandrum* in Western Himalayas.

Population	Population	Genetic Distance (Nei)	Gene Flow (Nm)
GHNP	Koksar	0.125	0.903
GHNP	Kukum-Seri	0.101	0.827
Koksar	Kukum-Seri	0.122	0.598
GHNP	Pangi	0.109	0.849
Koksar	Pangi	0.207	0.376
Kukum-Seri	Pangi	0.081	0.794
GHNP	Chamba	0.105	1.510
Koksar	Chamba	0.141	0.913
Kukum-Seri	Chamba	0.166	0.547
Pangi	Chamba	0.167	0.602

Propagation

Sexual method

Seed propagation through direct seed germination and excised embryo culture was attempted in the present study for attaining quick and high germination of *P. hexandrum* seeds.

Seed biology

The present study on seed biology of *P. hexandrum* indicates the existence of differential seed dormancy based on responses to various treatments and ontogenic variability in the five populations of HP.

Seed viability

Viability of seeds decreased with time when stored at 4°C for extended time periods (1–6 months). However, loss of viability was evident only by the excised embryo culture as the TTC test gave positive results even after 6months of storage. However, when the pods were stored at 4°C for up to 6 months, no loss of viability was recorded through the culture of excised embryos (Table 4.7).

Table 4.7. Embryo viability under different seed storage conditions.

Storage period (months)	Excised embryo culture of stored seeds	Excised embryo culture of seeds stored with pods at 4°C	TTC test of excised embryos of stored seeds
	Percentage germination		
0	100	100	100
1	100	100	100
2	93.3±5.7	100	100
3	63.3±5.7	100	100
4	40±0	100	100
5	26.6±11.5	100	93.3±5.7
6	0	100	90±0

Individual datum is the mean and s.d. of three replicates

Studies on water relations

When freshly collected seeds of *P. hexandrum* were subjected to water uptake study, it was revealed that seed absorbed very negligible amount (ca. 12%) of water even after 48h of inhibition.

Studies on seed dormancy

Different dormancy breaking treatments were employed in order to break the innate seed dormancy of *P. hexandrum* (Kharkwal *et al.*, 2002). Seeds treated with hot water at 60°C for 60 sec showed significant improvement in breaking the dormancy of otherwise dormant seeds of Chamba population for 2 consecutive years.

Comparative study on seed dormancy across five populations

Only three out of five populations responded to the standardized dormancy breaking treatments (Table 4.8). On an average the seeds from Chamba population showed maximum response, followed by Koksar and GHNP. Seeds collected from Kukum-Seri germinated (60–80%) after 10 months of treatment and seed germination was irrespective of treatment as the control seeds also germinated. The seeds from Pangi evoked very low germination.

Ontogeny

When the pattern of seed germination was studied, it was observed that while in the Kukum-Seri population seed germination was hypogeal (Fig. 4.6a) along with direct emergence of a true leaf (Fig. 4.6b), in the rest of the populations the germination was epigeal followed by emergence of the hypocotyl (Fig. 4.7) with two cotyledons forming a sheathing leaf base enclosing the plumular axis. In very few seeds of Chamba population's polyembryony was also observed as two hypocotyls bearing four cotyledonary leaves emerged from a single seed (Fig. 4.8).

Table 4.8. Relative germination response of different populations of *P. hexandrum* seeds to dormancy breaking treatments.

Treatment	Chamba	Koksar	GHNP	Pangi	Kukum-Seri*
	Percentage germination				
Control	20±0	16.6±5.7	20±10	16.6±11.5	83.3±5.7
Hot water	96.6±5.7	76.6±5.7	83.3±11.5	10±0	76.6±5.7
Nitrate	93.3±5.7	73.3±5.7	63.3±5.7	3.3±5.7	66.6±15.2
GA_3	90±0	73.3±5.7	63.3±5.7	6.6±5.7	86.6±15.2
Acid	86.6±5.7	73.3±5.7	63.3±5.7	10±5.7	80±10

* Data recorded after 10 months of seed sowing.
Individual datum is the mean and s.d. of three replicates

Emergence of functional leaves

Application of GA_3 (30, 45 and 60 ppm) as well as manual removal of both the cotyledons helped in reducing the time taken for leaf emergence considerably (Table 4.9).

Table 4.9. Effect of GA_3 on true leaf emergence in *P. hexandrum.*

Days	15	30	45	60	75	90	105	120
Treatment	Percentage emergence							
Control	0	0	0	22.2±9.6	22.2±9.6	22.2±9.6	27.7±9.6	27.7±9.6
GA_3 15ppm	0	0	5.5±9.6	5.5±9.6	5.5±9.6	5.5±9.6	5.5±9.6	5.5±9.6
GA_3 30ppm	5.5±9.6	16.6	16.6	22.2±9.6	22.2±9.6	33.3±9.6	38.8±25.4	55.5±9.6
GA_3 45ppm	0	5.5±9.6	11.1±9.6	16.6	16.6	38.8±9.6	50	61.1±19.2
GA_3 60ppm	0	0	0	0	0	16.6	33.3	50
GA_3 15ppm*	0	0	0	0	0	22.2±9.6	33.3	55.5±9.6
GA_3 30ppm*	0	22.2±9.6	16.6±	33.3±28.8	38.8±25.4	83.3±16.6	88.8±9.6	88.8±9.6
GA_3 45ppm*	11.1±19.6	22.2±25.4	33.3±16.6	38.8±9.6	49.9±28.8	66.6±16.6	77.7±25.4	88.8±9.6
GA_3 60ppm*	11.1±9.6	33.3±16.6	33.3±16.6	33.3±16.6	49.9±28.8	83.3	94.4±9.6	100
One cotyledon cut	0	0	5.5±9.6	5.5±9.6	5.5±9.6	38.8±9.6	38.8±9.6	38.8±9.6
Both cotyledons cut	16.6±25.4	61±34.6	72.2±34.6	72.2±34.6	72.2±34.6	77.7±25.5	83.3±16.6	83.3±16.6

Individual datum is the mean and s.d. of three replicates; *15 day interval

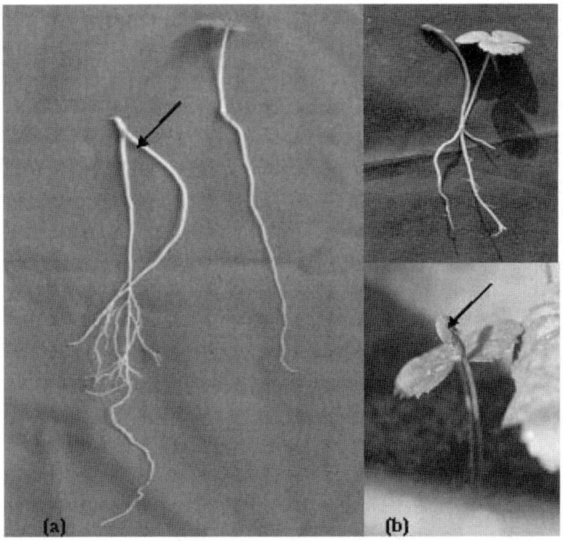

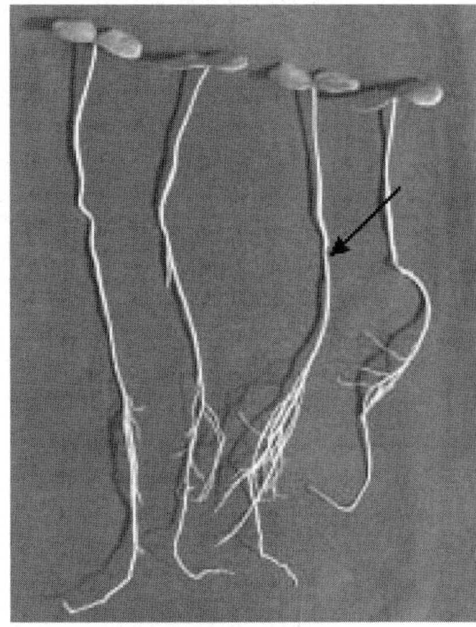

Fig. 4.6a–b. Hypogeal germination (a) and direct emergence of true leaf (b).
Fig. 4.7. Epigeal germination in *P. hexandrum* seeds.

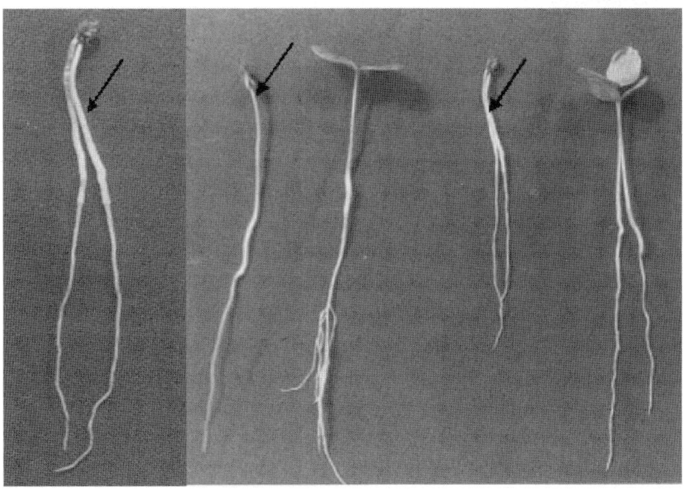

Fig. 4.8. Polyembryony in *P. hexandrum* seeds

Embryo culture

Effect of different media on the germination of embryos excised

While the immature embryos did not germinate at all, the partially mature and fully mature embryos germinated to the extent of 100% on the optimal medium within a week of their culture (Fig. 4.9; Table 4.10). The highest germination of both alpine and temperate zone embryos was recorded on B-5 as well as half strength B-5 media, and there were no significant differences in the percentage germination over the year.

Fig. 4.9. Excised embryo culture and mass propagation of *Podophyllum hexandrum.* (a) Excised embryos; (b) plantlets from germinated excised embryos; (c) plantlet raised from germinated embryos with two cotyledonary leaves; (d) embling with early emergence of true leaf from the base of the hypocotyls; (e) 1-year-old plants growing in a polytunnel.

Emergence of functional leaves

Prior to the setting of hypocotyl dormancy, i.e. 90 days after field transfer, percentage true leaf emergence in the temperate and alpine emblings was 38% and 54%, respectively (Fig. 4.10). In general, seeds collected above the tree line showed a higher response to true leaf emergence compared to those collected from the temperate zones.

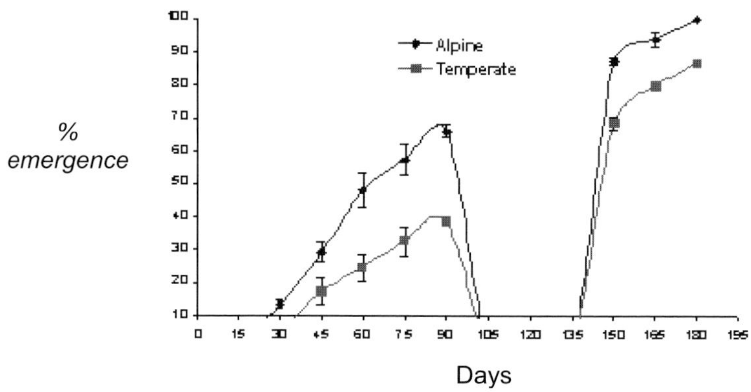

Fig. 4.10. True leaf emergence pattern of emblings from alpine and temperate zones.

Hardening of emblings derived from embryos germinated on different media

The effect of different media on the survival percentage of the emblings was observed when they were transferred to pots in the greenhouse (Table 4.10).

Table 4.10. Effect of media composition on germination, embling development, field survival and third leaf emergence in seed embryos from the alpine and temperate zones.

Media	Strength	Percentage germination after 1 week		Percent embling development after 4 weeks		Percent field survival after 12 weeks	
		ALP	TEMP	ALP	TEM	ALP	TEM
B-5	FS	100	100	100	100	83.3±5.7	100
	FS+G	100	100	100	100	100	100
	HS	100	100	100	83.3±5.7	76.6±5.7	83.3±5.7
	HS+G	100	100	100	100	80±10	86.6±5.7
WPM	FS	3.3±5.7	26.6±5.7	0	0	70±10	0
	FS+G	86.6±11.5	86.6±5.7	100	100	16.6±5.7	66.6±5.7
	HS	26.6±5.7	26.6±5.7	96.6±5.7	0	76.6±5.7	0
	HS+G	100	100	96.6±5.7	70± 0	86.6±5.7	0
MS	FS	60	3.33±5.7	66.6±5.7	0	86.6±5.7	0
	FS+G	93.3±11.5	3.3±11.5	83.3±5.7	83.3±5.7	26.6±5.7	53.3±5.7
	HS	100	6.6±25.1	26.6±5.7	0	16.66±5.7	0
	HS+G	100	96.6±5.7	83.3±5.7	83.3±5.7	93.33±5.7	0

ALP: alpine; TEMP: temperate; FS = full strength; HS = Half strength; G = GA_3 2.5 μM; individual datum is the mean and s.d. of three replicates

Establishment of emblings in the natural reserve

While 90.3% plantlets with true leaves grew vigorously after 1 year of transfer to the GHNP, nearly 81% plantlets survived and showed vigorous growth after the second year.

Asexual method

Somatic embryogenesis and rhizome cuttings were attempted as an alternate system for plant propagation.

Somatic embryogenesis

Explant response

Browning and death of the explants was observed on both basal B-5 and MS medium without the use of activated charcoal (600 mg/l) and ascorbic acid (10 mg/l). B-5 medium was more responsive in terms of time taken for callus induction and also proliferation.

Among the different explants selected for callus induction, hypocotyl gave the best response. Of the different PGRs tested, the maximum amount of proliferating callus (Fig 4.11b–d) was induced after 1 month of inoculation at 25°C only on basal B-5 medium containing picloram (5 μM) in combination with thidiazuron (TDZ) (1 μM). After subsequent removal of TDZ, picloram alone was not able to evince a friable callus growth response. However, when 2,4-D or NAA (2.5 μM each) in combination with BAP (1 μM) were supplemented to the picloram-containing medium, a fast sustainable growth of a friable callus was observed. This callus could be sub-cultured every 4 weeks over a prolonged period of over 18 months.

The liquid suspension culture growth could be obtained from this callus. This culture required a sub-culture interval of 10 days, having a cell doubling time of 6 days and can be gainfully employed in studies on production of secondary metabolites in culture.

Induction of somatic embryos

Globular embryos were formed on the yellow friable callus (Fig. 4.11e) after their transfer to medium containing 2.5 and 5 μM 2,4-D. These primary embryos when sub-cultured on the same medium gave rise to profuse secondary embryogenesis.

Development of the somatic embryos and their conversion

On transfer to hormone-free basal B-5 medium, the globular somatic embryos developed into heart, torpedo and cotyledonary stages (Fig. 4.11 f–h). These embryos could be finally converted into plantlets although their conversion frequency was very low.

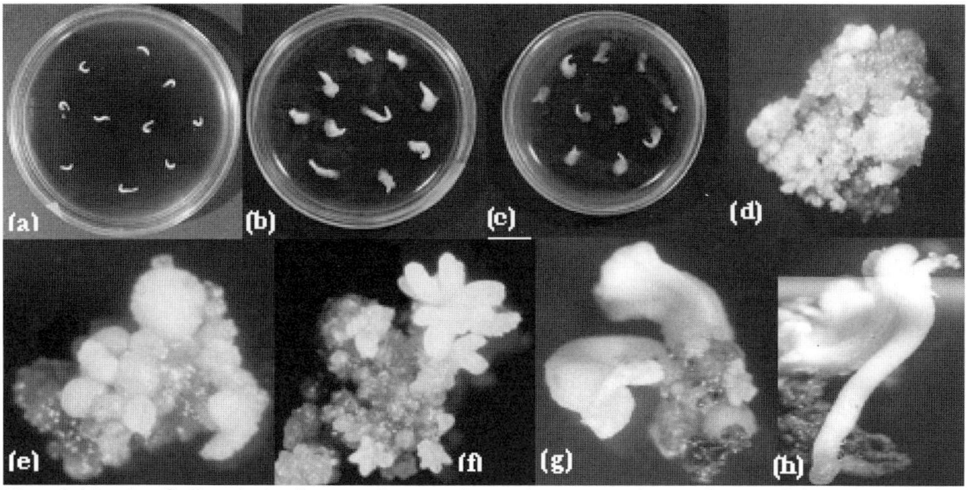

Fig. 4.11a–h. Callus induction and somatic embryogenesis; a: hypocotyl segments b–d: callus induction; e: globular embryos; f: heart shaped embryos; g: torpedo shaped embryos; h: cotyledon shaped embryos.

Propagation through rhizome cuttings

Out of the two types of rhizome segments used in the present study only the transversely cut segments with an apical bud were able to respond to the different treatments and produce plantlets (Table 4.11). The longitudinally cut rhizome did not respond to any treatment and turned necrotic.

Bud emergence

The emergence of buds in the segments started 2 weeks after sowing irrespective of the treatments. All the treatments involving auxins and cytokinins evoked better response in the number of buds/segment as compared to control (Fig. 4.12a–c).

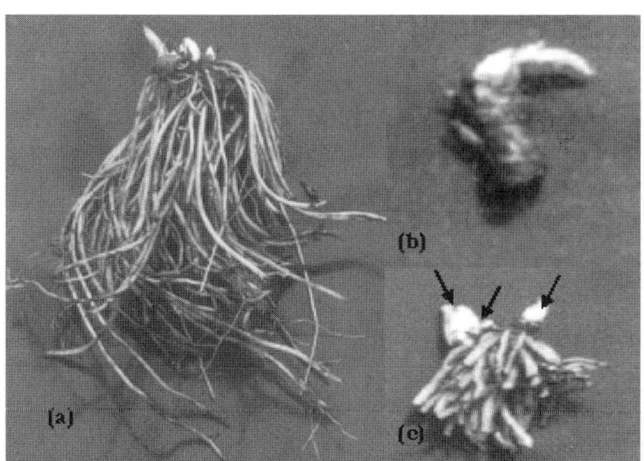

Fig. 4.12a–c. Mature rhizome (a), cut rhizome segment (b) and bud induction on rhizome segment (c).

Table 4.11. Effect of auxin and cytokinin treatment on rhizome cutting of *P. hexandrum.*

Treatment	No of Buds	No. of roots/rhizome	Root length (cm)
IBA 25ppm	2.45±0.07	10.66±0.28	10.43±0.50
IBA 50ppm	2.24±0.21	12.66±0.76	12.2±1.95
IBA 100ppm	2.37±0.21	23.1±4.16	16.63±0.66
IBA 200ppm	2.28±0.38	29.6±1.04	25.53±3.74
IBA 400ppm	2.37±0.45	31.5±0.70	24.9±2.16
NAA 25ppm	1.74±0.32	13.33±1.40	10.06±1.35
NAA 50ppm	1.87±0.37	17.83±0.76	13.66±1.26
NAA 100ppm	1.83±0.31	25.16±1.25	19.33±2.65
NAA 200ppm	2.03±0.38	27.66±0.57	25.96±3.23
NAA 400ppm	2.16±0.14	30.5±2.64	28.03±1.50
IAA 25ppm	1.7±0.40	13.0±1.32	11.46±1.18
IAA 50ppm	1.83±0.14	16.5±1.32	16.76±1.10
IAA 100ppm	2.12±0.21	20.16±2.25	20.7±0.72
IAA 200ppm	2.20±0.18	27.83±3.05	28.06±1.69
IAA 400ppm	2.41±0.14	32.5±2.0	32.53±1.05
Kinetin 25ppm	2.37±0.12	11.66±1.04	9.36±1.01
Kinetin 50ppm	2.0±0.0	13.66±1.75	10.06±1.95
Kinetin 100ppm	2.08±0.06	13.16±1.28	9.13±0.75
Kinetin 200ppm	2.04±0.06	13.0±2.64	10.26±1.65
BAP 25ppm	2.37±0.12	9.83±0.28	9.23±0.85
BAP 50ppm	2.12±0.12	10.0±0.5	10.66±2.15
BAP 100ppm	2.12±0.25	13.16±1.04	11.03±1.36
BAP 200ppm	2.33±0.06	13.16±0.76	11.06±1.87
Control	1.16±0.07	12.5±2.78	11.23±1.25

Individual datum is the mean and s.d. of three replicates

Rooting

The induction of rooting started about 6 weeks after sowing, much later than the bud emergence, irrespective of the treatment and by 12 weeks all the rhizome segments were able to root (Fig. 4.13a–b).

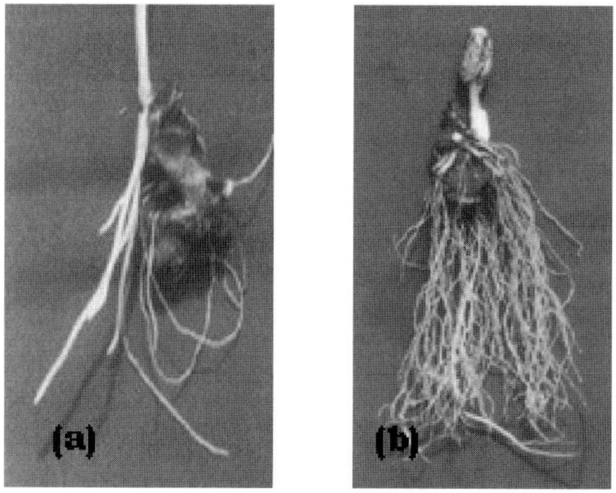

Fig. 4.13a–b. Rooting on cut rhizome segments of *P. hexandrum* (a: control, b: treatment).

Discussion

The present study on five populations from different locations shows a distinct pattern of variability whereby those populations that have been isolated over a period of time are relatively stable and have lower intra-population variability. The following discussion elaborates the outcome of key results on the nature of diversity and the germination responses following different techniques of excised embryo culture, somatic embryogenesis and direct seed germination so as to overcome the barriers of seed and hypocotyl dormancy.

Characterization

Estimated population density from the five different domains of Himachal Pradesh revealed small sized populations while only Kukum-Seri and Chamba populations exceeded minimum viable population (MVP) size required for survival and reproductive efficiency (Endels *et al.* 2002). A similar decline in plant populations of Garhwal Himalayas has also been reported by Bhadula *et al.* (1996). Following RAPD characterization, the five populations of *P. hexandrum* revealed high inter- and intra-population diversity (Kharkwal *et al.*, 2008). Earlier, molecular characterization of two populations from Himachal Pradesh revealed a high degree of polymorphism ranging between 25–73% (Sharma *et al.* 2000). Recently Sultan *et al.* (2008) have demonstrated high levels of genetic and chemical diversity in 12 nursery grown accessions from the western Himalayas.

Despite the fact that population size was low in *P. hexandrum*, the observed genetic diversity was reasonably high and as such it is not likely that reduction in population size will result in imminent loss of genetic variation (Widen and Andersson, 1993).

Among the populations, Kukum-Seri and Pangi showed narrow genetic base and they are distinct from other populations. Long seed dormancy periods of both the populations and a different ontogeny of Kukumeri support the clustering of populations into distinct groups.

Combining principal co-ordinate plot with Shanon index and DNA polymorphism, high genetic variability was found within Chamba and GHNP populations. Due to increased gene flow and more allelic similarity between the populations, the molecular variation among the populations was found to be 26%. The higher molecular variation within population could be attributed to the broad genetic base and diversity among individuals.

Comparison between geographical location and genetic variability detected by RAPD suggests a relationship between populations, genetic erosion and environmental conditions. The Chamba and GHNP populations have undergone high anthropogenic pressure resulting in the development of a broad genetic base, whereas, Koksar, Kukum-Seri and Pangi showed low genetic variability being isolated and comparatively undisturbed.

Though individuals of the Chamba population form a distinct group, even then they are distributed throughout the axis due to broad genetic base. Some of the individuals of the Koksar population were mingled with other populations. But the individuals of GHNP populations were distributed randomly throughout the axis suggesting a broad genetic base as well as relationship with other populations.

The GHNP and Chamba populations showed more genetic similarity with all other populations. The maximum genetic similarity was observed between Pangi and Kukumseri populations, followed by GHNP and Chamba populations. Though overall genetic distance among populations was below 21%, they form a distinct entity except few individuals.

The AMOVA analysis displayed low genetic structure and high genetic variation among individuals, which can be attributed to the divergence of the individuals suggesting risk of genetic erosion in the Pangi and Kukum-Seri populations and may lead to genetic drift also.

The relationship between Pangi and Kukum-Seri populations can be interpreted on their closeness of geographical locations (Table 4.1). Though the Koksar population was geographically close to Kukum-Seri since both represent the Lahul valley of Trans-Himalayan region, they are genetically distinct. However, the Chamba and GHNP populations representing the Greater Himalayan region have high genetic similarity and gene flow.

The findings of the work suggest strong gene flow between the populations, due to the seed dispersal under natural conditions through frugivory and agents like birds, resulting in long-range dispersal of seeds. Also, tribes of Western Himalayas carry the pods and plants from one place to another, which can be inferred through their grazing route, thus, resulting in the migration of germplasm. Such gene flow among populations either through seeds or ramets has been known to occur in other species also (Pleasants and Wendel, 1989). In comparison to others, the population from Pangi and Kukum-Seri were rather undisturbed. The GHNP population was greatly depleted due to anthropogenic pressures; moreover, there existed gene flow between GHNP and other populations suggesting the diversity of the population and correlation with the GHNP population.

In an earlier study based on AFLP analysis of seven populations of *P. hexandrum* from the Tibetan region of Sichuan Province, China, Xiao *et al.* (2006) reported a very low level of genetic diversity with limited gene flow (Nm=0.43). This low Nm value indicates genetic drift with distinct genetic differentiation as observed in the present study. The higher Nm value observed in the present study may be attributed to transfer of ramets and genets from one place to another.

Cytological characterization revealed similar chromosome types throughout the populations. But breakage of chromosome was observed only in the Kukum-Seri population, which may be due to the high concentration of podophyllotoxin present in the roots. The finding may be very well correlated with the cytotoxicity of podophyllotoxin known to induce DNA topoisomerase II-mediated chromosomal breakage (Osheroff, 1989). Interestingly, chromosome doubling as a result of cytomixis was also observed in the Pangi population. The event can be used to regenerate tetraploid plants with elite characters.

Analysis of genetic variation within the taxon is crucial for conservation purposes because of its implication on long term survival and continuous evolution of species (Young and Clarke, 2000; Kharkwal *et al.*, 2008). The reduced genetic variability in the Pangi and Kukum-Seri populations in the Western Himalaya could be the result of genetic isolation by habitat fragmentation and anthropogenic pressure. Most of these events have an additive effect with severe implications for survival of the population. The detected genetic variability among and between populations can be used to propose an appropriate management strategy.

Seed biology

Mass propagation of *P. hexandrum* has always been limited by a long juvenile phase (Rust and Roth, 1981), poor fruit setting, and long dormancy periods of 10 to 24 months, low and erratic seed germination and poor seedling establishment (Badhwar and Sharma, 1963; Arumugam and Bhojwani, 1990; Kharkwal *et al.*, 2004). Early attempts in domestication of *P. hexandrum* were by multiplication of rhizomes and seeds (Badhwar and Sharma, 1963). However, since rhizomes are also a source of podophyllotoxins, considerable loss is incurred in terms of propagules and harvestable material of economic importance (Sadowska *et al.*, 1997). Seed germination is not only erratic but is further limited by long periods of seed coat or mechanical dormancy and endosperm dormancy. Furthermore, it is difficult to break the dormancy by traditional methods (Badhwar and Sharma, 1963). More than seed dormancy, hypocotyl dormancy prevents the emergence of the true leaf, thereby reducing the chances of plant survival in nature (Nautiyal *et al.*, 1987). After germination hypocotyl dormancy is caused by the fusion of the two cotyledons in a sheathing base (Purohit and Nautiyal, 1986).

Work done on seed biology of *P. hexandrum* (Nautiyal *et al.*, 1987; Chaudhary *et al.*, 1996, 1998; Singh *et al.*, 1999; Nadeem *et al.*, 2000; Kharkwal *et al.*, 2004) indicates the extended time involved for overcoming seed dormancy.

The present study shows that the *P. hexandrum* seeds do not require any after-ripening treatment e.g. chilling. This is in contrast to earlier reports wherein Nautiyal *et al.* (1987), Chaudhary *et al.* (1996, 1998), Singh *et al.* (1999) and Nadeem *et al.* (2000) reported a need for after-ripening in this plant. From the result of the present study it is evident that freshly collected seeds germinate when sown immediately after pod harvest and after appropriate pre-germination treatment (Fig. 4.14). Thus, it can be easily concluded in the present study that there is no after-ripening period required in *P. hexandrum* seeds. The dormancy in *Podophyllum* seeds appeared to be more of chemico-mechanical type as also suggested by Arumugam and Bhojwani (1990). The seeds failed to imbibe water above 12%. Failure to imbibe water by seeds with a hard coat has been well documented in a number of plant species (Rolston, 1978). Badhwar and Sharma (1963) have also reported the occurrence of seed coat dormancy in *P. hexandrum*.

Fig. 4.14. Mass propagation of *P. hexandrum* through seeds.

Generally the seeds, which exhibit any kind of dormancy, also have prolonged seed viability (Tran and Cavanagh, 1984). However, from the present study the nature of *Podophyllum* seeds appears to be unique. The seeds exhibit some properties of recalcitrant seeds e.g. high moisture content (> 50%), sensitivity to desiccation and loss of viability on storage (Pammenter and Berjak, 1999), and concomitantly the seeds have a complex dormancy system involving both endosperm and seed coat, which is a property of orthodox seeds (Ellis *et al.,* 1989). As indicated by our studies and also by Pammenter and Berjak (1999) for recalcitrant seeds, the *Podophyllum* seeds maintain their viability when stored along with the pulpy fruit and not as seeds alone. In nature also the seeds germinate under moist conditions after tiding over the adverse winter. The simultaneous recalcitrance and dormant nature of *Podophyllum* appears to be an adaptation to its moist, cold and high altitude habitat. The classification and nature of *Podophyllum* seeds into recalcitrant and orthodox requires further investigation.

Consequent to germination of seedlings, hypocotyl dormancy has been considered to be a major constraint in plant establishment of *P. hexandrum* (Purohit and Nautiyal, 1986). After seed germination *P. hexandrum* requires one growing season to overcome hypocotyl dormancy (Purohit and Nautiyal, 1986). However, the present study demonstrates the successful establishment of *P. hexandrum* within the same growing season by a periodic GA_3 treatment. GA_3 is known to effect mobilization of storage reserves in cotyledons (Kaur *et al.* 1998); therefore, it may have evinced the true leaf emergence by making available the food reserves to the otherwise dormant plumular axis.

Differential germination in the seeds of the different populations also indicated differential dormancy. This is not surprising because the differences in seed dormancy within and between the populations could be due to many factors like genetic, maternal environment during maturation, age of the mother plant, position of seed on the plant (Gutterman, 1993), seed mass (Milberg *et al.* 1996) and size of the mother plant (Philippi, 1993).

Ontogenic studies on seed germination revealed that seeds from the Kukum-Seri population exhibited hypogeal germination, and those of the other populations showed epigeal germination. The Kukum-Seri seeds escaped the cotyledonary stage and hypocotyl dormancy through direct emergence of a true leaf. Polyembryony, which is the production of more than one embryo in a seed, was also observed in some of the seeds collected from Chamba populations at very low frequency, further suggesting a role of maternal factors in controlling plant development (Fenner, 1991).

Excised embryo culture

An effective protocol for mass propagation of *P. hexandrum* plants through embryo culture was also developed (Kharkwal *et al.* 2004). This technique offers an important solution to their domestication, which has been a persistent problem through conventional modes of propagation, i.e. seeds and vegetative propagation (Troup, 1915; Badhwar and Sharma, 1963).

Germination was significantly increased when excised embryos of partially mature and mature seeds, collected from various zones, were cultured under *in vitro* conditions. Therefore, it is evident that the excised embryo cultures have the potential for high and uniform germination and consequently high plant establishment and survival.

Embryo cultured raised plants were successfully established under field conditions at GHNP. The main utility of this technique lies in uniform and quick germination/development response across all the five populations studied, which showed differential germination response when propagated through seeds (Table 4.12).

Table 4.12. Comparison between germination time of seeds and excised embryos.

Accession	Days to germinate/Development	
	Seeds	Excised embryos
Chamba	60/90	7/30
GHNP	60/90	7/30
Koksar	60/90	7/30
Kukum-Seri	180/300	7/30
Pangi	-/-	7/30

Somatic embryogenesis

Plant regeneration through somatic embryogenesis was attempted in a *P. hexandrum* accession from the Chamba region to explore the *in vitro* route to propagation and conservation.

For the induction of somatic embryogenesis in *P. hexandrum*, only auxin (2.5 μM 2,4-D) was found to be suitable. Basal B-5 medium with additives, picloram and TDZ was used for initiation of callus on hypocotyls followed by their subsequent transfer to TDZ-free media containing picloram, 2,4-D or NAA and BAP. Although the medium used for callus production was more complex than the earlier reports, it was more superior in terms of sustainable callus proliferation and growth as was evident from the liquid culture studies where a cell doubling time of 6 days was achieved.

Like in most somatic embryogenesis systems (von Arnold *et al.*, 2002) subsequent development of the embryos occurred only on hormone-free medium. Generally during somatic embryogenesis a gradual reduction in the level of auxins is required for further development of the somatic embryos (Ammirato, 1983; von Arnold *et al.*, 2002).

Vegetative propagation

It is evident from the present study that the transversely cut segments with an apical bud could root, irrespective of the treatments. However, all the auxins at different concentration could evoke better response in all the three aspects i.e. early root initiation, higher number of roots and longer roots. Both auxins and cytokinins were able to induce the emergence of adventitious buds alike.

Multiplication of plants through rhizome segments can be used as an alternative method for propagation (Fig. 4.15). This method can help in the shortening of the cultivation cycle as compared to seeds that take more time for development to a marketable size of the rhizome as compared to rhizome cuttings (Troup, 1915).

The possible applications of different approaches to propagate *P. hexandrum* (Table 4.13) suggest their complementary uses so as to fully realize its sustainable use and the conservation of this important species.

Fig. 4.15. Mass propagation of *P. hexandrum* through rhizome segments.

Table 4.13. Possible applications of different approaches to propagate *P. hexandrum.*

Embryo culture	Seed propagation	Vegetative propagation	Somatic embryogenesis
It is used when there is a manifestation of strong innate dormancy as is shown in Kukum-Seri populations which cannot be broken within a short time span. The wide applicability of this method can help in conservation of valuable genetic diversity. This approach will also be critical in conventional breeding programmes.	Seed propagation can be used in the establishment of base nurseries to generate large planting material for cultivation and sustainable utilization.	Once the characterized base population is established, vegetative propagation can be used to generate uniform quality planting material.	Somatic embryogenesis should be used for rapid *in vitro* multiplication to bulk up elite varieties and as a tool for genetic manipulation. Furthermore, the approach will be useful for fundamental studies on embryo growth and development and its use in artificial seed production.

Conclusions

Due to the obvious threat of exploitation to *P. hexandrum* under natural conditions, *ex situ* nursery establishment and cultivation will go a long way to alleviate the existing pressures from nature. This will also help in restoration of viable populations under *in situ* conditions allowing continued evolution of the species. Furthermore, the policy of grazing in the high-altitude pastures needs to be reviewed from the point of view of restoration of regeneration capacity of the ecosystem for the overall conservation of species diversity prevailing in these zones for a long lasting solution. Application of such conservation strategy to other medicinally important but threatened plants of high altitudes will go a long way in the conservation of valued Himalayan biodiversity. There is a need today to complement traditional conservation strategies with modern ones. The present work on characterization and propagation of *Podophyllum hexandrum* is only a small step in the larger context of the problem. Nevertheless, it is a very important step.

Acknowledgements

ACK and DS are thankful to Council of Scientific and Industrial Research (CSIR), New Delhi for providing Senior Research Fellowship during the course of the study. We are thankful to Director, Institute of Himalayan Bioresource Technology, Palampur, Himachal Pradesh, India for providing necessary infrastructure. We are also thankful to Dr Ashok K. Chauhan, Founder President, RBEF, New Delhi, India for providing necessary support in writing the book chapter.

References

Airi, S., Rawal, R.S., Dhar, U. and Purohit, A.N. (1997) Population studies on *P. hexandrum* – a dwindling medicinal plant of the Himalaya. *Plant Genetic Resource Newsletter* 110, 29–34.

Ammirato, P.V. (1983) *Embryogenesis*. In: Evans, D.A., Sharp, W.R., Ammirato, P.V. and Yamada, Y. (eds) *Handbook of Plant Cell Culture*, Vol.1. *Techniques for Propagation and Breeding*. Macmillan Publishing Co., New York, USA, pp. 82–123.

Arumugam, N. and Bhojwani, S.S. (1990) Somatic embryogenesis in tissue cultures of *Podophyllum hexandrum*. *Canadian Journal of Botany* 68, 487–491.

Arora, R., Singh, S., Puri, S.C. and Sharma, R.K. (2008) Himalayan Mayapple (*Podophyllum hexandrum* Royle): Traditional Uses, Clinical Indications and Future Prospects. In: Watson, R. and Preedy, V.R. (eds) *Botanical Medicine in Clinical Practice*. CABI Publishing, Wallingford, Oxon, United Kingdom, pp. 71–84.

Badhwar, R.L. and Sharma, B.K. (1963) A note on the germination of Podophyllum seeds. *Indian Forester* 89, 445–447.

Bhadula, S.K., Singh, A., Lata, H., Kuniyal, C.P. and Purohit, A.N. (1996) Genetic resources of *Podophyllum hexandrum* Royle, an endangered medicinal species from Garhwal Himalaya. *Plant Genetic Resource Newsletter* 106, 26–29.

Chaudhary, D.K., Kaul, B.L. and Khan, S. (1996) Breaking seed dormancy of *Podophyllum hexandrum* Royle. Ex. Camb. (syn. *P. emodi* Wall ex. Honigberger). *Journal of Non Timber Forest Produce* 3, 10–12.

Chaudhary, D.K., Kaul, B.L. and Khan, S. (1998) Cultivation and conservation of *Podophyllum hexandrum* – an overview. *Journal of Medicinal and Aromatic Plant Sciences* 20, 1071–1073.

Ellis, R.H., Hong, T.D. and Roberts, E.H. (1989) A comparison of the low moisture content limit to the logarithmic relation between seed moisture longevity in twelve species. *Annals of Botany* 63, 601–611.

Endels, P., Jacquemyn, H., Brys, R., Hermy, M. and De Blust, G. (2002) Temporal changes (1986-1999) in populations of primrose (*Primula vulgaris* Huds.) in an agricultural landscape and implications for conservation. *Biological Conservation* 105, 11–25.

Fenner, M. (1991) The effects of the parent environment on seed germinability. *Seed Science Research* 1, 75–84.

Good, R. (1974) *The Geography of Flowering Plants*, 3rd edition, Longman, New York, USA.

Gutterman, Y. (1993) Seed germination in desert plants, adaptations of desert organisms. Springer, Berlin, p. 253

Kaplan, I.W. (1942) Condylomata acuminata. *New Orleans Medical Surgery Journal*. 94, 388–390.

Kaur, S., Gupta, A.K. and Kaur, N. (1998) Gibberellin A_3 reverses the effect of salt stress in chickpea (*Cicer arietinum* L.) seedlings by enhancing amylase activity and mobilization of starch in cotyledons. *Plant Growth Regulation* 26, 85–90.

Kharkwal, A.C., Prakash, O.M., Bhattacharya, A., Nagar, P.K. and Ahuja, P.S. (2002) Method for inducing improved seed germination in *Podophyllum hexandrum* Royle. United States Patent Number 6, 449, 899.

Kharkwal, A.C., Prakash, O., Bhattacharya, A. and Ahuja, P.S. (2004) Mass propagation and conservation of *Podophyllum emodi* Wall, an endangered medicinal plant of the Himalaya. *Plant Genetic Resources Characterization and Utilization* 2, 51–57.

Kharkwal, A.C., Singh, D., Rajkumar, S. and Ahuja, P.S. (2008) Genetic variation within and among the populations of *Podophyllum hexandrum* Royle (Podophyllaceae) in western Himalaya. *Plant Genetic Resources Newsletter FAO Rome* 156, 68–72.

Loike, J.K., Brewere, C.F., Sternlicht, H., Gensler, W.J. and Horwitz, S.B. (1978) Structure-activity study of the inhibition of microtubule assembly *in vitro* by podophyllotoxin and its congeners. *Cancer Research* 38, 2688.

Milberg, P., Andersson, L., Elfversson, C. and Regner, S. (1996) Germination characteristics of seeds differing in mass. *Seed Science Research* 6, 191–197.

Nadeem, M., Palni, L.M.S., Purohit, A.N., Pandey, H. and Nandi, S.K. (2000) Propagation and conservation of *Podophyllum hexandrum* Royle: an important medicinal herb. *Biological Conservation* 92, 121–129.

Nautiyal, M.C., Rawat, A.S., Bhadula, S.K. and Purohit, A.N. (1987) Seed germination in *Podophyllum hexandrum*. *Seed Research* 15, 206–209.

Osheroff, N. (1989) Effect of antineoplastic agents on the DNA cleavage/religation reaction of eukaryotic topoisomerase II: inhibition of DNA religation by etoposide. *Biochemistry* 28, 6157–6160.

Pammenter, N.W. and Berjak, P. (1999) A review of recalcitrant seed physiology in relation to desiccation-tolerance mechanisms. *Seed Science Research* 9, 13–37.

Philippi, T. (1993) Bet-hedging germination of desert annuals: variation among populations and maternal effects in *Lepidium lasiocarpum*. *American Naturalist* 142, 488–507.

Pleasants, J.M. and Wendel, J.F. (1989) Genetic diversity in clonal narrow endemic *Erythronium propullans* and its widespread progenitor *Erythronium albidum*. *American Journal of Botany* 76, 1136–1151.

Pugh, N., Khan, I., Moraes, R.M. and Pasco, D. (2001) Podophyllotoxin lignans enhance IL- 1β but suppress TNF-α mRNA expression in LPS-treated monocytes. *Immunopharmacology and Immunotoxicology* 23, 83–95.

Purohit, A.N. and Nautiyal, M.C. (1986) Inhibitory effect of cotyledons on plumular development in two alpine rosettes. *Canadian Journal of Botany* 66, 205–206.

Rolston, M.P. (1978) Water impermeable seed dormancy. *The Botanical Review* 44, 365–396.

Rust, R.W. and Roth, R.R. (1981) Seed production and seedling establishment in the mayapple *Podophyllum peltatum* L. *American Midland Naturalist* 105, 51–60.

Sadowska, A., Wiweger, M., Lata, B. and Obidoska, G. (1997) *In vitro* propagation of *Podophyllum peltatum* L. by the cultures of embrya and divided embrya. *Biologia Plantarum* 39, 331–336.

Sharma, T.R., Singh, B.M., Sharma, N.R. and Chauhan, R.S. (2000) Identification of high podophyllotoxin producing biotypes of *Podophyllum hexandrum* Royle from North-Western Himalaya. *Journal of Plant Biochemistry and Biotechnology* 9, 49–51.

Singh, A., Purohit, A.N., Bhadula, S.K., Lata, H., Kuniyal, C.P. and Chandra, S. (1999) Seed production potential and germination behaviour in populations of *Podophyllum hexandrum* Royle. *Journal of Plant Biology* 26, 51–57.

Stahelin, H.F. and von Warburg, A. (1991) The chemical and biological route from podophyllotoxin glucoside to etopside. *Cancer Research* 51, 5–15.

Sultan, P., Shawl, A.S., Ramteke, P.W., Kour, A. and Qazi, P.H. (2008) Assessment of diversity in *Podophyllum hexandrum* by genetic and phytochemical markers. *Scientia Horticulturae* 115, 398–408.

Tran, V.N. and Cavanagh, A.K. (1984) *Structural Aspects of Dormancy*. In: Murray, D.R. (ed.) *Seed Physiology*, Vol.2. Academic Press, Australia, pp. 1–44.

Troup, R.S. (1915) A note on the cultivation of *Podophyllum emodi*. *Indian Forester* 41, 361–365.

von Arnold, S., Sabala, I., Bozhkov, P., Dyachok, J. and Filonova, L. (2002) Developmental pathways of somatic embryogenesis. *Plant Cell, Tissue and Organ Culture* 69, 233–249.

Widen, B. and Andersson, S. (1993) Quantitative genetics of life-history and morphology in *Senecio integrifolius*. *Heredity* 70, 503–514.

Xiang, Q.-Y., Soltis, D.E., Soltis, P.S., Manchester, S.R. and Crawford, D.J. (2000) Timing the eastern Asian-eastern North American floristic disjunction: molecular clock corroborates paleontological data. *Molecular Phylogenetics and Evolution* 15, 462–472.

Xiao, M., Li, Q., Guo, L., Luo, T., Duan, W-X., He, W-X., Wang, Li. and Chen, F. (2006) AFLP analysis of genetic diversity of the endangered species *Sinopodophyllum hexandrum* in the Tibetan Region of Sichuan Province, China. *Biochemical Genetics* 44, 47–60.

Young, A.G. and Clarke, G.M. (2000) *Genetics, Demography and Viability of Fragmented Populations*. Cambridge University Press, London, UK.

Chapter 5

Microsatellite Markers: Potential and Opportunities in Medicinal Plants

Atul Grover, Ankit Jain and P.C. Sharma

Introduction

The global consumption of herbal medicines is increasing in response to the ever increasing pressures to fight the infectious and other diseases affecting the world population today (Joshi *et al.*, 2004; Canter *et al.*, 2005). The term 'medicinal plants' refers to a collection of approximately 50,000 plant species from which extracts can be obtained for cure or treatment of diseases and the majority of them grow as wild (Edwards, 2004). Some of the medicinal plants like *Cannabis* are cultivated, but their exploitation for the intoxicating properties of the plant products has put a restriction on their cultivation by farmers in most of the countries (Gilmore *et al.*, 2003). Compounds of medicinal value can be obtained from other plants as well, but the umbrella term 'medicinal plants' refers to the plant species whose primary economic importance relies on their therapeutic compounds. With the widespread and unplanned exploitation of these plant species, many are now threatened and their biodiversity is under serious threat (Canter *et al.*, 2005). Limited attempts for their domestication and subsequent cultivation have not proved very fruitful so far, primarily due to specific ecological requirements (Canter *et al.*, 2005).

The post genomic era offers vast opportunities for adapting medicinal plants to domestication and to undertake crop improvement programmes exploiting functional and comparative genomics approaches. For example, availability of genomic sequences of species like potato, tomato and tobacco may guide explorations into closely related medicinal plants within the family Solanaceae. The modern methods of crop improvement exploit genetic transformations, pathway engineering, genetic engineering and most importantly molecular breeding. However, to date there are very few reports of application and development of molecular markers in medicinal plants. Moreover, there have been almost no efforts towards construction of molecular genetic maps in medicinal plants.

Molecular Markers in Medicinal Plants

In order to ensure the efficacy of the drug, selection of the correct chemotype is critical (Joshi *et al.*, 2004). The use of molecular markers to establish the genetic identity is an important exercise in the interest of scientists and industrialists alike. Biochemical markers have excessively been used, but a trend in favour of DNA based molecular markers has started recently in medicinal plants also following other crop species. As little sequence data exist for medicinal plants, the genome characterization and analysis of genetic diversity in medicinal plants often rely on random amplified polymorphic DNA (RAPD; Williams *et al.*, 1990) and inter-simple sequence repeats markers (ISSR; Meyer *et al.*, 1993; Zietkiewicz *et al.*, 1994).

RAPD and ISSR markers are good choices as molecular markers when the genome under consideration is absolutely unexplored as an experimental system. However, comparatively poor reproducibility coupled with their dominant character restricts their use. The other systems like restriction fragment length polymorphism (RFLP; Botstein *et al.*, 1980) and amplified fragment length polymorphism (AFLP; Vos *et al.*, 1995) are costlier and time consuming, and suffer with the disadvantage of the requirement of a labelling system that may either be radioactive or fluorescent. Microsatellites, which are tandem repetitions of short nucleotide (1-6 bp) sequences, in comparison to major marker systems like RFLP, RAPD, AFLP, etc. offer several advantages. They are multi-allelic, highly polymorphic and reproducible, simple, co-dominant and amenable to automation (Powell *et al.*, 1996).

Microsatellite Hypermutability and Potential as Molecular Markers

The extensive microsatellite length polymorphism primarily emerges from the allelic variation in the number of tandemly arranged repeat motifs (VNTR) (Tautz, 1989). The molecular mechanisms governing microsatellite length variation are not completely understood. Higher microsatellite mutation rates (10^{-2}–10^{-6}) are thought to be responsible for their length variability (Ellegren, 2000). The most common mechanism involved is replication slippage at microsatellite sites (Schlotterer and Tautz, 1992). However, other molecular events like unequal crossing over, single nucleotide changes or duplications are also prevalent at these sites (Hancock, 1999). Microsatellite mutations also depend on factors like allele size, motif size, gender, age, sequence composition, neighbouring regions, etc. (Ellegren, 2000, 2002; Grover and Sharma, 2007; Grover *et al.*, 2007; Anmarkrud *et al.*, 2008; Roorkiwal *et al.*, 2009). Mutation patterns may also depend upon the effectiveness of mismatch repair enzymes (Modrich and Lahue, 1996; Harr *et al.*, 2002). Further, propensities to make three dimensional secondary structures and stabilization patterns also contribute to microsatellite mutation rates (Eckert *et al.*, 2002; Riley *et al.*, 2007). Recently, Eckert and Hile (2009) reviewed these mechanisms and hypothesized that different mechanisms might be occurring at different microsatellite loci within the same genome.

The prevalent hypermutability in microsatellites might or might not have a functional significance for the organism (Kashi and King, 2006), but has immensely been utilized to develop sequence tagged microsatellite sites (STMS) markers (Litt and Luty, 1989; Weber and May, 1989). In the following sections, we describe the development and applications of microsatellites in a variety of medicinal plants. Primarily, these markers have been utilized for assessing genetic diversity in natural populations of these species, but they also hold great potential for other uses like genome mapping, gene tagging, marker assisted selection, authentication and detection of adulteration of the medicinal plant material, protection of IPRs, etc.

Microsatellite Mining from Random Genomic Libraries

Conventionally, genomic microsatellites are isolated from the size selected genomic libraries (Zane *et al.*, 2002; Weising *et al.*, 2005). This approach requires screening of several thousand clones through colony hybridization with microsatellite repeat containing probes (Rassmann *et al.*, 1991). The first report of isolation of a microsatellite marker from a medicinal plant, *Helianthus annuus*, also used a genomic library (Brunel, 1994). Two microsatellite markers thus developed were tested for their polymorphism on more than 40 genotypes (Brunel, 1994).

Later, this approach was also found useful in isolation of microsatellites from other plant species of medicinal importance, for example *Aconitum napellus* (aconite; Le Cadre *et al.*, 2005) and *Acanthus ilicifolius* (sea holly; Geng *et al.*, 2008). In the latter report (Geng *et al.*, 2008), dual-suppression-PCR has been used to circumvent the need of radioactive screening of recombinant clones in the genomic libraries of *Acanthus ilicifolius* and *Lumnitzera racemosa* (black mangrove). To determine the region flanking one end of the microsatellite sequence, PCR amplifications followed by DNA sequencing were carried out using a microsatellite primer and an adapter sequence. The fragments thus obtained were used to design two primer pairs (IP1 and IP2) for nested PCR. Second PCR amplification was performed using the second adapter sequence at opposite end as one of the primer and any one of the primers designed in the previous step (e.g., IP1). The amplification product thus obtained was re-amplified using the second specific primer (e.g., IP2, as mentioned above) and second adapter primer. Amplification products at the end of each PCR step were sequenced. A single primer (IP3) was designed from this region as the flanking primer for microsatellite at the second end. Three flanking primers for each of the microsatellites thus provided two pairs of primers (IP1/IP3 and IP2/IP3) for developing markers (Geng *et al.*, 2008). Various modifications to this general protocol were also introduced by Geng *et al.* (2008). For example, a compound microsatellite repeat was used as a primer in the first step and primer from flanking region from only one of the ends was designed. Eight microsatellites thus obtained successfully detected polymorphisms in 40 individuals of mangroves representing *A. ilicifolius* and *L. racemosa.*

Screening of genomic libraries with radio-labelled microsatellite probes has routinely been used to isolate microsatellites and to exploit them for a variety of purposes. Six microsatellite markers developed by screening genomic libraries of *Aconitum napellus* were used to screen genetic diversity among 60 individuals representing six French natural populations to study inbreeding depression among these populations (Le Cadre *et al.*, 2005). Gilmore *et al.* (2003) used a cocktail of di- and tri-nucleotide repeat probes, as earlier described by Scott *et al.* (2001), for screening a genomic library of *Cannabis sativa.* Five microsatellite markers thus developed were used to genotype 93 plants representing drug and fibre accessions. A narrow genetic base of drug yielding accessions was reported in *C. sativa* (Gilmore *et al.*, 2003). For a species like *Cannabis*, which has a potential of being misused, development of microsatellite assays is an important activity, as the unique and accurate identification of genotypes can keep a watch on their illegitimate cultivation and production (Gilmore *et al.*, 2003).

Probably the most successful application of genomic libraries for development of microsatellite markers in a medicinal plant is displayed by Shokeen *et al.* (2007), developing 24 markers in *Catharanthus roseus* (Madagascar periwinkle) by screening genomic libraries with oligonucleotide probes of motifs CA, CT, GC and GCG. Genetic polymorphism was evaluated in 32 genotypes of *C. roseus* and transferability of these markers was tested on *C. trichophyllus*, *C. pusillus*, *Vinca minor*, *Thevetia peruviana* and *Nerium indicum* (Shokeen *et al.* 2007). Remarkably, this study demonstrated the use of GC probes for mining of microsatellites in a genomic library. Such microsatellite repeats are generally avoided for library screening considering problems expected due to self-complementarity of the single stranded probe. However, among the 38 microsatellites isolated by Shokeen *et al.* (2007) using this procedure, none contained a GC repeat. Our experience of analysing microsatellite repeats in rice and members of family Solanaceae suggest that GC repeats reveal low polymorphism (Grover *et al.*, 2007; Roorkiwal *et al.*, 2009), and hence their use must be avoided for the purposes of DNA fingerprinting and diversity assessment experiments.

Heterozygosity values and average number of alleles obtained in these studies showed marked variation. Particularly, lower heterozygosity and allele number was obtained in *Aconitum napellus* than in *Acanthus ilicifolius*. Though different methods were used to extract the microsatellite markers from these plants (Le Cadre *et al.*, 2005; Geng *et al.*, 2008), several other factors like mutability of the concerned loci or the genotypes under study, might have contributed to the development of microsatellites with different heterozygosity and allele numbers. Further, the inbreeding nature of *Aconitum napellus* accessions supports low heterozygosity (Le Cadre *et al.*, 2005). Another reason that might have contributed to the observed variation in allele number and heterozygosity is the microsatellite motifs used for screening genomic libraries. Repeats of family AC were probed by Brunel (1994) in sunflower, Geng *et al.* (2008) in *Acanthus,* and Le Cadre *et al.*, (2005) in *Aconitum*. Additionally, Geng *et al.* (2008) also targeted compound microsatellites of $(AC)_6(TC)_5$ and $(TC)_6(AC)_5$ type, and Le Cadre *et al.* (2005) also used motif TC and anchored tetranucleotide $CT(ATCT)_6$ and $(TGTA)_6TG$ probes.

Construction of genomic libraries for isolation of microsatellites is a simple approach and works well for the genomes which are richer in microsatellites, but for microsatellite poor genomes, this approach is tedious and inefficient (Zane *et al.*, 2002).

Microsatellite Mining from Enriched Libraries

Enriched libraries by far are the most popular way for isolation of microsatellites in medicinal plants. Different methods exist for enriching the libraries for microsatellites and most of them have been applied in medicinal plants for development of microsatellite markers.

Among the many different methods available for library enrichment for microsatellites, the most popular method is biotin-streptavadin capture protocol with its various modifications, successfully applied in the case of many medicinal plants for isolation of microsatellites (Table 5.1). Wardill *et al.* (2004) restricted the genomic DNA of *Acacia nilotica* ssp. *indica* (prickly acacia), ligated adapters and PCR amplified the entire size-selected genomic DNA. The amplified products were hybridized with biotin-labelled microsatellite probes. The eluted enriched genomic products were amplified again and cloned. Library enrichment protocol described by Fischer and Bachmann (1998) was used by Boontong *et al.* (2008) for isolation of eight polymorphic microsatellite markers in *Azadirachta indica* (Indian neem) and Gilmore and Peakall (2003) in *Cannabis sativa* (Table 5.1). Another variant of the same protocol as defined by Bloor *et al.* (2001) was applied by Crozier *et al.* (2007) for construction of AG- and TG-rich microsatellite libraries in *Ficus racemosa* (cluster fig) and *Ficus rubiginosa* (Port Jackson fig). Eleven microsatellite markers were tested for cross-species amplification and genetic diversity estimations in these species (Crozier *et al.*, 2007). Similarly, a selective hybridization procedure (Karagyozov *et al.*, 1993) was used for extraction of five and three microsatellite markers from *Ficus montana* (oak leaf fig) and *Ficus septica* (Noboloboi), respectively (Zavodna *et al.*, 2005). Enriched libraries were also constructed in *Hibiscus glaber* for isolation of CT based microsatellite repeats (Ohtani *et al.*, 2008) and ten of these proved highly polymorphic among natural populations and parentage analysis. Nine microsatellites could also be cross-amplified in closely related *Hibiscus tiliaceous*, another plant with medicinal importance and occurring in the same geographical and ecological regimes (Ohtani *et al.*, 2008). Such cross-amplifications are important and help to build the common resources for future development of these species into crop plants. Similarly, Takayama *et al.* (2006) discovered microsatellites in *Hibiscus tiliaceous* and reported their cross-amplification in *H. glaber*. Takayama *et al.* (2006) also mined microsatellites from an enriched library following FIASCO protocol (discussed below separately).

Microsatellites have been isolated from enriched libraries constructed using the biotin-streptavadin capture method in a number of medicinal plants described by traditional Chinese medicine system. Some of the reports include Wang *et al.* (2008) in *Hippophae rhamnoides* (Sea buckthorn), Wan *et al.* (2008) in *Przewalskia tangutica*, a Tibetan medicinal plant belonging to family Solanaceae, Aceto *et al.* (2003) in *Asparagus acutifolius* and Gu *et al.* (2007) in *Dendrobium officinale*, a Chinese medicinal herb. Ma *et al.* (2007) used yet another modification of this method as described by Dixit *et al.* (2005) for construction of enriched libraries, which were screened to develop 22 polymorphic microsatellite markers in *Panax ginseng* (ginseng).

The probability of finding a microsatellite, as described in most of these studies, was 50% on sequencing a clone after enrichment procedure (Table 5.1), much higher than that in our experience of handling genomic libraries for isolation of microsatellites (Grover *et al.*, 2009). Our attempts to isolate microsatellites from genomic libraries produced 1% positive clone at primary level of screening, and subsequently 2% following secondary screening.

Using the enrichment procedure of Fisher *et al.* (1996) based on 5'-anchored PCR protocol, Shokeen *et al.* (2005) developed seven microsatellite markers in *Catharanthus roseus*. The PCR required amplification using a degenerate primer with anchored region being KKVRVRV (K= G/T; V= G/C/A; R= G/A) followed by a simple sequence oligonucleotide of appropriate length. In original protocol, the 3' of the primer was constituted by $(CT)_6$, while Shokeen *et al.* (2005) substituted it by $(AG)_{10}$. A disadvantage of this technique is that the microsatellite is often present at the terminal position, and therefore only one flanking primer can be designed. For genetic diversity analysis, original degenerate primer has to be used in combination with a single flanking primer (Fisher *et al.*, 1996; Shokeen *et al.*, 2005). However, in certain cases, internal microsatellite markers could also be obtained, and they could be used in a conventional way by designing two flanking primers as usual (Fisher *et al.*, 1996; Shokeen *et al.*, 2005).

Edwards *et al.* (2007) initially amplified the size-selected DNA using *Sau*3AI linker primers, and then followed the enrichment procedures involving hybridization with biotin-labelled oligonucleotide probes. Radioactive screening was avoided by extracting the plasmid DNA and using it as a template in a PCR reaction that had microsatellite oligonucleotide as primer (Edwards *et al.*, 2007). This effort led to development of 19 microsatellite markers in *Hypericum cumulicola* (highlands scrub hypericum), which were tested on a population of 19 individuals.

Fagopyrum esculentum (common buckwheat) is an important crop plant in China and Japan, which is equally valued for its medicinal uses. Its importance as a crop has led to overall crop improvement efforts in this plant species, and thus more than 50 microsatellite markers have already been developed in common buckwheat using enrichment protocols (Iwata *et al.*, 2005; Konishi *et al.*, 2006). Five microsatellite markers were developed by Iwata *et al.* (2005), which were tested on a panel of 19 genotypes and compared with AFLP markers. The average heterozygosity values obtained for AFLP and microsatellite markers were 0.303 and 0.819, respectively. The microsatellite markers reported in this study were highly polymorphic, as 203 alleles were revealed for only five microsatellite loci (Iwata *et al.*, 2005). Another 48 microsatellite markers reported by Konishi *et al.* (2006) were also highly polymorphic with average PIC value of 0.79, which is only slightly less than that reported by Iwata *et al.* (2005) for five microsatellites. As the markers developed in cultivated common buckwheat also cross-amplified in seven wild relatives of buckwheat (Konishi *et al.*, 2006), the scope of improvement of common buckwheat for medicinal cultivation are promising. Molecular markers cross amplifying in related species is the foremost step towards introgressing useful traits from wild germplasm to the cultivated gene pool.

Fast Isolation by AFLP of Sequences COntaining repeats (FIASCO)

FIASCO protocol was first described by Zane *et al.* (2002) as a faster and simpler alternative to construction of libraries for isolation of microsatellites. As described in the original protocol, the technique relies on two important features of the primers: one, that adaptors are not phosphorylated to avoid self-ligation, and second that their sequence is designed in such a way so as not to restore *Mse*I restriction site, which is the usual choice in AFLP. Thereafter, the amplification step is carried out using a mixture of four primers with a different 3'-selective base. The amplification step is followed by the hybridization step as usual, except that a simple sequence oligonucleotide is taken as a hybridization probe. In the original protocol $(AC)_{17}$ was used as the hybridization probe. The hybridizing sequences are recovered using normal biotin-streptavadin capture, as in any other enrichment protocol. The sequences recovered thus are amplified using PCR performed with the primers described above. The amplification products are used for preparation of enriched genomic libraries (Zane *et al.*, 2002). FIASCO is essentially a method of library enrichment, but its popularity in construction of microsatellite enriched genomic libraries has made us consider its discussion separately here.

Table 5.1. Microsatellite development in medicinal plants using biotin-streptavadin capture or PCR based enrichment methods, and their applications.

Species	Enrichment procedure (modification)	Probe motifs for enrichment	Yield[a]	No. of primers designed	No. of truly polymorphic loci[b]	Polymorphism parameters[c]	Population/ Germplasm assayed	Reference
Acacia nilotica	De Barro *et al.* (2003)	Not known	50%	Not known	5	Alleles: 2–3 Heterozygosity: 0.000–0.458	3 Australian and 6 Indian individuals	Wardill *et al.* (2004)
Asparagus acutifolius	Tenzer *et al.* (1999)	GA, GTT	713 colonies screened. 28 picked up for sequencing	12	7	Alleles (per locus): 2–5; Heterozygosity: 0.20–0.73	15 individuals from a natural population from Pontecagnano, Italy	Aceto *et al.* (2003)
Azadirachta indica var. *indica*	Fischer and Bachmann (1998)	CT	68 colonies sequenced, 47 contained microsatellite	26	8	Alleles (avg.): 5.5 (within variety), 4.9 (across variety); Heterozygosity (avg.): 0.5 (within and between variety)	39 individuals from 2 accessions of Indian Neem (var. *indica*); cross-amplified in 39 samples of Thai Neem (var. *siamensis*)	Boontong *et al.* (2008)
Cannabis sativa	Modified Edwards *et al.* (1996) method	CT, GT, CAA, ATT, GCC, ACC, AGG, CTT, AGC, ACG, ACT, ATC	95 positive clones out of 192 sequenced (49.5%) in a library of 685 colonies	29	11	Alleles (avg.): 4.7 Heterozygosity (avg.): 0.529	41 *Cannabis* samples	Alghanim and Almirall (2003)

Cannabis sativa	Fischer and Bachmann (1998)	Not known	Not known	Not known	11	Alleles (avg.): 10.7 Heterozygosity (avg.): 0.68	48 samples representing 5 fibre crop accessions	Gilmore and Peakall (2003)
Catharanthus roseus	Fisher *et al.* (1996)	AG	< 50%	Not known	7	Alleles (avg.): 3.86 Heterozygosity (avg.): 0.7511	32 *C. roseus* accessions; cross-amplified in *C. trichophyllus*	Shokeen *et al.* (2005)
Fagopyrum esculentum	Tani *et al.* (2004)	CT, GT	1079 out of 1875 (57.5%) for CT enriched library; 404 out of 910 (44.4%) for GT enriched	237; out of which 54 were single locus	48	Alleles (avg.): 12.2; PIC (avg.): 0.79	Cross-amplified in seven *Fagopyrum* species and subspecies with different plcidy levels	Konishi *et al.* (2006)
Ficus montana	Karagyozov *et al.* (1993)	di-, tri-, tetra-nucleotide repeat probes	71 out of 308 clones sequenced	12	5	Alleles: 5–14 Heterozygosity: 0.23–0.87	24 *F. montana* individuals; 2 primers cross-amplified in *F. septica*	Zavodna *et al.* (2005)
Ficus racemosa	Bloor *et al.* (2001)	AG, TG	50%	86	11	Alleles: 2–8 Heterozygosity: 0.12–0.91	17–21 individuals of *F. racemosa*; cross-amplified in 16–24 individuals of *F. rubiginosa*	Crozier *et al.* (2007)

Ficus rubiginosa	Bloor *et al.* (2001)	AG, TG	50%	92	11	Alleles: 2–15 Heterozygosity: 0.12-0.91	16–24 individuals of *F. rubiginosa*; cross-amplified in 17–21 individuals of *F. racemosa*	Crozier *et al.* (2007)
Ficus septica	Karagyozov *et al.* (1993)	di-, tri-, tetra-nucleotide repeat probes	58 out of 223 clones sequenced	7	3	Alleles: 3-5 Heterozygosity: 0.36–0.49	36 *F.septica* individuals; 2 primers cross-amplified in *F. montana*	Zavodna *et al.* (2005)
Heracleum mantegazzianum	Schlotterer *et al.* (1997)	AT, GA	Not known	Not known	4 (nuclear) 1 (plastid)	Nucl. Alleles (avg.): 12.75 Cp Alleles: 4	13 British populations	Walker *et al.* (2003)
Hibiscus glaber	Tani *et al.* (2004)	CT	87 positive clones out of 208 sequenced	48	10	Alleles (avg.): 16.5 Heterozygosity (avg.): 0.854	78 individuals from Nishijama Island; cross-amplified in 12 individuals of *H. tiliaceous* from Chichijima Island	Ohtani *et al.* (2008)
Hippophae rhamnoides ssp. *sinensis*	Same as above	Same as above	26 microsatellites out of ~200 sequenced clones	26	9	Alleles: 3-12; Heterozygosity: 0.1397–0.2997	12 individuals from distantly related populations; cross-amplification in three species	Wang *et al.* (2008)

Hypericum cumulicola	Kandpal *et al.* (1994)	CA	88 positive clones out of 267 sequenced	46	19	Alleles (avg.): 2.21 Heterozygosity (avg.): 0.243	24 individuals from a natural population	Edwards *et al.* (2007)
Panax ginseng	Dixit *et al.* (2005)	GA, AGC, GGC, AAG, AAC, AGG	203 positive clones out of 504 sequenced	189	22	Alleles (avg.): 4.5 Heterozygosity (avg.): 0.554	10 Ginseng accessions	Ma *et al.* (2007)
Przewalskia tangutica	Biotinylation of probes not involving radioactive screening	AG, CT, GT, AC, CG, CCA	29 microsatellites out of ~200 sequenced clones	29	12	Alleles: 3-12; Heterozygosity: 0.1652–0.3183	17 individuals from distantly distributed populations	Wan *et al.* (2008)

[a] Yield refers to number of clones identified positive after screening;
[b] Truly polymorphic loci refers only to those loci which amplified in the expected size range and showed polymorphism reproducibly on the germplasm analysed
[c] Values indicated here take into account only the actually polymorphic loci. Heterozygosity values are the observed heterozy

There are at least seven published reports exemplifying the use of FIASCO for isolation of microsatellite markers, as indicated in Table 5.2. FIASCO has also been applied to other plants as well, which are primarily harvested for other purposes, but therapeutic uses also add to their economic importance. An example includes *Nelumbo nucifera* (sacred lotus), for which 24 microsatellite markers were reported using FIASCO (Pan *et al*., 2007).

Most importantly, the protocol has been used to extract eight microsatellite markers using AC probes in *Artemisia annua* (Huang *et al*., 2008). *A. annua* (Qinghao) is an important plant yielding valuable extracts with anti-malarial properties. A genotype of Qinghao, selected with the aid of SCAR markers for high yielding anti-malarial phenotypes has been patented, and named 'CIM-Arogya' (Khanuja *et al*., 2009).

Xu *et al*. (2006) introduced a few modifications in the original protocol of Zane *et al*., (2002) for extraction of 32 microsatellites from *Gastrodia elata* (Tian ma). The modifications introduced by Xu *et al*. (2006) include the initial amplification steps of digested-ligated fragments reduced to 17, hybridization of 5'-biotinylated AAG probes in double amount (150 pmol) as compared to the original protocol (Zane *et al*., 2002), streptavadin paramagnetic particles were prepared by washing with a weakly alkaline solution (Zane *et al*., 2002), and the number of amplification cycles were reduced to 23 cycles (Xu *et al*., 2008). The microsatellites developed thus when analysed for their polymorphism levels, had a high heterozygosity ranging from 0.000 to 1.000, but as most of the loci had low heterozygosities, average heterozygosity was comparatively lower. The FIASCO protocol followed by Xu *et al*. (2008) in *Epimedium brevicornum* (barrenwort) carries further modifications: the initial PCR step constituted of 20 cycles, amount of biotinylated probes further enhanced to 200 pmol, washing step for streptavadin paramagnetic particle preparation was restored as described in the original protocol (Zane *et al*., 2002), and final amplification was carried for 25 cycles.

Zane *et al*. (2002) favoured the use of FIASCO considering the costs and time saved by this protocol. The efficiency is such both in terms of time and cost that a single laboratory can prepare about ten libraries within a month.

A similar but technically distinct approach of constructing a microsatellite enriched library involves cloning ISSR-PCR products. Recently, Henry *et al*. (2008) demonstrated its use in *Heracleum mantegazzianum* (giant hogweed). The protocol requires cloning of ISSR-PCR products, followed by sequencing of some or all of the clones. Henry *et al*. (2008) sequenced 48 clones, with all of them containing the desired inserts. Nine of the clones had internal microsatellites also present within the insert, and thus two flanking primers were designed for these microsatellites as usual. Five of these proved useful in genotyping distant populations, and cross-species amplification (Henry *et al*., 2008). Technically speaking, this protocol is even simpler and faster than FIASCO. However, the success rate for developing true microsatellite markers with both the flanking primers available is lower in this case, compared to FIASCO (Table 5.2). Nevertheless, anchored primers in combination with a single flanking primer can be efficiently applied for any genetic analysis in the same way as a normal single locus microsatellite marker.

Table 5.2. Application of FIASCO protocol for development of microsatellites in medicinal plants.

Species	Probe motifs for enrichment	Yield[a]	No. of primers designed	No. of truly polymorphic loci[b]	Polymorphism parameters[c]	Population/ Germplasm assayed	Reference
Artemisia annua	AC	78 clones sequenced	32	8	Alleles (avg.): 3.1 Heterozygosity (avg.): 0.4328	54 samples from two populations of *A. annua* collected in Hebei and Guangxi, China.	Huang *et al.* (2008)
Epimedium brevicornum	AC	85 clones sequenced, 59 contained microsatellites	51	12	Alleles (avg.): 5.25 Heterozygosity (avg.): 0.38	38 individuals from wild population at Shanxi, China	Xu *et al.* (2008)
Eucommia ulmoides	AC	180 clones sequenced; 43 positive	29	19	Alleles (avg.): 7.21 Heterozygosity (avg.): 0.36	36 individuals from 10 populations from China	Deng *et al.* (2006)
Gastrodia elata	AAG	73 clones sequenced, 40 contained microsatellites	32	13	Alleles (avg.): 5.5 Heterozygosity (avg.): 0.123	32 individuals from 8 wild populations	Xu *et al.* (2006)
Hibiscus tiliaceus	GA, GT	Total colonies– 224; PCR amplified clones– 101 Microsatellite positive– 77	12	6	Alleles (avg.): 5.3 Heterozygosity (avg.): 0.424	37 individuals from South Africa; cross-amplified in four species of *Hibiscus*; also used for population studies (Takayama *et al.* 2008)	Takayama *et al.* (2006)
Hieracium pilosella	CT, CAA	461 colonies sequenced; 20% contained microsatellites	13	11	Alleles (avg.): 16.5 Heterozygosity (avg.): 0.424	127 individuals from six locations in Trentino, Italy	Zini and Komjanc (2008)

Trifolium repens	CA, ATG	6816 clones total; 4277 positive; 1689 contigs and 991 singletons constructed; 260 CA and 318 ATG repeats reported	578; 191 tested	116	Not known	GA02-15 and GA02-56 of genotypes SRVR and Durana as parents and six progenies from F1 in the mapping population	Zhang *et al.* (2008)

[a] Yield refers to number of clones identified positive after screening;
[b] Truly polymorphic loci refers only to those loci which amplified in the expected size range and showed polymorphism reproducibly on the germplasm analysed
[c] Values indicated here take into account only the actually polymorphic loci. Heterozygosity values are the observed heterozygosity.

Mining of Microsatellites from DNA Sequence Databases

Extracting microsatellite sequences from the publicly available DNA sequences deposited in major databases like GenBank is the most attractive source of developing microsatellite markers. However, the use of these data resources is restricted by the fact that these databases are over-represented by sequences of the model organisms, and whatever data are available for other organisms is either too miniscule or does not provide a meaningful coverage of the concerned genomes. Nevertheless, there are examples of use of sequence resources for development of microsatellite markers in medicinal plants. As described in Table 5.2, Zhang *et al.* (2008) made a genome-wide survey of *Trifolium repens* (white clover) using phpSSRMiner (available at http://bioinfo.noble.org/phpssrminer/) for the presence of microsatellites. Utility of these microsatellites was tested on a mapping population (Zhang *et al.*, 2008). Wide-scale integration of these markers into genome maps of *T. repens* is currently underway (Zhang *et al.*, 2008). Similarly, Acquadro *et al.* (2006) screened ESTs of *Apium graveolens* (celery) for the presence of microsatellites. Out of the 73 microsatellite loci discovered using sputnik (http://espressosoftware.com/sputnik/), marker potential for 39 were tested on 16 commercial varieties of celery, out of which 16 loci could be developed into markers (Acquadro *et al.*, 2006). Both *Trifolium repens* and *Apium graveolens* are domesticated plants, and thus availability of DNA sequence wealth is a benefit for developing markers in these species. Another species *Ipomoea trifida* (morning glory), though is not currently a domesticated species, is still an interesting subject of scientific investigation, being most likely ancestor of sweet potato (Austin, 1988). Screening of *I. trifida* sequence database led to the development of 12 microsatellite markers showing 100% transferability to *I. batatas* (sweet potato) (Hu *et al.*, 2004). These markers also displayed high transferability to three other *Ipomea* species.

On the similar lines, we are currently developing EST-derived microsatellite markers in sea buckthorn (*Hippophae rhamnoides*) and assessing their transferability to other species of *Hippophae*. The ESTs screened for development of markers in this study have been developed in our laboratory.

Exploiting Sequence Databases for Microsatellite Markers in Medicinal Plants

As discussed above, DNA sequence databases are attractive sources of microsatellite marker development. The disadvantage of representation of only a few species in these databases is partially overcome by the fact that microsatellites show a good rate of transferability among closely related species (reviewed by Sharma *et al.*, 2007). The transferability can sometimes be extended at genus level as well. Further, statistics available at the NCBI website indicate that complete genome sequences are available for two plants currently in addition to six draft assemblies and 43 genome sequencing projects in progress (http://www.ncbi.nlm.nih.gov/genomes/static/gpstat.html). Further, a number of other DNA sequence databases exist, and the data available at dbEST and unigene database are other potential sources for development of microsatellite markers in medicinal plant genomes. We suggest that wide scale explorations of microsatellite transferability within a family, especially in the genic regions may be helpful in developing useful microsatellite markers in orphaned medicinal plant genomes.

ISSR Markers in Medicinal Plants

Most of the medicinal plants provide new genetic systems to the biologists and are amenable to be characterized by RAPD and ISSR markers. ISSR markers (Meyer *et al.*, 1993; Zietkiewic *et al.*, 1994) rely on a primer containing simple repeat sequences as primer for PCR amplification to generate reproducible fingerprints. The primers may be the unanchored or anchored generally at the 5' end by selective nucleotides to prevent internal priming and to amplify only a subset of the targeted inter-repeat regions.

These markers have proved quite useful for genetic diversity analysis in medicinal plants (Joshi *et al.*, 2004; Yip *et al.*, 2007). ISSR fingerprinting has been used in the authentication of *Dendrobium officinale* (Tiepi shihu; Shen *et al.*, 2006) and in the studies of genetic variations and relationships of *Hippophae rhamnoides* (Tian *et al.*, 2004), *Houttuynia* (lizard tail; Wu *et al.*, 2005), *Codonopsis lanceolata* (bonnet bellflower; Guo *et al.*, 2006), *Cannabis sativa* (Hakki *et al.*, 2007), *Impatiens glandulifera* (Himalyana balsam; Provan *et al.*, 2007) and *Podophyllum hexandrum* (Indian mayapple; Alam *et al.*, 2008).

A disadvantage of ISSR-PCR is that the primers may anneal to sequences other than microsatellites of the desired repeats (Caldeira *et al.*, 2002). Moreover, the banding patterns can be affected by magnesium ion concentration, thermocycler and annealing temperature in use (Shen *et al.*, 2006), and even amount of amplification product loaded on the gel.

Selective Amplification of Microsatellite Polymorphic Loci (SAMPL) in Medicinal Plants

SAMPL can detect high levels of polymorphism between closely related genotypes, as they exploit the hypervariability of microsatellites, and do not require development of single locus individual microsatellite markers. Negi *et al.* (2006) used a pre-amplified AFLP restriction mixture of *Withania somnifera* for SAMPL analysis for genetic diversity analysis in *W. somnifera* and *W. coagulans* genotypes. SAMPL is a simple PCR-based assay that combines the benefits of both AFLP and microsatellite analysis.

Applications of Microsatellite Markers in Exploitation and Conservation of Medicinal Plants

Microsatellite markers have proved their utility in fields like genome mapping, marker assisted selection, genetic diversity estimations, taxonomy, evolutionary biology, etc. They have shown immense applications in crop plants, and have only recently started being employed for medicinal plants, as clear from the discussion above.

It is well recognized that geographical conditions affect the active constituents of the medicinal plants, and hence the activity of the extracts obtained from them vary (Oleszek *et al.*, 2002). Microsatellites have excessively been used for studying geographical variation at the genetic level. In fact, most of the microsatellite mining efforts described above had estimates of genetic diversity for the design of crop improvement programmes, or management of germplasm and evolving conservation strategies as an outcome. Such studies have also been made to characterize several population structures in wild. An interesting example has been presented by Wanke *et al.* (2006) who found an intronic microsatellite in chloroplast gene *trnK* of *Aristolochia*. Highly polymorphic microsatellite showed polymorphic banding patterns in 32 species of the genus *Aristolochia*. The microsatellite locus was also found useful in population studies in three species of *Aristolochia* including *A. pallida* in which it was originally discovered as a by-product of some evolutionary studies by the same group. A bulk of the studies described above also tested the transferability of microsatellite markers to other species. Such studies are important, as they not only make the basis to implement wide hybridization in future, they are also useful for meticulously planned introgressions, constructing common genomic maps and studying evolutionary relationships.

Microsatellites hold a lot of promise to be implemented in authentication of medicinal plants, especially to avoid adulteration and non-judicial use. Gilmore and Peakall (2003) demonstrated their use to distinguish between fibre and drug accessions of *Cannabis sativa*. Negi *et al.* (2006) used SAMPL to distinguish Kashmiri genotypes of *W. somnifera* from Nagori genotypes.

Marker-assisted selection (MAS) of desired chemotypes is another area where microsatellites markers are likely to find their utility. For the purpose of MAS and genome mapping, segregating populations or isogenic lines are required, and hence can be applicable only on those species which are presently under cultivation. Whatever MAS or gene tagging has been carried out in medicinal plants has been primarily based on the use of RAPD markers (reviewed by Joshi *et al.*, 2004).

Conclusions

Microsatellites are highly dynamic markers which can be useful for a variety of studies. Medicinal plants face the challenges of getting domesticated first, followed by selection of the most appropriate chemotypes. The ability to identify the desired genotypes at an early stage is the most remarkable characteristic of any molecular marker, and is implemented effectively using microsatellite markers. The chances of bringing the correct genotypes into cultivation and their effective breeding are thus greatly enhanced by employing microsatellites for genome mapping and marker assisted selection.

In general, there are three ways in which microsatellite analysis can contribute to the development and improvement of plant species with medicinal importance. Firstly, they can be employed for genetic diversity screening within the gene pool of a species, and characterizing the geographically diverse populations. The genetically diverse and agronomically suitable genotypes can thus be selected for improvement using plant breeding efforts. Secondly, microsatellite analysis can be used to establish linkage among different loci for development of genetic maps, and with different traits to facilitate marker assisted selection. Though the production of bioactive compounds is not frequently a quantitative trait, several QTLs may be required to be tagged when the species is under cultivation. Thirdly, microsatellites may be used for characterization of populations, so as to select suitable parents in future breeding efforts.

Further, microsatellites hold immense importance in authentication of accessions of medicinal plants, avoiding duplication and non-judicious use, and can be presented as evidence for IPRs. Thus, mining of more and more of the microsatellites for each of the species is desired. In fact, high-throughput efforts are required for pacing up the crop improvement efforts in these species. Beyond doubt, microsatellite markers are required for modernization, industrialization and internationalization of traditional medicines. Microsatellites can identify a sample in any physical form and provide reproducible results irrespective of age, tissue, environmental conditions, storage, etc. The requirement of small sample quantity is of particular importance for quality control.

References

Aceto, S., Sica, M., Gamba, G., Montieri, S., Farina, A. and Gaudio, L. (2003) Isolation and characterization of microsatellite loci from *Asparagus acutifolius* (Liliaceae). *Molecular Ecology Notes* 3, 242–243.

Acquadro, A., Magurno, F., Portis, E. and Lanteri, S. (2006) dbEST-derived microsatellite markers in celery (*Apium graveolens* L. var. Dulce). *Molecular Ecology Notes* 6, 1080–1082.

Alam, A., Naik, P.K., Gulati, P., Gulati, A.K. and Mishra, G.P. (2008) Characterization of genetic structure of *Podophyllum hexandrum* populations, an endangered medicinal herb of Northwestern Himalaya, using ISSR-PCR markers and its relatedness with podophyllotoxin content. *African Journal of Biotechnology* 7, 1028–1040.

Alghanim, H.J. and Almirall, J.R. (2003) Development of microsatellite markers in *Cannabis sativa* for DNA typing and genetic relatedness analyses. *Analytical and Bioanalytical Chemistry* 376, 1225–1233.

Anmarkrud, J.A., Kleven, O., Bachmann, L. and Lifjeld, J.T. (2008) Microsatellite evolution: Mutations, sequence variation, and homoplasy in the hypervariable avian microsatellite locus *HrU*10. *BMC Evolutionary Biology* 8, 138.

Austin, D.F. (1988) The Taxonomy, Evolution and Genetic Diversity of Sweet Potato and Related Wild Species. In: Gregory, P. (ed.) *Exploration, Maintenance and Utilization of Sweet Potato Genetic Resources.* International Potato Center, Lima, Peru, pp. 27–60.

Bloor, P.A., Barker, F.S., Watts, P.C., Noyes, H.A. and Kemp, S.J. (2001) Microsatellite Libraries by Enrichment. Protocol available at: http://www.genomics.liv.ac.uk/animal/Protocol1.html. University of Liverpool, Animal Genomics Laboratory, UK.

Boontong, C., Pandey, M. and Changtragoon, S. (2008) Isolation and characterization of microsatellite markers in Indian neem (*Azadirachta indica* var. *indica* A. Juss) and cross-amplification in Thai neem (*A. indica* var. *siamensis* Valenton). *Conservation Genetics* 10, 669–671.

Botstein, D., White, R.L., Skolnick, M. and Davis, R.W. (1980) Construction of genetic linkage map in man using restriction fragment length polymorphisms. *American Journal of Human Genetics* 32, 314–331.

Brunel, D. (1994) A microsatellite marker in *Helianthus annuus* L. *Plant Molecular Biology* 24, 397–400.

Caldeira, R.L., Carvalho, O.S., Cardoso, P.C.M., Lage, R.C.G. and Oliveira, G.C. (2002) Sequence characterization of *Biomphalaria glabrata* SSR-PCR products. *Memórias do Instituto Oswaldo Cruz* 97, 23–26.

Canter, P.H., Thomas, H. and Ernst, E. (2005) Bringing medicinal plants into cultivation: opportunities and challenges for biotechnology. *Trends in Biotechnology* 23, 180–185.
Crozier, Y.C., Jia, X.C., Yao, J.Y., Field, A.R., Cook, J.M. and Crozier, R.H. (2007) Microsatellite primers for *Ficus racemosa* and *Ficus rubiginosa*. *Molecular Ecology Notes* 7, 57–59.
De Barro, P.J., Scott, K.D., Graham, G.C., Lange, C.L. and Schutze, M.K. (2003) Isolation and characterization of microsatellite loci in *Bemisia tabaci*. *Molecular Ecology Notes* 3, 40–43.
Deng, J., Liu, Y. and Huang, H. (2006) Development and characterization of microsatellite markers in *Eucommia ulmoides* Oliver (Eucommiaceae). *Molecular Ecology Notes* 6, 496–498.
Dixit, A., Jin, M.H., Chung, J.W., Yu, J.W., Chung, H.K., Ma, K.H., Park, Y.J. and Cho, E.G. (2005) Development of polymorphic microsatellite markers in sesame (*Sesamum indicum* L.). *Molecular Ecology Notes* 5, 736–738.
Eckert, K.A. and Hile, S.E. (2009) Every microsatellite is different: Intrinsic DNA features dictate mutagenesis of common microsatellites present in the human genome. *Molecular Carcinogenesis* 48, 379–388.
Eckert, K.A., Yan, G. and Hile, S.E. (2002) Mutation rate and specificity analysis of tetranucleotide microsatellite DNA analysis in somatic human cells. *Molecular Carcinogenesis* 34, 140–150.
Edwards, K.J., Backer, J.H.A., Daly, A., Jones, C. and Karp, A. (1996) Microsatellite libraries enriched for several microsatellite sequences in plants. *Biotechniques* 20, 758–760.
Edwards, C.E., Arakaki, M., Ascencio, P.F.Q., Soltis, D.E. and Soltis, P.S. (2007) Isolation and characterization of microsatellite loci from the endangered highlands scrub hypericum (*Hypericum cumulicola*). *Molecular Ecology Notes* 7, 1135–1137.
Edwards, R. (2004) No remedy in sight for herbal ransack. *New Scientist* 181, 10–11.
Ellegren, H. (2000) Microsatellite mutations in the germline: implications for evolutionary inference. *Trends in Genetics* 16, 551–558.
Ellegren, H. (2002) Mismatch repair and mutational bias in microsatellite DNA. *Trends in Genetics* 18, 552.
Fischer, D. and Bachmann, K. (1998) Microsatellite enrichment in organisms with large genomes (*Allium cepa* L.). *BioTechniques* 24, 796–802.
Fisher, P.J., Gardner, R.C. and Richardson, T.E. (1996) Single locus microsatellites isolated using 5′-anchored PCR. *Nucleic Acids Research* 24, 4369–4371.
Geng, Q.F., Lian, C.L., Tao, J.M. and Hogetsu, T. (2008) Development of microsatellite markers for two nonviviparous mangrove species, *Acanthus ilicifolius* and *Lumnitzera racemosa*. *Molecular Ecology Resources* 8, 377–380.
Gilmore, S. and Peakall, R. (2003) Isolation of microsatellite markers in *Cannabis sativa* L. (marijuana). *Molecular Ecology Notes* 3, 105–107.
Gilmore, S., Peakall, R. and Robertson, J. (2003) Short tandem repeat (STR) DNA markers are hypervariable and informative in *Cannabis sativa*: implications for forensic investigations. *Forensic Science International* 131, 65–74.
Grover, A. and Sharma, P.C. (2007) Microsatellite motifs with moderate GC content are clustered around genes on *Arabidopsis thaliana* chromosome 2. *In Silico Biology* 7, 0021.
Grover, A., Aishwarya, V. and Sharma, P.C. (2007) Biased distribution of microsatellite motifs in the rice genome. *Molecular Genetics and Genomics* 277, 469–480.
Grover, A., Ramesh, B. and Sharma, P.C. (2009) Development of microsatellite markers in potato and their transferability in some members of Solanaceae. *Physiology and Molecular Biology of Plants* 15, 343–358.
Gu, S., Ding, X.Y., Wang, Y., Zhou, Q., Ding, G., Li, X.X. and Qian, L. (2007) Isolation and characterization of microsatellite markers in *Dendrobium officinale*, an endangered herb endemic to China. *Molecular Ecology Notes* 7, 1166–1168.
Guo, W.L., Gong, L., Ding, Z.F., Li, Y.D., Li, F.X., Zhao, S.P. and Liu, B. (2006) Genomic instability in phenotypically normal regenerants of medicinal plant *Codonopsis lanceolata* Benth. et Hook. f., as revealed by ISSR and RAPD markers. *Plant Cell Reports* 25, 896–906.

Hakki, E.E., Kayis, S.A., Pinarkara, E. and Sag, A. (2007) Inter simple sequence repeats separate efficiently hemp from marijuana (*Cannabis sativa* L.). *Electronic Journal of Biotechnology* 10, 570–581.

Hancock, J.M. (1999) Microsatellites and Other Simple Sequences: Genomic Context and Mutational Mechanisms. In: Goldstein, D.B. and Schlotterer, C. (eds) *Microsatellites: Evolution and Applications*. Oxford University Press, Oxford, UK, pp. 1–9.

Harr, B., Kauer, M. and Schlötterer, C. (2002) Hitchhiking mapping: a population-based fine-mapping strategy for adaptive mutations in *Drosophila melanogaster*. *Proceedings of National Academy of Sciences of the United States of America* 99, 12949–12954.

Henry, P., Provan, J., Goudet, J., Guisan, A., Jahodo, V. and Besnard, G. (2008) A set of primers for plastid indels and nuclear microsatellites in the invasive plant *Heracleum mantegazzianum* (Apiaceae) and their transferability to *Heracleum sphondylium*. *Molecular Ecology Resources* 8, 161–163.

Hu, J.J., Nakatani, M., Lalusin, A.G. and Fujimura, T. (2004) New microsatellite markers developed from reported *Ipomoea trifida* sequences and their application to sweet potato and its related wild species. *Scientia Horticulturae* 102, 375–386.

Huang, H.R., Zhou, G., Ge, X.J., Wei, X., Jiang, Y.S. and Tang, H. (2008) Eight polymorphic microsatellite loci for the Chinese medicinal plant *Artemisia annua* L. (Asteraceae). *Conservation Genetics* 10, 593–595.

Iwata, H., Imon, K., Tsumura, Y. and Ohsawa, R. (2005) Genetic diversity among Japanese indigenous common buckwheat (*Fagopyrum esculentum*) cultivars as determined from AFLP and SSR markers and quantitative agronomic traits. *Genome* 48, 367–377.

Joshi, K., Chavan, P., Warude, D. and Patwardhan, B. (2004) Molecular markers in herbal drug technology. *Current Science* 87, 159–165.

Kandpal, R.P., Kandpal, G. and Weissman, S.M. (1994) Construction of libraries enriched for sequence repeats and jumping clones, and hybridization selection for region-specific markers. *Proceedings of the National Academy of Sciences of the United States of America* 91, 88–92.

Karagyozov, L., Kalcheva, I.D. and Chapman, V. (1993) Construction of random small-insert genomic libraries highly enriched for simple sequence repeats. *Nucleic Acids Research* 21, 3911–3912.

Kashi, Y. and King, D.G. (2006) Simple sequence repeats as advantageous mutators in evolution. *Trends in Genetics* 22, 253–259.

Khanuja, S.P.S., Paul, S., Shashney, A.K., Darokar, M.P., Shukla, A.K., Gupta, M.M. and Kumar, A. (2009) Primers and a screening method for identification of artemisinin producing plants. *United States Patent 7473768*.

Konishi, T., Iwata, H., Yashiro, K., Tsumura, Y., Ohsawa, R., Yasui, Y. and Ohnishi, O. (2006) Development and characterization of microsatellite markers for common Buckwheat. *Breeding Science* 56, 277–285.

Le Cadre, S., Dubayle, M.C.B., Lambourdière, J., Machon, N., More, T.J. and Samad, S. (2005) Polymorphic microsatellites for the study of *Aconitum napellus* L. (Ranunculaceae), a rare species in France. *Molecular Ecology Notes* 5, 358–360.

Litt, M. and Luty, J.A. (1989) A hypervariable microsatellite revealed by *in vitro* amplification of a dinucleotide repeat within the cardiac muscle actin gene. *American Journal of Human Genetics* 44, 397–401.

Ma, K.H., Dixit, A., Kim, Y.C., Lee, D.Y., Kim, T.S., Cho, E.G. and Park, Y.J. (2007) Development and characterization of new microsatellite markers for ginseng (*Panax ginseng* C.A. Meyer). *Conservation Genetics* 8, 1507–1509.

Meyer, W., Mitchell, T.G., Freedman, E.Z. and Vilgalys, R. (1993) Hybridization probes for conventional DNA fingerprinting used as single primers in the polymerase chain reaction to distinguish strains of *Cryptococcus neoformans*. *Journal of Clinical Microbiology* 31, 2274–2280.

Modrich, P. and Lahue, R. (1996) Mismatch repair in replication fidelity, genetic recombination and cancer biology. *Annual Review of Biochemistry* 65, 101–133.

Negi, M.S., Sabharwal, V., Wilson, N. and Lakshmikumaran, M.S. (2006) Comparative analysis of the efficiency of SAMPL and AFLP in assessing genetic relationships among *Withania somnifera* genotypes. *Current Science* 91, 464–475.

Ohtani, M., Tani, N. and Yoshimaru, H. (2008) Development and characterization of microsatellite markers for *Hibiscus glaber* Matsum. ex Nakai, an endemic tree species of the oceanic Bonin Islands, Japan. *Molecular Ecology Resources* 8, 1316–1318.

Oleszek, W., Stochmal, A., Karolewski, P., Simonet, A.M., Macias, F.A. and Tava, A. (2002) Flavanoids from *Pinus sylvestris* needles and their variation in trees of different origin grown for nearly a century at the same area. *Biochemical Systematics and Ecology* 30, 1011–1022.

Pan, L., Quan, Z., Li, S., Liu, H., Huang, X., Ke, W. and Ding, Y. (2007) Isolation and characterization of microsatellite markers in the sacred lotus (*Nelumbo nucifera* Gaertn.). *Molecular Ecology Notes* 7, 1054–1056.

Powell, W., Morgante, M., Andre, C., Hanafey, M., Vogel, J., Tingey, S. and Rafalski, A. (1996) The comparison of RFLP, RAPD, AFLP and SSR (microsatellite) markers for germplasm analysis. *Molecular Breeding* 2, 225–238.

Provan, J., Love, H.M. and Maggs, C.A. (2007) Development of microsatellites for the invasive riparian plant *Impatiens glandulifera* (Himalayan balsam) using inter-simple sequence repeat cloning. *Molecular Ecology Notes* 7, 451–453.

Rassmann, K., Schlotterer, C. and Tautz, D. (1991) Isolation of simple sequence loci for use in polymerase chain reaction-based DNA fingerprinting. *Electrophoresis* 12, 113–118.

Riley, D.E., Jeon, J.S. and Kreiger, J.N. (2007) Simple repeat evolution includes dramatic primary sequence changes that conserve folding potential. *Biochemical and Biophysical Research Communications* 355, 619–625.

Roorkiwal, M., Grover, A. and Sharma, P.C. (2009) Genome-wide analysis of conservation and divergence of microsatellites in rice. *Molecular Genetics and Genomics* 282, 205–215.

Schlotterer, C. (1997) Microsatellites. In: Hoelzel, A.R. (ed.) *Molecular Genetic Analysis of Populations: A Practical Approach.* Oxford University Press, Oxford, pp. 237–260.

Schlotterer, C. and Tautz, D. (1992) Slippage synthesis of simple sequence DNA. *Nucleic Acids Research* 20, 211–215.

Scott, I.A., Hayes, C.M., Keogh, J.S. and Morrison, S.F. (2001) Isolation and characterisation of novel microsatellite markers from the Australian water skink *Eulamprus kosciuskoi* and cross-species amplification in other members of the species-group. *Molecular Ecology Notes* 1, 28–30.

Sharma, P.C., Grover, A. and Kahl, G. (2007) Mining microsatellites in eukaryotic genomes. *Trends in Biotechnology* 25, 490–498.

Shen, J., Ding, X., Liu, D., Ding, G., Ha, J., Li, X., Tang, F. and Chu, B. (2006) Inter-simple sequence repeats (ISSR) molecular fingerprinting markers for authenticating populations of *Dendrobium officinale* KIMURA et MIGO. *Biological and Pharmaceutical Bulletin* 29, 420–422.

Shokeen, B., Sethy, N.K., Choudhary, S. and Bhatia, S. (2005) Development of STMS markers from the medicinal plant Madagascar periwinkle (*Catharanthus roseus* (L.) G. Don.). *Molecular and Ecology Notes* 5, 818–820.

Shokeen, B., Sethy, N.K., Kumar, S. and Bhatia, S. (2007) Isolation and characterization of microsatellite markers for analysis of molecular variation in the medicinal plant Madagascar periwinkle (*Catharanthus roseus* (L.) G. Don). *Plant Science* 172, 441–451.

Takayama, K., Kajita, T., Murata, J. and Tateishi, Y. (2006) Isolation and characterization of microsatellites in the sea hibiscus (*Hibiscus tiliaceus* L., Malvaceae) and related hibiscus species. *Molecular Ecology Notes* 6, 721–723.

Takayama, K., Tateishi, Y., Murata, J. and Kajita, T. (2008) Gene-flow and population subdivision in a pantropical plant with sea-drifted seeds *Hibiscus tiliaceus* and its allied species: evidence from microsatellite analyses. *Molecular Ecology* 17, 2730–2742.

Tani, N., Takahashi, T., Ujino-Ihara, T., Iwata, H., Yoshimura, K. and Tsumura, Y. (2004) Development and characteristics of microsatellite markers for sugi (*Cryptomeria japonica* D. Don) derived from microsatellite-enriched libraries. *Annals of Forest Science* 61, 569–575.

Tautz, D. (1989) Hypervariability of simple sequences as a general source for polymorphic DNA markers. *Nucleic Acids Research* 17, 6463–6471.

Tenzer, I., Degli Ivanissevich, S., Morgante, M. and Gessler, C. (1999) Identification of microsatellite markers and their application to population genetics of *Venturia inaequalis*. *Phytopathology* 89, 748–753.

Tian, C.J., Nan, P., Shi, S.H., Chen, J.K. and Zhong, Y. (2004) Molecular genetic variation in Chinese populations of three subspecies of *Hippophae rhamnoides*. *Biochemical Genetics* 42, 259–266.

Vos, P., Hogers, R., Bleeker, M., Reijans, M., van de Lee, T., Hornes, M., Frijters, A., Pot, J., Pelemen, J. and Kuiper, M. (1995) AFLP: a new technique for DNA fingerprinting. *Nucleic Acids Research* 23, 4407–4414.

Walker, N.F., Hulme, P.F. and Hoelzel, A.R. (2003) Population genetics of an invasive species, *Heracleum mantegazzianum*: implications for the role of life history, demographics and independent introductions. *Molecular Ecology* 12, 1747–1756.

Wan, D., Wang, A., Wu, G. and Zhao, C. (2008) Isolation of polymorphic microsatellite markers from *Przewalskia tangutica* (Solanaceae). *Conservation Genetics* 9, 995–997.

Wang, A., Zhang, Q., Wan, D., Yang, Y. and Liu, J. (2008) Nine Microsatellite DNA primers for *Hippophae rhamnoides* ssp. *sinensis* (Elaeagnaceae). *Conservation Genetics* 9, 969–971.

Wanke, S., Quandt, D. and Neinhuis, C. (2006) Universal primers for a large cryptically simple cpDNA microsatellite region in *Aristolochia* (Aristolochiaceae). *Molecular Ecology Notes* 6, 1051–1053.

Wardill, T.J., Scott, K.D., Graham, G.C. and Zalucki, M.P. (2004) Isolation and characterization of microsatellite loci from *Acacia nilotica* ssp. *indica* (Mimosaceae). *Molecular Ecology Notes* 4, 361–363.

Weber, J.L. and May, P.E. (1989) Abundant class of human DNA polymorphisms which can be typed using the polymerase chain reaction. *American Journal of Human Genetics* 44, 388–396.

Weising, K., Nybom, H., Wolff, K. and Kahl, G. (2005) *DNA Fingerprinting in Plants: Principles, Methods, and Applications*, 2nd ed. CRC Press, Florida.

Williams, J.G.K., Kubelik, A.R., Livak, K.J., Rafalski, J.A. and Tingey, S.V. (1990) DNA polymorphisms amplified by arbitrary primers are useful as genetic markers. *Nucleic Acids Research* 18, 6531–6535.

Wu, W., Zheng, Y.L., Chen, L., Wei, Y.M., Yang, R.W. and Yan, Z.H. (2005) Evaluation of genetic relationships in the genus *Houttuynia* Thunb. in China based on RAPD and ISSR markers. *Biochemical Systematics and Ecology* 33, 1141–1157.

Xu, Y., Wang, Y., Li, Z., Bao, Z., Zhou, J. and Huang, H. (2006) Characterization of polymorphic microsatellite loci in a traditional Chinese medicinal plant, *Gastrodia elata*. *Molecular Ecology Notes* 6, 316–318.

Xu, Y., Huang, H., Li, Z. and Wang, Y. (2008) Development of 12 novel microsatellite loci in a traditional Chinese medicinal plant, *Epimedium brevicornum* and crossamplification in other related taxa. *Conservation Genetics* 9, 949–952.

Yip, P.Y., Chau, C.F., Mak, C.Y. and Kwan, H.S. (2007) DNA methods for identification of Chinese medicinal materials. *Chinese Medicine* 2, 9.

Zane, L., Bargelloni, L. and Patarnello, T. (2002) Strategies for microsatellite isolation: a review. *Molecular Ecology* 11, 1–16.

Zavodna, M., Arens, P., Van Dijk, J. and Vosman, B. (2005) Development and characterization of microsatellite markers for two dioecious Ficus species. *Molecular Ecology Notes* 5, 355–357.

Zhang, Y., He, J., Zhao, P.X., Bouton, J.H. and Monteros, M.J. (2008) Genome-wide identification of microsatellites in white clover (*Trifolium repens* L.) using FIASCO and phpSSRMiner. *Plant Methods* 4, 19.

Zietkiewicz, E., Rafalski, A. and Labuda, D. (1994) Genome fingerprinting by simple sequence repeat (SSR)-anchored polymerase chain reaction amplification. *Genomics* 20, 176–183.

Zini, E. and Komjanc, M. (2008) Identification of microsatellite markers in *Hieracium pilosella* L. *Conservation Genetics* 9, 487–489.

Chapter 6

In vitro Propagation of Medicinal Plants for Conservation and Quality Assurance

Christoph Wawrosch

Introduction

Mankind has been using plants since time immemorial for purposes such as food, timber, fodder, and as medicinal remedies. Worldwide, out of some estimated 422,000 plant species about 72,000 are used as medicinal and aromatic plants (MAPs) (Schippmann *et al.*, 2006). An estimation of the World Health Organization indicates that more than 80% of the world population still relies mainly on herbal medicine for their healthcare (Vines, 2004). Large amounts of MAPs are industrially processed for the preparation of pure compounds which are used as such, or further modified (Bonati, 1988). In Western countries there has been renewed interest for plant-based medicines in the past decades, including own traditional herbals and phytopharmaceuticals, but also foreign medicinal systems such as Ayurveda or Traditional Chinese Medicine.

Generally, the majority of plant species used for medicinal purposes are harvested from the wild. As a result of massive harvesting of MAPs from wild populations, a significant number of plants are endangered in their natural habitats in European countries (Lange, 1998). In South Africa, the rich resources of medicinally useful plants are decreasing at an alarming rate due to over-exploitation (Afolayan and Adebola, 2004). Deforestation has been regarded as one reason for the rapid loss of medicinal plant wealth in India (Rout *et al.*, 2000). The annual trade in wild collected MAPs and products thereof in South Asia is huge (Olsen, 2005), and greenhouse climate change will probably have major impact in the Himalayas as a future threat to the plant communities (Barnett *et al.*, 2005).

Globally, about 15,000 medicinal plant species are threatened, the reasons including loss of habitat, commercial over-harvesting, invasive species and pollution (Hamilton, 2008). It is by no doubt very necessary to push every possible effort into strategies for the conservation of our planet's invaluable resources. However, there are also other reasons why wild-collected medicinal plant material might be problematic. While there is always a chance for adulterations with unwanted, maybe even harmful species, companies are also interested in steady sources of reliably botanically identified raw material, genotypes with desired and improved traits, and controlled post-harvest handling (Leaman, 2001; Schippmann *et al.*, 2006).

Conservation and Quality Assurance

Two major concerns arise from the above mentioned facts: (i) an obvious and urgent need for measures aiming at protecting and conserving world-wide plant populations; and (ii) in the context of ongoing development of safe and high-quality plant-based medicines, quality assurance is a major issue (Chaturvedi *et al.*, 2007). As a possible solution, it has been recommended that wild species be brought into suitable cultivation systems. It is currently discussed whether or not this will be a general option in each and every situation, or whether sustainable harvest would be a more important conservation strategy for wild-harvested species (Schippmann *et al.*, 2006).

It seems likely that in many cases an approach tailored to the specific situation could be successful. However, it is undoubted that controlled field culture offers a number of advantages and would help in both conservation and quality assurance of medicinal plants. Such a trend has been observed in Germany where producers of phytomedicines are involved in the domestication of hitherto wild-collected species (Franke, 1999).

In vitro Propagation

The domestication of wild-growing species includes many steps ranging from studies of the plant's botanical features and ecological requirements, to the elaboration of the final cultivation process, and can last several years. At different stages *in vitro* techniques can be included, e.g. for storage of plant materials, propagation, or selection breeding for genetic improvement (Franz, 1991; Bohr, 1997). Certainly plant tissue culture is not to be used in each and every case, but offers clear advantages when species are difficult to propagate conventionally. This would be the case e.g. with seldomly or never flowering species, low seed germination rates or inhomogeneous seed quality, or with slow growing taxa like woody plants (Canter *et al.*, 2005; Wawrosch, 2005).

In vitro-propagation (micropropagation) offers a number of clear advances, as recently summarized by Debnath *et al.* (2006). Basically, these include: (i) the production of large numbers of plantlets in a comparably short time, due to usually high multiplication rates; (ii) micropropagation is feasible independently of the season; and (iii) plants produced *in vitro* are usually free from microorganism borne diseases, and valuable genotypes can be freed from plant viruses.

As generally also valid for any agronomically or horticulturally important plant species, it should be kept in mind that not every *in vitro* technique is equally suitable for the production of genetically homogeneous propagules. The avoidance of off-types is without doubt also important when *in vitro* techniques are applied within programmes for domestication and cultivation of medicinal plants. Thus, *in vitro* propagation starting from existing meristems should be the major strategy (i.e. shoot or nodal culture) because it is least likely to induce somaclonal variation. However, depending on the genotype, other techniques based on adventitious shoot formation or somatic embryogenesis may be used as long as the occurrence of genetic variation is acceptable or can be controlled (George and Debergh, 2008).

Medicinal Plants

Commercial *in vitro* propagation of plants has developed into a multi-billion-dollar industry since the 1960s. However, although available figures date back to more than a decade, it becomes apparent that MAPs play only a minor role (Pierik, 1991; Govil and Gupta, 1997; Hartman and Zimmerman, 1999). While there are a number of commercial and technological factors which generally limit the marketing of micropropagated plants (Ilan and Khayat, 1997), the increasing demand for medicinal drug material might well change the situation. Furthermore, the development of new bioreactor-based techniques is likely to reduce production costs and improve efficacy of *in vitro* propagation (Ziv, 2000; Etienne and Berthouly, 2002). Nevertheless, there has been substantial research in the development of propagation protocols for MAPs. For details we refer to the recent excellent reviews by Chaturvedi *et al.* (2007), Debnath *et al.* (2006), Afolayan and Adebola (2004), Tripathi and Tripathi (2003) and Rout *et al.* (2000). At the author's institution, *in vitro* propagation of medicinal plants has been dealt with for several years; Table 6.1 lists some species which have been studied.

Table 6.1. Some medicinal plant species for which *in vitro*-propagation protocols have been established at the Department of Pharmacognosy, University of Vienna.

Medicinal plant species	References
Achillea ceretanica Sennen	Wawrosch *et al.* (1997)
Allium wallichii Kunth	Wawrosch *et al.* (2001a)
Butea buteiformis (Voigt) Grierson and Long	Singh *et al.* (2006)
Charybdis numidica (Jour. & Fourr) Speta	Kongbangkerd *et al.* (2005); Kongbangkerd and Wawrosch (2003)
Crataegus monogyna Jacq. (Lindm.)	Wawrosch *et al.* (2007)
Dendrobium huoshanense C.Z. Tang et S.J. Cheng	Luo *et al.* (2009)
Drosera sp.	Wawrosch *et al.* (1996); Kopp *et al.* (2006)
Glycyrrhiza glabra L.	Wawrosch *et al.* (2009)
Lilium nepalense D.Don	Wawrosch *et al.* (2001b)
Picrorhiza kurroa Royle. ex Benth.	Wawrosch *et al.* (2002, 2003)
Stemona curtisii Hook f.	Montri *et al.* (2006)
Stemona tuberosa Lour.	Montri *et al.* (2009)
Swertia chirata Buch.-Ham. ex Wall	Wawrosch *et al.* (1999, 2005)
Tussilago farfara L.	Kopp *et al.* (1997); Wawrosch *et al.* (2000)

Conclusion

While 'theory and practice' concerning *in vitro* propagation of medicinal plants are still discussed, it is however a fact that protocols for the *in vitro* propagation of several MAP species are readily available. With improvements in *in vitro* techniques, economically feasible protocols are likely to be available soon. Certainly, *in vitro* propagation can contribute to (*ex situ*) conservation of endangered species, but also to quality assurance, as demonstrated.

References

Afolayan, A.J. and Adebola, P.O. (2004) *In vitro* propagation: A biotechnological tool capable of solving the problem of medicinal plants decimation in South Africa. *African Journal of Biotechnology* 3, 683–687.

Barnett, T.P., Adam, J.C. and Lettenmaier, D.P. (2005) Potential impacts of a warming climate on water availability in snow-dominated regions. *Nature* 438, 303–308.

Bohr, C. (1997) Inkulturnahme von bisher aus Wildsammlungen stammenden Wirkstoffpflanzen. *Drogenreport* 10, 37–39.

Bonati, A. (1988) Industry and the conservation of medicinal plants. *Fitoterapia* 59, 367–370.

Canter, P.H., Thomas, H. and Ernst, E. (2005) Bringing medicinal plants into cultivation: opportunities and challenges for biotechnology. *Trends in Biotechnology* 23, 180–185.

Chaturvedi, H.C., Jain, M. and Kidwai, N.R. (2007) Cloning of medicinal plants through tissue culture – a review. *Indian Journal of Experimental Biology* 45, 937–948.

Debnath, M., Malik, C.P. and Bisen, P.S. (2006) Micropropagation: A tool for the production of high quality plant-based medicines. *Current Pharmaceutical Biotechnology* 7, 33–49.

Etienne, H. and Berthouly, M. (2002) Temporary immersion systems in plant micropropagation. *Plant Cell, Tissue and Organ Culture* 69, 215–231.

Franke, R. (1999) Neuentstehung von Kulturpflanzen heute – Beispiele aus dem Bereich Arznei– und Gewürzpflanzen. *Zeitschrift für Arznei– und Gewürzpflanzen* 4, 24–38.

Franz, C. (1991) Kultivierung bedeutender wildwachsender Arznei– und Gewürzpflanzen. *entwicklung + ländlicher raum* 4, 3–7.

George, E.F. and Debergh, P.C. (2008) Micropropagation: Uses and Methods. In: George, E.F., Hall, M.A. and De Klerk, G.-J. (eds), *Plant Propagation by Tissue Culture* 3rd Edition, Vol. 1. Springer, Dordrecht, pp. 29–64.

Govil, S. and Gupta, S.C. (1997) Commercialization of plant tissue culture in India. *Plant Cell, Tissue and Organ Culture* 51, 65–73.

Hamilton, A.C. (ed.) (2008) *Medicinal Plants in Conservation and Development: Case Studies And Lessons Learnt*. Plantlife International, Salisbury.

Hartman, R.D. and Zimmerman, R.H. (1999) Commercial micropropagation in the United States, 1965–1998. In: Altman, A., Ziv, M. and Izhar, S. (eds) *Plant Biotechnology and In vitro Biology in the 21st Century*. Kluwer Academic Publishers, Dordrecht, pp. 699–707.

Ilan, A. and Khayat, E. (1997) An overview of commercial and technological limitations to marketing of micropropagated plants. *Acta Horticulturae* 447, 643–648.

Kongbangkerd, A. and Wawrosch, C. (2003) Improved shoot regeneration from nodules of *Charybdis numidica* in a temporary immersion system. *Journal of Horticultural Science & Biotechnology* 78, 650–655.

Kongbangkerd, A., Köpf, A., Allacher, P., Wawrosch, C. and Kopp, B. (2005) Micropropagation of Squill (*Charybdis numidica*) through nodule culture. *Plant Cell Reports* 23, 673–677.

Kopp, B., Wawrosch, C., Lebada, R. and Wiedenfeld, H. (1997) Pyrrolizidinalkaloid-freie Huflattich-Blätter (Farfarae folium DAB 10, Folium Tussilaginis ÖAB 90) bald am Markt. Teil I: *In vitro*–Kultivierung und Selektionszüchtung führten zum Erfolg. *Deutsche Apotheker Zeitung* 137, 4066–4069.

Kopp, B., Wawrosch, C., Buol, I. and Dorfer, T. (2006) Efficient production of Sundew (*Drosera rotundifolia* L.) *in vitro* using a temporary immersion system. *Planta Medica* 72, 1080.

Lange, D. (1998) *Europe's Medicinal and Aromatic Plants: Their Use, Trade and Conservation*. Traffic International, Cambridge, UK.

Leaman, D.J. (2001) Conservation, trade, sustainability and exploitation of medicinal plant species. In: Saxena, P.K. (ed.) *Development of Plant-Based Medicines: Conservation, Efficacy and Safety*. Kluwer Academic Publishers, Dordrecht, pp. 1–15.

Luo, J.-P., Wawrosch, C. and Kopp, B. (2009) Enhanced micropropagation of *Dendrobium huoshanense* C-Z. Tang et S.J. Cheng through protocorm-like bodies: The effects of cytokinins, carbohydrate sources and cold pretreatment. *Scientia Horticulturae* 123, 258–262.

Montri, N., Wawrosch, C. and Kopp, B. (2006) Micropropagation of *Stemona curtisii* Hook f., a Thai medicinal plant. *Acta Horticulturae* 725, 341–346.

Montri, N., Wawrosch, C. and Kopp, B. (2009) *In vitro* propagation of *Stemona tuberosa* Lour., an antitussive medicinal herb. *Acta Horticulturae* 812, 165–172.

Olsen, C.S. (2005) Quantification of the trade in medicinal and aromatic plants in and around Nepal. *Acta Horticulturae* 678, 29–35.

Pierik, R.L.M. (1991) *Commercial Micropropagation in Western Europe and Israel*. In: Debergh, P.C. and Zimmerman, R.H. (eds) *Micropropagation: Technology and Application*. Kluwer Academic Publishers, Dordrecht, pp. 155–165.

Rout, G.R., Samantaray, S. and Das, P. (2000) *In vitro* manipulation and propagation of medicinal plants. *Biotechnology Advances* 18, 91–120.

Schippmann, U., Leaman, D. and Cunningham, A.B. (2006) A Comparison of Cultivation and Wild Collection of Medicinal and Aromatic Plants Under Sustainability Aspects. In: Bogers, R.J., Craker, L.E. and Lange, D. (eds) *Medicinal and Aromatic Plants: Agricultural, Commercial, Ecological, Legal, Pharmacological and Social Aspects*. Springer, Dordrecht, pp. 75–95.

Singh, B.M., Wawrosch, C., Joshi, S.D. and Kopp, B. (2006) Micropropagation of *Butea buteiformis* (Voigt) Grierson and Long. *Nepal Journal of Science and Technology* 7, 119–122.

Tripathi, L. and Tripathi, J.N. (2003) Role of biotechnology in medicinal plants. *Tropical Journal of Pharmaceutical Research* 2, 243–253.

Vines, G. (2004) *Herbal Harvests With a Future: Towards Sustainable Sources for Medicinal Plants*. Plantlife International, Salisbury.

Wawrosch, C. (2005) *In vitro* Cultivation of Medicinal Plants. In: Yaniv, Z. and Bachrach, U. (eds) *Handbook of Medicinal Plants*. Haworth Press, Binghampton, pp. 261–278.

Wawrosch, C., Hugh-Bloch, A., Hostettmann, K. and Kopp, B. (2005) *Swertia chirata* Buch.–Ham. ex Wall. (Gentianaceae), an endangered Himalayan medicinal plant: Comparative study of the secondary compound patterns in market drug, *in vitro*-cultivated, and micropropagated field grown samples. *Scientia Pharmaceutica* 73, 127–137.

Wawrosch, C., Kazianka, C., Küchler, V., Lämmermayer, K., Winter, M. and Kopp, B. (2009) *In vitro* multiplication of *Glycyrrhiza glabra* L. through somatic embryogenesis. *Planta Medica* 75, 1036–1037.

Wawrosch, C., Kopp, B. and Wiedenfeld, H. (2000) Permanent monitoring of pyrrolizidine alkaloid content in micropropagated *Tussilago farfara* L.: a tool to fulfill statutory demands for the quality of coltsfoot in Austria and Germany. *Acta Horticulturae* 530, 469–472.

Wawrosch, C., Kopp, B., Stöckl, J., Glasl, S. and Kubelka, W. (1997) *In vitro* propagation of tetraploid *Achillea ceretanica* SENNEN. *Pharmaceutical & Pharmacological Letters* 7, 116–118.

Wawrosch, C., Malla, P.R. and Kopp, B. (2001a) Micropropagation of *Allium wallichii* KUNTH, a threatened medicinal plant of Nepal. *In vitro Cellular and Developmental Biology – Plant* 37, 555–557.

Wawrosch, C., Malla, P.R. and Kopp, B. (2001b) Clonal propagation of *Lilium nepalense* D. Don, a threatened medicinal plant of Nepal. *Plant Cell Reports* 20, 285–288.

Wawrosch, C., Markotai, J., Steinberger, B. and Kopp, B. (1996) *In vitro*–Vermehrung von Sonnentau–Arten. *Scientia Pharmaceutica* 64, 709–717.

Wawrosch, C., Maskay, N. and Kopp, B. (1999) Micropropagation of the threatened Nepalese medicinal plant *Swertia chirata* Buch.-Ham. Ex Wall. *Plant Cell Reports* 18, 997–1001.

Wawrosch, C., Prinz, S., Soleiman, Y. and Kopp, B. (2007) Clonal propagation of *Crataegus monogyna* Jacq. (Lindm.). *Planta Medica* 73, 1013.

Wawrosch, C., Rührlinger, S., Grauwald, B., Sturm, S., Stuppner, H. and Kopp, B. (2002) Clonal propagation of the Himalayan medicinal plant *Picrorhiza kurroa* (Scrophulariaceae). *Revista de Fitoterapia* 2, 330.

Wawrosch, C., Zeitlhofer, P., Grauwald, B. and Kopp, B. (2003) Effects of rooting auxins on the establishment of micro-propagated *Picrorhiza kurroa* plantlets in the greenhouse. *Acta Horticulturae* 616, 271–274.

Ziv, M. (2000) Bioreactor Technology for Plant Micropropagation. In: Janick, J. (ed.) *Horticultural Reviews*, Vol. 24. John Wiley & Sons, New York, USA, pp. 1–30.

Chapter 7

Propagation of Elite *Cannabis sativa* L. for the Production of Δ^9-Tetrahydrocannabinol (THC) using Biotechnological Tools

Suman Chandra, Hemant Lata, Ikhlas A. Khan and Mahmoud A. ElSohly

Introduction

Originally indigenous to temperate regions of Asia, *Cannabis sativa* L. is now widely distributed around the world. It grows in a variety of habitats ranging from sea level in tropical areas to alpine foot hills of the Himalayas. *Cannabis* has a long history of medicinal use in the Middle East and Asia, with references as far back as the 6th century BC. This species was introduced in Western Europe as medicine in the early 19th century to treat epilepsy, tetanus, rheumatism, migraine, asthma, trigeminal neuralgia, fatigue and insomnia (Doyle and Spence, 1995; Zuardi, 2006). *Cannabis sativa* contains a unique class of terpeno-phenolic compounds (cannabinoids or phytocannabinoids) which have been extensively studied since the discovery of the chemical structure of tetrahydrocannabinol (Δ^9-THC) commonly known as THC, which is the main constituent responsible for the psychoactive effects. The accumulation of THC is mainly found in glandular trichomes of the plant (Hammond and Mahlberg, 1977). A total of 537 *Cannabis* constituents including 109 phytocannabinoids have been reported in *Cannabis sativa* (Slade and ElSohly, 2009). The pharmacologic and therapeutic properties of preparations of *Cannabis sativa* L. and THC have been reported and reviewed by many researchers (Grinspoon and Bakalar, 1993; Mattes *et al.*, 1994; Brenneisen *et al.*, 1996). Beside Δ^9-THC, other major cannabinoids of *Cannabis* include Tetrahydrocannabivarin-THCV, Cannabidiol-CBD, Cannabichromene-CBC, Cannabigerol-CBG and Cannabinol-CBN, the chemical structures of which are shown in Fig. 7.1. Δ^8-THC is another closely related isomer of Δ^9-THC which is much less abundant (thought to be an artifact) and less potent than Δ^9-THC (Small and Marcus, 2003).

THC has a tremendous commercial value in the pharmaceutical area. Since *C. sativa* is a natural source of THC, efforts to select *Cannabis* varieties with the highest possible level of THC content are underway. However, due to the allogamous (cross fertilization) nature of the species, it is very difficult to maintain the chemical profile of selected high THC producing genotypes, grown from seeds, under field conditions. Since this plant is also used as an illicit drug and is highly regulated, its cultivation in open field must be done in secured areas. For the large scale production of *Cannabis*, being it outdoors or indoors, it is important to maintain genetic homogeneity of the cultivated plants. This is achieved by selecting plant material with the desired chemical profile followed by propagation of these plants either vegetatively or through micropropagation.

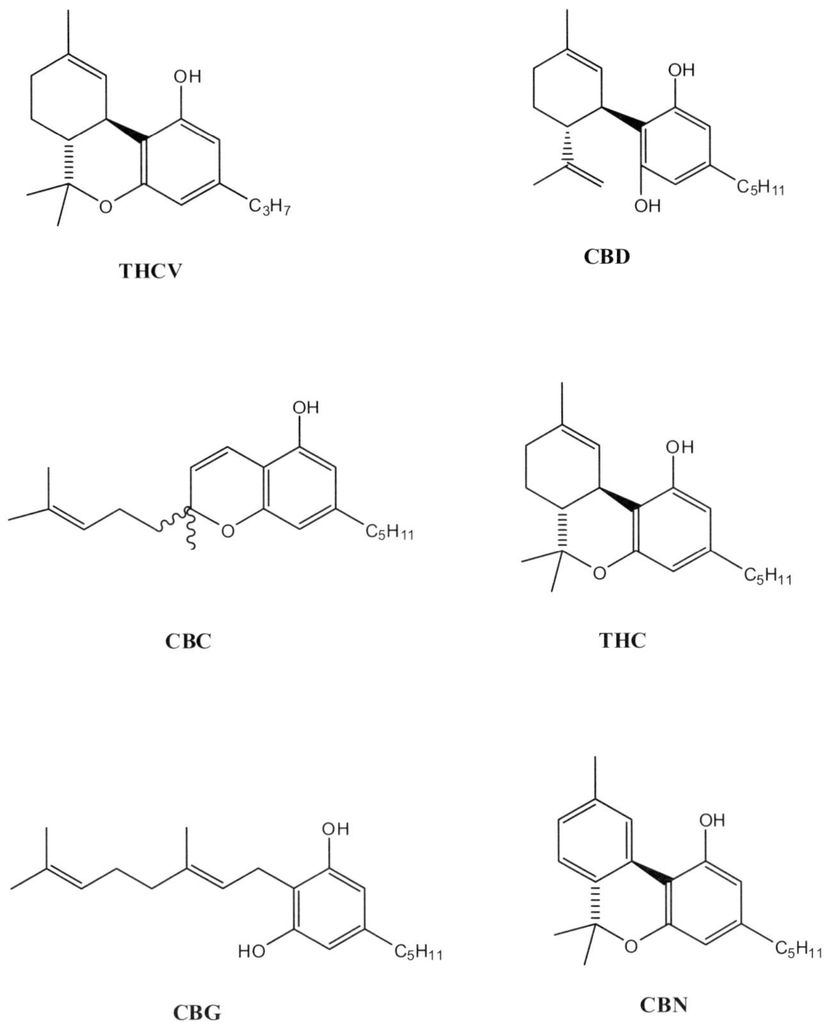

Fig. 7.1. Chemical structures of major phytocannabinoids in *Cannabis sativa*.

Selection of Elite Chemotypes

Based on its chemical constituents and their biological effects, *Cannabis* is considered a complex plant (ElSohly and Slade, 2005). Whether the genus *Cannabis* consists of one or more species is a matter of divided opinion (Small and Cronquist, 1976; Emboden, 1981). Using gas chromatography (GC), small differentiated *indica* strains from *sativa* strains based on their THC content (Small, 1979). Small and Cronquist (1976) have favoured a monospecific concept and assigned these two taxa to subspecies of *C. sativa*. Today, it is widely accepted that *Cannabis* is a monospecific genus (Small *et al.* 1976). However, a two-species concept of *Cannabis* is still advocated (Hillig and Mahlberg, 2004).

The analysis of the chemical profile of *Cannabis* using gas chromatography, particularly the concentration of cannabinoids in the dried inflorescence (leaves and buds) is considered to be the most objective measure to classify the plant according to psychoactive potency. The concentration of the most abundant cannabinoids, THC and CBD, in individual *Cannabis* parts can be characterized both quantitatively and qualitatively (Mandolino *et al.* 2003). Qualitative characterization involves determining the plant's THC/CBD ratio and assigning it to a discrete chemical phenotype (chemotype). Fetterman *et al.* (1971) have divided *Cannabis* plants into two distinct phenotypes, namely drug and fibre types. According to their classification, if THC/CBD ratio exceeded one, plants were classified as 'drug phenotype' otherwise as 'fibre phenotype'. Based on THC/CBD ratio, Small and Beckstead (1973a,b) distinguished three phenotypes, i.e. drug type (THC/CBD ratio>>1), intermediate type (THC/CBD ratio close to 1.0) and fibre type (THC/CBD ratio <<1). A rare, additional chemotype, characterized by a very low content of both THC and CBD and with CBG as the predominant constituent, was later identified by Fournier *et al.,* (1987). *Cannabis* can also be characterized quantitatively by measuring the cannabinoids content within its tissues. These levels are likely determined by the interaction of several genes and also influenced by the plant's growth environment (Hemphill *et al.*, 1980; de Meijer *et al.*, 1992; Chandra *et al.*, 2008, 2009a,b; Mendoza *et al.*, 2009). Numerous biotic and abiotic factors affect cannabinoid production including the sex and maturity of the plant (Small *et al.*, 1975), light cycle (Valle *et al.*, 1978), temperature (Bazzaz *et al.*, 1975; Chandra *et al.*, 2008), fertilization (Bo´Csa *et al.*, 1997) and light intensity (Pate, 1994). Variations in cannabinoids content among different tissues within a plant have been reported by Hemphill *et al.* (1980).

Considering all these variations, an efficient method for qualitative and quantitative determination of concentration of the major cannabinoids, namely Δ^9-THC, THCV, CBD, CBC, CBG and CBN was developed in our laboratory using gas chromatography–flame ionization detection, GC-FID (Ross *et al.* 1995). *Cannabis* samples were dried at 40°C and individually manicured by hand. Three 0.1 g samples were each extracted with 3 ml of internal standard/extracting solution (100 mg of 4-androstene-3, 17-dione + 10 ml chloroform + 90 ml methanol) at room temperature for 1h. The extracts were withdrawn into disposable transfer pipettes through cotton plugs for filtration and transferred into GC vials, which are then capped and placed on the auto sampler. One μl aliquots were injected. Gas chromatography analyses were performed using a Varian CP-3380 gas chromatograph equipped with a Varian CP-8400 automatic liquid sampler, a capillary injector and flame ionization detector. The column was a 15 m × 0.25 mm DB-1, 0.25 μ film (J&W Scientific, Inc.). Data are recorded using a Dell Optiplex GX1 computer and Varian Star (version 6.41) workstation software. Helium is used as the carrier gas. An indicating moisture trap and an indicating oxygen trap located in the helium line from upstream to downstream, respectively, were used. Helium was used as the 'make-up' gas at the detector. Hydrogen and compressed air were used as the combustion gases. The instrument parameters used for monitoring samples are: air: 30 psi (400 ml/min); hydrogen: 30 psi (30 ml/min); column head pressure: 14 psi (1.0 ml/min); split flow rate: 50 ml/min; split ratio:50:1; septum purge flow rate: 5 ml/min; make up gas pressure: 20 psi (20 ml/min); injector temp: 240°C; detector temp: 260°C; initial oven temp: 170 °C; initial temperature hold time: 1 min; temperature rate: 10°C/min; final oven temperature: 250°C and final temperature hold time: 3 min. The concentration of a specific cannabinoid is calculated as follows:

$$cannabinoid\% = \frac{GCarea(cannabinoid)}{GCarea(ISTD)} \times \frac{volume(ISTD)}{amount(sample)} \times 100$$

Currently, this method is being used by our group to analyse the confiscated marijuana samples submitted by the US Drug Enforcement Agency (DEA) and other US enforcement agencies under National Institute on Drug Abuse (NIDA) marijuana project, Potency monitoring programme at the University of Mississippi (ElSohly *et al.*, 2000; Mehmedic *et al.* 2009). A typical chromatogram of a screened and selected drug type *Cannabis sativa* clone is shown in Fig. 7.2.

Conventional Propagation

Once based upon gas chromatographic profile, the high THC yielding clone is screened and selected, a fresh nodal segment about 6–10 cm in length containing at least two nodes from the mother plant can be used for vegetative/conventional propagation. In our laboratory, we have developed simple and efficient vegetative propagation protocols for *Cannabis sativa* using solid (soil) and liquid medium (hydroponics).

Vegetative Propagation in Soil

Vegetative mass propagation of selected high yielding clones and their field plantation is shown in Figs 7.3B and 7.3E. Using a sterile tissue culture blade, a soft apical branch is cut at a 45° angle just below the node and dipped immediately in distilled water to avoid formation of any air bubble in the stem which might block fluid uptake later. About 2 cm of the base of the cuttings is dipped in green light rooting hormone (distributed by Green Light Co., San Antonio, TX, USA) for better rooting. Cuttings are placed in 2-inch jiffy pots containing coco natural growth medium and sterile potting mix fertilome, marketed by Canna Continental, Los Angeles, CA, USA) mixed in 1:1 ratio. At least one node is dipped in the soil for efficient rooting. Plants are irrigated regularly and kept under controlled environmental conditions (25 ± 3°C temperature and 55 ± 5% RH) in an indoor growing facility. Initially, the cuttings are kept under uniform flourescent light to avoid heating stress from regular grow lights. Rooting is initiated in 2–3 weeks, most of the cutting-raised plants are kept in jiffy pots for 6weeks and then transferred to bigger pots for better vegetative growth. Indoor light (~ 700 ± 24 μmol/m^2/s at plant canopy level measured by

LI-COR quantum meter, model LI-189) is then provided with seven full spectrum 1000 watts HID (high intensity discharge) lamps in combination with seven 1000 watt high pressure sodium bulbs (Sun Systems, CA), hung above plants covering 350 square foot area. A hot air suction fan is attached to each bulb assembly and about 3 to 4 feet distance between plants and bulb is maintained to avoid heating due to HID bulbs. These cuttings can be maintained in a constant vegetative stage under 18 or more light hours.

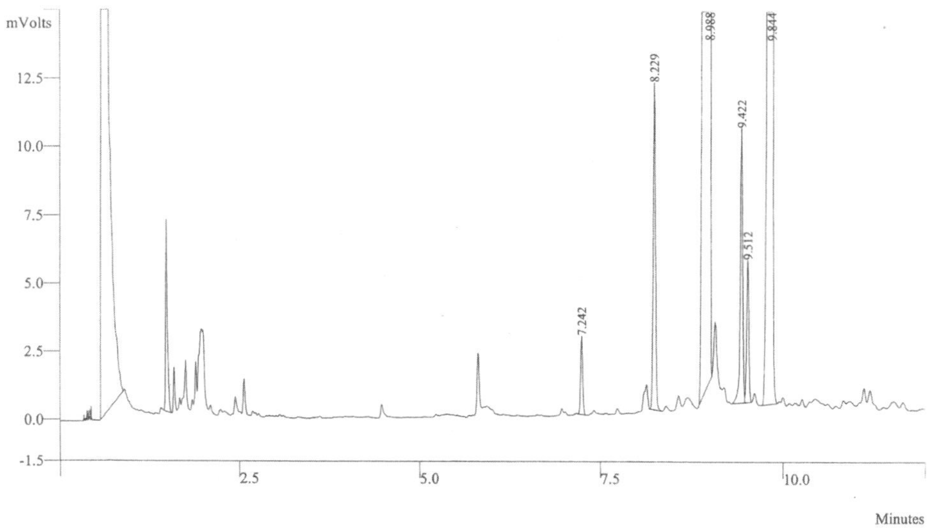

Peak No	Peak Name	% w/w	Ret Time (min)	Time Offset (min)	Status Codes	Peak Area (counts)	Sep. Code	Width 1/2 (sec)
1	THCV	0.0682	7.242	-0.016		5727	BB	1.8
2	CBC	0.3048	8.229	-0.018		25576	BB	1.9
3	DELTA-9-THC	10.4462	8.988	0.077		876646	BB	3.7
4	CBG	0.2844	9.422	0.016		23866	BV	2.1
5	CBN	0.1354	9.512	0.027		11366	VB	2.0
6	I.S.	Int Std	9.844	-0.045	S	251760	BB	3.4

Injection Notes:

Fig. 7.2. A representative chromatogram of a screened and selected high THC yielding clone of *Cannabis sativa* using gas chromatography–flame ionization detection (GC-FID). The amount of cannabidiol (CBD) in this clone was found to be insignificant (<0.10%, R_t ~8.16, appearing just before the CBC peak and not integrated).

Vegetative propagation in Hydroponics

Vegetative propagation of high yielding *Cannabis sativa* clones using a hydroponic system is shown in Figs 7.3C and 7.3D. Similar to propagation in soil, a small branch consisting of a growing tip with two or three leaves is cut. This clipping is then allowed to stand in water as the next clipping is cut. The procedure is repeated until the required number of clippings has been taken. A fresh cut is then made on each cutting just above the first cut. The clippings are then dipped in a rooting compound and inserted 1 inch deep into the rockwool rooting cube or in hydrotone clay ball supporting medium. Plants were supplied with vegetative fertilizer formula supplied by Advance nutrient (Abbotsford, BC, Canada). Plants were exposed to 18 h of diffused light and 6 h dark cycle for their vegetative growth in the hydroponic environment. The formation of roots started in 2–3 weeks. Once the roots are big enough plants can be transplanted into a bigger hydroponic system for further growth.

Micropropagation of High THC Yielding *Cannabis sativa*: Direct Organogenesis

Direct organogenesis is regarded as the most reliable method for clonal propagation as it upholds genetic uniformity among the progenies. In our laboratory, our goal in general is to develop a secure and stable *in vitro* clonal repository of elite *Cannabis* germplasm that will ensure future availability of desirable pharmacologically active chemotypes of *Cannabis sativa*. Thus, the first objective of our study was to develop a tissue culture system that will allow large scale clonal production of screened and selected high yielding elite clones without somaclonal variations. Although, few reports have been published describing the regeneration protocols developed for different *Cannabis* genotypes and explant sources (Mandolino and Ranalli, 1999; Slusarkiewicz-Jarzina *et al.*, 2005; Bing *et al.*, 2007), considerable variation has been reported in the response of cultures and in the morphogenic pathway. Production of roots through *Cannabis* calluses was reported by Fisse *et al.* (1981). However, these cultures were reported to be unreceptive to shoot formation. Occasional shoot regeneration from calluses was reported by Mandolino and Ranalli (1999), whereas Feeney and Punja (2003) failed to regenerate hemp plantlets, either directly or indirectly from callus or suspension cultures.

In our laboratory, we have refined a *Cannabis in vitro* propagation protocol for the micropropagation of high THC yielding elite clones for field plantation. This procedure will be a useful tool for multiplication and maintenance of high yielding elite plants, saving space and time in the selection programme. Apical nodal segments containing axillary buds (~1 cm in length) were used as an explant for initiation of shoot cultures. Explants were obtained from healthy branches of a screened and selected high yielding *Cannabis sativa* clone grown in an indoor cultivation facility housed at Coy-Waller laboratory, University of Mississippi. Explants were surface disinfected using 0.5% NaOCl (15% v/v bleach) and 0.1% Tween 20 for 20 min. The explants were washed in sterile distilled water three times for 5 min prior to inoculation on the culture medium. Micropropagation and hardening of micropropagated plants was done following the protocol described by Lata *et al.* (2009a).

Fig. 7.3. A: Micropropagated and well rooted plant of *Cannabis sativa*; **B:** Indoor grown vegetatively propagated crop of *Cannabis sativa* in soil, under controlled environmental

conditions; **C** and **D**: Hydroponic propagation; **E**: Field plantation of vegetatively propagated crop propagated from screened and selected high yielding mother plant.

Thidiazuron (TDZ), a synthetic phenylurea, is considered to be one of the most active cytokinins for shoot induction in plant tissue culture (Huetteman and Preece, 1993). Reports suggest that TDZ induces shoot regeneration better than other cytokinins (Thomas, 2003; Thomas and Puthur, 2004; Husain *et al.*, 2007). Optimum concentrations of growth regulators for the best effects on multiplication, shoot proliferation and root induction are shown in Table 7.1.

Best response for shoot induction was observed on Murashige and Skoog (MS) medium containing 0.5 μM TDZ in this species. Well developed shoots were then transferred to half strength MS medium activated charcoal supplemented by different concentrations of IAA, IBA and NAA for rooting. The highest percentage of rooting in micropropagated plants was achieved in half strength Murashige and Skoog (1/2 MS) salts with 500 mg/l activated charcoal supplemented with 2.5 μM indole-3-butyric acid (IBA). A well rooted micropropagated plant of *Cannabis sativa* is shown in Fig. 7.3A.

Since the preservation of genetic stability in germplasm collections and micropropagation of elite plants is of utmost importance, the propagation of *C. sativa* through nodal explants as compared to calluses would allow recovery of genetically stable and true to type progeny plants. The mass multiplication of high yielding varieties of *C. sativa* is feasible for field plantings to produce plants rich in specific cannabinoids for use as pharmaceutical raw materials.

Synthetic Seed Technology

Synthetic seed technology can be considered an important application of micropropagation that could be used for economical large-scale clonal propagation and germplasm conservation of the screened and selected elite germplasm of *C. sativa*. To further improve and refine the established protocol, the authors in our laboratory have reported a simple and efficient method for *in vitro* propagation of screened and selected high yielding varieties of *C. sativa* using synthetic seed technology (Lata *et al.*, 2009b).

Axillary buds of *Cannabis sativa* isolated from aseptic multiple shoot cultures were successfully encapsulated in calcium alginate beads. The best gel complexation was achieved using 5% sodium alginate with 50 mM $CaCl_2.2H_2O$.

Regrowth and conversion after encapsulation was evaluated both under *in vitro* and *in vivo* conditions on different planting substrates. The addition of antimicrobial substance (Plant Preservative Mixture (PPM)) had a positive effect on overall plantlet development. Encapsulated explants exhibited the best regrowth and conversion frequency on MS medium supplemented with TDZ (0.5 μM) and PPM (0.075%) under *in vitro* conditions. Under *in vivo* conditions, 100% conversion of encapsulated explants was obtained on 1:1 potting mix of fertilome with coco natural growth medium, moistened with full strength MS medium without TDZ, supplemented with 3% sucrose and 0.5% PPM. Plantlets regenerated from the encapsulated explants were hardened off and successfully transferred to the soil.

Table 7.1. Concentrations of growth regulators for the best response on multiplication and shoot proliferation, and root induction of *Cannabis sativa*.

Growth regulators	Optimum concentration (μM)	Average number	Average length (cm)	% explants producing
Shoot multiplication and proliferation				
			Shoots	
BA	0.5	3.00 ^{de}	3.20 ^{cde}	72.20
	2.5	5.20 ^{bc}	2.80 ^{de}	72.20
	5.0	6.30 ^{b}	3.40 ^{cde}	61.10
KN	0.5	2.60 ^{def}	2.50 ^{e}	77.70
	2.5	5.60 ^{bc}	3.50 ^{cd}	83.30
	5.0	2.20 ^{cd}	2.80 ^{de}	50.00
TDZ	0.5	12.60 ^{a}	7.10 ^{a}	100.00
	2.5	11.10^{a}	6.10 ^{b}	94.40
	5.0	5.80 ^{b}	3.20 ^{cde}	83.30
BA + GA_3	0.5 + 7.0	3.20 ^{f}	2.50 ^{e}	50.00
	2.5 + 7.0	3.60 ^{ef}	3.50 ^{d}	66.60
	5.0 + 7.0	4.50 ^{def}	3.60 ^{d}	33.30
Kn + GA_3	0.5 + 7.0	3.20 ^{f}	2.60 ^{e}	44.40
	2.5 + 7.0	5.50 ^{d}	2.10 ^{e}	66.60
	5.0 + 7.0	4.80 ^{de}	2.50 ^{e}	55.50
TDZ + GA_3	0.5 + 7.0	13.80 ^{a}	7.90 ^{a}	83.30
	2.5 + 7.0	11.20 ^{b}	7.00 ^{b}	94.40
	5.0 + 7.0	7.30 ^{c}	5.00 ^{c}	77.70
Root Induction				
			Roots	
IAA*	2.5	1.25 ^{b}	1.40 ^{b}	33.30
	5.0	1.60 ^{b}	1.30 ^{b}	44.40
IBA*	2.5	4.80 ^{a}	4.80 ^{a}	94.40
	5.0	3.80 ^{a}	4.70 ^{a}	83.30
NAA*	2.5	2.20 ^{b}	2.30 ^{b}	55.50
	5.0	2.30 ^{b}	2.40 ^{b}	44.40

Data represent the mean of three replicates with six explants for each treatment. Means followed by the same letter do not differ statistically at P= 0.05 according the Tukey test. *Medium used is half strength Murashige and Skoog (1/2 MS) salts supplemented with 500 mg/l activated charcoal.

From Lata, H. *et al.* Thidiazuron-induced high-frequency direct shoot organogenesis of *Cannabis sativa* L. Reproduced with permission from *In vitro* Cellular and Developmental Biology – Plant, 2009, Vol. 45, pp. 12–19. Copyright © 2009 by the Society for *In vitro* Biology, formerly the Tissue Culture Association.

Synthetic seed technology, one of the most promising biotechnological tools, would allow production of mass propagation material in the form of a synthetic seed possessing the ability of regrowth and rooting and developing into plantlet for *in vitro* and ex vitro usage. This technique has a tremendous potential in scaling up the micropropagation procedure while at the same time economizing upon time, space and cost (Mathur *et al.*, 1989; Bornman, 1993). Synthetic seeds production and plantlet regeneration have been reported for various types of plants including cereals, vegetables, fruits, ornamentals, aromatic grass and conifers (Ganapathi *et al.*, 2001; Brischia *et al.*, 2002; Hao and Deng, 2003). However, in most cases somatic embryos were used in the encapsulation process. Few reporters (Mathur *et al.*, 1989; Pattnaik and Chand, 2000) described the encapsulation of vegetative propagules such as axillary buds or shoot tips which could be used for mass clonal propagation as well as in long-term conservation of germplasm. The inclusion of antimicrobial substances in the planting substrate would allow possible future sowing of synthetic seeds in greenhouse (*ex vitro*), without sterility requirements. Conversion of encapsulated nodal segments into plantlets in extremely simplified medium such as 1:1 potting mix- fertilome with coco natural growth medium indicates that this method could be used in developing a cost effective propagation protocol for the mass propagation of elite varieties of *C. sativa*. Other than assuring a high degree of genetic uniformity and stability, the use of vegetative propagules minimizes the occurrence of somaclonal variations.

Indirect Organogenesis

The regeneration of plants by *in vitro* cell and tissue culture is a fundamental step for the genetic manipulation and improvement of crops (Vasil and Vasil, 1986). The callus mediated plant regeneration can provide an easy way to obtain somaclonal variation as has been emphasized in several plants (Lata *et al.* 2002; Faisal and Anis, 2005). Thus, indirect organogenesis could be exploited in *Cannabis sativa* for the induction of somaclonal variation. Introduction of valuable variation induced by callus cultures and subsequent regeneration may help in programmes designed to improve characteristics of the crop. These variants could be further screened for the desirable traits. The authors in our laboratory have developed a high frequency plant regeneration system from leaf tissue derived callus of a high yielding variety of *C. sativa* (Lata *et al.*, 2010). Calli were introduced from leaf explant on Murashige and Skoog medium supplemented with different concentrations (0.5, 1.0, 1.5 and 2.0 μM) of indole-3-acetic acid (IAA), indole-3-butyric acid (IBA), naphthalene acetic acid (NAA) and dichlorophenoxyacetic acid (2,4-D) in combination with 1.0 μM thidiazuron (TDZ) for the production of callus. The optimum callus growth and maintenance was in 0.5 μM NAA plus 1.0 μM TDZ. The 2-month old calli were subcultured to MS media containing different concentrations of cytokinins (BAP, KN, TDZ). The rate of shoot induction and proliferation was highest in 0.5 μM TDZ. Regenerated shoots rooted best on half strength MS medium supplemented with 2.5 μM IBA of the various auxins (IAA, IBA and NAA) tested. Further screening of the somaclonal variants of *C. sativa* is in progress.

Assessment of Stability of Micropropagated Plants of *Cannabis sativa*

The occurrence of physiological and genetic variation is a matter of great concern where commercial success in micropropagation depends solely on clonal uniformity. Therefore, it is extremely important to ascertain the suitability of a particular micropropagation protocol developed for a particular species, in terms of the production of qualitatively and quantitatively identical plants. Clonal fidelity of micropropagated *Cannabis sativa* plants in our laboratory was tested by comparing them with mother plants and conventionally propagated plants from the same mother plant, for genetic stability using inter sample sequence repeat (ISSR) marker, for its cannabinoids contents using GC-FID and for growth and physiological traits using gas and water vapour characteristics.

Genetic Stability

True-to-type clonal fidelity is one of the most important prerequisites in the micropropagation of any crop species. To use plant tissue culture as a tool for commercial utilization, periodic monitoring of the degree of genetic stability among *in vitro* grown plants is of utmost importance. Particularly, the maintenance of clonal fidelity is an important issue in developing a secure and stable *in vitro* clonal repository of elite *C. sativa* germplasm.

We have successfully assessed the genetic stability of the micropropagated plants with over 30 passages in culture and hardened in soil for 8 months, using ISSR DNA fingerprinting (Lata *et al.*, 2009c). Inter simple sequence repeat profiles of micropropagated and hardened plantlets were compared with the mother plant grown indoors under similar environmental conditions. A total of 15 ISSR primers resulted in 115 distinct and reproducible bands. All the ISSR profiles from micropropagated plants were monomorphic and comparable to mother plants. Furthermore, no variation was detected within the micropropagated plants. Our results suggest that the micropropagation protocol developed by us for the rapid *in vitro* multiplication is appropriate and applicable for clonal mass propagation of *C. sativa.* A similar approach to test the genetic stability in other plants has been reported by Modgil *et al.* (2005), Lakshmanan *et al.* (2007), Chandrika and Rai (2009) and Huang *et al.* (2009). This type of study is of high significance if plants are being propagated for the production and isolation of biologically active secondary metabolites for pharmaceutical use.

Stability of Cannabinoid Profile: Qualitatively and Quantitatively

In general, a problem that has been observed with the plants produced via *in vitro* propagation regimes is a high rate of somatic mutation (Rani *et al.*, 1995; Damasco *et al.*,1996; Wang *et al.*, 2000; Salvi *et al.*, 2001). Because the major goal of our research was to develop an efficient *in vitro* propagation method for the mass production of high yielding *Cannabis* plants as a consistent source of biomass for the extraction of THC, we wanted to be certain that the method did not introduce mutations that could lead to alterations in metabolism. To evaluate whether such mutation occurred, we used gas chromatography–flame ionization detection (GC-FID) to assess the chemical profile and quantification of cannabinoids to identify the differences, if any, in the chemical constituents of *in vitro* propagated plants (IVP), conventionally grown plants (VP) and indoor grown mother plant (MP-Indoor) of a high THC yielding variety of *Cannabis sativa* L. during different developmental stages of growth. In general, THC content in all groups increased with plant age up to a highest level during budding stage where the THC content reached a plateau before the onset of senescence. The pattern of changes observed in the concentration of other cannabinoids content with plant age has followed a similar trend in all groups of plants. Qualitatively, the cannabinoid profiles obtained using GC-FID, in MP-indoor, VP and IVP plants were found to be similar to each other and to that of the field grown mother plant (MP field) of *C. sativa*. Minor differences were observed in cannabinoid concentration within and among the groups which were not statistically significant (Lata *et al.*, 2009c; Chandra *et al.*, 2009b).

Our results confirm the clonal fidelity of IVP plants of *C. sativa* and suggest that the biochemical mechanism used in this study to produce the micropropagated plants does not affect the metabolic content and can be used for mass propagation of true to type plants of this species for commercial pharmaceutical use. The study also serves as a tool for quality control of the micropropagated crop of this variety of *Cannabis sativa* and is of high value, since these plants are selected as high yielding elite clones for the production of THC for pharmaceutical use. Working on *in vitro* micropropagated and conventionally greenhouse grown ginger (*Zingiber officinale*) and turmeric (*Curcuma longa*) plants, Ma and Gang (2006a, b) also reported no significant differences in useful secondary metabolite content between both types of plants.

Physiological Stability: Gas and water vapour exchange

Plant survival, growth and productivity are intimately coupled with the aerial environment through processes such as energy exchange, loss of water vapour in transpiration and uptake of carbon dioxide in photosynthesis (Stoutjesdijk and Barkman, 1992). The water vapour exchange rate affects the energy budget and transpiration of leaves and consequently the physiology of the whole plant (Chandra and Dhyani, 1997; Chandra, 2004). Therefore, data on physiological parameters such as gas and water vapour exchange are likely to provide valuable information regarding the suitability of tissue culture raised plants for field plantation.

We have evaluated the performance of *in vitro* propagated and hardened plants, on the basis of selected physiological parameters, in comparison to those of *ex vitro* vegetatively grown plants from the same mother plant of the same age. The effects of different light intensities on photosynthesis and water vapour exchange characteristics have also been examined. Following acclimatization of the micropropagated plants, growth performance of 4-months-old *in vitro* propagated plants was compared with *ex vitro* vegetatively grown plants of the same age. The photosynthesis and transpiration characteristics were studied under different light levels (0, 500, 1000, 1500, or 2000 $\mu mol/m^2/s$). An increase in photosynthesis (P_N) was observed with increase in the light intensity up to 1500 $\mu mol/m^2/s$ and then decreased subsequently at higher light levels in both types of plants. However, the increase was more pronounced at lower light intensities below 500 μmol/m²/s. Stomatal conductance (g_s) and transpiration (E) increased with light intensity up to the highest level tested (2000 μmol/m²/s). Intercellular CO_2 concentration (C_i) and the ratio of intercellular CO_2 concentration to ambient CO_2 (C_i/C_a) decreased with the increase in light intensity in both *in vitro* as well as *ex vitro* raised plants. The results show that *in vitro* propagated and hardened plants were functionally comparable to *ex vitro* plants of the same age in terms of gas and water vapour exchange characteristics (Lata *et al.*, 2009a). In a similar study, Bag *et al.* (2000) reported comparable photosynthetic characteristics and growth in seed raised and micropropagated plants of *Thamnocalamus spathiflorus*, a temperate bamboo from Central Himalayas.

Conclusion

The *Cannabis* plant has acquired much interest over the last few years, not only because of the problems associated with its abuse, but also because of the therapeutic potential of many of its compounds. In this chapter, an attempt has been made to elucidate the role of biotechnology and our efforts to screen and propagate high THC yielding elite *Cannabis sativa* for the production of Δ^9-THC. Since it is very important to have consistency in plant material used for the production of any natural product of pharmaceutical interest, plants raised through micropropagation were compared with mother plants and vegetatively propagated plants from the same mother plant in terms of their genetic stability, cannabinoid profile and physiological traits, i.e. gas and water vapour characteristics. Our results confirm that plants raised through micropropagation were highly comparable to the mother plant and the vegetatively propagated plants, and therefore the micropropagation protocol followed can be used for the mass propagation of true to type plants of this species for the commercial production of natural constituents for pharmaceutical use.

Acknowledgement

This work was supported in part by The National Institute on Drug Abuse (NIDA), USA, Contract No. N01DA-5-7746.

References

Bag, N., Chandra, S., Palni, L.M.S. and Nandi, S.K. (2000) Micropropagation of Dev ringal (*Thamnocalamus spathiflorus* (Trin.) Munro), a temperate bamboo and comparison between *in vitro* propagated plants and seedlings. *Plant Science* 156, 125–135.

Bazzaz, F.A., Dusek, D., Seigler, D.S. and Haney, A.W. (1975) Photosynthesis and cannabinoid content of temperate and tropical populations of *Cannabis sativa*. *Biochemical Systematics and Ecology* 3, 15–18.

Bing, X., Ning, L., Jinfeng, T. and Nan, G. (2007) Rapid tissue culture method of *Cannabis sativa* for industrial uses. CN 1887043 A 20070103 Patent (pp. 9).

Bo´Csa, I., Ma´ The´, P. and Hangyel, L. (1997) Effect of nitrogen on tetrahydrocannabinol (THC) content in hemp (*Cannabis sativa* L.) leaves at different positions. *Journal of the International Hemp Association* 4, 80– 81.

Bornman, C.H. (1993) Maturation of Somatic Embryos. In: Redenbaugh, K. (ed.) *Synseeds: Application of Synthetic Seeds to Crop Improvement.* CRC Press, Boca Raton, Florida, pp.105–114.

Brenneisen, R., Egli, A., ElSohly, M.A., Henn, V. and Spiess, Y. (1996) The effect of orally and rectally administered Δ^9-tetrahydrocannabinol on spasticity: a pilot study with 2 patients. *International Journal* of *Clinical Pharmacology and Therapeutics* 34(10), 446–452.

Brischia, R., Piccioni, E. and Standardi, A. (2002) Micropropagation and synthetic seed in M.26 apple rootstock (II): A new protocol for production of encapsulated differentiating propagules. *Plant Cell Tissue and Organ Culture* 68, 137–141.

Chandra, S. (2004) Effect of altitude on energy exchange characteristics of some alpine medicinal crops from Central Himalayas. *Journal of Agronomy and Crop Science* 190, 13–20.

Chandra, S. and Dhyani, P.P. (1997) Diurnal and monthly variation in leaf temperature, water vapour transfer and energy exchange in the leaves of *Ficus glomerata* during summer. *Physiology and Molecular Biology of Plants* 3, 135–147.

Chandra, S., Lata, H., Khan, I.A. and ElSohly, M.A. (2008) Photosynthetic response of *Cannabis sativa* L. to variations in photosynthetic photon flux densities, temperature and CO_2 conditions. *Physiology and Molecular Biology of Plants* 14, 299-306.

Chandra, S., Lata, H., Khan, I.A. and ElSohly, M.A. (2009a) Variations in temperature response of photosynthesis in drug and fiber type varieties of *Cannabis sativa* L. Eighth annual Oxford international conference on the science of botanicals. *Planta Medica* 75, p. 5.

Chandra, S., Lata, H., Khan, I.A., Mehmedic, Z. and ElSohly, M.A. (2009b) Assessment of cannabinoids content in micropropagated plants of *Cannabis sativa* L. and their comparison with conventionally propagated plants and mother plant during developmental stages of growth. *Planta Medica* Advance online publication; DOI 10.1055/s-0029-1240628.

Chandrika, M. and Rai, V.R. (2009) Genetic fidelity in micropropagated plantlets of *Ochreinauclea missionis* an endemic, threatened and medicinal tree using ISSR markers. *African Journal of Biotechnology* 8, 2933–2938.

Damasco, O.P., Graham, G.C., Henry, R.J., Adkins, S.W., Smith, M.K. and Godwin, I.D. (1996) Random amplified polymorphic DNA (RAPD) detection of dwarf off types in micropropagated Cavendish (*Musa* spp. AAA) bananas. *Plant Cell Reports* 16, 118–123.

De Meijer, E.P.M., Van Der Kamp, H.J. and Van Eeuwijk, F.A. (1992) Characterization of *Cannabis* accessions with regard to cannabinoid content in relation to other plant characters. *Euphytica* 62, 187–200.

Doyle, E. and Spence, A.A. (1995) *Cannabis* as a medicine? *British Journal of Anesthesia* 74, 359–361.

ElSohly, M.A. and Slade, D. (2005) Chemical constituents of marijuana: the complex mixture of natural cannabinoids. *Life Science* 78, 539–548.

ElSohly, M.A., Ross, S.A, Mehmedic, Z., Arafat, R., Yi, B. and Banahan, B.F. (2000) Potency trends of Δ^9-THC and other cannabinoids in confiscated marijuana from 1980–1997. *Journal of Forensic Science* 45, 24–30.

Emboden, W.A. (1981) The genus Cannabis and the correct use of taxonomic categories. *Journal of Psychoactive Drugs* 13, 15–21.

Faisal, M. and Anis, M. (2005) An efficient *in vitro* method for mass propagation of *Tylophora indica. Biologia Plantarum* 49(2), 257–260.

Feeney, M. and Punja, Z.K. (2003) Tissue culture and Agrobacterium-mediated transformation of hemp (*Cannabis sativa* L.) *In vitro Cellular and Developmental Biology – Plant* 39, 578-585.

Fetterman, P.S., Keith, E.S., Waller, C.W., Guerrero, O., Doorenbos, N.J. and Quimby, M.W. (1971) Mississippi-grown *Cannabis sativa* L. Preliminary observation on chemical definition of phenotype and variations in tetrahydrocannabinol content versus age, sex, and plant part. *Journal of Pharmaceutical Sciences* 60, 1246–1249.

Fisse, J., Braut, F., Cosson, L. and Paris, M. (1981) Etude *in vitro* des capacities organogenetiques de tissues de *Cannabis sativa* L. Effet de differentes substances de croissance. *Planta Medica* 15, 217–223.

Fournier, G., Dumanois, C.R., Duvezin, Mathieu, J.P. and Paris, M. (1987) Identification of a new chemotype in *Cannabis sativa*: cannabigerol-dominant plants, biogenetic and agronomic prospects. *Planta Medica* 53, 277–280.

Ganapathi, T.R., Srinivas, I., Suprasanna, P. and Bapat, V.A. (2001) Regeneration of plants from alginated-encapsulated somatic embryos of banana cv. Rasthali (*Musa* spp. AAB group). *Biologia Plantarum* 37, 178–181.

Grinspoon, L. and Bakalar, J.B. (1993) *Marihuana, the Forbidden Medicine*. Yale University Press, New Haven.

Hammond, C.T. and Mahlberg, P.G. (1977) Morphogenesis of 1980 Breeding of Cannabis. Determination of cannabinoids by capitate glandular hairs of *Cannabis sativa* (Cannabaceae). *American Journal of Botany* 64, 1023–1031.

Hao, Y.J. and Deng, X.X. (2003) Genetically stable regeneration of apple plants from slow growth. *Plant Cell Tissue and Organ Culture* 72, 253–260.

Hemphill, J.K.; Turner, J.C. and Mahlberg P.G. (1980) Cannabinoid content of individual plant organs from different geographical strains of *Cannabis sativa* L. *Journal of Natural Products* 43, 112–122.

Hillig, K. and Mahlberg, P.G. (2004) A chemotaxonomic analysis of cannabinoid variation in Cannabis (Cannabaceae). *American Journal of Botany* 91, 966–975.

Huang, W.J., Ning, G.G., Liu G.F. and Bao, M.J. (2009) Determination of genetic stability of long-term micropropagated plantlets of *Platanus acerifolia* using ISSR markers. *Biologia Plantarum* 53, 159–163.

Huetteman, C.A. and Preece, J.E. (1993) Thidiazuron: a potent cytokinin for woody plant tissue culture. *Plant Cell Tissue Organ Culture* 33, 105–119.

Husain, M.K., Anis, M. and Shahzad, A. (2007) *In vitro* propagation of Indian Kino (*Pterocarpus marsupium* Roxb.) using Thidiazuron. *In vitro Cellular and Developmental Biology-Plant* 43, 59–64.

Lakshmanan, V., Reddampalli, S., Reddy, V. and Neelwarne, B. (2007) Molecular analysis of genetic stability in long-term micropropagated shoots of banana using RAPD and ISSR markers. *Electronic Journal of Biotechnology* 10, 106–113.

Lata, H., Bedir, E., Hosick, A., Ganzera, M., Khan, I. and Moraes, R.M. (2002) *In vitro* plant regeneration from leaf-derived callus in Black cohosh (*Cimicifuga racemosa*). *Planta Medica* 68, 912–915.

Lata, H., Chandra, S., Khan, I.A. and ElSohly, M.A. (2009a) Thidiazuron induced high frequency direct shoot organogenesis of *Cannabis sativa* L. *In vitro Cellular and Developmental Biology-Plant* 45, 12–19.

Lata, H., Chandra, S., Khan, I.A. and ElSohly, M.A. (2009b) Propagation through alginate encapsulation of axillary buds of *Cannabis sativa* L. – An important medicinal plant. *Physiology and Molecular Biology of Plants* 15(1), 79–86.

Lata, H., Chandra, S., Khan, I.A. and ElSohly, M.A. (2010) High frequency plant regeneration from leaf derived callus of high Δ^9- tetrahydrocannabinol yielding *Cannabis sativa* L. *Planta Medica* Advance online publication; DOI 10.1055/s-0030-1249773.

Lata, H., Chandra, S., Techen, N., Khan, I.A. and ElSohly, M.A. (2009c) Assessment of genetic stability of micropropagated plants of *Cannabis sativa* L. by ISSR markers. *Planta Medica* 76, 97–100.

Ma, X. and Gang, D.R. (2006a) Metabolic profiling of *in vitro* micropropagated and conventionally greenhouse grown ginger (*Zingiber officinale*). *Phytochemistry* 67, 2239–2255.

Ma, X. and Gang, D.R. (2006b) Metabolic profiling of turmeric (*Curcuma longa* L.) plants derived from *in vitro* micropropagation and conventional greenhouse cultivation. *Journal of Agriculture and Food Chemistry* 54, 9573–9583.

Mandolino, G. and Ranalli, P. (1999) Advances in Biotechnological Approaches for Hemp Breeding and Industry. In: Ranalli, P. (ed.) *Advances in Hemp Research*. Haworth Press, New York, pp. 185–208.

Mandolino, G., Bagatta, M., Carboni, A., Ranalli, P. and De Meijer, E. (2003) Qualitative and quantitative aspects of the inheritance of chemical phenotype in *Cannabis*. *Journal of Industrial Hemp* 8, 51–72.

Mathur, J., Ahuja, P.S., Lal, N. and Mathur, A.K. (1989) Propagation of *Valeriana wallichii* DC using encapsulated apical and axial shoot buds. *Plant Science* 60, 111–116.

Mattes, R.D., Egelman, K., Shaw, L.M. and ElSohly, M.A. (1994) Cannabinoids appetite stimulation. *Pharmacology Biochemistry and Behavior* 49(1), 187

Mehmedic, Z., Chandra, S., Slade. D., Denham, H., Foster, S., Patel, A.S., Ross, S.A., Khan, I.A. and ElSohly, M.A. (2009) Potency trends of Δ^9-THC and other cannabinoids in confiscated *Cannabis* preparations from 1993–2008. *Journal of Forensic Science* 55(5).

Mendoza, M., Mills, D.E., Lata, H., Chandra, S., ElSohly, M.A. and Almirall, J. (2009) Genetic Individualization of *Cannabis sativa* by an STR Multiplex. *Analytical and Bioanalytical Chemistry* 393, 719–726.

Modgil, M., Mahajan, K., Chakrabarti, S.K., Sharma, D.R. and Sobti, R.C. (2005) Molecular analysis of genetic stability in micropropagated apple rootstock MM106. *Scientia Horiculture* 104, 151–160.

Murashige, T. and Skoog, F. (1962) A revised medium for rapid growth and bioassays with tobacco tissue cultures. *Plant Physiology* 15, 473– 497.

Pate, D.W. (1994) Chemical ecology of Cannabis. *Journal of the International Hemp Association* 1, 29, 32–37.

Pattnaik, S. and Chand, P.K. (2000) Morphogenic response of the alginate encapsulated axillary buds from *in vitro* shoot cultures of six mulberries. *Plant Cell Tissue and Organ Culture* 64, 177–185.

Rani, V., Parida, P. and Raina, S.N. (1995) Random amplified polymorphic DNA (RAPD) markers for genetic analysis in micropropagated plants of *Populus deltoides* Marsh. *Plant Cell Reports* 14, 459–462.

Ross, S.A., Parker, M., Arafat, R., Lovett, K. and ElSohly, M.A. (1995) The analysis of confiscated marijuana samples for different cannabinoids using GC/FID. *American Labs* 27(16), 16–17.

Salvi, N.D., George, L. and Eapen, S. (2001) Plant regeneration from leaf callus of turmeric and random amplified polymorphic DNA analysis of regenerated plants. *Plant Cell Tissue and Organ Culture* 66, 113–119.

Slade, D. and ElSohly, M.A. (2009) New cannabinoids from high potency *Cannabis*: A short review (2005-2009). *Bioorganic & Medicinal Chemistry Letters* (communicated).

Slusarkiewicz-Jarzina, A., Ponitka, A. and Kaczmarek, Z. (2005) Influence of cultivar, explant source and plant growth regulator on callus induction and plant regeneration of *Cannabis sativa* L. *Acta Biologica Cracoviensia Series Botanica* 47(2), 145–151.

Small, E. (1979) *The Species Problem in Cannabis*, vol. 1, Science. Corpus Information Services, Toronto, Ontario, Canada.

Small, E. and Beckstead, H.D. (1973a) Cannabinoid phenotypes in *Cannabis sativa*. *Nature* 245, 147–148.

Small, E. and Beckstead, H.D. (1973b) Common cannabinoid phenotypes in 350 stocks of *Cannabis*. *Lloydia* 36, 144–165.

Small, E., Backstead, H.D. and Chan, A. (1975) The evolution of cannabinoids phenotypes. Cannabis. *Economic Botany* 29, 219–232.

Small, E. and Cronquist, A. (1976) A practical and natural taxonomy for *Cannabis*. *Taxon* 25, 405–435.

Small, E. and Marcus, D. (2003) Tetrahydrocannabinol levels in hemp (*Cannabis sativa*) germplasm resources. *Economic Botany* 57, 545–558.

Small, E., Backstead, H.D. and Chan, A. (1975) The evolution of cannabinoids phenotypes. Cannabis. *Economic Botany* 29, 219–232.

Small, E.; Jui, P.Y. and Lefkovitch, L.P. (1976) A numerical taxonomy analysis of *Cannabis* with special reference to species delimitation. *Syst. Bot.* 1, 67–84.

Stoutjesdijk, P.H. and Barkman, J.J. (1992) *Microclimate, Vegetation and Fauna*. Opulus Press Publ., Sweden.

Thomas, T.D. (2003) Thidiazuron induced multiple shoot induction and plant regeneration from cotyledonary explants of mulberry. *Biologia Plantarum* 46, 529–533.
Thomas, T.D. and Puthur, J.T. (2004) Thidiazuron induced high frequency shoot organogenesis in callus from *Kigelia pinnata* L. *Botanical Bulletin of Academia Sinica* 45, 307–313.
Valle, J.R., Vieira, J.E.V., Auce'Lio, J.G. and Valio, I.F.M. (1978) Influence of photoperiodism on cannabinoid content of *Cannabis sativa* L. *Bulletin on Narcotics* 30, 67–68.
Vasil, I.K. and Vasil, V. (1986) Regeneration in cereal and other grass species. In: Vasil, I.K. (ed.) *Cell Culture and Somatic Cell Genetics of Plants.* Vol. 3. Plant Regeneration and Genetic Variability. Academic Press, Orlando, USA, pp. 121–150.
Wang, X.R., Szmidt, A.E. and Nguyen, H.N. (2000) The phylogenetic position of the endemic flat-needle pine *Pinus krempfii* (Lec., Pinaceae) from Vietnam, based on PCR-RFLP analysis of chloroplast DNA. *Plant Systematics and Evolution* 220, 21–36.
Zuardi, A.W. (2006) History of *Cannabis* as a medicine: a review. *Brazilian Journal of Psychiatry* 28, 153–157.

Chapter 8

In vitro Saponin Production in Plant Cell and Tissue Cultures

Archana Mathur and Ajay K. Mathur

Introduction

Phytosaponins are a group of plant secondary metabolites with a diverse range of biological activities. Major problems associated with agri-based production of plant saponins have been their low productivity *in planta*, exclusive occurrence in only a few angiospermous families and in general the extremely slow growing nature of the taxa that harbour them. Poor understanding of their chemistry (biogenesis, extraction, processing and identification) and complex chirality, further narrow down the option of their industrial production through an economically viable chemical route. Clearly, there are only two options to address this issue: (i) to improve *in planta* yields of desired saponin by producing designer crops; or (ii) to explore the possibility of using cultured plant cells and tissues as a continuous renewable resource for their commercial production in bioreactors (Verpoorte *et al.* 2002; Horn *et al.* 2004). Modern tools of plant biotechnology are, therefore, being extensively explored to test these options.

Employment of cell culture approaches to produce plant saponins has received a lot of momentum in the last two decades because of a host of biological activities that have now been ascribed to them (Ustundag and Mazza, 2007). Most of these *in vitro* production systems have employed cell suspensions or transformed hairy roots (Hayashi *et al.*, 2005; Guillon *et al.*, 2006; Thanh *et al.*, 2006; Chapagain *et al.*, 2008; Xu *et al.*, 2008). Cells or tissues in liquid medium predominantly provide uniform conditions for growth and metabolite synthesis, rapid biomass gain, better amenability for precursor feeding, biotic/abiotic elicitation, higher feasibility for bioreactor scaling up and easy extraction/downstream processing. This chapter summarizes the salient developments made in the area of *in vitro* phytosaponin production using the cell/tissue culture approach, taking *Panax* ginsenosides as a successful case study.

What are saponins?

The word 'saponin' has been derived from the soapwort plant (Genus *Saponaria*, Family Caryophyllaceae), the root of which was used historically as a soap. Saponins are a class of complex glycosides mainly found in a variety of higher plants as secondary metabolites (Ustundag and Mazza, 2007). Saponins are amphiphilic molecules consisting of a hydrophobic aglycone linked to one or more hydrophilic sugar moieties. The presence of saponins has been frequently documented in more than 100 plant families (Table 8.1).

Broadly, the monocotyledonous taxa are a rich source of steroidal saponins whereas dicotyledonous taxa are major accumulators of triterpene saponins. Plants such as soybean, chickpea, mungbean, lentils, oats, garlic, Asiatic pennywort, asparagus, bacopa, yams, licorice, yucca, ginseng, soap-bark tree, foxglove, psoralea, fenugreek etc. are the major sources of phytosaponins for industrial usages (Oleszek and Marston, 2000; Ustundag and Mazza, 2007).

These compounds are basically classified into three groups depending on the structure of the aglycone moiety which can be a triterpenoid, a steroid or a steroidal glycoalkaloid. They are further classified as monodesmosidic, bidesmosidic or tridesmosidic according to the number of sugar moieties attached to the aglycone. The sugar moiety is linked to the aglycone through an ether or ester glycosidic linkage. The triterpenoid backbone undergoes various modifications (oxidation, substitution and glycosylation), mediated by cytochrome P450 dependent monooxygenases, glycosyltransferases and other enzymes (Haralampidis *et al.*, 2002). Squalene synthase (SQS) and squalene epoxidase (SQE) are key enzymes of saponin synthesis in plants (Osbourn *et al.*, 1997). Saponins have historically been understood to be plant-derived, but they have also been isolated from several marine organisms. Saponins have been found to possess a wide range of biological activities (Table 8.1).

Chemical extraction and purification of phytosaponin is still a challenge to plant chemists due to their structural complexity, strong polar amphiphilic nature, lack of chromophore, low concentration in plants and non-availability of marker standards for quantitative and qualitative profiling (Marston *et al.* 2000; Ustundag and Mazza, 2007).

For plants, saponins mainly serve as anti-feedants, and anti-microbial. Some plant saponins (e.g. from oat and spinach) may enhance nutrient absorption and thus aid in animal digestion. However, saponins are usually bitter in taste, and thus have low palatability in livestock feeds, or may even cause life-threatening toxicity. Antifungal, antibacterial, insecticidal, antihelminthic, molluscicidal, piscicidal, anticancerous, immunomodulatory, hypolipidaemic actions of saponins are well documented in literature (Ustundag and Mazza, 2007).

Production of Plant Saponins by *In vitro* Cultures

Studies concerning the *in vitro* production of plant saponins in cultured cells, tissues and organs are summarized in Table 8.2. The majority of these represent the optimization of cell types and culture conditions for their sustained production (Liu and Zhong, 1997; Gangwar, 2003; Hayashi *et al.*, 2005; Thanh *et al.*, 2006; Chapagain *et al.*, 2008; Xu *et al.*, 2008). Some of the experimental strategies employed for improved saponin production in cell, tissue and organ cultures are highlighted in succeeding sections. References have been frequently drawn from the work carried out on *Panax* species (including the work done in the author's own laboratory) that represent the most comprehensively studied system in this field of research.

Ginseng – The green gold crop for phytosaponins

Ginseng (*Panax*) species are the primary non-food sources of phytosaponins used in health-care systems (Hostettmann and Marston, 1995; Balandrin, 1996). Interest in the use of ginseng as a health food comes from the strong adaptogenic and anti-aging activities

(Schultz *et al.*, 1998; Attele *et al.*, 1999; Varshney *et al.*, 2001). These 'health tonic' activities are believed to improve the body's tolerance to stress, resulting in better physical and mental performance (Schultz *et al.*, 1998; Nocerino *et al.*, 2000; Johannsen, 2006). Ginsengs are slow growing perennial herbs of the family Araliaceae. The four major *Panax* species that are in commercial use comprise *P. ginseng* (Korean ginseng), *P. quinquefolium* (American ginseng), *P. japonicus* (Japanese ginseng) and *P. notoginseng* (Chinese ginseng). Recently, an Indian species of *Panax*, *P. sikkimensis* has also been included in this list based on its chemotypic and genotypic profiles (Mathur *et al.*, 2002b, 2003a,b). The plant is valued for its storage roots, which are the source of a group of triterpene saponins, collectively called as 'ginsenosides' (Proctor and Bailey, 1987; Dewick, 1997; Huang, 1999; Haughton, 1999; Ngan *et al.*, 1999; Liang and Zhao, 2008). Chemically, the ginsenosides are glycosylated derivatives of two aglycones: panaxadiol and panaxatriol (Bruneton, 1995; Dewick, 1997). More than 40 ginsenosides have been identified in ginseng roots/flower buds/leaves (Dou *et al.*, 2001; Park, I.J. *et al.*, 2002; Park, I.H. *et al.*, 2002; Yoshikawa *et al.*, 2007).

Amongst these, Rb and Rg groups of ginsenosides namely Rb1, Rb2, Rc, Rd, Re, Rf, Rg1 and Rg2 are considered to be most potent CNS-stimulant (WHO, 1999; Liang and Zhao, 2008). Relative occurrence of these different ginsenosides is genotype, plant age and sesason dependent (Li and Mazza, 1999; Varshney *et al.*, 2001; Schlag and McIntosh, 2006).

Ginseng species normally need nutrient enriched soils for their growth and hence, the chemoprofiles of plant roots also vary with the nutritional status of the soil in which they grow (Li and Mazza, 1999). Agriculture-based production of ginseng roots is expensive and difficult due to prolonged seed dormancy phase of 18–22 months, coupled with an extended juvenile vegetative growth phase of 5–7 years for ginsenoside accumulation. Presently, ginseng roots (powder as well as extracts) are consumed throughout the world with an estimated market size of US$3.5 billion (Hong *et al.*, 2006). The market demand of ginseng roots is likely to expand further as more pharmacological activities like anti-cancer, anti-diabetic, radioprotection, anti-oxidant, anti-fatigue etc. have been elucidated recently (Furuya and Ushiyama, 1994; Sticher, 1998; Huang, 1999; Attele *et al.*, 1999; Shibata, 2001). Therefore, biotechnological interventions to generate ginseng root biomass and/or *in vitro* culture-based production systems for ginsenosides have been extensively explored at the levels of callus (Furuya *et al.*, 1983a; Furuya, 1988; Odnevall *et al.*, 1989; Mathur *et al.*, 1994, 1999, 2000, 2003 a,b; Gangwar, 2003), cell suspensions (Mathur *et al.*, 1994; Zhong and Yue, 2005; Thanh *et al.*, 2005, 2006) and adventitious/transformed hairy root cultures (Yoshikawa and Furuya, 1987; Inomata *et al.*, 1993; Washida *et al.*, 1998; Choi *et al.*, 2000; Yu *et al.*, 2000; Mallol *et al.*, 2001; Kim, Y.S. *et al.*, 2004; Palazón *et al.*, 2003a,b; Woo *et al.*, 2004; Ali *et al.*, 2005; Sivakumar *et al.*, 2005; Mathur *et al.*, 2010b). These studies are summarized in Table 8.3 and Table 8.4.

Factors affecting saponin production *in vitro*

Secondary metabolism in cultured cells and tissues is a dynamic process. The net accumulation of desired metabolites represents an equilibrium between its biogenesis, storage and degradation within the cellular compartments or specialized tissues (Rao, 2000).

Table 8.1. List of some saponin-yielding plants and biological activities of their active constituents.

Plant species	Family	Plant part used	Active constituents	Biological activity
Acacia auriculiformis	Fabaceae	Fruit	Acaciaside (A,B)	Sperm immobilizing activity
Aesculus hippocastanum	Sapindaceae	Seeds	Escins	Anti-inflammatory, antioedema, venotonic
Allium cepa	Liliaceae	Bulb	Alliospirosides (A,B,C,D)	Tonic, stimulant, stomachic, expectorant, aphrodisiac, diuretic
Allium sativum	Liliaceae	Bulb	Allin, Allinase, Sativoside R2	Hypocholesterolaemic, hypolipaedemic
Anagallis arvensis	Primulaceae	Root, aerial parts	Anagallisins, anagallogenin, cyclamin	Antiviral, antifungal, expectorant, vulnerary
Aralia elata	Araliaceae	Root bark	Aralia saponins	Tonic, anti-arthritic, hypoglycemic, hepatoprotective
Asparagus officinalis	Liliaceae	Tuberous root	Saponin (As-1)	Antifunal, antioxytocic, galactagogue, demulcent, aphrodisiac
Bacopa monnieri	Scrophulariaceae	Whole plant	Bacoside (A,B), Bacosaponin (A,B,C,D)	Nerve tonic, cardiotonic, diuretic
Bupleurum falcuactum	Apiaceae	Root	Saikosapnoin (triterpenoid saponins)	Anti-allergic, analgesic, anti-inflammatory
Camellia chinensis	Theaceae	Leaves	Caffeine, polyphenols	Antioxidant, anti-inflammatory
Centella asiatica	Apiaceae	Whole plant	Asiaticosides (A,B)	Psychotropic, for skin diseases
Chlorophytum borivilianum	Liliaceae	Root	Saponin, Sapogenin, (hecogenin)	Nerve tonic, aphrodisiac
Daucus carota	Apiaceae	Seeds, root	Diosgenin (steroidal saponins)	Diuretic, deobsturent, cardiac stimulant
Digitalis purpurea	Scrophulariaceae	Leaves	Digitoxigenin, digoxin, digoxigenin	Cardiotonic
Glycine max	Fabaceae	Seeds	Soysaponins	Antioxidant, anticarcinogenic, hepatoprotective, hypocholesterolaemic
Glycyrrhiza glabra	Fabaceae	Roots	Glycyrrhizin, glycyrrhizic acid	Expectorant, demulcent, spermicidal, laxative
Panax species	Araliaceae	Root	Ginsenosides (triterpene glycosides)	Immunomodulatory, adaptogenic, CNS stimulant
Terminalia bellerica	Combretaceae	Stem, bark	Bellericoside	Cardiotonic
Yucca species	Agavaceae	Leaves	Steroidal sapogenins	Precursor of oral contraceptive, sex hormones and other useful steroids

Table 8.2. Studies concerning phytosaponin production in cell, tissue and organ cultures.*

Plant species	Culture type	Saponin biosynthesized *in vitro*	Reference(s)
Agastache rugosa	C, CS	Rosmarinic acid	Xu *et al.*, 2008
Akebia quinata	C	Nortriterpenoid saponins	Ikuta and Itokawa,1989
A. trifoliata	C	Trifosides A,C, mubenoside	Ikuta,1995
Astragalus membranaceous	HR	Triglycosidic triterpene, astragalosides	Zhou *et al.*, 1995
Balanites aegyptiaca	C	Saponins	Chapagain *et al.*, 2008
Bupleurum falcatum	AR	Saikosaponins	Yamamoto and Kamura, 1997
Centella asiatica	WP, C, SH	Asiaticoside, Madecassoside	Kim, O.T. *et al.*, 2004a; Aziz *et al.*, 2007; Yadav *et al.*, 2007
Cistanche desrticola	CS	Phenylethanoidglycosides	Liu and Cheng, 2008
Gypsophilla paniculata	C, SH, AR, SC	Gypsogenin-3,0-glucuronide, Triterpene saponin	Hanafy and Setta, 2007; Fulcheri *et al.*, 1998; Pauthe-Dayde *et al.*, 1990
Glycyrrhiza glabra	CS , HR HR	Betulinic acid, β-glycyrrhetinic acid, glycyrrhizin, glycyrrhizic acid and triterpene saponin	Hayashi *et al.*, 1992, 2005; Kovalenko and Maliuta, 2003; Mehrotra *et al.*, 2008; Tenea *et al.*, 2008
G. uralensis	HR	Glycyrrhizin	Kovalenko and Maliuta, 2003
Gymnema sylvestre	C, CS	Gymnemic acid, gymnemagenin	Gopi and Vatsala, 2006
Gynostemma pentaphyllum	HR	Gypenosides	Chang *et al.*, 2005
Ruscus aculeatus	ARS	Ruscogenin, neuruscogenin	Palazón *et al.*, 2006
Saponaria officinalis	CS	Triterpene saponin	Fulcheri *et al.*, 1998
Solanum aculeatissimum	HR	Steroidal saponins (aculeatiside A,B)	Ikenaga *et al.*, 1995
S. eleagnifolium	CS	Solasodine	Nigra *et al.*, 1990
S. khasianum	HR	Solasodine	Jacob and Malpathak, 2005
S. paludosum	SH	Solamargine	Badaoui *et al.*, 1996
Trigonella foenum-graecum	C	Steroidal sapogenins	Jain and Agrawal, 1994
Uncaria tomentosa	CS	Ursolic and oleanolic acid	Isvett *et al.*, 2002

*Studies carried out in *Panax* species are separately given in Tables 8.3 and 8.4.
AR, Adventitious roots; ARS, Aerial shoots; C, Callus; CS, cell suspension; HR, Hairy roots; SH, Multiple shoots; WP, Whole plantlets

Table 8.3. Studies concerning *in vitro* saponin production in *Panax* species.

Panax species	Culture type	Medium used	Saponin produced *in vitro* (d. wt. unless specified)	References
Panax ginseng	C	INP	Panaxacol and dihydropanaxacol	Fujimoto *et al.*, 1990
	C	MS + 4.5μM 2,4-D + 0.5 μM Kn	Ginsenosides (1.46 mg/g)	Asaka *et al.*, 1994
	SE	MS + 4.65 μM Kn + 8% sucrose	Ginsenosides (7.21 mg/g)	Asaka *et al.*, 1994
	ASRP	MS basal medium	Ginsenosides (21.35 mg/g)	Asaka *et al.*, 1994
	ASL	MS basal medium	Ginsenosides (18.78 mg/g)	Asaka *et al.*, 1994
	SRP	MS basal medium	Ginsenosides (2.96 mg/g)	Asaka *et al.*, 1994
	SSL	MS basal medium	Ginsenosides (2.27 mg/g)	Asaka *et al.*, 1994
	C	MS + 5 μM 2,4D + 0.5 μM Kn + 60 mM K^+	Ginsenosides (15.4 mg/l/d), poly-saccharides (21.0 mg/l/d)	Liu and Zhong, 1996
	CS	MS + 5 μM 2,4-D + 0.5 μM Kn + 60 mM total initial nitrogen with NO_3^-/NH_4^+ ratio of 2:1	Ginsenosides (230 mg/l), polysaccharides (1190 mg/l)	Liu and Zhong, 1997
	CS	INP	Saponins production	Hu *et al.*, 2003a,b; Thanh *et al.*, 2006; Lu *et al.*, 2001
	CS	MS + 4.5 μM 2,4-D + 0.3% YE + 0.05% CH	Ginsenosides (1.61%)	Lu *et al.*, 2001
		MS + 0.05% CH + 500 μM MJ	Ginsenosides (2.08%)	Lu *et al.*, 2001
	AR	SH medium + 0.15% gelrite + 24.6 μM IBA with or without MJ and ethaphon elicitation	Ginsenosides (1% d. wt.) with >60 mg/l Rb and Rg content	Choi *et al.*, 2000; Kim, J.S.*et al.*, 2005; Yu *et al.*, 2005; Ali *et al.*, 2005; Bae *et al.*, 2006; Kim *et al.*, 2007; Jeong *et al.*, 2008
Panax notoginseng	CS	MS + 10 μM 2,4-D + 3.5 μM Kn + 6 μM Cu^{++}	Ginsenosides (65.5 mg l/d) and Polysaccharides (62.6 mg/l/d)	Zhong and Wang, 1996

		MS +10 µM 2,4-D + 3.5 µM Kn	Ginsenosides (136 mg/l/d)	Zhong *et al.*, 1997
		MS + 10 µM 2,4-D + 3.5 µM Kn + 1 µM Cu^{++} + 3.75 µM PO_4^{---}	Ginsenosides (55 mg/l/d), polysaccharides (186 mg/l/d)	Zhang and Zhong, 1997
		MS + 10 µM 2,4-D + 3.5 µM Kn	Polysaccharides (78.0 mg/l/d)	Yao and Zhong, 1999
		MS + 10 µM 2,4-D + 3.5 µM Kn + 3.75 µM PO_4^{---} + 1 µM Cu^{++} + 5% sucrose	Ginsenosides (75 mg/l/d), polysaccharides (140 mg/l/d)	Zhong, 2000
		Modified MS media + 50% conditioned medium	Ginsenosides (94 mg/l/d), polysaccharides (245 mg/l/d)	Zhong, 2000
		Elicitation by MJ & HMJA	Rb group ginsenoside enhanced by 9 fold	Wang and Zhong , 2002
Panax pseudoginseng	C, CS	MS + 5 µM 2,4-D + 1.2 µM Kn + 3% CH	Crude ginsenosides (1.1% fresh wt.)	Mathur *et al.*, 1999, 2003a,b
	C, CS	MS + 5 µM 2,4-D + 1.2 µM Kn + 3% CH	Crude ginsenosides (1.1% fresh wt.)	Mathur *et al.*, 1999, 2003a,b
Panax quinquefolium	C, CS	MS + 5 µM 2,4-D + 1.2 µM Kn	Ginsenosides (0.56–1.2 % f.wt.)	Mathur *et al.*, 1994, 1999
		MS+ 5 µM 2,4-D	Ginsenosides (1.9 mg/g)	Wang *et al.*, 1999
	CS	MS + 12.5 µM IBA + 2 µM 2,4-D + 0.5 µM Kn	Ginsenosides (33.9 mg/l/d), Polysaccharides (61.3 mg/l/d)	Zhong and Wang, 1998
	SL	½ MS salts + B_5 vitamins + 1.5% Sucrose + 3% gelrite	Ginsenosides (24.4 mg/g)	Wang *et al.*, 1999
	SEP	MS+1.45 µM GA_3 + 2.2 µM BAP	Ginsenosides (6.6 mg/g)	Wang *et al.*, 1999
	RPL	½MS + 1.45 µM GA_3	Ginsenosides (12.8 mg/g)	Wang *et al.*, 1999
	AR	MJ elicitation	INP	Ali *et al.*, 2005
Panax sikkimensis	C	MS + 5 µM 2,4-D + 1.2 µM Kn	Ginsenosides (0.95% fresh wt.)	Mathur *et al.*, 1999

AR, Adventitious roots; ASRP, Aerial part of regenerated plants; ASL, Aerial part of seedlings; C, Callus; CH, Casein hydrolysate; HMJA, Hemimethyl jasmonic acid; IBA, Indole-3-butyric acid; INP, Information not provided; Kn, Kinetic; MJ, Methyl jasmonate; MS, Murashige & Skoog medium; RPL, Roots of regenerated plantlets; SE, Somatic embryoids; SEP, Somatic embryo-derived plantlets; SH, Schenk and Hildebrandt Medium; SL, Seedlings; SRP- Subterranean part of regenerated plants; SSL, Subterranean part of seedling; YE, Yeast extract; 2,4-D, 2,4-dichlorophenoxyacetic acid.

Table 8.4. Studies concerning *in vitro* saponin production in *Agrobacterium rhizogenes*-mediated transformed hairy root cultures of *Panax* species.

Panax sps.	Bacterial strain	Explant	Medium used	Saponins produced *in vitro*	References
P. ginseng	A_4	RC	MS basal (Liquid)	Ginsenosides (0.95 % dry wt.)	Yoshikawa and Furuya, 1987
	ATCC-15834	R	LS basal (liquid)	Ginsenoside 7.4 mg/l/d in bioreactor and 11.9 mg/l/d in shake flask cultures	Inomata *et al.*, 1993
	ATCC-15834	P	½ MS (semi-solid)	INP	Yoshimatsu *et al.*, 1996
	A4	S	Modified MS (liquid)	Ginsenosides (13.23 -21.27 mg/g/ d.wt.)	Bulgakov *et al.*, 1998
	ATCC-15834	R	MS (semi-solid)	INP	Yang and Choi, 2000
	KCTC- 2703	R	MS + 23.8 µM JA	Ginsenosides (504.39 mg/l)	Yu *et al.*, 2000
			MS + 0.03% peptone	Ginsenosides (244.0 mg/l)	Yu *et al.*, 2000
	A4	R	SH (liquid)	Ginsenosides (5.4 mg/g dry wt.)	Mallol *et al.*, 2001
	ATCC- 15834	R	MS + 2.5 µM IBA + 2.5 µM NAA	INP	Washida *et al.*, 2004
	KCTC- 2703	R	MS + JA(4.8-23.8 µM)	Rb1 and Rb2 increased by 4.6 and 7.7 times	Yu *et al.*, 2002
	A4	R	JA, vanadyl sulphate and chitosan elicitation	Ginsenosides (3 fold increase)	Palazón et al., 2003a, b
	A4	R	INP	Ginsenosides	Woo *et al.*, 2004; Yu, K.W. *et al.*, 2005
	A4	R	Elicitation by oligosaccharides of *Paris polyphylla*	Ginsenosides	Zhou *et al.*, 2007
P. quinquefolium	LBA-9402	EP	¼ B5 (liquid)	Ginsenosides (1.0% fresh wt.)	Mathur *et al.*, 2010b
Panax hybrid	ATCC-15834	P	B5 (liquid)	Ginsenosides (2.87% dry wt.)	Washida *et al.*, 1998

ATCC, American Type Culture Collection; B5, Gamborg et al (1969) medium; EP, Epicotyl; INP,Information not provided; JA,Jasmonic acid; KCTC, Korean Collection for Type Cultures LS, Linsmaier & Skoog medium; MS,Murashige & Skoog medium; P,Petiole; R,Roots; RC,Root callus; Schenk and Hildebrandt Medium

Optimization of a culture-based production system for a plant product requires standardization of various techniques ranging from cell line selection to media manipulation, pathway elicitation, precursor feeding, and genetic transformations (Mathur and Ahuja, 1990; Dixon and Steele, 1999; Giri *et al.*, 2001; Rao and Ravishankar, 2002; Liang and Zhao, 2008). The role of some of these factors in relation to *in vitro* production of saponins is discussed below.

Selection of hyper-productive cell lines

Cell cultures may offer better selectivity and yield for the desired bioactive product since the cell strains may be selected from tissues or organs, which show more productivity than other parts of the plant (Dicosmo and Misawa, 1995). Because different levels of secondary metabolite production can be found within a cell line, significant headway in the productivity of plant cell cultures has followed on the heels of intensive selection for high-producing cell lines (Watanabe *et al.*, 1982; Dicosmo and Misawa, 1995).

Plant cell lines can be recurrently selected to amplify the productivity of the cell culture as has been done in American ginseng, *Panax quinquefolium* (Wang *et al.*, 1999; Mathur *et al.*, 2001; US Patent No. 6326202; 2002b, 2003b). Similar amplification of yield has followed selection in highly pigmented cell culture line of *Panax sikkimensis,* an Indian species of ginseng (Mathur *et al.*, 2002a, b; US Patent No. 6368860, 2010a).

Manipulation of culture conditions

Composition of the media including the types and amounts of plant growth regulators, mineral salts, carbon sources, as well as culture conditions such as temperature, pH, illumination condition, aeration etc. may affect the production of secondary metabolites. The effects of some medium components especially micro- and macro-nutrients have been studied on growth and saponin production in cell cultures of many plants (Nigra *et al.*, 1990; Liu and Zhong, 1997; Mathur *et al.*, 2000). Cu^{++}, K^{+}, PO_4^{---} and total nitrogen concentration in the media have been found to significantly affect the cell growth and saponin accumulation in cell cultures of various *Panax* species (Table 8.3). Optimal mineral element ratio is required to increase the growth and biomass production in callus and cell suspension cultures of *Panax notoginseng* (Zhang *et al.*, 1996a) and *Panax quinquefolium* (Mathur *et al.*, 2000; Gangwar, 2003). Manipulation of other media constituents including NO_3^- concentration (Liu and Zhong, 1997; Gangwar, 2003), sucrose (Akalezi *et al.*,1999; Zhang *et al.*, 1996b), PO_4 (Zhong and Zhu, 1995) and plant growth promoting substances (Zhang *et al.*,1996a,b) has also been shown to affect ginsenoside production in different species of *Panax.* Ginsenoside production in callus and cell suspension cultures of *P. quinquefolium* was also found to be affected by culture age (Mathur *et al.*, 1994, 2001) and medium replenishment/exchange strategy in *Panax* (Jeong *et al.*, 2008). In *P. ginseng*, oxygen supplementation to bioreactor-based cell suspension culture improved biomass accumulation and saponin production (Thanh *et al.*, 2006). Media component optimization for growth and secondary metabolite production was also studied for hairy roots of *Solanum khasianum* and it was observed that while MS media supported the growth, Gamborg's B5 media supported secondary metabolite production (Jacob and Malpathak, 2005). Nutrient feeding and osmotic shocks have also been demonstrated to be beneficial for biomass accumulation and ginsenoside biosynthesis in *P. ginseng* (Wu *et al.*, 2005).

Organ cultures

While research to date has succeeded in producing a wide range of valuable phytosaponins in unorganized callus or suspension cultures, in some cases the production requires a more differentiated state of tissues like multiple shoots or root cultures (Dörnenberg and Knorr, 1997). This is particularly required when the metabolite of interest is only produced in specialized plant tissues or glands in the parent plant. Adventitious or transformed hairy root cultures are the focus of such an approach. Hairy root cultures are genetically more stable because of a higher degree of tissue differentiation and organization than isolated cells in culture and usually grow faster than normal plant roots without any exogenous phytohormone (Giri and Narasu, 2000; Rao and Ravishankar, 2002; Sevon and Oksman-Caldentey, 2002; Verpoorte *et al*., 2002). *A. rhizogenes*-mediated hairy root formation also provides a transgenic system for introducing foreign pathway genes through Ri-plasmid of *A. rhizogenes* into the cultured cells. The induction and establishment of hairy roots after the infection of *P. ginseng* roots with *A. rhizogenes* has been successfully performed (Yoshikawa and Furuya, 1987; Ko *et al*., 1989). However, these roots required phytohormones for satisfactory growth. The roots grow more rapidly and produce higher levels of saponins than the non-transformed adventitious roots grown *in vitro*. Inomata *et al.* (1993), Bulgakov *et al.* (1998) and Washida *et al.* (1998), however showed the ability of *P. ginseng* hairy roots to grow rapidly in a phytohormone-free medium. Inomata *et al.,* (1993) and Yoshimatsu *et al.* (1996) established hairy root cultures by infecting petiole segments with *A. rhizogenes* strain 15834 and studied their storage at 4°C and cryopreservation at -196°C. The hairy roots regenerated from cryopreserved root tips grew well, having the same ginsenoside productivity pattern as those of the control hairy roots maintained continuously at 25°C. Hwang *et al.* (1996) found that the growth in light conditions favours ginsenoside synthesis, while darkness favours growth of the hairy roots. Yu *et al.* (2000) reported the effect of jasmonic acid, phenylalanine, caffeic acid, catechin, chitin, gum karaya, fucoidan and peptone on metabolite productivity of hairy root cultures of *P. ginseng*. They observed that only jasmonic acid strongly improved total ginsenoside production in these hairy roots. Peptone also gave a few positive signals but it was less efficacious than jasmonic acid.

Hairy root cultures, in addition to being an excellent system for *in vitro* saponin production, are also being used as expression systems to test the efficacy of 'pathway gene constructs of saponin biosynthesis' (Woo *et al*., 2004; Liang and Zhao, 2008). During upscaling, most remarkable developments of scale-up in large vessels were reported for *Panax ginseng* hairy root biomass in a 20-ton cultivation tank (Sivakumar *et al.,* 2005).

Employment of multiple shoot cultures for *in vitro* saponin production is also reported in *Centella asiatica* (Kim, T. *et al.* 2004 a,b; Susan *et al*., 2006; Yadav *et al*., 2007), *Galphimia glauca* and *Ruscus aculeatus* (Susan *et al*., 2006). The growth versus asiaticoside production in shoot cultures of *C. asiatica* has been studied over a culture period of 90 days. The results indicated that optimal growth with high asiaticoside production occurs between 30 and 40 days of culture. The shoots can be efficiently cultured in liquid medium supplemented with kinetin. An important point to mention here is that there is differential expression of asiaticoside production in leaf and petiole of the cultured shoots. Leaves produce more asiaticoside as compared to the petiole. Some biotic elicitors (*Trichoderma* exudates) have shown enhancement in asiaticoside production in these cultures (Yadav *et al*., 2007; Mathur, unpublished results).

Precursor feeding, biotransformation and permeabilization

One of the important strategies to enhance the natural product accumulation is feeding the cell cultures with inexpensive and abundantly available pathway precursors. The concept is based on the idea that any compound that is an intermediate in or at the beginning of a secondary biosynthetic route stands a good chance of increasing the yield of the final product (Rao and Ravishankar, 2002). Addition of known or putative precursors in a pathway may enhance the production of a desired metabolite in plant cell cultures if the endogenous level of such precursors is a limiting factor. This approach has been found to be very effective in the case of *Solanum lyratum* (Lee *et al.*, 2007) for the production of solanine, solanidine and solasodine and in *Cistanche salsa* for the production of a phenylethanoid glycoside (Liu *et al.*, 2007). Furuya *et al.* (1983b) studied the effect of some precursors (like sodium acetate, mevalonic acid and farnesol and some inhibitors of the side chain reactions of saponin synthesis like semicarbazide and thiosemicarbazide and found that these could significantly enhance saponin production in callus cultures of *P. ginseng*. Linsefors *et al.* (1989) also reported a significant increase of nearly 40% in the saponin accumulation in ginseng cell cultures with the addition of mevalonic acid in the medium. In *C. asiatica* callus cultures various triterpene precursors like squalene, isopentenyl pyrophosphate (IPP), farnesyl pyrophosphate (FPP) and leucine were tested by Kiong *et al.* (2005) to increase the asiaticoside production. Squalene was found to be the best in promoting the biomass and saponin production in such callus cultures. Feeding metabolic precursors in some instances can also help to identify steps in a pathway at which the activity of an enzyme or the availability of its substrate might limit the overall flux through the pathway (Han *et al.*, 2002).

In plant cell cultures the secondary metabolites are mostly stored within the cell vacuole which becomes a major limiting factor for continuous operation. Secretion of metabolites into the medium either naturally or by induction treatment offers easy removal of the product and better biomass re-use. Using *P. notoginseng* cell cultures Zhong *et al.* (1997) studied the effective release of saponins by short- and long-term treatment with DMSO. They noticed that long-term DMSO treatment can release much more saponin with high biological activity than those under short-term permeabilization treatment.The amount of saponin leached out into the medium was 136 mg/l on day 24 successively for five subcultures when the cultures were treated with 1% DMSO. In *P. quinquefolium* cell suspension cultures an appreciable amount of ginsenosides were found to auto-leach into the medium (Mathur *et al.*, 1994, 2001).

Elicitation

Biotic and abiotic elicitation of a metabolic pathway represents a very effective strategy to realize hyper-accumulation of a desired metabolite in cell, tissue and organ cultures (Roberts and Shuler, 1997; Rao and Ravishankar, 2002; Radman *et al.*, 2003; Namdev, 2007). This approach primarily works on the hypothesis that accumulation of secondary metabolites in plants is a part of its defence response to biotic and abiotic stresses that act as triggering signals for their synthesis. Some of the frequently used elicitation agents are carbohydrate fractions of fungal or bacterial cell walls, glycoproteins, polyamines, fatty acids, hepta-ß-glucosides, methyl jasmonate, chitosan, yeast extract, heavy metal ions (Cu^{++},Cd^{++},Ca^{++}, Ni^{+++}) and UV radiations (Radman *et al.*, 2003; Veersham, 2004; Kim, T. *et al.*, 2004b). Osmotic shocks (dehydration) and low energy ultrasound waves have also been frequently employed as elicitation factors (Lin and Wu, 2002; Wu *et al*., 2005; Liu and Cheng, 2008). Elicitors mostly trigger the signal transduction pathway in the cells through a cascade of biochemical events like change in membrane potential, ion flux through tonoplast, production of reactive oxygen species (ROS) and synthesis of phytoalexins (Giri *et al.*, 2001; Zhao *et al.*, 2005). Their positive influence on expression of secondary metabolic pathways are considered to be mediated by activation of pathway specific transcription factors that in turn, switch on or off several genes of a targeted biogenetic route.

Various elicitation strategies have been utilized to stimulate saponin biosynthesis in cell, shoot and root cultures of *Panax*, *Glycyrrhiza*, *Solanum* and *Centella asiatica* (Hu *et al.*, 2003a, b, c; Choi *et al.*, 2005; Kim, J.S. *et al.*, 2005; Kim, O.T. *et al.*, 2005b; Thanh *et al.*, 2005; Yu, C.J. *et al.*, 2005; Yu, K.W. *et al.*, 2005; Wang *et al.*, 2005 a,b; Wang *et al.*, 2006). Saponin accumulation in cultured ginseng cells was found to be stimulated by various elicitors such as yeast extract, peptone, methyl jasmonate, jasmonic acid, chitosan, yeast cell wall fractions and an oligo-galacturonic acid elicitor (Yu *et al.*, 2000; Lu *et al.*, 2001; Hu *et al.*, 2003b, c; Wang *et al.*, 2006). The saponin and polysaccharide synthesis in cell cultures of *P. notoginseng* was also enhanced by the use of conditioned/left-over medium (Yao and Zhong, 1999; Zhong, 2000).

There are several reports on positive influence of methyl jasmonate elicitation on ginsenoside production in cell suspension and root cultures of *P. ginseng* and *P. notoginseng* (Lu *et al.*, 2001; Palazón *et al.*, 2003a; Choi *et al.*, 2005; Bae *et al.*, 2006; Wang *et al.,* 2006; Hu and Zhong, 2007, 2008). Since methyl jasmonate has very low solubility in water besides having a cytotoxic effect, several ester derivatives of methyl jasmonate like 2-hydroxy-ethyl-jasmonate and trifluoroethyl jasmonate have been synthesized to achieve a much better elicitation of ginenoside pathway in *P. notoginseng* (Wang *et al.,* 2006; Hu and Zhong, 2008). Kim *et al.* (2009) showed that levels of Rb and Rg types of ginsenosides in hairy roots of *P. ginseng* increased by 5.5–9.7 and 1.85–3.82 fold within 7 days of MJ treatment. The Rg1 ginsenoside content was, however, negatively affected upon MJ treatment in these cultures. These workers also used the MJ-based elicitation approach to study the expression of several pathway genes associated with ginsenoside synthesis. Three genes of glucosyltransferase family involved in hydroxylation and glycosidation steps of the pathway were found to be upregulated in the elicited cultures. This study has opened up a fresh avenue for metabolic engineering in *Panax* species. MJ elicitation of saponin pathway in organized plantlet cultures of *Galphimia glauca* and *Centella asiatica* have also been reported (Susan *et al.*, 2006). The elicited cultures showed enhancement in their triterpenoid saponin synthesized from 2,3-oxidosqualene. However, in the case of *Ruscus aculeatus* which synthesizes steroidal saponins (mainly spirostane type) indirectly from 2,3-oxidosqualene, the growth rate and free sterol content decreased but spirostane content remained unchanged in aerial parts and decreased in the roots in presence of MJ. These results suggest that MJ may be used as an inducer of enzymes involved in the triterpene biosynthesis downstream from 2,3-oxidosqualene in both *Galphimia glauca* and *C. asiatica* plantlets, but in *C. asiatica* and *R. aculeatus* it inhibited the enzymes involved in sterol synthesis downstream from cycloartenol (Susan *et al.*, 2006). Responses of defence signals and transcription of squalene synthase (SQS), squalene epoxidase (SQE) and farnesyl diphosphate genes have also been reported in cultures of *P. ginseng* (Liang and Zhao, 2008), *C. asiatica* (Kim, J.S. *et al.*, 2005; Kim, O.T. *et al.*, 2005b), *G. glabra* (Hayashi *et al.*, 2003) and *M. truncatula* (Iturbe *et al.*, 2003) treated with MJ (Hu *et al.*, 2003c; Choi *et al.*, 2005). Induction of saponin biosynthesis was also observed under the influence of fungal cell wall extract of

Colletotrichum lagenarium by generation of H_2O_2 (Hu *et al.*, 2003b, c), singlet O_2 and ethylene (Xiaojie *et al.*, 2005) in cell cultures of *P. ginseng*. Elicitation treatment triggers the activation of plasma membrane NADPH oxidase and 1-aminocyclopropane-1-carboxylic acid oxidase (ACO), leading to ethylene release and increased mRNA expression of squalene synthase (SQS) and squalene epoxidase (SQE), and accumulation of β-amyrin synthase (β-AS). Heavy metal elicitors such as $CdCl_2$ and $CuCl_2$, were also used to elicit asiaticoside production in whole plant cultures of *C. asiatica* for stimulation of asiaticoside production (Kim *et al.*, 2004a). Another novel and special elicitation stimulus in the form of low energy ultrasound (US) has been found effective in stimulating saponin production in *P. ginseng* (Wu and Lin, 2002). This stimulating effect was ascribed to the mechanical stress created by ultrasound waves in the liquid medium.

Bioreactor upscaling

There are very few reports on bioreactor upscaling of cell cultures for phytosaponin production. The only commercially successful venture has been documented for *Panax ginseng* cell culture by Nitto Denko Corporation (Japan) wherein the production is scaled up to a level of 20 g/l/week biomass in a 20,000 l capacity fermenter (Ushiyama and Hibino, 1997). The cell lines of *P. notoginseng* and *P. quinquefolium* have also been cultivated in air-lift and stirred tank bioreactors for ginsenoside production (Hu and Zhong, 2001; Zhang and Zhong, 2004). Clearly, there are several technological/engineering gaps to be addressed in this area. Plant cells in culture have different morphologic and growth characteristics from those of microorganisms such as their relatively large size, formation of large aggregates, shear sensitivity and slow growth. Therefore, the conventional bioreactors that are used for microbial cells are not suitable for plant cells. Initially only the airlift reactors were used (Wagner and Vogelmann, 1977) due to their low shear forces. Nowadays, however, reactors with stirrers, like those of the ship impellor type, are preferred because of their better mixing potential (Shuler, 2001; Zhong, 2003). The fed batch mode has been efficiently utilized in cases where high concentration of the substrate addition affects the growth as reported by Wu and Ho (1999) in the case of *P. ginseng*. Using different agitation speeds during different growth stages would be a more rational aproach for successful cultivation of plant cell suspension cultures. *P. ginseng* cell cultures have been successfully grown in a balloon-type bubble bioreactor with an improved quality of inlet air via an appropriate level of oxygen supplementation up to a level of 40% (Thanh *et al.*, 2006). A fast growing and high ginsenoside producing cell line of *Panax quinquefolium* has also been upscaled by us in a 5 l bioreactor (A. Mathur, unpublished data).

It is evident that several operational parameters like optimum oxygen transfer, uniform media percolation, online growth monitoring, protection against cellular injury and *in situ* retrieval of the product are yet to be optimized in this direction (Shuler, 2001; Zhong, 2003; Ionkova, 2007). Intense efforts towards these are underway for several plant systems and rapid progress is expected in the near future (Guillon *et al.*, 2006; Srivastava and Srivastava, 2007; Cloutier *et al.*, 2008; Misra and Ranjan, 2008).

Metabolic Engineering of Saponin Biosynthetic Pathways

Plant saponins are in focus of metabolic engineering (ME) efforts for their enhanced expression in cultured cells and corresponding plants. Unfortunately, our current knowledge on their biogenetic pathways is still linear and scanty (Ustundag and Mazza, 2007; Liang and Zhao, 2008). Initial efforts to apply the techniques of genome-wide profiling to identify pathway genes for saponin synthesis have begun in the last few years in plants like *Panax*, *Centella*, *Glycyrrhiza* and *Medicago* (Hayashi *et al.,* 2003; Iturbe *et al.*, 2003; Jenner *et al.*, 2005; Kim, J.S. *et al.,* 2005; Kim O.T. 2005a,b,c; Liang and Zhao, 2008). Efforts are also on in several laboratories to dissect the saponin biosynthetic pathways in plants (Kushiro *et al.*, 1998; Lee *et al.*, 2004; Kim *et al.,* 2005; Han *et al.*, 2006). Many genes associated with triterpene biosynthesis have now been characterized and cloned. To create a ginseng gene resource that contains the genes involved in ginsenoside biosynthesis, Choi *et al.* (2005) generated 3134 ESTs from MJ-treated ginseng hairy roots. A novel oxidosqualene cyclase (OSC) gene was also identified by this analysis. RNA gel blot analysis showed that transcription of this OSC and squalene esterase (SQE) gene transcription increased by MJ treatment. Three genes of glucosyltransferase that regulate glycosidation in ginseng were also selected as candidates, although their characterization in terms of their functions in a heterogenetic system is yet to be worked out. In future, if the function of the gene associated with hydroxylation and glycosidation steps can be clearly elucidated, it will be possible to manipulate the ginsenoside pathway for the large-scale production of a target saponin in ginseng hairy root cultures (Kim *et al.*, 2009). Likewise three major pathway genes coding for branch point enzymes like farnesyl diphosphate synthase, oxidosqualene cyclase and amyrin synthase involved in synthesis of asiaticosides in *C. asiatica* have been cloned (Kim, J.S. *et al.,* 2005; Kim O.T. 2005a) and further efforts in this direction will pave the way of designing better asiaticoside yielding clones.

Conclusion and Future Directions

The last three decades have witnessed notable progress in the development of various *in vitro* approaches for the production of plant saponins from cultured cells and tissues. These knowhows now deserve a cross-talk for their synergism and translation for industrial exploitation. Plant scientists and chemical engineers must join hands to address the scaling up issues. It is hoped that amalgamation of combinatorial genomic and metabolic engineering tools with available knowledge will further facilitate the fine tuning of the various bio-processes involved in the *in vitro* saponin synthesis, storage, catabolism and recovery from the cultured tissues. Need for further research on molecular understanding, redesigning and obtaining better regulatory controls of saponin pathways *in planta* and in cultured cells will continue to motivate and excite the scientific community engaged with phytosaponins.

References

Akalezi, C.O., Liu, S., Li, Q.S., Yu, J.T. and Zhong, J.J. (1999) Combined effects of initial sucrose concentration and inoculum size on cell growth and ginseng saponin production by suspension cultures of *Panax ginseng*. *Process Biochemistry* 34, 639–642.

Ali, M.B.,Yu, K.W., Hahn, E.J. and Paek, K.Y. (2005) Differential responses of antioxidant enzymes, lipoxygenase activity, ascorbate content and the production of saponins in tissue cultured root of mountain *Panax ginseng* C.A. Mayer and *Panax quinquefolium* L. in bioreactor subjected to methyl jasmonate stress. *Plant Science* 169, 83–92.

Asaka, I., Ii, I., Hirotani, M., Asada, Y. and Furuya, T. (1994) Ginsenoside contents of plantlets regenerated from *Panax ginseng* embryoids. *Phytochemistry* 36, 61–63.

Attele, A.S., Wu, J.A. and Yuan, C.S. (1999) Ginseng pharmacology: multiple constituents and multiple actions. *Biochemical Pharmacology* 54, 1–8.

Aziz, Z.A., Davey, M.R., Power, J.B., Anthony, P., Smith, R.M. and Lowe, K.C. (2007) Production of asiaticoside and madecassoside in *Centella asiatica in vitro* and *in vivo*. *Biologia Plantarum (Prague)* 51, 34–42.

Badaoui, H.E., Morard, P. and Henry, M. (1996) Stimulation of growth and solamargine production by *Solanum paludosum* multiple shoot cultures using a new culture medium. *Plant Cell Tissue and Organ Culture* 45, 153–158.

Bae, K.H., Choi, Y.E., Shin, C.G., Kim, Y.Y. and Kim, Y.S. (2006) Enhanced ginsenoside productivity by combination of ethephon and methyl jasmonate in ginseng (*Panax ginseng* C.A. Meyer) adventitious root cultures. *Biotechnology Letters* 28, 1163–1166.

Balandrin, M.F. (1996) Commercial utilization of plant-derived saponins: An overview of medicinal, pharmaceutical and industrial applications. *Advances in Experimental Medicine and Biology* 404, 1–14.

Bruneton, J. (1995) *Pharmacognosy, Phytochemistry, Medicinal Plants.* Lavisior, Paris.

Bulgakov, V.P., Khodakovskaya, M.V., Labetskaya, N.V., Chernoded, G.K. and Zhuravlev, Y.N. (1998) The impact of plant *rol C* oncogene on ginsenoside production by ginseng hairy root cultures. *Phytochemistry* 49, 1929–1934.

Chang, C.K., Chang, K.S., Lin, S.Y. and Chen, C.Y. (2005) Hairy root cultures of *Gynostemma pentaphyllum* (Thunb.) Makino: a promising approach for the production of gypenosides as an alternative of ginseng saponins. *Biotechnology Letters* 27, 1165–1169.

Chapagain, B.P., Saharan, V. and Weisman, Z. (2008) Larvicidal activity of saponins from *Balanites aegyptiaca* callus against *Aedes aegypti* mosquito. *Bioresource Technol*ogy 99, 1165–1168.

Choi, D.W., Jung, J.D., Ha, Y.I., Park, H.W., In, D.S., Chung, H.J. and Liu, J.R. (2005) Analysis of transcripts in methyl jasmonate-treated ginseng hairy roots to identify genes involved in the biosynthesis of ginsenosides and other secondary metabolites. *Plant Cell Reports* 23, 557–566.

Choi, S.M., Son, S.H., Yun, S.R., Kwon, O.W., Seon, J.H. and Paek, K.Y. (2000) Pilot-scale culture of adventitious roots of ginseng in a bioreactor system. *Plant Cell Tissue and Organ Culture* 62, 187–193.

Cloutier, M., Bouchard-Marchaud, É., Perrier, M. and Jolicoeur, M. (2008) A predictive nutritional model for plant cells and hairy roots. *Biotechnol Bioeng* 99, 189–200.

Dewick, P.M. (1997) *Medicinal Natural Products*. John Wiley & Sons, West Sussex, UK.

Dicosmo, F. and Misawa, M. (1995) Plant cell and tissue culture: Alternatives for metabolite production. *Biotechnology Advances* 13, 425–453.

Dixon, R.A. and Steele, C.L. (1999) Flavonoids and isoflavonoids– A gold mine for metabolic engineering. *Trends Plant Sciences* 4, 394–400.

Dörnenburg, H. and Knorr, D. (1997) Strategies for the improvement of secondary metabolite production in plant cell cultures. *Enzyme and Microbial Technology* 17, 674–684.

Dou, D.Q., Chen, Y.J., Liang, L.H., Pang, F.G., Shimizu, N. and Takada, T. (2001) Six new dammarane-type triterpene saponin from the leaves of *Panax ginseng*. *Chemical and Pharmaceutical Bulletin (Tokyo)* 49, 442–446.

Fujimoto, Y., Satoh, M., Takeuchi, N. and Kirisawa, M. (1990) Synthesis and absolute configurations of the cytotoxic polyacetylenes isolated from the callus of *Panax ginseng*. *Chemical and Pharmaceutical Bulletin* 35, 1447–1450.

Fulcheri, C., Morard, P. and Henry, M. (1998) Stimulation of the growth and the triterpenoid saponin accumulation of *Saponaria officinalis* cell and *Gypsophila paniculata* root suspension cultures by improvement of the mineral composition of the media. *Journal of Agricultural and Food Chemistry* 46 (5), 2055–2061.

Furuya, T. (1988) Saponins (Ginseng saponins). In: Constabel, F. and Vasil, I.K. (eds) *Cell Culture and Somatic Cell Genetics of Plants*, Vol. 5. *Phytochemicals in Plant Cell Cultures*. Academic Press, New York, pp. 213–234.

Furuya, T. and Ushiyama, K. (1994) Ginseng production in cultures of *Panax ginseng* cells. In: Shargool, P.D. and Ngo, T.T. (eds) *Current Topics in Plant Molecular Biology: Biotechnological Application of Plant Cultures*. CRC Press, Boca Raton, Florida, USA, pp. 1–22.

Furuya, T., Yoshikawa, T., Ishii, T. and Kajii, K. (1983a) Effects of auxins on growth and saponin production in callus cultures of *Panax ginseng*. *Planta Medica* 47, 183–187.

Furuya, T., Yoshikawa, T., Ishii, T. and Kajii, K. (1983b) Regulation of saponin production in callus cultures of *Panax ginseng*. *Planta Medica* 47, 200–204.

Gangwar, A. (2003) Morphogenetic studies and *in vitro* secondary metabolite production in Panax species. PhD thesis, Lucknow University, India.

Giri, A. and Narasu, M.L. (2000) Transgenic hairy roots: recent trends and applications. *Biotechnology Advances* 18, 1–22.

Giri, A., Dhingra, V., Giri, C.C., Singh, A., Ward, O.P. and Narasu, M.L. (2001) Biotransformation using plant cells, organ cultures and enzyme system: current trends and future prospects. *Biotechnology Advances* 19, 175–199.

Gopi, C. and Vatsala, T.M. (2006) *In vitro* studies on effects of plant growth regulators on callus and suspension culture biomass yield from *Gymnema sylvestre* R.Br. *African Journal of Biotechnology* 5(12), 1215–1219.

Guillon, S., Trémouillaux-Guiller, J., Kumar, P.P, Rideau, M. and Gantet, P. (2006) Harnessing the potential of hairy roots: dawn of a new era. *Trends in Biotechnology* 24, 403–409.

Han, J.Y., Kwon, Y.S., Yang, D.C., Jung, Y.R. and Choi, Y.E. (2006) Expression and RNA interference-induced silencing of the dammarenediol synthase gene in *Panax ginseng*. *Plant Cell Physiology* 47, 1653–1662.

Han, Y.S., van der Heijden, R. and Verpoorte, R. (2002) Improved anthraquinone accumulation in cell cultures of *Cinchona* 'Robusta' by feeding of biosynthetic precursors and inhibitors. *Biotechnology Letters* 24, 705–710.

Hanafy, M.S. and Abou-Setta, L.M. (2007) Saponin production in shoot and callus cultures of *Gypsophila paniculata*. *Journal of Applied Sciences Research* 3, 1045–1049.

Haralampidis, K., Trojanowska, M. and Osbourn, A.E. (2002) Biosynthesis of triterpenoid saponins in plants. *Advances in Biochemical Engineering Biotechnology* 75, 31–49.

Haughton, P. (1999) Roots of remedies: Plants, people and pharmaceuticals. *Chemistry and Industry* 1, 15–19.

Hayashi, H., Yamada, K., Fukui, H. and Tabata, M. (1992) Metabolism of exogenous 18 β glycyrrhetinic acid in cultured cells of *Glycyrrhiza glabra*. *Phytochemistry* 31, 2729–2733.

Hayashi, H., Huang, P. and Inoue, K. (2003) Upregulation of soyasaponin biosynthesis by methyl jasmonate in cultured cells of *Glycyrrhiza glabra*. *Plant Cell Physiology* 44, 404–411.

Hayashi, H., Hiraoka, N. and Ikeshiro, Y. (2005) Differential regulation of soyasaponin and betulinic acid production by yeast extract in cultured licorice cells. *Plant Biotechnology* 22, 241–244.

Hong, S.G., Lee, K.H., Kwak, J. and Bae, K.S. (2006) Diversity of yeasts associated with *Panax ginseng*. *Journal of Microbiology* 44, 674–679.

Horn, M.E., Woodard, S.L. and Howard, J.A. (2004) Plant molecular farming: systems and products. *Plant Cell Reports* 22,711–720.

Hostettmann, K. and Marston, A. (1995) *Saponins*. Cambridge University Press, Cambridge.

Hu, W.W. and Zhong J.-J. (2001) Effect of bottom clearance on performance of airlift bioreactor in high-density culture of *Panax notoginseng* cells. *Journal of Bioscience and Bioengineering* 92(4), 389–392.

Hu, F.X. and Zhong, J.J. (2007) Role of jasmonic acid in alteration of ginsenoside heterogeneity in elicited cell cultures of *Panax notoginseng*. *Journal of Bioscience and Bioengineering* 104, 513–516.

Hu, F.X. and Zhong, J.J. (2008) Jasmonic acid mediates gene transcription of ginsenoside biosynthesis in cell cultures of *Panax notoginseng* treated with chemically synthesized 2-hydroxyethyl jasmonate. *Process Biochemistry* 43, 113–118.

Hu, X., Neill, S., Fang, J., Cai, W. and Tang, Z. (2003a) Nitric oxide mediates elicitor-induced saponin synthesis in cell cultures of *Panax ginseng*. *Functional Plant Biology* 30, 901–907.

Hu, X., Neill, S., Fang, J., Cai, W. and Tang, Z. (2003b) Hydrogen peroxide and jasmonic acid mediate oligogalacturonic acid-induced saponin synthesis in suspension cultured cells of *Panax ginseng*. *Physiologia Plantarum* 118, 414–421.

Hu, X., Neill, S., Fang, J., Cai, W. and Tang, Z. (2003c) The mediation of defence responses of ginseng cells to an elicitor from cell walls of *Colletotrichum lagenarium* by plasma membrane NAD(P)H oxidase. *Acta Botanica Sinica* 45, 32–39.

Huang, K.C. (1999) *Pharmacology of Chinese Herbs*. CRC Press, Boca Raton, Florida, USA.

Hwang, B., Yang, D.C., Park, J.C., Choi, K.J., Min, K.K. and Choi, K.T. (1996) Mass culture and ginsenoside production of ginseng hairy root by two-step culture process. *Journal of Plant Biology* 39, 63–69.

Ikenaga, T., Oyama, T. and Muranaka, T. (1995) Growth and steroidal saponin production in hairy root cultures of *Solanum aculeatissimum*. *Plant Cell Reports* 14, 413–417.

Ikuta, A. (1995) Saponins and triterpenes from callus tissues of *Akebia trifoliata* and comparison with the constituents of other lardizabalaceous callus tissues. *Journal of Natural Products* 58, 1378–1383.

Ikuta, A. and Itokawa, H. (1989) 30-noroleanane saponins from callus tissues of *Akebia quinata*. *Phytochemistry* 28, 2663–2665.

Inomata, S., Yokoyama, M., Gozu, Y., Shimizu, T. and Yanagi, M. (1993) Growth pattern and ginsenoside production of *Agrobacterium*-transformed *Panax ginseng* roots. *Plant Cell Reports* 12, 681–686.

Ionkova, I. (2007) Biotechnological approaches for the production of lignans. *Pharmacognosy Reviews* 1, 57–68.

Isvett, J., Sanchez, F., Lopez, J.O. and Montes-Horcasitas, M.C. (2002) Biosynthesis of sterols and triterpenes in cell suspension cultures of *Uncaria tomentosa*. *Plant Cell Physiology* 43, 1502–1509.

Iturbe-Ormaetxe, I., Haralampidis, K., Papadopoulou, K. and Osbourn, A.E. (2003) Molecular cloning and characterization of triterpene synthases from *Medicago truncatula* and *Lotus japonicus*. *Plant Molecular Biology* 51, 731–743.

Jacob, A. and Malpathak, N. (2005) Manipulation of MS and B5 components for enhancement of growth and solasodine production in hairy root cultures of *Solanum khasianum* Clarke. *Plant Cell Tissue and Organ Culture* 80, 247–257.

Jain, S.C. and Agrawal, M. (1994) Effect of mutagens on steroidal sapogenins in *Trigonella foenum-graecum* tissue cultures. *Fitoterapia* 65, 367–370.

Jenner, H., Townsend, B. and Osbourn, A. (2005) Unravelling triterpene glycoside synthesis in plants: phytochemistry and functional genomics join forces. *Planta* 220, 503–506.

Jeong, C.S., Murthy, H.N., Hahn, E.J. and Paek, K.Y. (2008) Imporoved production of ginsenosides in suspension cultures of ginseng by medium replenishment strategy. *Journal of Bioscience and Bioengineering* 105, 288–291.

Johanssen, K. (2006) *Ginseng Dreams: The Secret World of America's most Valuable plant*. The University Press of Kentucky, USA.

Kim, J.S., Chang, E.J. and Oh, H. (2005) Saponin production in submerged adventitious root culture of *Panax ginseng* as affected by culture conditions and elicitors. *Asia Pacific Journal of Molecular Biology and Biotechnology* 13, 87–91.

Kim, O.T., Kim, M.Y., Hong, M.H., Ahn, J.C. and Hwang, B. (2004a) Stimulation of asiaticoside accumulation in the whole plant cultures of *Centella asiatica* Linn. Urban by elicitors. *Plant Cell Reports* 23, 339–344.

Kim, O.T., Kim, M.Y., Huh, S.M., Ahn, J.C., Seong, N.S. and Hwang, B. (2004b) Effect of growth regulators on asiaticoside production in whole plant cultures of *Centella asiatica* (L.) Urban. *Journal of Plant Biology* 47(4), 361–365.

Kim, O.T., Kim, M.Y., Huh, S.M., Bai, D.G., Ahn, J.C. and Hwang, B. (2005a) Cloning of cDNA probably encoding oxidosqualene cyclase associated with asiaticoside biosynthesis from *Centella asiatica* Linn. *Plant Cell Reports* 24, 304–311.

Kim, O.T., Ahn, J.C., Hwang, S.J. and Hwang, B. (2005b) Cloning and expression of a farnesyl diphosphate synthase in *Centella asiatica* (L.) Urban. *Molecular Cell* 19, 294–299.
Kim, O.T., Kim, M.Y., Hwang, S.J., Ahn, J.C. and Hwang, B. (2005c) Cloning and molecular analysis of cDNA encoding cycloartenol synthase from *Centella asiatica* (L.). *Biotechnology and Bioprocess Engineering* 10, 16–22.
Kim, O.T., Bang, K.H., Kim, Y.C., Hyun, D.Y., Kim, M.Y. and Cha, S.W. (2009) Upregulation of ginsenoside and gene expression related to triterpene biosynthesis in ginseng hairy root cultures elicited by methyl jasmonate. *Plant Cell Tissue and Organ Culture* 98, 25–33.
Kim, Y., Wyslouzil, B.E. and Weathers, P.J. (2002) Secondary metabolism of hairy root cultures in bioreactors. *In vitro Cellular and Developmental Biology– Plant* 38, 1–10.
Kim, Y.S., Hahn, E.J., Murthy, H.N. and Paek, K.Y. (2004) Adventitious root growth and ginsenoside accumulation in *Panax ginseng* cultures as affected by methyl jasmonate. *Biotechnology Letters* 26, 1619–1622.
Kim, Y.S., Yeung, E.C., Hahn, E.J. and Paek, K.Y. (2007) Combined effects of phytohormone, indolebutyric acid and methyl jasmonate on root growth and ginsenoside production in adventitious root cultures of *Panax ginseng* C.A. *Biotechnology Letters* 29, 1789–1792.
Kiong, L.P., Mahmood, M., Fadzillah, N.M. and Daud, S.K. (2005) Effects of precursor supplementation on the production of triterpenes by *Centella asiatica* callus cultures. *Pakistan Journal of Biological Sciences* 8(8), 1160–1169.
Ko, K.S., Noguchi, H., Ebizuka, Y. and Sankawa, U. (1989) Oligoside production by hairy root cultures transformed by Ri plasmids. *Chemical and Pharmaceutical Bulletin (Tokyo)* 37, 245–248.
Kovalenko, P.G. and Maliuta, S.S. (2003) An effect of transformation by Ri– plasmids and elicitors on licorice cells and secondary metabolites production. *Ukrainskii Bioorganic Acta* 1, 50–60.
Kushiro, T., Shibuya, M. and Ebizuka, Y. (1998) β-amyrin synthase-cloning of oxidosqualene cyclase that catalyzes the formation of the most popular triterpene among higher plants. *European Journal of Biochemistry* 256, 238–244.
Lee, M.H., Jeong, J.H., Seo, J.W., Shin, C.G., Kim, Y.S., In, J.G., Yang, D.C., Yi, J.S. and Choi, Y.E. (2004) Enhanced triterpene and phytosterol biosynthesis in *Panax ginseng* overexpressing squalene synthase gene. *Plant and Cell Physiology* 45, 976–984.
Lee, M.H., Cheng, J.J., Liu, C.Y., Chen, Y.J. and Lu, M.K. (2007) Precursor feeding strategy for the production of solanine, solanidine, and solasodine by a cell culture of *Solanum lyratum*. *Process Biochemistry* 42, 899–903.
Li, T.S.C. and Mazza, G. (1999) Correlations between leaf and soil mineral concentrations and ginsenoside contents in American ginseng. *Hortscience* 34, 85–87.
Liang, Y. and Zhao, S. (2008) Progress in understanding of ginsenoside biosynthesis. *Plant Biology* 10, 415–421.
Lin, L. and Wu, J. (2002) Enhancement of shikonin production in single and two-phase suspension cultures of *Lithospermum erythrorhizon* cells using low-energy ultrasound. *Biotechnology and Bioengineering* 78(1), 81–88.
Linsefors, L., Björk, L. and Mosbach, K. (1989) Influence of elicitors and mevalonic acid on the biosynthesis of ginsenosides in tissue cultures of *Panax ginseng*. *Biochem. Physiol. Pflanzen.* 184, 413–418.
Linsmaier, E.M. and Skoog, F. (1965) Organic growth factor requirements of tobacco tissue cultures. *Physiologia Plantarum* 18, 100–127.
Liu, C.Z. and Cheng, X.Y. (2008) Enhancement of phenylethanpid glycosides biosynthesis in cell cultures of *Cistanche deserticola* by osmotic stress. *Plant Cell Reports* 27, 357–362
Liu, S. and Zhong, J.J. (1996) Effects of potassium ion on cell growth and production of ginseng saponin and polysaccharide in suspension cultures of *Panax ginseng*. *Journal of Biotechnology* 52, 121–126.
Liu, S. and Zhong, J.J. (1997) Simultaneous production of ginseng saponin and polysaccharide by suspension cultures of *Panax ginseng*: Nitrogen effects. *Enzyme and Microbial Technology* 21, 518–524.

Liu, Y.J., Gao, Z.G. and Zeng, Z.I. (2007) Improved accumulation of phenylethanoid glycoside by precursor feeding to suspension culture of *Cistanche salsa*. *Biochem. Eng. J.* 33, 88–95.

Lu, M.B., Wong, H.L. and Teng, W.L.(2001) Effects of elicitation on the production of saponin in cell culture of *Panax ginseng*. *Plant Cell Reports* 20, 674–677.

Mallol, A., Cusidó, R.M., Palazón, J., Bonfill, M., Morales, C. and Piňol, M.T. (2001) Ginsenoside production in different phenotypes of *Panax ginseng* transformed roots. *Phytochemistry* 57, 365–371.

Marston, A., Wolfender, J.L. and Hostettmann, K. (2000) Analysis and isolation of saponins from plant material. In: Oleszek, W. and Marston, A. (eds) *Saponins in Food, Foodstuffs and Medicinal Plants. Proceedings of the Phytochemical Society of Europe* (45) Kluwer Academic Publishers, Dordrecht-Boston/London, pp. 171–179.

Mathur, A. and Ahuja, P.S. (1990) Production of secondary metabolites in plant tissue cultures. *Current Research on Medicinal and Aromatic Plants* 12, 37–50.

Mathur, A., Shukla, Y.N., Pal, M., Ahuja, P.S. and Uniyal, G.C. (1994) Saponin production in callus and cell suspension cultures of *Panax quinquefolium*. *Phytochemistry* 35, 1221–1225.

Mathur, A., Mathur, A.K., Pal, M. and Uniyal, G.C. (1999) Comparison of qualitative and quantitative *in vitro* ginsenoside production in callus cultures of three *Panax* species. *Planta Medica* 65, 484–486.

Mathur, A., Mathur, A.K. and Gangwar, A. (2000) Saponin production by cell/callus cultures of *Panax* species. In: Oleszek, W. and Marston, A. (eds) *Saponins in Food, Foodstuffs and Medicinal Plants. Proceedings of the Phytochemical Society of Europe* (45) Kluwer Academic Publishers, Dordrecht-Boston/London, pp.171–179.

Mathur, A., Mathur, A.K., Uniyal, G.C., Pal, M. and Sangwan, R.S. (2001) A process for the development of a stable high yielding callus line of *Panax quinquefolium*. (a) US Patent No. 6326202; b) Japan Patent No. 11302760; (c) Korean Patent No. 9952236.

Mathur, A., Gangwar, A., Mathur, A.K., Sangwan, R.S. and Jain, D.C. (2002a) A procedure for the development of an anthocyanin producing callus line of *Panax sikkimensis* (Indian species of ginseng). *US Patent No. 6368860.*

Mathur, A., Mathur, A.K. and Gangwar, A. (2002b) Efficacy of some cell culture approaches applied for micropropagation, conservation and secondary metabolite production in ginseng (*Panax* spp.). In: Nandi, S.K., Palani, L.M.S. and Kumar, A. (eds) *Role of Plant Tissue Culture in Biodiversity Conservation and Economic Development.* Gyanodaya Prakashan, Nainital, India, pp. 267–275.

Mathur, A., Mathur, A.K., Sangwan, R.S., Gangwar, A. and Uniyal, G.C. (2003a) Differential responses, ginsenoside metabolism and RAPD pattern of three *Panax* species. *Genetic Resources and Crop Evolution* 50, 245–252.

Mathur, A., Mathur, A.K. and Gangwar, A. (2003b) Development of high saponin and anthocyanin yielding callus lines of Panax for industrial exploitation. In: Mathur, A.K. (ed.) *Proceedings of the First National Interactive Meet on Medicinal and Aromatic plants*, Central Institute of Medicinal and Aromatic Plants, Lucknow, pp. 399–406.

Mathur, A., Mathur, A.K., Gangwar, A., Verma, P. and Sangwan, R.S. (2010a) Anthocyanin production in a callus line of *Panax sikkimensis* Ban. *In vitro Cellular and Developmental Biology–Plant* 46(1), 13–21.

Mathur, A., Gangwar, A., Mathur, A.K., Verma, P., Uniyal, G.C. and Lal, R.K. (2010b) Growth kinetics and ginsenosides production in transformed hairy roots of American ginseng–*Panax quinquefolium* L. *Biotechnol Letters* 32(3), 457–461.

Mehrotra, S., Kukreja, A.K., Khanuja, S.P.S. and Misra, B.N. (2008) Genetic transformation studies and scale up of hairy root culture of *Glycyrrhiza glabra* in bioreactor. *Electronic Journal of Biotechnology* 11, 1–9.

Misra, B.N. and Ranjan, R. (2008) Growth of hairy root cultures in various bioreactors for the production of secondary metabolites. *Biotechnology and Applied Biochemistry* 49, 1–10.

Murashige, T. and Skoog, F. (1962) A revised medium for rapid growth and bioassays with tobacco tissue cultures. *Physiologia Plantarum* 15, 473–497.

Namdev, A.G. (2007) Plant cell elicitation for production of secondary metabolites: A review. *Pharmacognosy Reviews* 1, 69–79.

Ngan, F., Shaw, P., But, P. and Wang, J. (1999) Molecular authentication of *Panax* species. *Phytochemistry* 50, 787–791.
Nigra, H.M., Alvarez, M.A. and Giulietti, A.M. (1990) Effect of carbon and nitrogen sources on growth and solasodine production in batch suspension cultures of *Solanum eleagnifolium* Cav. *Plant Cell Tissue and Organ Culture* 21, 55–60.
Nocerino, E., Amato, M. and Izzo, A.A. (2000) The aphrodisiac and adaptogenic properties of ginseng. *Fitoterapia* 71, S1–S5.
Odnevall, A., Björk, L. and Berglund, T. (1989) Rapid establishment of tissue cultures from seeds of *Panax ginseng* and *Panax pseudoginseng*. *Biochem. Physiol. Pflanzen.* 185, 131–134.
Oleszek, W. and Marston, A. (ed.) (2000) *Saponins in Food, Foodstuffs and Medicinal Plants. Proceedings of the Phytochemical Society of Europe* (45). Kluwer Academic Publishers, Dordrecht–Boston/London.
Osbourn, A.E., Wubbern, J.P. and Daniels, M. (1997) Saponin detoxification by phytopathogenic fungi. In: Stacey, G. and Keen, N.T.(eds) *Plant Miocrobe Interactions.* Vol. 2. International Thomson Publishing, Chapman and Hall, New York, NY, USA, pp. 99–104.
Palazón, J., Cusidó, R.M., Bonfill, M., Mallol, A., Moyano, E., Morales, C. and Piñol, M.T. (2003a) Elicitation of different *Panax ginseng* root phenotypes for an improved ginsenoside production. *Plant Physiology and Biochemistry (Paris)* 41, 1019–1025.
Palazón, J., Mallol, A., Eibl, R., Lettenbauer, C., Cusidó, R.M. and Piñol, M.T. (2003b) Growth and ginsenoside production in hairy root cultures of *Panax ginseng* using a novel bioreactor. *Planta Medica* 69, 344–349.
Palazón, J., Moyano, E., Bonfill, M., Osuna, L.T., Cusidó, R.M. and Piñol, M.T. (2006) Effect of organogenesis on steroidal saponin biosynthesis in calli cultures of *Ruscus aculeatus. Fitoterapia* 77, 216–220.
Park, I.H., Kim, N.Y., Han, S.B., Kim, J.M., Kwon, S.W., Kim, H.J., Park, M.K. and Park, J.H. (2002) Three new dammarane glycosides from heat processed ginseng. *Archives of Pharmacol Research (Seoul)* 25, 428– 432.
Park, I.J., Piao, L.Z., Kwon, S.W., Lee, Y.J., Cho, S.Y. and Park, M.K. (2002) Cytotoxic dammarane glycosides from processed ginseng. *Chemical and Pharmaceutical Bulletin (Tokyo)* 50, 538–540.
Pauthe-Dayde, D., Rochd, M. and Henry, M. (1990) Triterpenoid saponin production in callus and multiple shoot cultures of *Gypsophila* spp. *Phytochemistry* 29, 483–487.
Proctor, J.T.A. and Bailey, W.G. (1987) Ginseng: Industry, botany and culture. *Horticultural Reviews* 9, 187–236.
Radman, R., Saez, T., Bucke, C. and Keshavarz, T. (2003) Elicitation of plants and microbial cell systems. *Biotechnology and Applied Biochemistry.* 37, 91–102.
Rao, R.S. (2000) Biotechnological production of phytopharmaceuticals. *Journal of Biochemistry Molecular Biology and Biophysics* 4, 73–112.
Rao, R.S. and Ravishankar, G.S. (2002) Plant cell cultures: chemical factories of secondary metabolites. *Biotechnology Advances* 20, 101–153.
Roberts, S.C. and Shuler, M.L. (1997) Large scale plant cell culture. *Current Opinion in Biotechnology* 8, 154–159.
Schlag, E.M. and McIntosh, M.S. (2006) Ginsenoside content and variation among and within American ginseng (*Panax quinquefolius* L.) populations. *Phytochemistry* 67, 1510–1519.
Schultz, V., Hänsel, R. and Tyler, V. (1998) *Rational Phytotherapy: A Physician's Guide to Herbal Medicine.* Springer-Verlag, Berlin, Germany.
Sevon, N. and Oksman-Caldentey, K.M. (2002) *Agrobacterium rhizogenes*-mediated transformation: root cultures as source of alkaloids. *Planta Medica* 68, 859–868.
Shibata, S. (2001) Chemistry and cancer preventing activities of ginseng saponins and some related triterpenoid compounds. *Journal of Korean Medical Science* 16, S28–S37.
Shuler, M. (2001) *Bioprocess Engineering: Basic Concepts.* 2nd edn. Prentice Hall, Englewood Cliffs, NJ, USA.
Sivakumar, G., Yu, K.W., Hahn, E.J. and Paek, K.Y.(2005) Optimisation of organic nutrients for ginseng hairy roots production in large scale bioreactors. *Current Science* 89, 641–649.

Srivastava, S. and Srivastava, A.K. (2007) Hairy root culture for mass-production of high-value secondary metabolites. *Critical Reviews in Biotechnology* 27, 29–43.
Sticher, O. (1998) Getting to the root of ginseng. *Chemtech* 28, 26–32.
Susan, M., Merce, B., Lidia, O., Elisabeth, M., Jaime, T., Rosa, C., Teresa, P.M. and Javier, P. (2006) The effect of methyl jasmonate on triterpene and sterol metabolisms of *Centella asiatica*, *Ruscus aculeatus* and *Galphimia glauca* cultured plants. *Phytochemistry* 67, 2041–2049.
Tenea, G.N., Calin, A. and Cucu, N. (2008) Manipulation of root biomass and biosynthetic potential of *Glycyrrhiza glabra* L. plants by *Agrobacterium rhizogenes* mediated transformation. *Romanian Biotechnology Letters* 13, 3922–3932.
Thanh, N.T., Murthy, H.N., Yu, K.W., Hahn, E.J. and Paek, K.Y. (2005) Methyl jasmonate elicitation enhanced synthesis of ginsenoside by cell suspension cultures of *P.ginseng* in 5-l balloon type bubble bioreactors. *Applied Microbiology and Biotechnology* 67, 197–201.
Thanh, N.T., Murthy, H.N., Yu, K.W., Jeong, C.S., Hahn, E.J. and Paek, K.Y. (2006) Effect of oxygen supply on cell growth and saponin production in bioreactor cultures of *Panax ginseng*. *Journal of Plant Physiology* 163, 1337–1341.
Ushiyama, K. and Hibino, K. (1997) Commercial production of ginseng by plant cell cultures. *American Chemical Society* 213 *National Meeting*, San Francisco, CA, Abstract Pt.1, AGFD057.
Ustundag, O.G. and Mazza, G. (2007) Saponins: Properties, applications and processing. *Critical Reviews in Food Science and Nutrition* 47, 231–258.
Varshney, A., Dhawan, V. and Srivastava, P.S. (2001) Ginseng: wonder drug of the world. In: Khan, I.A. and Khanum, A. (eds) *Role of Biotechnology in Medicinal and Aromatic Plants* .Vol. IV. Ukaaz Publications, Hyderabad, India, pp. 26–41.
Veersham, C. (2004) In: *Elicitation Medicinal Plant Biotechnology*, CBS Publishers, India, pp. 270–293.
Verpoorte, R., Contin, A. and Memelink, J. (2002) Biotechnology for the production of plant secondary metabolites. *Phytochemical Reviews* 1, 13–25.
Wagner, F. and Vogelmann, H. (1977) Cultivation of plant tissue cultures in bioreactors and formation of secondary metabolites. In : Barz, W., Reinhard, E.and Zenk, M.H. (eds) *Plant Tissue Culture and its Biotechnological Application*, Springer, New York, pp. 245–252.
Wang, W. and Zhong, J.J. (2002) Manipulation of ginsenoside heterogeneity in cell cultures of *Panax notoginseng* by addition of jasmonates. *Journal of Bioscience and Bioengineering* 93, 48–53.
Wang, W., Zhang, Z.Y. and Zhong, J.J. (2005a) Enhancement of ginsenoside biosynthesis in high-density cultivation of *Panax notoginseng* cells by various strategies of methyl jasmonate elicitation. *Applied Microbiology and Biotechnology* 67, 752–758.
Wang, W., Zhao, Z.J., Xu, Y., Qian, X. and Zhong, J.J. (2005b) Efficient elicitation of ginsenoside biosynthesis in cell cultures of *Panax notoginseng* by using self-chemically synthesized jasmonates. *Biotechnology and Bioprocess Engineering* 10, 162–165.
Wang, W., Zhao, Z.J., Xu, Y., Qian, X. and Zhong, J.J. (2006) Efficient induction of ginsenoside biosynthesis and alteration of ginsenoside heterogeneity in cell cultures of *Panax notoginseng* by using chemically synthesized 2-hyroxylethyl jasmonate. *Applied Microbiology and Biotechnology* 70, 298–307.
Wang, X., Proctor, J.T.A., Kakuda, Y., Krishnaraj, S. and Saxena, P.K. (1999) Ginsenosides in American ginseng: Comparison of *in vitro* derived and field-grown plant tissues. *Journal of Herb, Spices and Medicinal Plants* 6, 1–10.
Washida, D., Shimomura, K., Nakajima, Y., Takido, M. and Kitanaka, S. (1998) Ginsenosides in hairy roots of a *Panax* hybrid. *Phytochemistry* 49, 2331–2335.
Washida, D., Shimomura, K., Takido, M. and Kitanaka, S. (2004) Auxins affected ginsenoside production and growth of hairy roots in *Panax* hybrid. *Biological and Pharmaceutical Bulletin* 27, 657–660.
Watanabe, K., Yano, S.Y. and Yamada, Y. (1982) Selection of cultured plant cell lines producing high levels of biotin. *Phytochemicals* 21, 513–516.
WHO (1999) *Monographs on Selected Medicinal Plants,* Vol. 1. World Health Organization, Geneva

Woo, S.S., Song, J.S., Lee, J.Y., In, D.S., Chung, H.J., Liu, J.R. and Choi, D.W. (2004) Selection of high ginsenoside producing hairy root lines using targeted metabolic analysis. *Phytochemistry* 65, 2751–2761.

Wu, J. and Ho, K.P. (1999) Assessment of various carbon sources and nutrient feeding strategies for *Panax ginseng* cell culture. *Applied Biochemistry and Biotechnology* 82, 17–26.

Wu, J. and Lin, L. (2002) Elicitor like effects of low energy ultrasound on plant (*Panax ginseng*) cells: induction of plant defense responses and secondary metabolite production. *Applied Microbiology and Biotechnology* 59, 51–57.

Wu, J.Y., Wong, K., Ho, K.P. and Zhou, L.G. (2005) Enhancement of saponin production in *Panax ginseng* cell culture by osmotic stress and nutrient feeding. *Enzyme and Microbial Technology* 36, 133–138.

Xiaojie, X., Xiangyang, H., Neill, S.J., Jianying, F. and Weiming, C. (2005) Fungal elicitor induces singlet oxygen generation, ethylene release and saponin synthesis in cultured cells of *Panax ginseng* C.A. Meyer. *Plant and Cell Physiology* 46, 947–954.

Xu, H., Kim, Y.K., Jin, X., Lee, S.Y. and Park, S.U. (2008) Rosmarinic acid biosynthesis in callus and cell cultures of *Agastache rugosa* Kuntze. *Journal of Medicnal Plant Research* 2, 237–241.

Yadav, S., Mathur, A., Mishra, P., Singh, M., Gupta, M.M., Alam, M. and Mathur, A.K. (2007*) In vitro* growth and asiaticoside production in shoot cultures of *Centella asiatica* L.. In: Kukreja, A.K. *et al.* (eds) *Proceedings of National Symposium on Plant Biotechnology: New Frontiers*, pp. 373–378.

Yamamoto, O. and Kamura, K. (1997) Production of saikosaponin in cultured roots of *Bupleurum falcatum*. *Plant Tissue Culture and Biotechnology* 3, 138–147.

Yang, D.C. and Choi, Y.E. (2000) Production of transgenic plants of *Panax ginseng* from *Agrobacterium*-transformed hairy roots. *Plant Cell Reports* 19, 491–496.

Yao, H. and Zhong, J.J. (1999) Improved production of ginseng polysaccharide by adding conditioned medium to *Panax notoginseng* cell cultures. *Biotechnological Technology* 13, 347–349.

Yoshikawa, M., Sugimoto, S., Nakamura, S. and Matsuda, H. (2007) Medicinal flowers.XI. Structures of new dammarane-type triterpene diglycosides with hydroperoxide group from flower buds of *Panax ginseng*. *Chemical and Pharmaceutical Bulletin (Tokyo)* 55, 571–576.

Yoshikawa, T. and Furuya, T. (1987) Saponin production by cultures of *Panax ginseng* transformed with *Agrobacterium rhizogenes*. *Plant Cell Reports* 6, 449–453.

Yoshimatsu, K., Yamaguchi, H. and Shimomura, K. (1996) Traits of *Panax ginseng* hairy roots after cold storage and cryopreservation. *Plant Cell Reports* 15, 555–560.

Yu, C.J., Murthy, H.N., Jeong, C.S., Hahn, E.J. and Paek, K.Y. (2005) Organic germanium stimulates the growth of ginseng adventitious roots and ginsenoside production. *Process Biochemistry* 40, 2959–2961.

Yu, K.W., Gao, W.Y., Son, S.H. and Paek, K.Y. (2000) Improvement of ginsenoside production by jasmonic acid and some other elicitors in hairy roots culture of ginseng (*Panax ginseng* C.A. Meyer). *In vitro Cellular and Developmental Biology– Plant* 36, 424–428.

Yu, K.W., Gao, W., Han, E.J. and Paek, K.Y. (2002) Jasmonic acid improves ginsenoside accumulation in adventitious root culture of *Panax ginseng* C.A. Meyer. *Biochemical Engineering* 11, 211–215.

Yu, K.W., Murthy, H.N., Hahn, E.J. and Paek, K.Y. (2005) Ginsenoside production by hairy root cultures of *Panax ginseng*: influence of temperature and light quality. *Biochemical Engineering Journal* 23, 53–56.

Zhang, Y.H. and Zhong, J.J. (1997) Hyperproduction of ginseng saponin and polysaccharide by high density ultivation of *Panax ginseng* cells. *Enzyme and Microbial Technology* 21, 59–63.

Zhang, Y.H., Zhong, J.J. and Yu, J.T. (1996a) Effect of nitrogen source on cell growth and production of ginseng saponin and polysaccharide in suspension cultures of *Panax notoginseng*. *Biotechnology Progress* 12, 567–571.

Zhang, Y.H., Zhong, J.J. and Yu, J.T. (1996b) Enhancement of ginseng saponin production in suspension cultures of *Panax notoginseng*: Manipulation of medium sucrose. *Journal of Biotechnology* 51(1), 49–56.

Zhang, Z.Y. and Zhong, J.J. (2004) Scale-up of centrifugal impeller bioreactor for hyperproduction of ginseng saponin and polysaccharide by high-density cultivation of *Panax notoginseng* cells. *Biotechnology Progress* 20, 1076–1081.

Zhao, J., Davis, L.C. and Verpoorte, R. (2005) Elicitor signal transduction leading to production of plant secondary metabolites. *Biotechnology Advances* 23, 283–333.

Zhong, J.J. (2000) Production of ginseng saponins by cell suspension cultures of *Panax notoginseng* in bioreactors. In: Oleszek, W. and Marston, A. (eds) *Saponins in Food Feedstuffs and Medicinal Plants. Proc. Phytochemical Society of Europe*, Vol. 45. Kluwer Academic Publishers, Dordrecht, Boston, London, pp. 163–170.

Zhong, J.J. (2003) Biochemical engineering of the production of plant specific secondary metabolites of cell suspension cultures. *Advances in Biochemical Engineering Biotechnology* 72, 1–26.

Zhong, J.J. and Wang, D.J. (1996) Improvement of cell growth and production of ginseng saponin and polysaccharide in suspension cultures of *Panax notoginseng*: Cu^{2+} effect. *Journal of Biotechnology* 46, 69–72.

Zhong, J.J. and Wang, S.J. (1998) Effects of nitrogen source on the production of ginseng saponin and polysaccharide by cell cultures of *Panax quinquefolium*. *Process Biochemistry* 33, 671–675.

Zhong, J.J. and Yue, C.J. (2005) Plant cells: secondary metabolite heterogeneity and its manipulation. *Advances in Biochemical Engineering Biotechnology* 100, 53–88.

Zhong, J.J. and Zhu, Q.X. (1995) Effect of initial phosphate concentration on cell growth and ginseng saponin production by suspended cultures of *Panax notoginseng*. *Applied Biochemistry and Biotechnology* 55, 241–247.

Zhong, J.J., Meng, X.D., Zhang, Y.H. and Liu, S. (1997) Effective release of ginseng saponin from suspension cells of *Panax notoginseng*. *Biotech. Tech.* 11, 241–243.

Zhou, L., Cao, X., Zhang, R., Peng, Y., Zhao, S. and Wu, J. (2007) Stimulation of saponin production in *Panax ginseng* hairy roots by two oligosaccharides from *Paris polyphalla* var. yunnanensis. *Biotechnology Letters* 29, 631–634.

Zhou, Y., Hirotani, M., Rui, H. and Furuya, T. (1995) Two triglycosidic triterpene astragalosides from hairy root cultures of *Astragalus membranaceus*. *Phytochemistry* 38, 1407–1410.

Chapter 9

Podophyllotoxin and Related Lignans: Biotechnological Production by *In vitro* Plant Cell Cultures

Iliana Ionkova

Introduction

The aryltetralin lignan podophyllotoxin (PTOX) is a natural occurring lignan derived from the roots and rhizomes of the Himalayan *Podophyllum hexandrum* and the American *Podophyllum peltatum* L. (Podophyllaceae ~Berberidaceae) (Ionkova, 2007; Arora *et al.*, 2008). *Podophyllum* is a genus of six species of herbaceous perennial plants in the family Berberidaceae, native to eastern Asia (five species) and eastern North America (one species, *P. peltatum*). They are woodland plants, typically growing in colonies derived from a single root. This small group of perennials (commonly called May Apples) is originally from North America, the Himalayas and western China. They grow from 12 to 18 inches high and have large, deeply lobed leaves on long, fleshy stems, which rise straight up from the soil. The name *Podophyllum* is taken from podos, a foot, and phyllon, a leaf, and refers to the resemblance of the leaves to a duck's foot. A drug known as podophyllin is made from the rhizomes of these plants. *P. hexandrum* has pretty leaves that are divided into three lobes. They completely unfurl after the plant has bloomed and are dark green splotched with brown. In the spring, white or pale pink, six-petalled flowers are borne at the ends of stout stems; these are followed by fleshy, oval, red berries. The perennial herb *Podophyllum hexandrum* (syn. *P. emodi*), bearing the common names Himalayan mayapple or Indian Mayapple, is native to the lower elevations in and surrounding the Himalaya (Gupta and Sethi, 1983; Arora *et al.*, 2008). It is low to the ground with glossy green, drooping, lobed leaves on its few stiff branches, and it bears a pale pink flower and bright red-orange bulbous fruit. The ornamental appearance of the plant makes it a desirable addition to woodland-type gardens. It can be propagated by seed or by dividing the rhizome. It is very tolerant of cold temperatures, as would be expected of a Himalayan plant, but it is not tolerant of dry conditions.

Medicinal Use

Lignans have a long history of medicinal use as the first records date back over 1000 years (Kelly and Hartwell, 1954). The roots of wild Chervil (*Anthriscus sylvestris* L. Apiaceae), containing several lignans, including deoxypodophyllotoxin, were used in a salve for treating cancer (Cockayne, 1961). Another source from 400–600 years ago reveals the use of the resin, derived from an alcoholic extract of the roots and rhizomes of *Podophyllum* perennials as a catharctic and poison, both by the natives of the Himalayas and the American Penobscot Indians of Maine (Ayres and Loike, 1990). Throughout the years, lignan-containing plant products were used for a wide number of ailments in Chinese medicine – roots of *Kadsura coccinea* Hance. ex Benth. (Schizandraceae) for treatment of rheumatoid arthritis, gastric and duodenal ulcers (Tu, 1977), Japanese – *Fraxinus japonica* Blume ex K. Koch. (Oleaceae) (Kariyone and Kimurta, 1976; Kodaira, 1983) – diuretic, antipyretic, an analgesic and antirheumatic agent. The bark of *Olea europaea* L. (Oleaceae) has been studied (Tsukamoto *et al.*, 1984) for its antipyretic, antirheumatic, tonic and scrofula remedy actions.

The Podophyllum plant is poisonous but when processed has medicinal properties. The rhizome of the plant contains a resin, known generally and commercially as Indian Podophyllum Resin, which can be processed to extract podophyllotoxin, or podophyllin, a neurotoxin. It has been historically used as an intestinal purgative and emetic, salve for infected and necrotic wounds, and inhibitor of tumour growth. The North American variant of this Asian plant contains a lower concentration of the toxin but has been more extensively studied. All the parts of the plant, excepting the fruit, are poisonous. Even the fruit, though not dangerously poisonous, can cause unpleasant indigestion. *Podophyllum* gets its name from the Greek words podos and phyllon, meaning foot shaped leaves. *Podophyllum* rhizomes have a long medicinal history among native North American tribes who used a rhizome powder as a laxative or an agent that expels worms (anthelmintic). A poultice of the powder was also used to treat warts and tumourous growths on the skin.

Podophyllotoxin is a plant-derived compound used to produce two cytostatic drugs, etoposide and teniposide. The substance has been primarily obtained from the American mayapple (*Podophyllum peltatum*). The Himalayan mayapple (*Podophyllum hexandrum* or *P. emodi*) contains this constituent in a much greater quantity, but is endangered in the wild. The substance they contain (podophyllotoxin or podophyllin) is used as a purgative and as a cytostatic. Posalfilin is a drug containing podophyllin and salicylic acid that is used to treat the plantar wart. Several podophyllotoxin preparations are on the market for dermatological use to treat genital warts. Since the total synthesis of podophyllotoxin is an expensive process, availability of the compound from natural renewable resources is an important issue for pharmaceutical companies that manufacture these drugs (Moraes *et al.*, 2002). In recent years, *P. hexandrum* has been extensively investigated for its potent radioprotective properties (Arora *et al.*, 2005, 2007, 2010a,b; Chawla *et al.*, 2006; Singh *et al.*, 2009).

The lignan podophyllotoxin, occurring in *Podophyllum emodi* Wall. ex Royle and *Podophyllum peltatum* L., is the starting compound for the semi-synthesis of the anticancer drugs. It is currently being used as a lead compound for the semi-synthesis of anticancer drugs etoposide, teniposide and etopophos (Fig. 9.1), which are used for the treatment of lung and testicular cancers and certain leukaemias (Stahelin and Wartburg, 1991; Imbert,

1998). The drug etoposide (VePesid®) is the semisynthetic derivative of podophyllotoxin, and is approved by the US Food and Drug Administration (FDA) for various types of cancer. Currently, extracts of the podophyllum plant are used also in topical medications for genital warts, HIV-related oral hairy leukoplakia, and some skin cancers. Preliminary research also shows that CPH 82, an oral form of *Podophyllum emodi* composed of two purified semisynthetic lignan glycosides, may be useful in treating rheumatoid arthritis. However, when used orally, *Podophyllum* can be lethal and should be avoided.

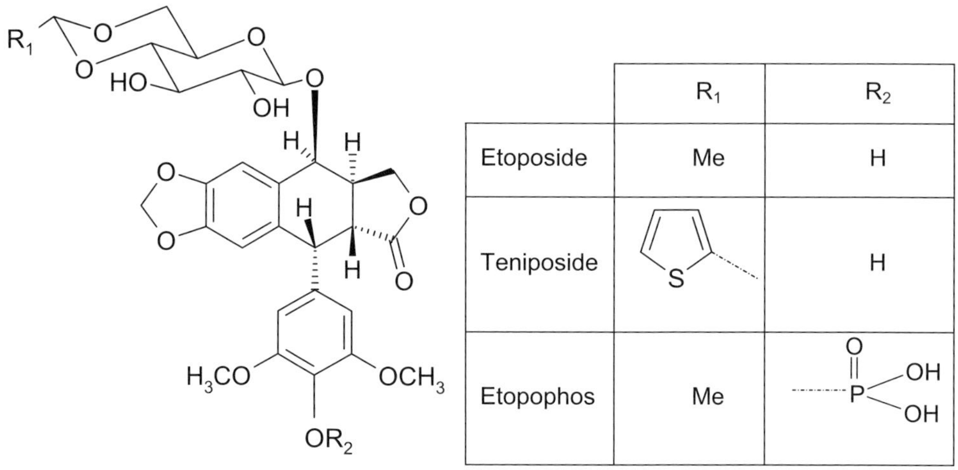

Fig. 9.1. Structures of etoposide, teniposide and etopophos.

Supply of podophyllotoxin

The supply of podophyllotoxin depends mainly on its extraction from roots and rhizomes of *Podophyllum hexandrum* Royle (from Himalayas region) and *Podophyllum peltatum* L. (North America), which contain 4% and 0.2% of the active substance on a dry mass basis, respectively. Those resources are, however limited, because of the intensive collection of the plants, lack of cultivation and the long juvenile phase and poor reproduction capacities of the plant (Van Uden, 1992). Podophyllotoxin and related compounds (Fig. 9.2) are not only present in Podophyllaceae, but also in e.g. Juniperaceae, Lamiaceae and Linaceae (Petersen and Alfermann, 2001). A detailed phytochemical analysis of the lignans in the Linaceae has been done in the groups of Ionkova (Sofia, Bulgaria), T.J. Schmidt (Münster, Germany) and A.W. Alfermann (Düsseldorf, Germany). Genera in which abundance of PTOX has been reported are *Linum* (Linaceae) (Ionkova, 2007; Broomhead and Dewick, 1990a; Konuklugil, 1996a; Vasilev *et al.*, 2005a,b; Konuklugil *et al*, 2007). *Juniperus* (Cupressaceae) (Kupchan, 1965; San Feliciano *et al.*, 1989a,b), *Hyptis* (Lamiaceae) (Kuhnt *et al.*, 1994), *Thymus* (Lamiaceae), *Teucrium* (Lamiaceae), *Nepeta* (Lamiaceae) (Konuklugil, 1996b), *Dysosma* (Berberidaceae) (Yu *et al.*, 1991), *Diphylleia* (Berberidaceae) (Broomhead and Dewick, 1990a), *Jeffersonia* (Berberidaceae) (Bedir *et al.*, 2002) and *Thuja* (Cupressaceae) (Muranaka *et al*, 1998).

General approaches to the chemical synthesis of podophyllotoxin derivatives (Canel *et al.*, 2000) and chemical synthesis of lignans (Ward, 2003) have been proposed, however, an efficient commercially viable route to the synthesis of podophyllotoxin is still to be sought.

Need for production of podophyllotoxin and related lignans by plant *in vitro* cultures

The supply of podophyllotoxin depends mainly on its extraction from roots and rhizomes of *Podophyllum hexandrum* Royle (from Himalayas region) and *Podophyllum peltatum* L. (North America), which contain 4% and 0.2% of the active substance on a dry mass basis, respectively. Those resources are, however limited, because of the intensive collection of the plants, lack of cultivation and the long juvenile phase and poor reproduction capacities of the plant (Van Uden, 1992).

In the coming decades, several new enabling technologies will be required to develop the next generation of advanced plant-based pharmaceuticals. With modern biotechnology, it has become possible to use plant cells for the production of specific pharmaceuticals. Using the right culture medium and appropriate phytohormones it is possible to establish *in vitro* cultures of almost every plant species. Starting from callus tissue, cell suspension cultures can be established that can even be grown in large bioreactors. Moreover, the biotechnological production of these plant products is more environmentally friendly way than is currently occurring.

Malignant diseases are the second leading cause of mortality within the human population. Despite the serious progress in establishing and introducing novel specifically-targeted drugs the therapy of these diseases remains a severe medical and social problem. Some of the most effective cancer treatments to date are natural products or compounds derived from plant products. Seven of the most consumed anticancer drugs are of plant origin: etoposide, teniposide, taxol, vinblastine, vincristine, topotecan, and irinotecan. They are some of the most vigorous products in cancer therapy and are still derived from plants since the chemical synthesis of the chiral molecules is not economic (Dechamp, 1999).

Market prices for the plant-derived anticancer drugs are quite high: 1 kg of vincristine (*Catharanthus* alkaloid) costs about US$20,000 and the annual world market is about US$5 million per year. Isolation of pharmaceuticals from plants is difficult due to their extremely low concentrations. The industry currently lacks sufficient methods for producing all of the desired plant-derived pharmaceutical molecules. Some substances can only be isolated from extremely rare plants. The biotechnological approach offers a quick and efficient method for producing these high-value medical compounds in cultivated cells. In the future, a new production method may also offer alternatives to other highly expensive drugs. Biotechnological production in plant cell cultures is an attractive alternative but has so far had only limited commercial success (for example, paclitaxel or Taxol), due to a lack of understanding of the complex multistep biosynthetic events leading to the desired end-product. Discoveries of cell cultures capable of producing specific medicinal compounds at a rate similar or superior to that of intact plants have accelerated in the last few years in Bulgaria and other countries in Europe.

Lignans are a large group of phenolic compounds defined as ββ′-dimers of phenylpropane (C6C3) units. This widely spread group of natural products possesses a long and remarkable history of medicinal use in the ancient cultures of many peoples. The first unifying definition of lignans was made by Howarth in 1936, who described them as a group of plant phenols with a structure determined by the union of two cinnamic acid residues or their biogenetic equivalents (Howarth, 1936). According to IUPAC nomenclature, lignans are 8,8′-coupled dimmers of coniferyl alcohol or other cinnamyl alcohols (Moss, 2000). Lignans occur in many plant species, but only in low concentrations. The biotechnological part focuses on alternative production systems for these natural compounds, because the plant *in vitro* cultivation has several advantages over collecting plants from fields (Alfermann *et al.*, 2003). Growing plant cells permit a stricter control of the quality of the products as well as their regular production without dependence on the variations of natural production resulting from climate and socio-political changes in their countries of origin. Problems connected with gathering, storing (in special conditions), processing and disposal of huge amounts of biomass, connected with extraction of active substances from *in vivo* plants are also solved. Suspension cultures are of special interest due to their high growth rate and short cycle of reproduction. Another advantage is the fact that undifferentiated plant cells, maintained in a liquid medium, possess a high metabolic activity due to which considerably high yields of secondary products can be achieved in short terms (from 1 to 3 weeks of cultivation). This raises the question of investigation of *in vitro* cultures of new plant species for the production of podophyllotoxin derivatives (Fuss, 2003). In Table 9.1, the accumulation of lignans in plant tissue and organ cultures has been summarized.

Although there are many examples for the synthesis of podophyllotoxin and its derivatives in plant cell and tissue cultures, the *in vitro* production still has to cope with multiple tasks for the purpose of finding economically feasible paths for enhancing production. The plant-specific secondary products as a podophyllotoxin and its derivatives were long considered as a major limitation for an extensive use of plant-made pharmaceuticals in human therapy.

Production advantages provided by plant* in vitro *cultures

The two principal advantages of plant-based production systems are:

1. Scalability: no other production system offers the potential scalability of plant products. High-value products could be produced in sufficient amounts in plant cell culture and will allow product manufacture on a massive scale that can match global demand.
2. Adaptability: In the post-genomic era, it has become feasible to engineer plant cell and tissue cultures, not only to produce complex proteins but also to produce high-value secondary metabolites or entirely novel structures (such as new lead compounds for pharmaceutical industry) (Stakeholders proposal, 2005).

Additional major advantages of a cell culture system over the conventional cultivation of whole plants are:

1. Useful compounds can be produced under controlled conditions independent of climatic changes or soil conditions;
2. Cultured cells would be free of microbes and insects;
3. The cells of any plants, tropical or alpine, could easily be multiplied to yield their specific metabolites;
4. Automated control of cell growth and rational regulation of metabolite processes would reduce labour costs and improve productivity;

5. Organic substances are extractable from callus cultures. Production by cell cultures could be justified, for rare products that are costly and difficult to obtain through other means.

Table 9.1. Podophyllotoxin and related lignans in plant *in vitro* cultures*.

Species	*In vitro* culture	Lignans synthesized
Callitris drummondii	Callus, Suspension	PTOX
Daphne odora	Callus, Suspension	Matairesinol, Lariciresinol, Pinoresinol, Secoisolariciresinol, Wikstromol
Forsythia x intermedia	Callus, Suspension	Epipinoresinol
Forsythia x intermedia	Callus, Suspension	Matairesinol
Forsythia x intermedia	Suspension	Pinoresinol, Matairesinol
Forsythia sp.	Callus, Suspension	Matairesinol, Epipinoresinol, Phillyrin, Arctigenin
Haplophyllum patavinum	Callus	Justicidin B, Diphyllin, Tuberculation, Arctigenin
Ipomea cairica	Callus	Trachelogenin, Arctigenin
Ipomea cairica	Callus	Pinoresinol
Jamesoniella autumnalis	Gametophyte	8′, 8,2′-Tricarboxy-5,4-dihydroxy-7′ (5′)-6′-pyranonyl-7′,8′-dihydronaphthalene and its two monomethylesters
Juniperus chinensis	Callus	PTOX
Larix leptolepis	Callus	Pinoresinol; 2,3-dihydro-2-(4-hydroxy-3-metoxyphenyl)-3-hydroxymethyl-5-(ω-hydroxypropyl)-7-metoxybenzofuran, Lariciresinol, Secoisolariciresinol, Iso-lariciresinol
Linum album	Suspension	PTOX, 6MPTOX, DPTOX, Pinoresinol, Matairesinol, Lariciresinol, β-peltatin, α-peltatin
Linum altaicum	Cell cultures	Justicidin B, Isojusticidin B
Linum austriacum	Callus, Suspension, Root, Hairy root	Justicidin B, Isojusticidin B
Linum austriacum ssp. *euxinum*	Cell cultures	Justicidin B, Iisojusticidin B
Linum africanum	Callus, Suspension	PTOX, DPTOX
Linum campanulatum	Callus, Suspension	Justicidin B
Linum cariense	Suspension	6MPTOX 5′-demethoxy-6-methoxypodophyllotoxin, and the corresponding 8′-epimers 6-methoxypicropodophyllin, 5′-demethoxy-6-methoxypicropodophyllin
Linum flavum	Root	6MPTOX
Linum flavum	Suspension, Embryogenic Suspension	6MPTOX

Linum flavum	Suspension, Root like tissue	6MPTOX, 5′-demethoxy-6-methoxy-PTOX
Linum flavum	Hairy roots	6MPTOX
Linum flavum	Hairy roots	Coniferin
Linum elegans	Callus, Suspension	6MPTOX, β-peltatin
Linum leonii	Callus	Justicidin B
Linum leonii	Hairy roots	Justicidin B
Linum lewisii	Cell cultures	Justicidin B, Isojusticidin B
Linum linearifolium	Callus, Suspension, Shoots, Hairy roots	PTOX, 6MPTOX, Justicidin B
Linum mucronatum ssp. *armenum*	Shoot, Suspension	
Linum narbonese	Callus	Justicidin B
Linum nodiflorum	Suspension	6MPTOX
Linum nodiflorum	Suspension	6MPTOX, PTOX, DPTOX
Linum persicum	Callus, Cell cultures	PTOX, 6MPTOX, α- and β-peltatin
Linum tauricum	Callus, Suspension, Shoots, Hairy roots	6MPTOX 4′-demethyl-6MPTOX
Linum serbicum	Callus, Suspension	6MPTOX, β-peltatin
Picea glechni	Suspension	Pinoresinol, Dihydrodehydrodiconiferil alcohol
Podophyllum hexandrum	Callus, Suspension	PTOX
Podophyllum hexandrum	Embryogenic callus	PTOX
Podophyllum peltatum	Callus	PTOX
Podophyllum peltatum	Embryogenic suspension	PTOX, DPTOX, 4′- DPTOX
Podophyllum sp.	Callus	PTOX
Sesamum indicum	Callus, Suspension, Hairy roots	Sesamin, Sesamolin

* (The data are cited in Ionkova, 2007; Konuklugil *et al.*, 2007; Ionkova *et al.*, 2006, 2007; Vasilev *et al.*, 2008; Vasilev and Ionkova, 2005).

Development of *in vitro* cultures

The lignans are a large group of natural substances, which occur in a range of plant species, but only in low concentrations. Their natural abundance, however, is scarce and their chemical synthesis is not yet economically feasible. Podophyllotoxin (PTOX) is a lignan, which is used as adduct for the semisynthesis of etoposide and tenoposide. Both antineoplastic drugs are of importance for therapy and today and in the future will be a high demand for the commercial drug and its precursors for the treatment of leukaemia. For the production of podophyllotoxin different routes are known: extraction and isolation from *Podophyllum hexandrum*, *Linum* and other species, production in plant cell cultures (Van Uden *et al.*, 1990c) and organic synthesis (Wink *et al.*, 2005). Extraction and isolation of podophyllotoxin from Indian *Podophyllum hexandrum* species is not economical and is characterized by a high price for the final product.

Another problem is cultivation of the plant, which is why mostly the natural product is extracted from wild collected species. The main reason for this is the limited growth rate and culture conditions for accelerated growth. As a consequence, *P. hexandrum* is actually endangered in the wild in India, especially in the Himalayan region. PTOX and related compounds are present in the plant families Juniperaceae, Berberidaceae, Lamiaceae and Linaceae (Ionkova, 2007).

***Podophyllum* species**

Up to this moment aryltetralin lignans are actually derived exclusively from plant sources. Due to their restricted natural abundance and important pharmacological application, identification of new sources or rational *in vitro* synthesis is very important for the production of therapeutic candidates for cancer chemotherapy.

Podophyllotoxin was produced by cell culture of *Podophyllum hexandrum* under *in vitro* culture conditions. A maximum of 4.26 mg/l of podophyllotoxin was produced when *P. hexandrum* was cultivated in a 3 l stirred tank bioreactor. The compound extracted from the cell culture was applied to the human breast cancer cell line (MCF-7) and 1 nM podophyllotoxin was able to inhibit the growth of the cancer cells by 50%. The most profound inhibitory effect of podophyllotoxin was observed when it was applied in the beginning of cell growth (Chattopadhyay *et al.*, 2004).

Submerged cultivation of *Podophyllum hexandrum* for the production of podophyllotoxin was carried out in a 3 l stirred tank bioreactor fitted with a low-shear Setric impeller. The specific requirements of the medium, such as carbon source (sugar) and light, were established for the growth of and podophyllotoxin production by *P. hexandrum* in suspension cultures. Substitution of sucrose by glucose resulted in higher growth and podophyllotoxin production. The biosynthesis of podophyllotoxin was favoured when plant cells were cultivated in the dark. An agitation speed of 100 rpm was sufficient to mix the culture broth in the bioreactor without causing any significant cell damage. Biomass and podophyllotoxin accumulation in the 3 l bioreactor under batch growth conditions were 6.5 g/l and 4.26 mg/l, respectively, in 22 days. This resulted in an overall podophyllotoxin productivity of 0.19 mg/l/d), which represented an increase of 27% in comparison to its productivity in a shake flask. Podophyllotoxin production was found to be a combined growth-associated and non-growth associated process (Chattopadhyay *et al.*, 2002).

***Linum* species**

Antitumour activity will undoubtedly continue to be the most clinically relevant property of lignans. The production of anticancer compounds, such as lignan podophyllotoxin, by plant *in vitro* cultures from different plant species has been demonstrated. Cell cultures of different *Linum* species of section Syllinum are shown to produce considerable amounts of lignans, mainly MPTOX. Although both PTOX and MPTOX have comparable cytotoxic activity, due to the different substitution in position 6 MPTOX is not used for the production of anticancer drugs (Kuhlmann *et al.*, 2002). Results, summarized in Table 9.2, show that a broad range of experiments have been carried out, resulting in enhancement of lignan production of *in vitro* cultures from *Linum* species. Since PTOX is the preferred precursor for the semi-synthesis of anticancer drugs like Etoposide and Etopophos®, the accumulation of predominantly PTOX is especially interesting. *L. linearifolium* is now beside *L. album* and *L. persicum* the third *Linum* species of section Syllinum with PTOX (ca 0.8% DW) as the main lignan (Kuhlmann *et al.*, 2002).

As a new biotechnological alternative is our success in stereoselective hydroxylation of deoxypodophyllotoxin by recombinant CYP3A4 in *Escherichia coli* to epipodophyllotoxin (Vasilev *et al.*, 2006a), the direct precursor for the semi-synthesis of anti-cancer drugs etoposide (VP-16) and teniposide (VM-26).

A large number of aryltetralin lignans were identified from *in vivo* and *in vitro* species of the *Linum* genus in our research group (*L. tauricum* Willd. ssp. *bulgaricum* (Podp.) Petrova; *L. tauricum* Willd. ssp. *tauricum*; *L. tauricum* Willd. ssp. *serbicum* (Podp.) Petrova; *L. tauricum* Willd. ssp. *linearifolium* (Lindem.) Petrova; *L. elegans* Sprun. ex Boiss.; *L. flavum* L. ssp. *sparsiflorum* (Stoj.) Petrova; *L.capitatum* Kit. ex Schult. var. *laxiflorum* (Stoj.) Petrova; *L. cariense* Boiss; *L. altaicum* Ledeb.; *L. austriacum* var *euxinum* Juz.; *L. lewissi* Pursh.; *L. campanulatum* L.; *L. setaceum* Brot.; *L. africanum, L. strictum* L.; *L. leonii* F. W. Schulz.; *L. narbonense* L.). Lignans in different samples of *Linum* species, mainly occurring in Bulgaria, were analysed by HPLC-ESI/MS and HPLC-UV/DAD. The ESI/MS fragmentation pathways recently established for aryltetralin lignans are now extended to ester and glycoside derivatives. In total, ca. 40 different lignans, mainly of the aryltetralin type, were identified (Vasilev *et al.*, 2008). 6-Methoxypodophyllotoxin and its glucoside were present as major constituents in all samples. Differences between the investigated taxa were observed especially with respect to the accumulation of 6-deoxy-7-hydroxy-aryltetralins such as podophyllotoxin and of 6-hydroxy-7-deoxy-aryltetralin lignans of the peltatin type. Studies on extracts of *in vitro* cultures of *L. narbonese* and *L. leonii* have shown cytotoxic activity, due to the presence of arylnaphthalene lignan Justicidin B (Vasilev *et al.*, 2006b).

Increasing the yields of plant cell cultures for the productions of podophyllotoxin and related lignans

Different methods and approaches have been applied to *in vitro* cultures producing PTOX, its derivatives and other lignans (Fig. 9.2 (Ionkova, 2007)).

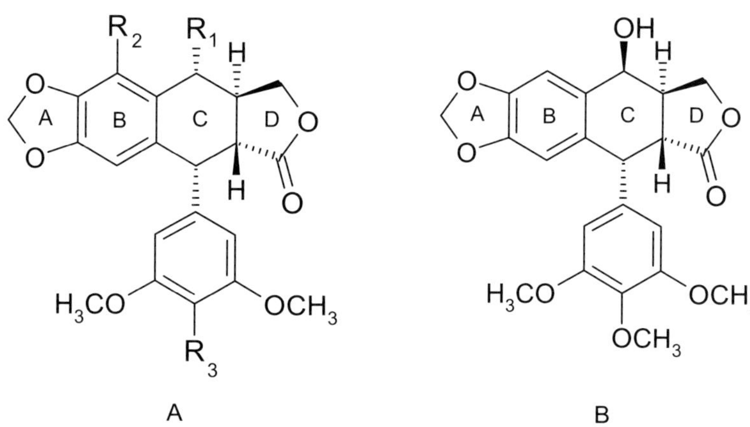

Fig. 9.2. Podophyllotoxin (R_1=OH, R_2=H, R_3=OCH_3), Deoxypodophyllotoxin (R_1=H, R_2=H, R_3=OCH_3), β-peltatin (R_1=H, R_2=OH, R_3=OCH_3) – A, Epipodophyllotoxin– B.

Optimization of conditions of culturing, of plant in vitro *cultures, for enhancement of lignan production.*

Multiple experiments from different working groups have been done for the enhancement of lignan production from *in vitro* cultures with established synthesis (Table 9.2).

Table 9.2. Optimization of lignan *in vitro* production.

Species, (*in vitro* culture)	Lignans synthesized	Varying factors with impact on the culture	Results of optimization
Callitris drummondii (Suspension)	PTOX-beta-D-glucoside	Illumination	0.02% in the dark 0.11% in light
Forsythia x intermedia (Suspension)	Pinoresinol Matairesinol	Carbon source	2% sucrose: 0.001% lignans 6% sucrose: 0.07% pinoresinol; 0.1% matairesinol
Ipomoea cairica (Callus)	Arctigenin, trace-Logenin	Phytohormones, Carbon source	4 mg/l 2,4-D, 3% maltose, pH 6.4 - 0.03 % lignans
Juniperus chinensis (Immobilized cells)	PTOX	Various calcium alginate concentrations	1.8% Ca-alginate gel: 5-fold increased levels, compared to free cell suspension; maximum excretion of PTOX in the medium 3% Ca-alginate gel: (0.21–0.025 mg/g dw) 4–5-fold increase 6% Ca-alginate gel: small amount
Linum africanum (Suspension)	PTOX DPTOX	Source of explant material, light, IAA, NAA, 2,4-D, Cytokinin/Kinetin ratios	Highest synthesis: Kinetin 10ml/l, IAA: 0.4ml/l, 2,4-D: 0.2ml/l; PTOX: increases in dark conditions (no difference between callus and suspension) DPTOX: increases when cultivated in the dark and has higher levels in callus than suspension
Linum album (Suspension)	PTOX and related lignans	Illumination	0.2% - dark 0.5% - light
Linum album (Suspension, shake-flasks: 1000ml – 50ml medium, 300ml – 50	PTOX	Oxygen supply Increasing of shaker speed	Enhancement of PTOX accumulation

ml medium, establishment of bioreactor)			
L. altaicum (Callus, Suspension)	Justicidin B Isojusticidin B	Dark	Justicidin B between 0.92–0.96%.
L. austriacum ssp. *euxinum* (Callus, suspension)	Justicidin B	Dark	Justicidin B between 0.50–0.96%
Linum campanulatum (Callus, Suspension)	Justicidin B	Illumination, Kinetin, IAA, 2,4-D, Succrose	Higher lignan content: dark conditions 1.41 % in the dark 0.40 % in light
L. leonii Hairy roots		Agrobacterium strains	
L. lewisii (Callus, Suspension)	Justicidin B	Dark	Justicidin B between 0.16–0.30%
Linum nodiflorum (Suspension)	6MPTOX	Illumination	Light: 0.6 % Dark: trace amounts
Linum tauricum ssp. *tauricum* (Suspension, Shoots, Callus, Hairy roots)	4′-Demethyl-6-MPTOX 6MPTOX	Phytohormones Methyl jasmonate *Agrobacterium* strains	Suspension: 4 mg/l NAA, maximal production of 4′-demethyl-6-methoxy-podophyllotoxin (2.16 mg/g dw) Callus: 0.1mg/l 2,4-D, 0.2mg/l IAA, 2.0mg/l Kinetin: maximal production of 6-methoxy-podophyllotoxin (3.99 mg/g dw)
Podophyllum hexandrum (Suspension)	PTOX	Illumination	Light: 0.03% Dark: 0.09%
Podophyllum hexandrum (Callus)	PTOX	Phytohormones	2,4-D + Kinetin: 0.077%
Podophyllum hexandrum (Suspension, 3l-stirred-tank bioreactor)	PTOX	Mathematical model for developing nutritient-feeding strategies, low shear conditions in fed-batch modes of operation, prolonging the productive log-phase of cultivation.	Improvement to 48.8 mg/l PTOX, corresponding volumetric productivity of 0.80 mg/l per day
Podophyllum hexandrum (Suspension)	PTOX	pH, Phytohormones, Carbon source, Inoculum	pH 6.0 IAA: 1.25 mg/l Glucose 72 g/l Inoculum level: 8 g/l

Podophyllum hexandrum (Suspension)	PTOX	Sugar, Nitrogen source, Phosphate	MS medium, NH_4^+Salts: Nitrate: 1:2, 60 g/l Glucose: highest growth and PTOX accumulation
Podophyllum peltatum (Callus)	PTOX	Red light (660 nm) Carbon source Phytohormones	Enhancement of production in red light Sucrose: 0.057% Maltose: 0.023% 2,4-D + Kinetin: 0.57%
Rollinia mucosa (Jacq.) Bail. (Callus)	Furofuranic lignans: Epigambin, Magnolin, Epiyangambin	Origin of plant material, Explant type, Growth regulators (2,4-D, NAA, BA, PIC)	Foliar blade explants: biomass, synthesis enhancement; PIC: best biomass production; NAA, 2,4-D: Epiyangambin, Magnolin (dependent on explant source); PIC: Epiyangambin: Calli from foliar blade

* (The Data are cited in Vasilev and Ionkova, 2005; Ionkova *et al.*, 2006, 2007; Konuklugil *et al.*, 2007; Vasilev *et al.*, 2008).

Selection of high producing strains

A cell line of *Linum album*, accumulating mainly PTOX at a level of 0.2% on dry mass basis was reported by Empt *et al.* (2000). With this cell line, about 28 mg PTOX can be produced in 1l of culture medium in 11 days (Petersen and Alfermann, 2001).

Development of differentiated and transformed cultures

A 14-day old hairy root culture of *Linum album* Kotschy. (Linaceae) was reported to produce about 2.5% of 6MPTOX on dry mass basis (Windhövel *et al.*, 2003). A hairy root culture of *L. austriacum* L. was reported by Mohagheghzadeh *et al.* (2002), which synthesized the arylnaphthalene lignans justicidin B and isojusticidin B. Hairy root cultures of *Linum leonii*, obtained by genetic transformation using the agropine-type strain *Agrobacterium rhizogenes* 15834, accumulate justicidin B (Vasilev *et al.*, 2006b). The products encoded by *rol A* and *rol C* genes were found to have a synergistic effect on root induction and to induce increased sensitivity to auxin. The transformation of these genes from *A. rhizogenes* into the hairy root was checked by PCR (polymerase chain reaction). Proof of transformation was given by the PCR products showing that *rol A* and *rol C* genes were present in the hairy roots of *L. leonii*. Genetically modified hairy roots produced 5-fold higher yields of justicidin B (10.8 mg/g DW) compared to untreated callus (Vasilev and Ionkova, 2005) and show that differentiated roots produce higher amounts of secondary compound. This suggests that this technique may be used to enhance the accumulation of justicidin B. In addition to the production, we investigated the cytotoxic effect of justicidin B in three chronic human myeloid leukaemia-derived cell lines (LAMA-84, K-562 and SKW-3), that show a lower responsiveness to cytotoxic drugs due to a strong expression of the fusion oncoprotein BCR-ABL (a non-receptor tyrosine kinase). IC_{50} values of justicidin B were 1.1, 6.1 and 1.5 μM for the chronic myeloid leukaemia

(LAMA-84), pre-B-cell lymphoma (K-562) and chronic lymphoid leukaemia (SKW) cell lines respectively. These IC_{50} were comparable to the anticancer drug etoposide (a semi-synthetic lignan derivative).

Influence of stress factors

The influence of methyl jasmonate has been reported on pinoresinol and matairesinol production in *Forsythia* × *intermedia* Zab. (Oleaceae) suspension culture (Dewick, 1994). In a medium, containing 2% sucrose, the amount of pinoresinol was increased eightfold (0.058±0.015mg/g dry mass), matairesinol about fivefold (0.029±0.026 mg/g dry mass). In a medium containing 6% sucrose, pinoresinol was enhanced to 0.086±0.19 mg/g dry mass and matairesinol to 2.24±1.00mg/g dry mass. The accumulation of PTOX and 6MPTOX was enhanced about twofold, expressed on dry mass basis in a given cell line of suspension culture of *Linum album* after the addition of methyl jasmonate, by 7.69±1.45 mg/g dry mass and 1.11±0.09mg/g dry mass respectively (Van Furden *et al.*, 2005). The PTOX production of *Juniperus chinensis* L. (Cupressaceae) callus culture was stimulated to 6.4-fold by addition of COS (chito-oligosaccharides), a biotic elicitor with most active component chitopentaose, detected by HPLC (Premjet *et al.*, 2002). The concentration 100 μM of the elicitor methyl jasmonate in the nutrition medium of suspension *of Linum tauricum ssp. tauricum*, leads to substantial increase of the levels of lignans 4'DM-6MPTOX and 6MPTOX from traces, reaching a maximum of 0.1180 mg/g dw and 0.1250 mg/g dw respectively. The results (Ionkova *et al.*, 2006) indicate that addition of extracellular methyl jasmonate can not only increase the biosynthesis of both 4'DM-6MPTOX and 6MPTOX in a *L. tauricum* ssp. *tauricum* suspension culture, but also change the ratio of both compounds, in comparison with the intact plant and callus cultures, towards the more pharmacologically valuable 4'-DM-6MPTOX (Vasilev *et al.*, 2005a).

Feeding of precursors and/or biotransformation

Coniferin feeding to *in vitro* culture to produce PTOX derivatives has been reported in literature (Van Uden *et al.*, 1990a; Woerdenbag *et al.*, 1990). Levels of PTOX increased 13- to 56-fold after application of coniferin. The initiation of *Linum flavum* L. hairy roots was reported by Han-wei Lin *et al.* (2003a) as a source of coniferin. Significant variation was reported for its accumulation between hairy root lines, originating from different *L. flavum* seedlings and/or *A. rhizogenes* strains. After culturing the roots in Linsmaier and Skoog medium (LS) with 2,4-D (2,4-dichlorophenoxyacetic acid) and NAA (naphthalenacetic acid) as growth regulators coniferin reached 58 mg/g dry mass. Another experiment is the cross species co-culture of *L. album* hairy roots, as a source of coniferin, and *Podophyllum hexandrum* cell suspensions (Lin *et al.*, 2003b). Increasing PTOX concentrations in coculture was observed.

Feeding of coniferin led to fast increase of pinoresinol content, and no influence on matairesinol synthesis in *Forsythia* × *intermedia* suspension culture (Schmidt and Petersen, 2002). Feeding of phenylalanine and coniferyl alcohol in *Juniperus chinensis* callus culture, led to 3.6- and 2.2- fold increase of PTOX, respectively (Premjet *et al.*, 2002).

Certain enzymes of podophyllotoxin biosynthetic pathway have been isolated and characterized and podophyllotoxin and 6MPTOX have been established to be stored in the vacuole (Schmidt and Petersen, 2002). Elucidation in details of the biosynthetic pathway of

PTOX, isolating all enzymes and genes responsible for their encoding, would lead to new strategies for enhancement of the yields of this valuable substance by means of a combination of biotechnological and biochemical methods.

A novel system for the production of 2,7'-cyclolignans was recently demonstrated. Deoxypodophyllotoxin is stereoselectively converted into epipodophyllotoxin by recombinant human cytochrome P450 3A4 (CYP3A4). CYP3A4 is the main human metabolizing enzyme. As a new biotechnological alternative, recently we described the successful stereoselective hydroxylation of deoxypodophyllotoxin by recombinant CYP3A4 in *Escherichia coli*. Epipodophyllotoxin (Fig. 9.3) has been detected as the only metabolite in yields up to 90% (Vasilev *et al.*, 2006a). Therefore, the heterologous expression of CYP3A4 in *E. coli* presents an interesting alternative for the large-scale production of epipodophyllotoxin.

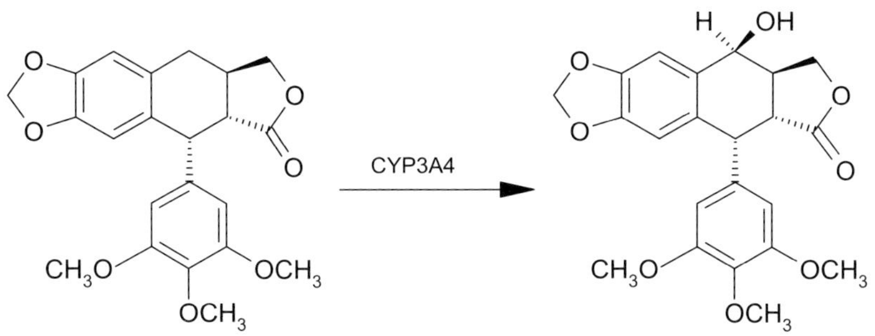

Fig. 9.3. Stereoselective hydroxylation of deoxypodophyllotoxin by recombinant CYP3A4 in *Escherichia coli* to epipodophyllotoxin.

Conclusion

Results, summarized in Table 9.1 and Table 9.2, show that a broad range of experiments have been carried out, resulting in enhancement of lignan production in *in vitro* cultures. Cell cultures of different plant species are shown to produce considerable amounts of lignans. Although both PTOX and MPTOX have comparable cytotoxic activity, due to the different substitution in position 6 MPTOX is not used for the production of anticancer drugs (Kuhlmann *et al.*, 2002). Since PTOX is the preferred precursor for the semi-synthesis of anticancer drugs like etoposide and etopophos®, the accumulation of predominantly PTOX is especially interesting. *L. tauricum* ssp. *linearifolium* is now beside *L. album* and *L. persicum* the third *Linum* species of section Syllinum with PTOX as the main lignan (Ionkova *et al.*, 2007).

Development of differentiated cultures as a general rule results in higher production of active substances. This approach however is not economically feasible for scale-up of production, as it encounters problems of *in vitro* cultivation and processing of great biomass and longer growth periods than undifferentiated cultures. For this reason, differentiated cultures are usually not used for *in vitro* production of lignans. An efficient alternative of differentiated cultures are the genetically transformed, hairy roots cultures. Their advantage is the shorter cycle (10–14 day), combined with the state of differentiation which makes possible the stable production of active substances. Stress factors as biotic (chito-oligosaccharides (COS) in *Juniperus chinensis*) and abiotic elicitors (methyl jasmonate in *Forsythia* × *intermedia, Linum tauricum* ssp. *tauricum*) have been demonstrated to enhance production of lignans. The feeding of a cheaper precursor as phenylalanine, coniferin and coniferil alcohol in plant cell and tissue cultures of lignin-producing plants, results in higher levels of production, as plant cells represent a 'ready' and organized system for bioprocessing and synthesis of target compounds.

A new biotechnological alternative is the successful stereoselective hydroxylation of deoxypodophyllotoxin by recombinant CYP3A4 in *Escherichia coli* to epipodophyllotoxin (Vasilev *et al.*, 2006a), the direct precursor for the semi-synthesis of anticancer drugs etoposide (VP-16) and teniposide (VM-26).

Due to the pharmaceutical importance and the low content in the plants the present review focuses on alternative production systems for podophyllotoxin and related lignans. Accumulation of lignans in plant tissue and organ cultures has been discussed based on the work of different authors. A survey of literature data has shown positive results in experiments with optimization of conditions of culturing, selection of high producing cell lines, influence of stress factors and feeding of precursors in plant *in vitro* cultures for the enhancement of production of lignans.

Recently, the interest of international pharmaceutical industries has been directed more and more to plant based anticancer compounds. In this case, it has been shown that production of podophyllotoxin and related lignans in cell cultures is possible. We believe that cell cultures of different plants as a source of biologically active lignans can play a role in this respect.

Abbreviations

PTOX – podophyllotoxin; 6MPTOX – 6-methoxypodophyllotoxin; DPTOX – deoxypodophyllotoxin; 4′DM-6MPTOX – 4′-demethyl-6-MPTOX; HR – hairy root cultures.

Acknowledgements

Financial support from Ministry of Education and Science, Sofia, Bulgaria (grant D002-128/2008 Iliana Ionkova) is acknowledged.

References

Alfermann, A.W., Petersen, M. and Fuss, E. (2003) Production of natural products by plant cell biotechnology: results, problems and perspective. In: Laimer, M. and Rücker, W. (eds) *Plant Tissue Culture* (100 Years Since Gottlieb Haberlandt). Springer, New York, pp. 153–166.

Arora, R., Gupta, D., Chawla, R., Prasad, J., Singh, S., Sharma, A.K., Kumar, R., Sagar, R.K., Samanta, N. and Sharma, R.K. (2005) Radioprotection by Plant Products: Present Status and Future Prospects. *Phytotherapy Research* 19, 1–22.

Arora, R., Sagar, R.K., Singh, S., Kumar, R., Singh, S., Prasad, J., Puri, S.C., Qazi, G.N., Krishan, B., Sharma, R.K. and Tripathi, R.P. (2007) Hemoprotective effect of *P. hexandrum* against γ-radiation: Role of heme oxygenase-1 and BCl-2. *Integrative Cancer Therapies* 6(1), 54–65.

Arora, R., Chawla, R., Dhaker, A.S., Adhikari, M., Sharma, J., Singh, S., Gupta, D., Kumar, R., Sharma, A.K., Puri, S.C., Sharma, R.K. Tripathi, R.P. (2010a) *Podophyllum hexandrum* as a potential botanical supplement for the medical management of nuclear and radiological emergencies. *J Dietary Supplements* 7(1), 31–50.

Arora, R., Gupta, D., Chawla, R., Adhikari, M., Sharma, J., Dhaker, A.S., Goyal, V., Kumar, R., Sharma, A., Sharma, R.K., Tripathi, R.P. and Puri, S.C. (2010b) Antioxidant, Anticancer, Cytoprotective and Radioprotective Properties of Indian *P. hexandrum*. In: Arora, R. (ed.) *Herbal Medicine: A Cancer Chemopreventive and Therapeutic Perspective*, Jaypee Brothers Medical Publishers (P) Ltd., Delhi, India, pp. 616-630.

Arora, R., Singh, S., Puri, S.C. and Sharma, R.K. (2008) Himalayan Mayapple (*Podophyllum hexandrum* Royle): Traditional Uses, Clinical Indications and Future Prospects. In: Watson, R. and Preedy, V.R. (eds) *Botanical Medicine in Clinical Practice*. CABI Publishing, Wallingford, Oxon, United Kingdom, pp. 71–84.

Ayres, D.C. and Loike, J. (1990) *Lignans, Chemical, Biological and Clinical Properties*. Cambridge University Press, Oxford, pp. 85–90.

Bedir, E., Khan, I. and Moraes, R.M. (2002) Bioprospecting for podophyllotoxin. In: Janick J. and Whipkey, A. (eds) *Trends in New Crops and New Uses*. ASHS Press, Alexandria, VA, pp. 545–549.

Broomhead, A.J. and Dewick, P.M. (1990a) Aryltetralin lignans from *Linum flavum* and *Linum capitatum*. *Phytochemistry* 29, 3839–3844.

Broomhead, A.J. and Dewick, P.M. (1990b) Tumour-inhibitory aryltetralin lignans in *Podophyllum versipelle*, *Diphylleia cymosa* and I. *Phytochemistry* 29, 3831–3837.

Canel, C., Moraes, R.M., Dayan, F.E. and Ferreira, D. (2000) Molecules of interest 'Podophyllotoxin'. *Phytochemistry* 54, 115–120.

Chattopadhyay, S., Bisaria, V.S., Panda, A.K. and Srivastava, A.K. (2004) Cytotoxicity of *in vitro* produced podophyllotoxin from *Podophyllum hexandrum* on human cancer cell line. *Natural Product Research* 18, 51–57.

Chattopadhyay, S., Srivastava, A.K., Bhojwani, S.S. and Bisaria, V.S. (2002) Production of Podophyllotoxin by plant cell cultures of *Podophyllum hexandrum* in bioreactor. *Journal of Bioscience and Bioengineering* 93(2), 215–220.

Chawla, R., Arora, R., Singh, S., Sagar, R.K., Sharma, R.K., Kumar, R., Sharma, A., Tripathi, R.P., Puri, S.C., Khan, H.A., Shawl, A.S., Sultan, P.S., Krishen, T. and Qazi, G.N. (2006) *Podophyllum hexandrum* offers radioprotection by modulating free radical flux: Role of aryl-tetralin lignans. *Evidence Based Complementary and Alternative Medicine* 3(4), 503–511.

Cockayne, T.O. (1961) Leech book of Bald, In: Cockayne, T.O. (ed) *Leechdom, Wortcunnings and Starcraft of Early England*. The Holland Press, London, pp. 313.

Dechamp, J., (1999) Herbal medicinal products and patents need in Europe. *Drug Information Journal* 33, 309–313.

Dewick, P.M. (1994) Forsythia species: *in vitro* culture and the production of lignans and other secondary metabolites. In: Bajaj, Y.P.S. (ed.) *Biotechnology in Agriculture and Forestry*. Springer Verlag, Berlin/Heidelberg, pp. 236–256.

Empt, U., Alfermann, A.W., Pras, N. and Petersen, M., (2000) The use of plant cell cultures for the production of podophyllotoxin and related lignans. *Journal of Applied Botany* 74, 145–150.

Fuss, E. (2003) Lignans in plant cell and organ cultures. *Phytochemistry Review* 2, 307–320.

Gupta, R. and Sethi, K.L. (1983) Conservation of medicinal plant resources in the Himalayan region. In: Jain, S.K. and Mehra, K.L. (eds) *Conservation of Tropical Plant Resources*. Botanical Survey of India, Howrah, India, pp. 101–107.

Howarth, R.D. (1936) Natural resins. *Annual Report Progress Chemistry* 33, 266–279.

Imbert, F. (1998) Discovery of podophyllotoxins. *Biochimie* 80, 207–222.

Ionkova, I. (2007) Biotechnological approaches for the production of lignans. *Pharmacognosy Reviews* 1(1), 57–68.

Ionkova, I., Antonova, I., Momekov, G. and Fuss, E. (2007) Cytotoxic activity of extracts from Bulgarian *Linum* species. *Pharmacognosy Magazine* 3(1), 192–195.

Ionkova, I., Goermova, B., Vasilev, N. and Fuss, E. (2006) Lignan production in cell cultures of *Linum tauricum* ssp. Tauricum. *Proceedings of IV Balkan Botanical Congress,* 661–664.

Kariyone, T. and Kimurta, Y. (eds) (1976) *Latest Sino-Japanese Herbal Plants.* Hirokawas Shoten, Tokyo, pp. 105.

Kelly, M.G. and Hartwell, J.L. (1954) The biological effects and the chemical composition of podophyllin: a review. *Journal of the National Cancer Institute* 14, 967–1010.

Kodaira, H. (1983) Isolation and identification of anti-platelet aggregation principles from the bark of *Fraxinus japonica* BLUME. *Chemical and Pharmaceutical Bulletin (Tokyo)* 31, 2262–2268.

Konuklugil, B. (1996a) Aryltetralyn lignans from genus *Linum. Fitoterapia* 67, 379–381.

Konuklugil, B. (1996b) Investigation of podophyllotoxin in some plants in Lamiaceae using HPLC. *Journal Faculty of Pharmacy Ankara* 25, 23–27.

Konuklugil, B., Ionkova, I., Vasilev, N., Schmidt, T., Windhövec, J., Fuss, E. and Alfermann, A. (2007) Lignans from *Linum* species of sections *Sylinum* and *Linum. Natural Product Research* 21, 1–6.

Kuhlmann, S., Kranz, K., Lücking, B., Alfermann, A.W. and Petersen, M. (2002) Aspects of cytotoxic lignan biosynthesis in suspension cultures of *Linum nodiflorum. Phytochemistry Review* 1, 37–43.

Kuhnt, M., Rimper, H. and Heinrich, M. (1994) Lignans and other compounds from the mixed Indian medicinal plant *Hyptis verticillata. Phytochemistry* 36, 485–489.

Kupchan, S.M. (1965) Tumour inhibitors VII. Podophyllotoxin, the active principle of *Juniperus virginiana. Journal of Pharmaceutical Sciences* 54, 659–660.

Lin, H.W., Kwok, K.H. and Doran, P.M. (2003a) Production of podophyllotoxin using cross-species coculture of *Linum flavum* hairy roots and *Podophyllum hexandrum* cell suspensions. *Biotechnology Progress* 19, 1417–1426.

Lin, H.W., Kwok, K.H. and Doran, P.M. (2003b) Development of *Linum flavum* hairy root cultures for production of coniferin. *Biotechnology Letters* 25, 521–525.

Mohagheghzadeh, A., Schmidt, T.J. and Alfermann, A.W. (2002) Arylnaphtalene lignans from *in vitro* cultures of *Linum austriacum* L. *Journal of Natural Products* 65, 69–71.

Moraes, R.M., Hemant, L., Ebru, B., Muhammad, M. and Cushman, K. (2002) The American Mayapple and its potential for podophyllotoxin production. In: Janick, J. and Whipkey, A. (eds) *Trends in New Crops and New Uses.* ASHS Press, Alexandria, VA, USA.

Moss, G.P. (2000) Nomenclature of lignans and neolignans (IUPAC Recommendations 2000). *Pure and Applied Chemistry* 72, 1493–1523.

Muranaka, T., Miyata, M., Ito, K. and Taschinaba, S. (1998) Production of podophyllotoxin in *Juniperus chinensis* callus cultures treated with oligosaccharides and a biogenic precursor. *Phytochemistry* 49, 491–496.

Petersen, M. and Alfermann, A.W. (2001) The production of cytotoxic lignans by plant cell cultures, *Applied Microbiology and Biotechnology* 55, 135–142.

Premjet, D., Itoh, K. and Tachibana, S. (2002) Stimulation of production of podophyllotoxin by biogenetic precursors and an elicitor in *Juniperus chinensis* stem-derived callus cultures. *Pakistan Journal of Biological Science* 5, 313–316.

San Feliciano, A., Del Corral, J.M.M., Gordaliza, M. and Castro, M.A. (1989a) Acetylated lignans from *Juniperus sabinai. Phytochemistry* 28, 659–660.

San Feliciano, A., Medarde, M., Lopez, J.L., Puebla, P., Del Corral, J.M.M. and Barrero, A.F. (1989b) Lignans from *Juniperus thurifera. Phytochemistry* 28, 2863–2866.

Schmitt, J. and Petersen, M. (2002) Influence of methyl jasmonate and conyferyl alcohol on pinoresinol and matairesinol accumulation in a *Forsythia* × *intermedia* cell suspension culture. *Plant Cell Reports* 20, 885–889.

Singh, P., Kumar, R., Sharma, A., Arora, R., Chawla, R., Jain, S. and Sharma, R. (2009) *Podophyllum hexandrum* fraction (REC-2006) shows higher radioprotective efficacy in p53 carrying hepatoma cell line: A role of cell cycle regulatory proteins. *Integrative Cancer Therapies* 8(3), 261–272.

Stahelin, H.F. and Wartburg, A.V. (1991) The chemical and biological route from podophyllotoxin glucoside to etoposide. *Cancer Research* 51, 5–15.

Stakeholders Proposal for a Strategic Research Agenda to 2025, including Draft Action Plan 2010 - 3.2.1. WP 2-2. Natural plant pharmaceutical products. http://europa.eu.int/comm/research/biosociety/library/glossaryfind_en.cfm) (2005).

Tsukamoto, H., Hisada, S. and Nishibe, S. (1984) Lignans from the bark of *Olea* plants. *Chemical and Pharmaceutical Bulletin* 32, 2730–2735.

Tu, G. (ed.) (1977) *Pharmacopoeia of the Peoples Republic of China*, Beijing. Guangdong Science and Technology Press Vol.1. p. 594.

Van Furden, B., Hamburg, A. and Fuss, E. (2005) Influence of methyl jasmonate on podophyllotoxin and 6-methoxypodophyllotoxin accumulation in *Linum album* cell suspension cultures. *Plant Cell Reports* 24, 312–317.

Van Uden, W. (1992) *The production of podophyllotoxin and related cytotoxic lignans by plant cell cultures.* PhD Thesis, University of Groningen, Groningen, the Netherlands.

Van Uden, W., Pras, N. and Malingré, T.M. (1990a) On the improvement of podophyllotoxin production by phenylpropanoid precursor feeding to cell cultures of *Podophyllum hexandrum* Royle. *Plant Cell Tissue and Organ Culture* 23, 217–224.

Van Uden, W., Pras, N. and Malingré, T.M. (1990b) The accumulation of podophyllotoxin-β-D-glucoside by cell suspension cultures derived from the conifer *Callitris dummondii. Plant Cell Reports* 9(5), 257–260.

Vasilev, N. and Ionkova, I. (2005) Cytotoxic activity of extracts from *Linum* cell cultures. *Fitoterapia* 76, 50–53.

Vasilev, N., Ebel, R., Edrada, R., Alfermann, W., Fuss, E., Ionkova, I., Petrova, A. and Schmidt, T. (2008) Metabolic profiling of lignan variability in *Linum* species of section Syllinum native to Bulgaria. *Planta Medica* 74, 273–280.

Vasilev, N., Elfahmi, B, R., Kaiser, O., Momekov, G., Konstantinov, S. and Ionkova, I. (2006b), Production of Justicidine B, a cytotoxic arylnaphthalene lignan from genetically transformed root cultures of *Linum leonii. Journal of Natural Products* 69(7), 1014–1017.

Vasilev, N., Julsing, M., Koulman, A., Clarkson, C., Woenderbag, H., Ionkova, I., Boss, R., Jaroszewski, J. and Kaiser, O. (2006a) Bioconversion of deoxypodophyllotoxin into epipodophyllotoxin in *E. coli* using human cytochrome P450 3A4. *Journal of Biotechnology* 126, 383–393.

Vasilev, N., Momekov, G., Zaharieva, M., Konstantinov, S., Bremner, P., Heinrich, M. and Ionkova, I. (2005a) Cytotoxic activity of podophyllotoxin-like lignan from *Linum tauricum* Willd, *Neoplasma* 52, 425–429.

Vasilev, N.P., Gibbons, S., Ionkova, I., Kayzer, O., Arroo, R. and Woolley, J. (2005b) Phytochemical investigation of lignans from *Linum tauricum* ssp. Bulgaricum. International Congress and 53rd Annual Meeting of the Society for Medicinal Plant Research with Societa Italiana di Fitochimica, August 21–25, Florence, Italy, Book of Abstracts, Society for Medicinal Plant Research Publisher, Florence, Italy, pp. 366.

Ward, R.S. (2003) Different strategies for the chemical synthesis of lignans. *Phytochemistry Review* 2, 391–400.

Windhövel, J., Mohagheghzadeh, A., Godt, D., Wink, M., Wildi, E. and Alfermann, A.W. (2003) Lignan production in hairy roots cultures of *Linum* species. *International Meeting*, Phytochemistry and Biology of Lignans, April 6–9, Bornheim-Walberberg Germany, Book of Abstracts, The Phytochemical Society of Europe Publisher, Dusseldorf, Germany, p. 29.

Wink, M., Alfermann, W., Franke, R., Wetterauer, B., Distl, M., Windhövel, J., Krohn, O., Fuss, E., Garden, H., Mohagheghzadeh, A., Wildi, E. and Ripplinger, P. (2005) Sustainable bioproduction of phytochemicals by plant *in vitro* cultures: anticancer agents. *Plant Genetic Resources* 3, 90–100.

Woerdenbag, H.J., Van Uden, W., Frijlink, H.W., Lerk, C.F., Pras N. and Malingré, T.M. (1990) Increased podophyllotoxin production in *Podophyllum hexandrum* cell suspension cultures after feeding coniferyl alcohol as a β-cyclodextrin complex. *Plant Cell Reports 9,* 97–100.

Yu, P., Wang, L. and Chen, Z. (1991) A new podophyllotoxin-type lignan from *Dysosma versipellis* var. Tomentosa. *Journal of Natural Products* 54, 1422–1424.

Chapter 10

Hairy Root Culture: Copying Nature in New Bioprocesses

Milen I. Georgiev, Jutta Ludwig-Müller and Thomas Bley

Copying Nature: Transformation with *Agrobacterium rhizogenes*

Conventional cell cultures need to be sustained using plant growth hormones, but this is not necessary after transformation of plants with *Agrobacterium rhizogenes*, which leads to the production of 'hairy roots' that do not normally need exogenous hormones and can grow indefinitely. In addition, such cultures may produce different metabolites from cell cultures, and they have much greater resemblance to the root system of a normal plant, i.e. morphologically *A. rhizogenes*-induced hairy roots are very similar to wild-type roots, except that they are more branched, agravitropic and their root hairs are longer and more numerous (Veena and Taylor, 2007). In addition, since there is still concern about transgenic plants, if the *Agrobacteria* used for the transformation are not transformed with foreign DNA (see the section on metabolic engineering), these cultures may be more readily accepted than possible transgenic alternatives. Furthermore, production of hairy roots is less time consuming than most transformation procedures for whole plants.

A wide range of plants can be transformed with *Agrobacterium*, including numerous dicot and monocot angiosperm species (Gelvin, 2003; Li and Leung, 2003) and gymnosperms (Li and Leung, 2003; McAfee *et al.*, 1993; Yibrah *et al.*, 1996). However, the outcome of transformation/infection in nature varies between *Agrobacterium* species. While *A. tumefaciens* causes the widely distributed crown gall disease, *A. rhizogenes* infection leads (as mentioned above) to hairy root formation. In the laboratory, so-called disarmed vectors without the tumour-inducing genes, but containing the genes of interest to be introduced into the target plant, are used for *A. tumefaciens* transformation (Tzfira and Citovsky, 2006). In contrast, when plants are transformed with *A. rhizogenes*, the *rol* genes (see below) are still present, which facilitates the generation of the root cultures (either with or without foreign DNA, see section on metabolic engineering).

The molecular mechanism of hairy root induction is not completely understood (Veena and Taylor, 2007; Georgiev *et al.*, 2008). However, the transferred DNA (T-DNA) is known to be a segment of the *Agrobacterium* Ri-plasmid. Hairy root formation in plants infected with *A. rhizogenes* is not thought to be dependent on changes in the phytohormonal balance in plant cells, probably because the cells become more sensitive to auxins (Nilsson and Olsson, 1997; Altamura, 2004). It should be noted that Park and Facchini (2000) found that although hairy root cultures should, in theory, be able to grow without exogenous plant hormones, the growth rates of transformed roots of two species, opium poppy and Californian poppy, was improved by the addition of auxin to the liquid culture medium. Exogenous application of auxin can also reportedly stimulate growth in hairy root cultures of *Lippia dulcis* Trev. (Sauerwein *et al.*, 1991). The minimum doubling times, of approximately 2 weeks for California poppy and 3–4 weeks for opium poppy hairy root cultures are comparable to those of other species (Loyola-Vargas and Miranda-Ham, 1995).

The products of the four *rol* (*A*, *B*, *C* and *D*) genes play key roles in hairy root formation, but the *rolB* gene appears to be the most important in hairy root induction, since loss-of-function mutation at this locus renders the plasmid avirulent (Nilsson and Olsson, 1997; Altamura, 2004). Further, when *rolB* is introduced into the host plant genome as a single gene, it is capable of hairy root induction (Altamura, 2004). Auxin has been proposed to be synthesized from tryptophan (Trp) by a two-step reaction in hairy roots, because Trp can be converted by the *aux1* gene product tryptophan monooxygenase to indole-3-acetamide (IAM), which is then converted by indoleacetamide hydrolase, the *aux2* gene product, to auxin (indole-3-acetic acid; IAA) (Thomashow *et al.*, 1984; Offringa *et al.*, 1986). However, very recently it was shown that introduction of the *rolABCD* genes alone does not affect the auxin requirement of tobacco BY-2 cells. Instead, the *aux1* and *aux2* genes, rather than the *rolABCD* genes, appear to be responsible for the auxin autotrophy of these transgenic lines (Nemoto *et al.*, 2009). These findings, together with older reports (see above), imply that although the *rolABCD* genes may be sufficient to generate hairy-root tumours, they are irrelevant to auxin autotrophy. A further interesting observation was made by Nemoto *et al.* (2009), that the BY-2 cells were able to grow without supplementary auxin even when only the *aux1* gene was present, suggesting that there might be plant enzymes capable of catalyzing the amidase reaction. Accordingly, Pollmann *et al.* (2003) found an amidase in *Arabidopsis* and Ishikawa *et al.* (2007) reported that amidase genes are present in *Brassica* sp., so such genes may be quite ubiquitous, and could contribute to the increased auxin levels in hairy root cultures. Understanding the mechanisms that lead to rhizogenesis may also be useful for establishing faster growing cultures in sub-optimal culture conditions, e.g. in bioreactors (see the section on bioprocessing). However, auxin production can also have effects on secondary metabolite production, which need to be taken into account. For example, Rischer *et al.* (2006) found that addition of auxin to cell cultures of *Catharanthus roseus* inhibited accumulation of the terpenoid indole alkaloid ajmalicine, but stimulated the accumulation of other alkaloid metabolites, such as tabersonine and catharanthine, and when methyl jasmonate (MeJA) was added to auxin-treated cells it restored their ability to produce all alkaloids. Such responses may also occur in hairy roots, which already produce auxin via the *aux* genes, and possibly change the desired metabolic pattern. Since the *rol* genes alone are sufficient to induce hairy roots, the *aux* genes may warrant further consideration before planned transformations. However, the *rol* genes themselves can also influence secondary metabolite production. For instance, Palazon *et al.* (1997) observed striking differences in nicotine production between transgenic root lines of tobacco transformed with the *rolA*, *rolB* and *rolC* genes singly and in combination. While the *rolC* gene strongly stimulated alkaloid production indirectly, by increasing growth, the *rolA* gene stimulated it mainly by a direct effect. Further, the presence of *rolA*, *rolB* and *rolC* together dramatically enhanced both the growth rate and nicotine production, showing that the effects of these three *rol* genes were synergistic.

Besides transformation via *A. rhizogenes*, it should be noted that there are alternative approaches for producing hairy root cultures, i.e. establishment of root cultures from mutants with a hairy root phenotype. For example, Suzuki *et al.* (2008) developed an alternative root culture system using an allelic mutant of *Arabidopsis thaliana superroot 1* (*sur1*). When the mutant was cultured in liquid medium developed for hairy root cultures, it grew rapidly, developing a globular, rooty phenotype and was easily subcultured in the medium. This mutant is advantageous because the rooty phenotype is recessive, hence the desired characteristics can likely be maintained via seeds. Crossing heterozygous *sur1* plants with fatty acid desaturase mutants in the cited study by Suzuki *et al.* (2008) resulted in the accumulation of high amounts of oleic acid and linolenic acid, indicating that the *sur1* root culture system is an alternative tool for the production of useful compounds in roots.

The long-term maintenance of induced hairy root strains is of both practical and financial significance. Usually, transformed root cultures, grown on solid media are subcultured every 3–4 weeks. However this is expensive in terms of time, labour and resources, with consequently high risks of contamination and eventual loss of the original strains. Conservation of the original strains by cryopreservation could avoid these problems. However, despite the progress made in the last decade on cryopreservation of plant material, there have been few reports on the cryopreservation of hairy root cultures (Lambert *et al.*, 2009). Generally the main approaches used for cryopreservation of plant tissues and organs include slow cooling, vitrification, encapsulation-dehydration and encapsulation-vitrification, the last two of these appear to be most effective (Xue *et al.*, 2008). Lambert *et al.* (2009) studied the potential applicability of encapsulation in calcium-alginate beads (containing 0.1 M sucrose), followed by dehydration (in air streams, to 35% of the initial mass) for hairy root cultures of *Maesa lanceolata* and *Medicago truncatula* and established survival rates of 90% and 53%, respectively, which was concluded to be sufficient for large-scale purposes.

Applications of Transformed Root Cultures

Currently, the main applications of transformed root cultures include production of high-value metabolites and biotransformation, phytoremediation, regeneration of whole plants and production of artificial seeds (Giri *et al.*, 2001; Guillon *et al.*, 2006). Plant regeneration and artificial seed production are beyond the scope of this chapter and, thus, will not be further considered here. Further applications include their use in detailed studies of fundamental molecular, genetic and biochemical aspects of genetic transformation and hairy root induction.

Production of metabolites

The enormous biosynthetic potential of hairy root cultures was largely neglected for years, in which investigations mainly focused on the mechanisms underlying the hairy root syndrome. Nowadays, transformed root cultures receive more attention as technological objects for production of valuable metabolites, for which they have several attractive features, including high genetic and biochemical stability, and relatively fast growth rates in hormone-free media (Georgiev *et al.*, 2007). Figure 10.1 shows the structures of some high-value metabolites that can be produced by transformed root cultures, including anticancer (paclitaxel, podophyllotoxin, camptothecin, justicidine B and berberine), antimalarial (artemisinin), and anti-inflammatory (harpagoside) substances.

To illustrate the economic potential of substances from natural sources, Potterat and Hamburger (2008) noted that over 60% of all anti-cancer drugs are of natural origin, and that annual sales of such drugs exceed several billion US dollars. In 1971 Wall and Wani reported the isolation of paclitaxel from the stem bark of the pacific yew *Taxus brevifolia*. Paclitaxel (a complex diterpenoid that is the active ingredient of Taxol®) has received considerable attention because of its activity against various kinds of cancer, *inter alia* breast cancer, ovarian cancer, AIDS-related Kaposi's sarcoma, and non-small cell lung cancer. The annual sales of paclitaxel and related taxanes in 2004 exceeded 2 billion US dollars (Tabata, 2004). Obtaining commercial supplies of paclitaxel from natural sources is not economically feasible because the source *Taxus* plants grow very slowly, and their paclitaxel contents are very low (~0.01% of the dry weight of the bark, where levels are highest). Tabata (2004) calculated that a 100-year-old yew tree would yield just 300 mg of the compound. The total synthesis of paclitaxel has been achieved, but this too is not economically viable. Therefore, paclitaxel is now produced commercially in large-scale plant cell suspension cultures (by Phyton and Samyang Genex Corporation), in one of the most remarkable examples of the potential value of plant *in vitro* culture. Hairy root cultures of *Taxus* x *media* var. *Hicksii* have also been found to produce paclitaxel at amounts of 69 μg/g dry weight. Further, Furmanowa and Syklowska-Baranek (2000) found that the content of paclitaxel in transformed roots can be enhanced through elicitation with MeJA; a week after addition of 100 μM MeJA the paclitaxel content increased to 210 μg/g dry weight. Another compound that is produced by transformed root cultures (of *Ophiorrhiza mungos*) is camptothecin, an alkaloid with anticancer activities, in amounts (~0.3% per unit dry weight) significantly exceeding those obtained from cell suspension cultures (~0.01% per unit dry weight) of the same species (Wink *et al.*, 2005).

Podophyllotoxin, a lignan with antimitotic activity, and the closely related 6-methoxypodophyllotoxin are also accumulated in hairy root cultures, of *Linum album* and *L. persicum* (Wink *et al.*, 2005). Therapeutic application of podophyllotoxins is hampered due to severe side effects. However, they serve as useful bases for production of the semisynthetic etoposide and teniposide, which have found clinical use for the treatment of various cancers (Potterat and Hamburger, 2008).

In addition, Vasilev *et al.* (2006) recently reported the production of justicidine B (an arylnaphthalene lignan with cytotoxic, antiviral, fungicidal, antiprotozoal and antiplatelet properties) by transformed root cultures of *L. leonii*. After two weeks of submerged cultivation the justicidine B content reached 10.8 mg/g dry; 5-fold higher than yields from a callus culture of the same species. Berberine (an alkaloid with anti-inflammatory and antimicrobial activities, which was recently also found to possess anticancer properties) was also found to accumulate at higher amounts in transformed root cultures of *Phellodendron chinense* than in the callus cultures of the same species. All of these studies support the hypothesis that the biosynthesis of some metabolites (e.g. alkaloids) is associated with organogenesis.

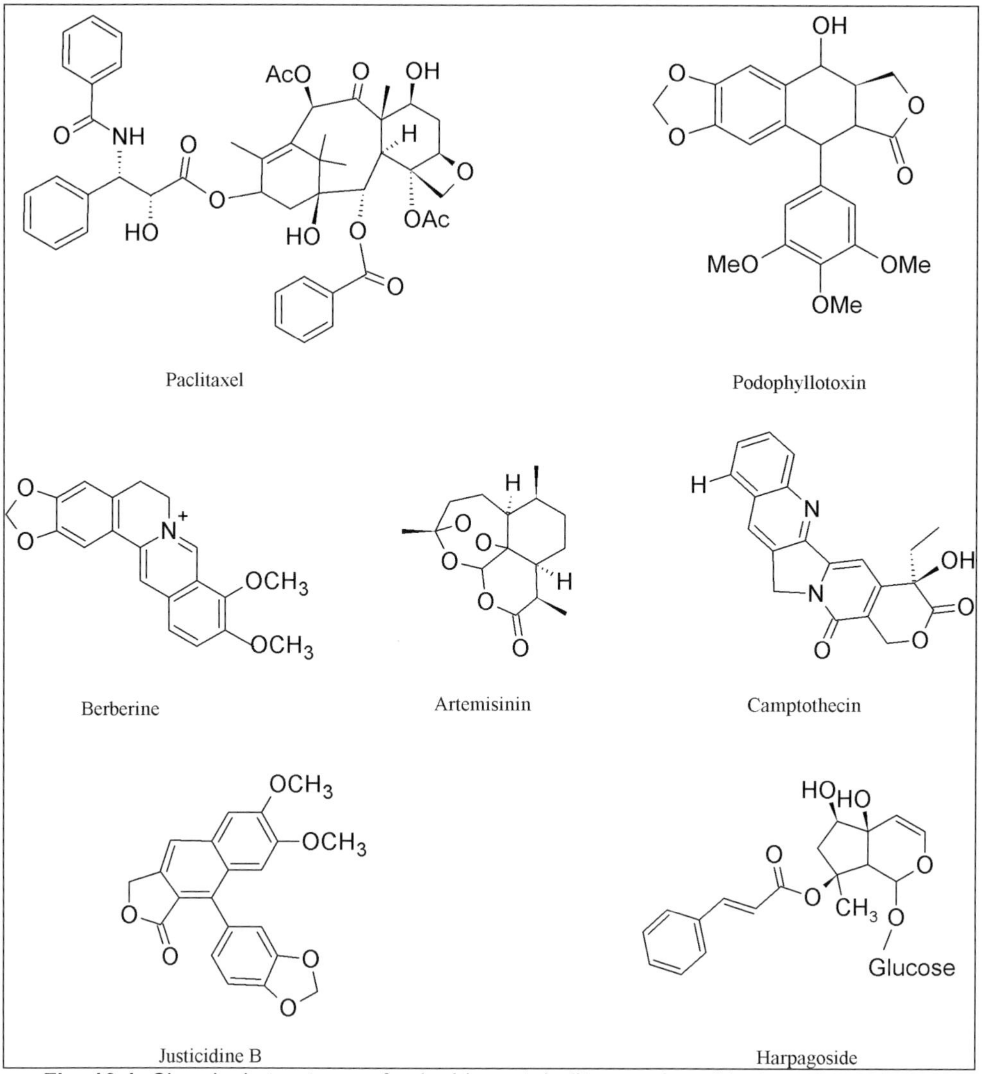

Fig. 10.1. Chemical structures of valuable metabolites produced by transformed root cultures.

Finally, for this section, artemisinin (a sesquiterpene lactone produced by *Artemisia annua*; Fig. 10.1) is highly active against the parasitic protozoa of the genus *Plasmodium* that cause malaria, but it has low toxicity to humans. Malaria is a very common and virulent tropical disease, causing ~2 million deaths (mainly of infants and young children) amongst 300–500 million cases per year (Potterat and Hamburger, 2008). When extracted from the plant yields of artemisinin are very low (0.01–0.8%), making large-scale production costly (Potterat and Hamburger, 2008). However, Kim *et al.* (2001) examined artemisinin production by transformed root cultures of *A. annua*, and obtained yields up to 2.64 μg/g, and ca. ~3-fold higher levels from cultivation in mist reactors than in bubble column reactors.

Biotransformation

Biotransformation refers to the chemical modification of the molecular structure of organic compounds, by microorganisms, plant *in vitro* systems or enzymes. Plant *in vitro* systems can perform a wide range of reactions, such as glucosylation, hydroxylation, reduction of carbon double bonds (C = C), hydrolysis, epoxidation and many others (Giri *et al.*, 2001). Glucosylation reactions are of particular interest in this respect because they facilitate the conversion of water-insoluble compounds (e.g. phenols, phenylpropanoid acid or their analogues) to water-soluble compounds (Giri *et al.*, 2001). More specifically, in an analysis by Faria *et al.* (2009) of the capacity of *Anethum graveolens* hairy roots to transform menthol and geraniol, neither of these monoterpenes adversely affected growth of the roots, menthyl acetate was formed following menthol addition and ten new biotransformation products appeared following geraniol addition (including the alcohol linalool and the aldehyde geranial). In another example of the biotransformation capacity of hairy root cultures, Chen *et al.* (2008) observed regiospecific glucosylation of three isomers of hydroxybenzoic acid (*p*-, *m*- and *o*-hydroxybenzoic acid) in hairy root cultures of *Panax ginseng*. *p*-Hydroxybenzoic acid and *m*-hydroxybenzoic acid were converted into their corresponding glycosides and glycosyl esters, but no metabolite of *o*-hydroxybenzoic acid was detected.

Phytoremediation

In phytoremediation plants and/or plant *in vitro* systems (e.g. hairy root cultures) are used to remove, or reduce the concentrations of, toxic organic pollutants or to absorb and accumulate heavy metals (Oller *et al.*, 2005; Guillon *et al.*, 2006). Use of plants or plant cultures has considerable advantages for these purposes, including the scope to use them *in situ* or *ex situ* and their environmental acceptability (Oller *et al.*, 2005). Most successful applications of hairy root cultures for phytoremediation have been for removing phenol (Table 10.1), since the process is mainly catalyzed by peroxidases, which are produced abundantly by hairy roots. Furthermore, Alderete *et al.* (2009) recently reported the production of transgenic *Nicotiana tabacum* hairy roots, expressing heterologous basic peroxidases, with removal efficiencies that were significantly higher than those of control 'wild type' hairy roots.

Another major application of phytoremediation with hairy roots is for heavy metal detoxication. The exploitation of plants and plant *in vitro* systems to remove toxic heavy metals from the environment is currently of considerable commercial interest. Of particular importance for such uses are hyperaccumulator plants/*in vitro* systems, which are capable of taking up and storing elevated concentrations of heavy metals (Cd, Hg, Ag, Pb, Cu, Zn etc.), without displaying symptoms of severe metal toxicity or cell damage (Boominathan and Doran, 2003). Hairy roots of *Daucus carota* have been found to be sensitive to uranium exposure by Straczek *et al.* (2009), but their study provides further insights for application of hairy root cultures for phytomining purposes.

In confirmation, Boominathan and Doran (2003) observed that hairy roots of *Thlaspi caerulescens* are hyperaccumulators of Cd ions, since exposure to 178 μM Cd neither suppressed their growth, nor significantly enhanced their H_2O_2 levels. In addition, 300-fold greater constitutive catalase activity was detected in the hairy roots of *T. caerulescens* (than in *N. tabacum*), illustrating the further scope for overexpressing catalase and/or superoxide dismutase in order to improve Cd tolerance (Boominathan and Doran, 2003).

Table 10.1. Selected examples of phytoremediation and metal accumulation by hairy root cultures.

Hairy root culture	Organic or inorganic pollutant	Enzyme(s) involved	Reference
Lycopersicon esculentum	Phenol	Peroxidase	Gonzales *et al.*, 2006
Brassica napus	Phenol	Peroxidase	Coniglio *et al.*, 2008
Daucus carota, *Ipomoea batatas*, *Solanum aviculare*	Phenol and chlorophenols	Peroxidase	Santos de Araujo *et al.*, 2006
Armoracia rusticana	Acetaminophen (paracetamol)	P450 monooxygenase; Glutathione S-transferases	Huber *et al.*, 2009
Brassica napus	Prochiral diketones	Dehydrogenases	Orden *et al.*, 2006
Solanum nigrum	Polychlorinated bisphenyls	-	Rezek *et al.*, 2007
Helianthus annuus	Tetracycline, Oxytetracycline	-	Gujarathi *et al.*, 2005
Daucus carota	Uranium	-	Straczek *et al.*, 2009
Thlaspi caerulescens	Cadmium	Catalase	Boominathan and Doran, 2003

Metabolic Engineering

While production of naturally occurring metabolites in tissue cultures can be elicited by various signalling molecules, or simply by the presence of *A. rhizogenes* genes from the TL-DNA (see the section on transformation with *A. rhizogenes*), it may also be desirable to modify the genetic features of hairy root cultures so that a targeted metabolite can be produced in higher amounts if there is a rate-limiting step in its production pathway. However, such a strategy requires thorough knowledge of the pathway(s) leading to the desired metabolite, and other pathways that may compete for precursors or intermediates to avoid potentially adverse secondary effects if a necessary precursor such as an amino acid is depleted (see, for instance, Verpoorte *et al.*, 2000). Other potential factors to be considered include: (i) feedback inhibition of enzymes in the pathway; (ii) the availability of cofactors; and (iii) differences in the compartmentation of enzymes and substrates, and hence the possible need to transport an intermediate to another compartment (see Fig. 10.2A). However, it is also possible to block pathways competing for intermediates. In the schematic diagram presented in Fig. 10.2A, for instance, the precursor is converted to intermediate A which is not rate-limiting. However, if the next step (conversion of A to B) would be rate limiting, then increasing the levels of enzymes that catalyze the conversion of A to B would increase the amount of intermediate B. It might also be helpful to increase levels of intermediate A by adding the compound. Intermediate B is then converted to C (not rate-limiting), but B is also a precursor for metabolite K. Hence, it would be desirable to reduce production of K, by either blocking the reaction chemically or using antisense approaches. Alternatively, the expression (or dosage) of the gene encoding the protein required for conversion of B to C could be increased. Metabolite C might have to be transported to a different compartment, which could lead to a drop in the concentration of C. A solution could be to start engineering the pathway after the transport step. In addition, high rates of synthesis of the end product D could require an additional cofactor. It is also possible that the end product (or an intermediate) could reduce production of one of the important rate-limiting intermediates (e.g. B) by feedback inhibition, if so steps should be taken to prevent this happening.

Strategies for genetic engineering of more complex pathways could also include: (i) heterologous gene expression to produce compounds that are not normally produced from a precursor molecule present in the given species (see below); (ii) antisense inhibition or co-suppression (RNAi) of alternative pathways; (iii) expression of multiple genes in a pathway; and (iv) expression of genes encoding transcription factors that regulate multiple steps in a pathway (Fig. 10.2B).

The potential utility of the last of these approaches has been demonstrated in a study where the efficiency of the maize *Sn* gene, which transactivates the anthocyanin pathway in various tissues, for regulating anthocyanin biosynthesis in several dicotyledonous species was tested using transformed root cultures (Damiani *et al.*, 1998). The *Sn* gene was found to be capable of inducing anthocyanin biosynthesis in some heterologous roots, such as lucerne (*Medicago sativa* L.) and lotus (*Lotus angustissimus* L.), but not in others, such as petunia (*Petunia hybrida* L.).

Knowledge of control factors that could be easily transferred to hairy root cultures has been obtained in several studies, including the following. Conserved regulatory elements in the strictosidine synthase promoter have been used to isolate various regulatory genes that control JA-inducible expression of monoterpene indole alkaloid (MIA) biosynthesis in *C. roseus* cell cultures (Memelink and Gantet, 2007). In another study, T-DNA activation tagging led to the identification of the cDNA encoding the transcription factor ORCA3 (van der Fits and Memelink, 2000), which regulates several early steps in monoterpene indole alkaloid biosynthesis, and over-expression of ORCA3 in transformed *C. roseus* cell cultures increased the expression of several MIA pathway genes. Interestingly, treating *C. roseus* cell cultures with JA resulted in a post-translational modification of ORCA3, possibly via phosphorylation (Memelink and Gantet, 2007).

Genetically engineered root cultures have also been used as model systems to study various aspects of the metabolic and molecular regulation of several natural product pathways. For example, expression of a cDNA encoding *Antirrhinum majus* L. dihydroflavonol reductase in hairy roots of *Lotus corniculatus* L. caused an increase in the content, and alteration in the structure, of condensed tannins in a manner consistent with the substrate specificity of the transgene product (Bavage *et al.*, 1997).

In addition, the same precursor can sometimes be used for the production of several different metabolites, e.g. tryptophan can be converted to either indole glucosinolates or tryptamine in canola (De Luca, 2000), the latter leading to the desired terpenoid indole alkaloids, while coumaryl-CoA and malonyl-CoA can be substrates for either stilbene synthase or chalcone synthase, leading to resveratrol and flavonol/anthocyanidin production, respectively (Hain and Grimmig, 2000). Sometimes it is sufficient to overexpress genes encoding a single key enzyme in the pathway (see above), notably genes encoding hyoscyamine 6β-hydroxylase (Table 10.2), which catalyzes the conversion of hyoscyamine to the medicinally important scopolamine (Yun *et al.*, 1992).

An engineered *A. rhizogenes* strain, harbouring the 35S-H6H gene construct, was introduced into *Hyoscyamus muticus* by Jouhikainen *et al.* (1999), and into hyoscyamine-rich hairy roots of *Atropa belladonna* L. by Hashimoto *et al.* (1993), leading to almost exclusive production of scopolamine and increased yields in both species. Using the same approach, Palazon *et al.* (2003b) transformed *Duboisia* hybrids, selected for initially high scopolamine contents, using *A. rhizogenes* carrying the 35S-H6H construct and further regenerated plants with even higher scopolamine contents.

Similarly, Moyano *et al.* (2007) reported the establishment of an *N. tabacum* cell suspension culture, containing the 35S-H6H gene from *Hyoscyamus niger*, which showed considerable capacity for the bioconversion of hyoscyamine into scopolamine. This approach has also been used to alter levels of alkaloids normally produced in roots via the expression of yeast, rather than plant, genes. In one such study, a yeast cDNA encoding ornithine decarboxylase (ODC) was introduced into *Nicotiana rustica* L. hairy roots using an *Agrobacterium*-derived vector (Hamill *et al.*, 1990). Although the hairy root cultures produced more nicotine than wild-type roots, the increase in ODC activity was proportionately higher than the increase in nicotine levels, suggesting that ornithine decarboxylation is not a rate-determining step in the nicotine pathway. In addition, transformed *Lithospermum erythrorhizon* root cultures expressing a bacterial gene encoding chorismate pyruvate-lyase, which converts chorismate to the shikonin precursor 4-hydroxybenzoate (4HB), did not produce higher levels of shikonin (Sommer *et al.*, 1999). These results suggest that the availability of 4HB does not normally limit the biosynthesis of shikonin.

Cinchona officinalis hairy roots constitutively expressing constructs of cDNAs encoding the enzymes tryptophan decarboxylase and strictosidine synthase from *Catharanthus roseus*, two key enzymes in terpenoid indole and quinoline alkaloid biosynthesis, have high amounts of tryptamine and strictosidine, as well as quinine and quinidine (Geerlings *et al.*, 1999). These findings show that genetic engineering with multiple genes is certainly feasible in hairy roots of *C. officinalis*. Similarly, the introduction of a heterologous tryptophan decarboxylase gene into hairy root cultures of *Peganum harmala* (Table 10.2) resulted in the accumulation of two biosynthetically related, tryptamine-derived secondary metabolites: serotonin and β-carboline alkaloids (Berlin *et al.*, 1993). Although serotonin accumulation in transgenic root cultures with elevated TDC activity was higher than in control cultures, β-carboline alkaloid levels were not affected; thus, the tryptamine supply was shown to be limiting for serotonin, but not for β-carboline alkaloid, biosynthesis. In some cases it might therefore be necessary to augment the medium with a precursor to obtain higher yields. For example, terpenoid indole alkaloid production in transgenic *Catharanthus roseus* cells can be stimulated by addition of the precursor loganin (Canel *et al.*, 1998; Whitmer *et al.*, 1998), and transformed hairy root cultures would likely respond in a similar manner. In accordance with this hypothesis, *C. roseus* hairy root cultures expressing various forms of anthranilate synthase and/or TDC exhibited higher fluxes through the tryptophan branch of the monoterpene indole alkaloid pathway in a study by Hughes *et al.* (2004). Hairy root cultures expressing these genes under the control of constitutive or inducible promoters accumulated higher levels of tryptamine than alkaloids. In another attempt to alter the isoprenoid pathway, a hamster 3-hydroxy-3-methylglutaryl CoA reductase gene was expressed in hairy roots and a line was generated that accumulated up to seven-fold higher levels of serpentine than controls. However, no effects on catharanthine levels were detected (Ayora-Talavera *et al.*, 2002). In addition, the genetic engineering and expression of the enzyme catalyzing the terminal step of vindoline biosynthesis, deacetylvindoline-4-O-acetyltransferase (DAT), in *Catharanthus roseus* hairy root cultures has been recently reported (Magnotta *et al.*, 2007). *C. roseus* produces the powerful anticancer drugs vinblastine and vincristine, both of which are derived from dimerization of the monoterpenoid indole alkaloids, vindoline and catharanthine. Metabolite analysis of transgenic hairy root cultures in the cited study established that hairy root extracts had an altered alkaloid monoterpenoid indole profile, possibly useful for the production of the desired compounds.

Overall, there is still little detailed knowledge of complete biochemical pathways involved in the synthesis of secondary metabolites in plants. In addition, the regulatory features are not well understood. Therefore, progress in metabolic engineering remains slow and is frequently hampered (Verpoorte *et al.*, 2002; Oksman-Caldentey and Inze, 2004). For example, relatively few ESTs are available for alkaloid-producing plants, except for tobacco, compared with the numbers available for model plants such as *Arabidopsis*, rice and tomato (Facchini and De Luca, 2008). However, extensive EST databases have now been compiled for opium poppy and *C. roseus*. In addition, functional genomic approaches (including transcriptomics, proteomics and metabolomics) are being used to accelerate comprehensive investigations of biological systems (Oksman-Caldentey and Inze, 2004). For example, Rischer *et al.* (2006) recently attempted to combine genome-wide transcript profiling and metabolic profiling (by cDNA-amplified fragment-length polymorphism and LC-MS analysis, respectively) of MeJA-elicited *Catharanthus roseus* cell cultures and identified sets of both known and previously undescribed transcript tags (417 gene sequence tags) and metabolites (178) associated with terpenoid indole alkaloids. The analysis and subsequent modification of such gene-to-metabolite networks seem to provide a promising approach for adjusting plant biochemical machinery to improve the production of useful compounds and/or new compounds with novel and/or more desirable biological activities (Goossens and Rischer, 2007).

In addition to modifying naturally occurring pathways, the expression of completely foreign genes can lead to the production of desired secondary metabolites. For instance, *de novo* synthesis of poly(3-hydroxybutyrate) has been reported from hairy root cultures of *Beta vulgaris*, transformed with *A. tumefaciens*, harbouring the root-inducing plasmid pRi15834 together with the pBI ABC plasmid containing the *phaA*, *phaB* and *phaC* genes of *Cupriavidus necator* (Menzel *et al.*, 2003). High amounts of the mammalian immunomodulator murine interleukin-12 have also been produced from transgenic *Nicotiana tabacum* hairy root cultures (Table 10.2) grown in mist bioreactors (see the section on bioprocessing).

The reports described above all present more or less successful attempts to genetically engineer hairy root cultures in order to improve the production of targeted metabolites, but in numerous cases overexpression of a gene encoding a specific enzyme involved in a biosynthetic pathway of interest has not led to accumulation of the desired metabolite, and sometimes it has even led to reductions in its production (Palazon *et al.*, 2008 and references cited therein). This could be desirable for a side metabolite (Fig. 10.2), but not for the main targeted compound. The cited authors list several examples related to the production of tropane alkaloids. One problem associated with homologous gene expression, or even expression of a heterologous gene with high homology to native genes, is co-suppression, as mentioned above. Moreover, secondary metabolite synthesis is highly compartmented, thus tissue-specific expression may be required, which is not always achieved even in hairy root cultures. In conclusion, more factors need to be considered for successful transgenic metabolite production than merely the required genes.

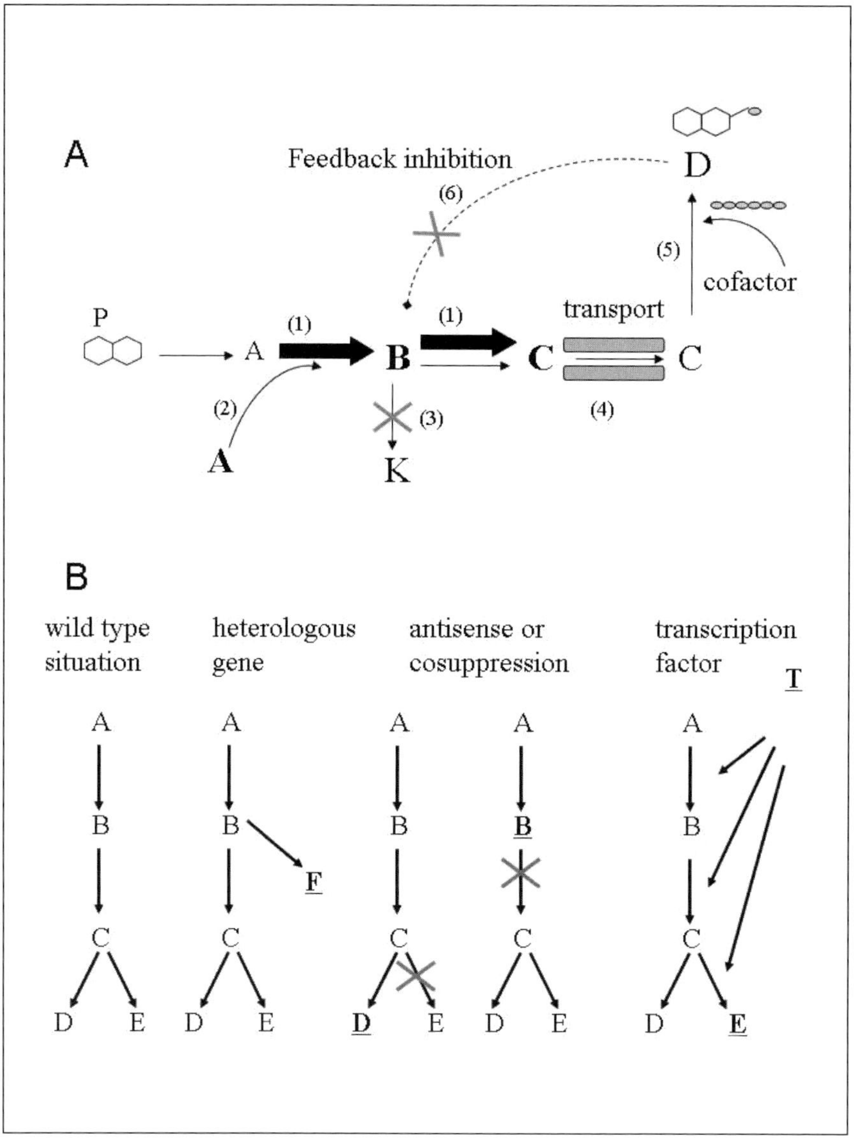

Fig. 10.2. (A) Aspects of metabolic pathways that need to be considered when manipulating a specific pathway, starting with precursor P, by metabolic engineering, including: (1) the rate limiting step(s), (2) addition of rate-limiting intermediates, (3) alternative pathways using the same intermediates, (4) transport to different compartments, (5) availability of cofactors, and (6) feedback inhibition. Bold arrows indicate overexpression of genes encoding proteins required for the indicated steps (modified after Verpoorte *et al.*, 2000). (B) Possibilities to change pathways in plants by heterologous gene expression, antisense inhibition or co-suppression of undesired pathways or the expression of genes encoding transcription factors regulating one or several steps in the pathway. The changes compared to the wild type situation are shown underlined.

Table 10.2. Summary of engineered hairy root cultures that have been used for secondary metabolite production.

Plant species	**Gene(s) transformed**	**Product**	**Reference**
	Homologous genes		
Hyoscyamus	Hyoscyamine 6β-hydroxylase	Scopolamine	Jouhikainen *et al.*, 1999
Atropa	Hyoscyamine 6β-hydroxylase	Scopolamine	Yun *et al.*, 1992
Catharanthus	Deacetylvindoline-4-O-acetyltransferase	Horhammericine	Magnotta *et al.*, 2007
Catharanthus	Anthranilate synthase and/or tryptophan decarboxylase	Tryptamine	Hughes *et al.*, 2004
	Foreign plant genes		
Duboisia	*Hyoscyamus* h6h	Scopolamine	Palazon *et al.*, 2003b
Nicotiana	*Hyoscyamus* h6h	Scopolamine	Moyano *et al.*, 2007
Lotus	*Antirrhinum* dihydroflavonol reductase	Condensed tannins	Bavage *et al.*, 1997
Peganum	*Catharanthus* tryptophan decarboxylase	Serotonin	Berlin *et al.*, 1993
Cinchona	*Catharanthus* tryptophan decarboxylase, strictosidine synthase	Tryptamine, strictosidine, quinine, quinidine	Geerlings *et al.*, 1999
Medicago	Maize Sn gene (TF) [a]	Anthocyanin	Damiani *et al.*, 1998
Lotus	Maize Sn gene (TF)	Anthocyanin	Damiani *et al.*, 1998
	Foreign genes from other kingdoms		
Lithospermum	Bacterial ubiC	Shikonin	Sommer *et al.*, 1999
Nicotiana	Yeast ornithine decarboxylase	Nicotine	Hamill *et al.*, 1990
Beta	Bacterial poly(3-hydroxybutyrate) synthesis genes	Poly(3-hydroxybutyrate)	Menzel *et al.*, 2003
Catharanthus	Hamster 3-hydroxy-3-methylglutaryl CoA reductase gene	Serpentine	Ayora-Talavera *et al.*, 2002
Nicotiana	Sequences, encoding a single-chain form of murine IL-12	Murine Interleukin-12	Liu *et al.*, 2009
Trichosanthes	Ribosome-inactivating protein gene	Ribosome-inactivating protein	Thorup *et al.*, 1994

[a] TF - transcription factor

Bioprocessing of Hairy Root Cultures

Bioprocessing aspects of transformed root cultures for mass production of desired metabolites have been intensively studied during the last two decades. Developing economically feasible commercial processes requires detailed knowledge of the physiology of the hairy root strains, understanding of the interactions of the roots with the culture environment, and tight control of the whole process (from the solid culture stage to the end product; Fig. 10.3).

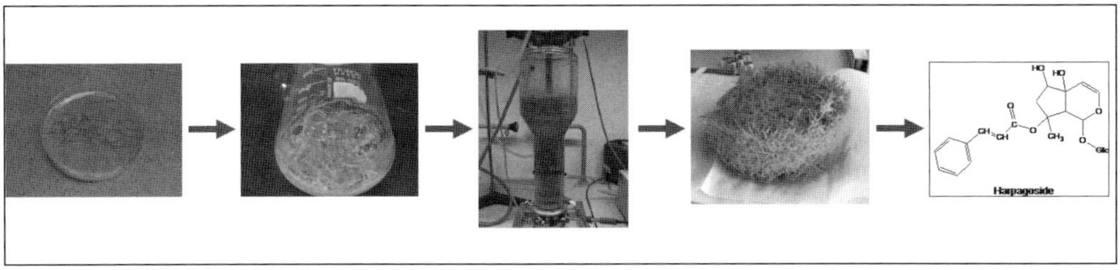

Fig. 10.3. Hairy root culture of *Harpagophytum procumbens* (Devil's claw, Pedaliaceae): from Petri dish to product formation.

An important, and in several cases limiting, stage is the bioreactor cultivation of transformed root cultures. Diverse bioreactor configurations have been used for cultivating transformed root cultures, including (*inter alia*) mechanically driven reactors (e.g. stirred tank reactors, wave reactors and rotating drum reactors), pneumatically driven reactors (e.g. bubble column reactors and airlift reactors) and bed reactors (e.g. trickle bed reactors and mist reactors). It would be difficult, if not impossible, to pick any single, ideal type of bioreactor for cultivating transformed root cultures for all purposes, although use of ordinary, mechanically agitated reactors is not recommended because of the high stress-sensitivity of the transformed roots (Georgiev *et al.*, 2008).

However, slight changes in the stirred tank reactor design, such as separating the impeller from the culture (using a stainless steel mesh) or simply reducing the agitation speed has allowed successful cultivation of transformed root cultures of *Atropa belladonna* and *Beta vulgaris* (Lee *et al.*, 1999; Georgiev *et al.*, 2006). Table 10.3 lists various types of bioreactor that have been used to cultivate transformed root cultures of several plant species, the final densities obtained and the biomasses productivity.

The disposable (often called single-use) wave bioreactors (which have a wave-induced working motion that ensures mixing and aeration), originally developed for cultivating highly stress-sensitive animal cell cultures, have been successfully modified by Prof Eibl's group from Switzerland for cultivating transformed root cultures from *Hyoscyamus muticus*, *Panax ginseng* (see Table 10.3) and *Harpagophytum procumbens* (Eibl and Eibl, 2008; Eibl *et al.*, 2009). The advantages of the disposable reactors over classical reactors include tremendous flexibility, ease of handling and high biosafety (contamination levels <1% are guaranteed; Eibl and Eibl, 2006). Thus, use of disposable bioreactors in current Good Manufacturing Practices (cGMP) production processes can minimize complex cleaning, sterilization, validation and process costs, with consequent reductions in development times and times-to-market for new products (Eibl and Eibl, 2006). The commercially available Wave reactor™ and BioWave® reactors (with 250 and 300 l working volumes, respectively) may also facilitate the establishment of large-scale processes (sources: www.wavebiotech.com and www.wavebiotech.net).

Mist and trickle-bed reactors seem to offer perfect environments for cultivating transformed root cultures. Mist reactors are types of gas-phase reactors in which ultrasonic systems are used to spray a fine mist of water and nutrients on surfaces of hairy root cultures (in the form of a thin film of liquid medium), which are suspended in a plastic bag. The nutrient solution is collected at the bottom of the bag and recycled through the system. In this way, all of the materials are completely contained and isolated from the environment. Gas-phase reactors can virtually eliminate any oxygen deficiency in dense root beds, while providing a low shear stress environment (Kim *et al.*, 2002). In experiments performed in Prof Weathers' lab in the USA with hairy roots of *A. annua* and *N. tabacum*, grown in 1.5-l and 4-l mist reactors (respectively), biomasses of 14.4 g/l and 5.2 g/l have been attained, respectively (Table 10.3). Although, in the case with transgenic *N. tabacum* the accumulated dry biomass was not considered as high, the amount of murine interleukin-12 was higher, compared to amounts produced in other tested cultivation systems. Ramakrishnan and Curtis (2004) reported very high growth of *H. muticus* hairy roots in 14-l trickle-bed reactors, with final biomass densities significantly exceeding previously reported values (Table 10.3). The bioreactor characterization (experiments performed at 1.6 l and 14 l scales) was sufficient for the cited authors to carry out preliminary design calculations and they concluded that scale-up to at least 10,000 l would be feasible (Ramakrishnan and Curtis, 2004).

Table 10.3. Selected examples of bioreactor configurations (operated in batch mode) used for cultivating hairy root cultures.

Bioreactor type[a]	**Hairy root culture**	**Final density (g DW/l)/ Productivity (g DW/l/day)** [b]	**Reference**
Mechanically driven			
Stirred tank reactor (5 l)	*Beta vulgaris*	12.9 / 0.68	Georgiev et al., 2006
Stirred tank reactor with separate impeller (25 l)	*Atropa belladonna*	6.02 / 0.20	Lee *et al.*, 1999
Wave reactor (0.5 l)	*Panax ginseng*	11.6 / 0.41	Palazon et al., 2003a
Pneumatically driven			
Bubble column reactor (2 l)	*Harpagophytum procumbens*	6.6 / 0.31	Ludwig-Müller *et al.*, 2008
Bubble column reactor (2 l)	*Beta vulgaris*	12.7 / 0.79	Pavlov *et al.*, 2007
Airlift reactor (2 l)	*Nicotiana tabacum*	4.8 / 0.34	Liu *et al.*, 2009
Bed reactors			
Mist reactor (4 l)	*Nicotiana tabacum*	5.2 / 0.37	Liu *et al.*, 2009
Mist reactor (1.5 l)	*Artemisia annua*	14.4 / 0.38	Kim *et al.*, 2002
Trickle bed (14 l)	*Hyoscyamus muticus*	36.2 / 1.45	Ramakrishnan and Curtis, 2004

[a] Bioreactor working volumes are given in parenthesis; [b] DW – dry weight

It should be noted that transformed root cultures have not been used in any commercial production process to date, although the large-scale cultivation of transformed roots (at 500 l scale) was achieved more than a decade ago (Wilson, 1997). A major issue, which should be addressed before commercialization, is the inoculation of large-scale reactors. Unlike in plant cell suspension-based processes, in which the inoculum can be easily transferred pneumatically, transferring inocula of hairy root cultures from small reactors to large-scale reactors might be difficult.

A possible solution was offered by Pavlov *et al.* (2007), who applied fed-batch operation mode for cultivating *Beta vulgaris* hairy roots in a bubble column bioreactor. The recently developed commercial system (involving use of 10,000 l balloon type reactors) for biomass and ginsenoside production from normal (adventitious) roots of *Panax ginseng* in South Korea (Sivakumar *et al.*, 2005), may also provide a solution to the inoculum problem; the inoculation steps prior to the 10,000 l reactors are likely to be readily applicable to hairy roots given the high morphological similarities of adventitious root and transformed root cultures.

Conclusion and Future Prospects

In recent years transformed root cultures have become increasingly serious alternatives for the mass production of high-value plant-derived metabolites, as well as suitable systems for producing recombinant proteins. Today, hairy root cultures can be induced from any plant species, although some are more difficult to transform than others (e.g. monocotyledons). This raises the possibility of inducing transformed root cultures from rare and threatened plants (*e.g.* some medicinal plant species), which would contribute to the preservation of global biodiversity. In addition, transformed root cultures show promising potential for other ecologically valuable applications in phytoremediation, and metabolic engineering offers powerful tools for deliberate adjustment of desired metabolites' biosynthetic pathways. Finally, recent trends in bioreactor design (e.g. the development of single-use bioreactor systems and novel methods for agitating cultures) should lead to significant reductions in process costs and facilitate fulfilment of cGMP requirements. Thus, hairy root cultures have high and increasing potential applicability for diverse purposes.

Acknowledgements

This work has been supported by grants from National Science Fund of Bulgaria (under contract number DO-02-261/2008).

References

Alderete, L.G.S., Talano, M.A., Ibanez, S.G., Purro, S., Agostini, A., Milrad, S.R. and Medina, M.I. (2009) Establishment of transgenic tobacco hairy roots expressing basic peroxidases and its application for phenol removal. *Journal of Biotechnology* 139, 273–279.

Altamura, M.M. (2004) *Agrobacterium rhizogenes rolB* and *rolD* genes: regulation and involvement in plant development. *Plant Cell, Tissue and Organ Culture* 77, 89–101.

Ayora-Talavera, T., Chappell, J., Lozoya-Gloria, E. and Loyola-Vargas, V.M. (2002) Overexpression in *Catharanthus roseus* hairy roots of a truncated hamster 3-hydroxy-3-methylglutaryl-CoA reductase gene. *Applied Biochemistry and Biotechnology* 97, 135–145.

Bavage, A.D., Davies, I.G., Robbins, M.P. and Morris, P. (1997) Expression of an *Antirrhinum* dihydroflavonol reductase gene results in changes in condensed tannin structure and accumulation in root cultures of *Lotus corniculatus* (bird's foot trefoil). *Plant Molecular Biology* 35, 443–458.

Berlin, J., Rügenhagen, C., Dietze, P., Fecker, L.F., Goddijn, O.J.M. and Hoge, J.H.C. (1993) Increased production of serotonin by suspension and root cultures of *Peganum harmala* transformed with a tryptophan decarboxylase cDNA clone from *Catharanthus roseus*. *Transgenic Research* 2, 336–344.

Boominathan, R. and Doran, P.M. (2003) Cadmium tolerance and antioxidative defenses in hairy roots of the cadmium hyperaccumulator, *Thlaspi caerulescens*. *Biotechnology and Bioengineering* 83, 158–167.

Canel, C., Lopes-Cardoso, I., Whitmer, S., van der Fits, L., Pasquali, G., van der Heijden, R., Hoge, J.H.C. and Verpoorte, R. (1998) Effects of over-expression of strictosidine synthase and tryptophan decarboxylase on alkaloid production by cell cultures of *Catharanthus roseus*. *Planta* 205, 414–419.

Chen, X., Zhang, J., Liu, J.-H. And Yu, B.-Y. (2008) Biotransformation of *p*-, *m*-, and *o*-hydroxybenzoic acids by *Panax ginseng* hairy root cultures. *Journal of Molecular Catalysis B: Enzymatic* 54, 72–75.

Coniglio, M.S., Busto, V.D., Gonzales, P.S., Medina, M.I., Milrad, S. and Agositini, E. (2008) Application of *Brassica napus* hairy root cultures for phenol removal from aqueous solutions. *Chemosphere* 72, 1035–1042.

Damiani, F., Paolocci, F., Consonni, G., Crea, F., Tonelli, C. and Arcioni, S. (1998) A maize anthocyanin trans activator induces pigmentation in hairy roots of dicotyledonous species. *Plant Cell Reports* 17, 339–344.

De Luca, V. (2000) Metabolic engineering of crops with the tryptophan decarboxylase of *Catharanthus roseus*. In: Verpoorte, R. and Alfermann, A.W. (eds), *Metabolic Engineering of Plant Secondary Metabolism*. Kluwer Academic Publishers, the Netherlands, pp. 179–194.

Eibl, R. and Eibl, D. (2006) Disposable bioreactors for pharmaceutical research and manufacturing. In: Sorvari, S. and Toldi, O. (eds) *Proceedings of Second International Congress on Bioreactor Technology in Cell, Tissue Cultures and Biomedical Applications*. Karhukopio, OY, Turku, Finland, pp. 12–21.

Eibl, R. and Eibl, D. (2008) Design of bioreactors for plant cell and tissue cultures. *Phytochemistry Reviews* 7, 593–598.

Eibl, R., Werner, S. and Eibl, D. (2009) Disposable bioreactors for plant liquid cultures at litre-scale: review. *Engineering in Life Sciences* 9, 156–164.

Facchini, P.J. and De Luca, V. (2008) Opium poppy and Madagascar periwinkle: model non-model systems to investigate alkaloid biosynthesis in plants. *Plant Journal* 54, 763–784.

Faria, J.M.S., Nunes, I.S., Figueiredo, A.Ch., Pedro, L.G., Trindade, H. and Barroso, J.G. (2009) Biotransformation of menthol and geraniol by hairy root cultures of *Anethum graveolens*: effect on growth and volatile components. *Biotechnology Letters* 31, 897–903.

Furmanowa, M. and Syklowska-Baranek, K. (2000) Hairy root cultures of *Taxus* x *media* var. *Hicksii* Rehd. as a new source of paclitaxel and 10-deacetylbaccatin III. *Biotechnology Letters* 22, 683–686.

Geerlings, A., Hallard, D., Martinez Caballero, A., Lopes Cardoso, I., van der Heijden, R. and Verpoorte, R. (1999) Alkaloid production by a *Cinchona officinalis* Ledgeriana hairy root culture containing constitutive expression constructs of tryptophan decarboxylase and strictosidine synthase cDNAs from *Catharanthus roseus*. *Plant Cell Reports* 19, 191–196.

Gelvin, S.B. (2003) *Agrobacterium*-mediated plant transformation: The biology behind the 'Gene-Jockeying' tool. *Microbiology and Molecular Biology Reviews* 67, 16–37.

Georgiev, M.I., Pavlov, A.I. and Bley, Th. (2006) Betalains by transformed *Beta vulgaris* roots in stirred tank bioreactor: batch and fed-batch processes. In: Sorvari, S. and Toldi, O. (eds), *Proceedings of Second International Congress on Bioreactor Technology in Cell, Tissue Cultures and Biomedical Applications*. Karhukopio OY, Turku, Finland, pp. 22–28.

Georgiev, M.I., Pavlov, A.I. and Bley, Th. (2007) Hairy root type plant *in vitro* systems as sources of bioactive substances. *Applied Microbiology and Biotechnology* 74, 1175–1185.

Georgiev, M., Georgiev, V., Weber, J., Bley, Th., Ilieva, M. and Pavlov, A. (2008) *Agrobacterium rhizogenes*-mediated genetic transformations: A powerful tool for the production of metabolites. In: Wolf, T. and Koch, J. (eds.), *Genetically Modified Plants*. Nova Science Publishers, New York, USA, pp. 99–126.

Giri, A., Dhingra, V., Giri, C.C., Singh, A., Ward, O.P. and Narasu, M.L. (2001) Biotransformations using plant cells, organ cultures and enzyme systems: current trends and future prospects. *Biotechnology Advances* 19, 175–199.

Gonzales, P.S., Capozucca, C.E., Tigier, H.A., Milrad, S.R. and Agositini, E. (2006) Phytoremediation of phenol from wastewater, by peroxidases of tomato hairy root cultures. *Enzyme and Microbial Technology* 39, 647–653.

Goossens, A. and Rischer, H. (2007) Implementation of functional genomics for gene discovery in alkaloid producing plants. *Phytochemistry Reviews* 6, 35–49.

Guillon, S., Tremouillaux-Guiller, J., Pati, P.K., Rideau, M. and Gantet, P. (2006) Hairy root research: recent scenario and exciting prospects. *Current Opinion in Plant Biotechnology* 9, 341–346.

Gujarathi, N.P., Haney, B.J., Park, H.J., Wickramasinghe, S.R. and Linden, J.C. (2005) Hairy roots of *Helianthus annuus*: a model system to study phytoremediation of tetracycline and oxytetracycline. *Biotechnology Progress* 21, 775–780.

Hain, R. and Grimmig, B. (2000) Modification of plant secondary metabolism by genetic engineering. In: Verpoorte, R. and Alfermann, A.W. (eds), *Metabolic Engineering of Plant Secondary Metabolism*. Kluwer Academic Publishers, the Netherlands, pp. 217–213.

Hamill, J.D., Robins, R.J., Parr, A.J., Evans, D.M., Furze, J.M. and Rhodes, M.J.C. (1990) Overexpressing a yeast ornithine decarboxylase gene in transgenic roots of *Nicotiana rustica* can lead to enhanced nicotine accumulation. *Plant Molecular Biology* 15, 27–38.

Hashimoto, T., Yun, D.-J. and Yamada, Y. (1993) Production of tropane alkaloids in genetically engineered root cultures. *Phytochemistry* 32, 713–718.

Huber, Ch., Barhta, B., Harpaintner, R. and Schroeder, P. (2009) Metabolism of acetaminophen (paracetamol) in plants – two independent pathways result in the formation of a glutathione and a glucose conjugate. *Environmental Science and Pollution Research* 16, 206–213.

Hughes, E.H., Hong, S.B., Gibson, S.I., Shanks, J.V. and San, K.Y. (2004) Metabolic engineering of the indole pathway in *Catharanthus roseus* hairy roots and increased accumulation of tryptamine and serpentine. *Metabolic Engineering* 6, 268–276.

Ishikawa, T., Kuroda, H., Okazaki, K., Itoh, K., Mitsui, T. and Hori, H. (2007) Evaluation of roles of amidase which converts indole-3-acetamide to indole-3-acetic acid, in formation of clubroot in turnip. *Bulletin of the Faculty of Agriculture Niigata University* 60, 53–60.

Jouhikainen, K., Lindgren, L., Jokelainen, T., Hiltunen, R., Teeri, T.H. and Oksman-Caldentey, K.-M. (1999) Enhancement of scopolamine production in *Hyoscyamus muticus* L. hairy root cultures by genetic engineering. *Planta* 208, 545–551.

Kim, Y., Wyslouzil, B.E. and Weathers, P.J. (2001) A comparative study of mist and bubble column reactors in the *in vitro* production of artemisinin. *Plant Cell Reports* 20, 451–455.

Kim, Y.J., Weathers, P.J. and Wyslouzil, B.E. (2002) Growth of *Artemisia annua* hairy roots in liquid- and gas-phase reactors. *Biotechnology and Bioengineering* 80, 454–464.

Lambert, E., Goossens, A., Panis, B., Van Labeke, M.C. and Geelen, D. (2009) Cryopreservation of hairy root cultures of *Maesa lanceolata* and *Medicago truncatula*. *Plant Cell, Tissue and Organ Culture* 96, 289–296.

Lee, K.-T., Suzuki, T., Yamakawa, T., Kodama, T., Igarashi, Y. and Shimomura, K. (1999) Production of tropane alkaloids by transformed root cultures of *Atropa belladonna* in stirred tank bioreactors with a stainless steel net. *Plant Cell Reports* 18, 567–571.

Li, M. and Leung, D.W.M. (2003) Root induction in radiata pine using *Agrobacterium rhizogenes*. *Electronic Journal of Biotechnology* 6, 251–258.

Liu, C., Towler, M.J., Medrano, G., Cramer, C.L. and Weathers, P.J. (2009) Production of mouse interleukin-12 is greater in tobacco hairy roots grown in a mist reactor than in an airlift reactor. *Biotechnology and Bioengineering* 102, 1074–1086.

Loyola-Vargas, V.M. and Miranda-Ham, M.L. (1995) Root culture as a source of secondary metabolites of economic importance. In: Arnason, J.T., Mata, R. and Romeo. J.T. (eds) *Recent Advances in Phytochemistry,* Vol. 29. *Phytochemistry of Medicinal Plants*. Plenum Press, New York, pp. 217–248.

Ludwig-Müller, J., Georgiev, M. and Bley, Th. (2008) Metabolite and hormonal status of hairy root cultures of Devil's claw (*Harpagophytum procumbens*) in flasks and in a bubble column bioreactor. *Process Biochemistry* 43, 15–23.

Magnotta, M., Murata, J., Chen, J.X. and De Luca, V. (2007) Expression of deacetylvindoline-4-O-acetyltransferase in *Catharanthus roseus* hairy roots. *Phytochemistry* 68, 1922–1931.

McAfee, B.J., White, E.E., Pelcher, L.E. and Lapp, M.S. (1993) Root induction in pine (*Pinus*) and larch (*Larix*) spp. using *Agrobacterium rhizogenes. Plant Cell, Tissue and Organ Culture* 34, 53–62.

Memelink, J. and Gantet, P. (2007) Transcription factors involved in terpenoid indole alkaloid biosynthesis in *Catharanthus roseus*. *Phytochemistry Reviews* 6, 353–362.

Menzel, G., Harloff, H.-J. and Jung, C. (2003) Expression of bacterial poly(3-hydroxybutyrate) synthesis genes in hairy roots of sugar beet (*Beta vulgaris* L.). *Applied Microbiology and Biotechnology* 60, 571–576.

Moyano, E., Palazon, H., Bonfill, M., Osuna, L., Cusido, R.M., Oksman-Caldentey, K.-M. and Pinol, M.T. (2007) Biotransformation of hyoscyamine into scopolamine in transgenic tobacco cell cultures. *Journal of Plant Physiology* 164, 521–524.

Nemoto, K., Hara, M., Goto, S., Kasai, K., Seki, H., Suzuki, M., Oka, A., Muranaka, T. and Mano, Y. (2009) The *aux1* gene of the Ri plasmid is sufficient to confer auxin autotrophy in tobacco BY-2 cells. *Journal of Plant Physiology* 166, 729–738.

Nilsson, O. and Olsson, O. (1997) Getting to the root: The role of the *Agrobacterium rhizogenes rol*-genes in the formation of hairy roots. *Physiologia Plantarum* 100, 463–473.

Offringa, I.A., Melchers, L.S., Regensburg-Tuink, A.J.G., Costantino, P., Schilperoort, R.A. and Hooykaas, P.J.J. (1986) Complementation of *Agrobacterium tumefaciens* tumour-inducing aux mutants by genes from the TR-region of the Ri plasmid of *Agrobacterium rhizogenes*. *Proceedings of the National Academy of Sciences of the United States of America* 83, 6935–6939.

Oksman-Caldentey, K.-M. and Inze, D. (2004) Plant cell factories in the post-genomic era: new ways to produce designer secondary metabolites. *Trends in Plant Science* 9, 433–440.

Oller, A.L.W., Agostini, E., Talano, M.A., Capozucca, C., Milrad, S.R., Tigier, H.A. and Medina, M.I. (2005) Overexpression of a basic peroxidase in transgenic tomato (*Lycopersicon esculantum* Mill. Cv. Pera) hairy roots increases phytoremediation of phenol. *Plant Science* 169, 1102–1111.

Orden, A.A., Bisogno, F.R., Cifuente, D.A., Giordano, O.S. and Sanz, M.K. (2006) Asymmetric bioreduction of natural xenobiotic diketones by *Brassica napus* hairy roots. *Journal of molecular Catalysis B: Enzymatic* 42, 71–77.

Palazon, J., Cusido, R.M., Roig, C. and Pinol, M.T. (1997) Effect of *rol* genes from *Agrobacterium rhizogenes* TL-DNA on nicotine production in tobacco root cultures. *Plant Physiology and Biochemistry* 35, 155–162.

Palazon, J., Mallol, A., Eibl, R., Lettenbauer, C., Cusido, R.M. and Pinol, M.T. (2003a) Growth and ginsenoside production in hairy root cultures of *Panax ginseng* using a novel bioreactor. *Planta Medica* 69, 344–349.

Palazon, J., Moyano, E., Cusido, R.M., Bonfill, M., Oksman-Caldentey, K.-M. and Pinol, M.T. (2003b) Alkaloid production in *Duboisia* hybrid hairy roots and plants overexpressing the *h6h* gene. *Plant Science* 165, 1289–1295.

Palazon, J., Navarro-Ocaña, A., Hernandez-Vazquez, L. and Mirjalili, M.H. (2008) Application of metabolic engineering to the production of scopolamine. *Molecules* 13, 1722–1742.

Park, S.-U. and Facchini, P.J. (2000) *Agrobacterium rhizogenes*-mediated transformation of opium poppy, *Papaver somniferum* L., and California poppy, *Eschscholzia californica* Cham., root cultures. *Journal of Experimental Botany* 51, 1005–1016.

Pavlov, A.I., Georgiev, M.I. and Bley, Th. (2007) Batch and fed-batch production of betalains by Red beet (*Beta vulgaris*) hairy roots in a bubble column reactor. *Zeitschrift fuer Naturforschung C* 62, 439–446.

Pollmann, S., Neu, D. and Weiler, E.W. (2003) Molecular cloning and characterization of an amidase from *Arabidopsis thaliana* capable of converting indole-3-acetamide into the plant growth hormone, indole-3-acetic acid. *Phytochemistry* 62, 293–300.

Potterat, O. and Hamburger, M. (2008) Drug discovery and development with plant-derived compounds. In: Petersen, F. and Amstutz, R. (eds), *Progress in Drug Research*. Birkhäuser Verlag, Basel, Switzerland, pp. 47–118.

Ramakrishnan, D. and Curtis, W.R. (2004) Trickle-bed root culture bioreactor design and scale-up: growth, fluid-dynamics and oxygen mass transfer. *Biotechnology and Bioengineering* 88, 248–260.

Rezek, J., Macek, T., Mackova, M. and Triska, J. (2007) Plant metabolites of polychlorinated biphenyls in hairy root culture of black nightshade *Solanum nigrum* SNC-90. *Chemosphere* 69, 1221–1227.

Rischer, H., Oresi, M., Seppänen-Laakso, T., Katajamaa, M., Lammertyn, F., Ardiles-Diaz, W., Van Montagu, M.C.E., Inzé, D., Oksman-Caldentey, K.M. and Goossens, A. (2006) Gene-to metabolite networks for terpenoid indole alkaloid biosynthesis in *Catharanthus roseus* cells. *Proceedings of the National Academy of Sciences of the United States of America* 103, 5614–5619.

Santos de Araujo, B., Dec, J., Bollag, J.M. and Pletsch, M. (2006) Uptake and transformation of phenol and chlorophenols by hairy root cultures of *Daucus carota*, *Ipomoea batatas* and *Solanum aviculare*. *Chemosphere* 63, 642–651.

Sauerwein, M., Yamazaki, T. and Shimomura, K. (1991) Hernandulcin in hairy root cultures of *Lippia dulcis*. *Plant Cell Reports* 9, 579–581.

Sivakumar, G., Yu, K.W. and Paek, K.Y. (2005) Production of biomass and ginsenosides from adventitious roots of *Panax ginseng* in bioreactor cultures. *Engineering in Life Sciences* 5, 333–342.

Sommer, S., Köhle, A., Yazaki, K., Shimomura, K., Bechthold, A. and Heide, L. (1999) Genetic engineering of shikonin biosynthesis in hairy root cultures of *Lithospermum erythrorhizon* transformed with the bacterial *ubiC* gene. *Plant Molecular Biology* 39, 683–693.

Straczek, A., Wannijn, J., Van Hees, M., Thijs, H. and Thiry, Y. (2009) Tolerance of hairy roots of carrots to U chronic exposure in a standardized *in vitro* device. *Environmental and Experimental Botany* 65, 82–89.

Suzuki, M., Kobayashi, K., Nagata, N., Matsuda, O., Iba, K., Yoshida, S. and Muranaka, T. (2008) Novel root culture system using a recessive mutant with a rooty phenotype. *Plant Biotechnology* 25, 197–200.

Tabata, H. (2004) Paclitaxel production by plant-cell-culture technology. *Advances in Biochemical Engineering/Biotechnology* 87, 1–23.

Thomashow, L.S., Reeves, S. and Thomashow, M.F. (1984) Crowngall oncogenesis: evidence that a T-DNA gene from the Agrobacterium Ti plasmid pTiA6 encodes an enzyme that catalyzes synthesis of indoleacetic acid. *Proceedings of the National Academy of Sciences of the United States of America* 81, 5071–5075.

Thorup, J.E., McDonald, K.A., Jackman, A.P., Bhatia, N. and Dandekar, A.M. (1994) Ribosome-inactivating protein production from *Trichosanthes kirilowii* plant cell cultures. *Biotechnology Progress* 10, 345–352.

Tzfira, T. and Citovsky, V. (2006) *Agrobacterium*-mediated genetic transformation of plants: Biology and biotechnology. *Current Opinion in Biotechnology* 17, 147–154.

Van der Fits, L. and Memelink, J. (2000) ORCA3, a jasmonate responsive transcriptional regulator of plant primary and secondary metabolism. *Science* 289, 295–297.

Vasilev, N., Elfahmi, Bos, R., Kayser, O., Momekov, G., Konstantinov, S. and Ionkova, I. (2006) Production of justicidin B, a cytotoxic arylnaphthalene lignan from genetically transformed root cultures of *Linum leonii*. *Journal of Natural Products* 69, 1014–1017.

Veena, V. and Taylor, C.G. (2007) *Agrobacterium rhizogenes*: recent developments and promising applications. *In vitro Cellular and Developmental Biology - Plant* 43, 383–403.

Verpoorte, R., van der Heijden, R. and Memelink, J. (2000) General strategies. In: Verpoorte, R. and Alfermann, A.W. (eds), *Metabolic Engineering of Plant Secondary Metabolism*. Kluwer Academic Publishers, the Netherlands, pp. 31–50

Verpoorte, R., Contin, A. and Memelink, J. (2002) Biotechnology for the production of plant secondary metabolites. *Phytochemistry Reviews* 1, 13–25.

Whitmer, S., Canel, C., Hallard, D., Goncalves, C. and Verpoorte, R. (1998) Influence of precursor availability on alkaloid accumulation by a transgenic cell line of *Catharanthus roseus*. *Plant Physiology* 116, 853–857.

Wilson, P.D.G. (1997) The pilot-scale cultivation of transformed roots. In: Doran, P.M. (ed.) *Hairy Roots and Culture Applications*. Harwood Academic Publishers, Amsterdam, pp. 179–190.

Wink, M., Alfermann, A.W., Franke, R., Wetterauer, B., Distl, M., Windhövel, J., Krohn. O., Fuss, E., Garden, H., Mohagheghzadeh, A., Wildi, E. and Ripplinger, P. (2005) Sustainable bioproduction of phytochemicals by plant *in vitro* cultures: anticancer agents. *Plant Genetic Resources* 3, 90–100.

Xue, S.-H., Luo, X.-J., Wu, Z.-H., Zhang, H.-L. and Wang, X.-Y. (2008) Cold storage and cryopreservation of hairy root cultures of medicinal plant *Eruca sativa* Mill., *Astragalus membranaceus* and *Gentiana macrophylla* Pall. *Plant Cell, Tissue and Organ Culture* 92, 251–260.

Yibrah, H.S., Grönroos, R., Lindroth, A., Franzen, H., Clapham, D. and von Arnold, S. (1996) *Agrobacterium rhizogenes*-mediated induction of adventitious rooting from *Pinus contorta* hypocotyls and the effect of 5-azacytidine on transgene activity. *Transgenic Research* 5, 75–85.

Yun, D.-J., Hashimoto, T. and Yamada, Y. (1992) Metabolic engineering of medicinal plants: transgenic *Atropa belladonna* with an improved alkaloid composition. *Proceedings of the National Academy of Sciences of the United States of America* 89, 11799–11803.

Chapter 11

Genetic Transformation of *Catharanthus roseus* (L.) G. Don. for Augmenting Secondary Metabolite Production

Rajesh Arora, Archana Mathur, Ajay K. Mathur and C.M. Govil

Introduction

Catharanthus is a perennial short-lived plant that commonly grows in tropical countries and is mainly cultivated as an ornamental plant in the tropical and subtropical world. The plant grows in warm regions of the world and especially in the Southern United States. The plant normally grows 8 to 18 inches in length and a 1-foot spread, although the trailing types can spread up to 2 feet. It is highly branched and develops a woody base. The leaves of the plant are petiolate, oblong in shape, thick and leathery in texture, while the arrangement is opposite or alternate (Virmani *et al.*, 1978). The plant is referred to by a variety of names in various countries (Table 11.1.) and the various species are mildly poisonous.

Flowers of the plant are single and never double and most modern varieties show overlapping petals. The colour of the flowers ranges from pink, white or a mixture of both to pale pink in colour with a dark violet dot in the centre. The chromosome number for all Catharanthus species is 2n=16. However, tetraploid plants grow faster and the flowers are reported to be bigger in size (Virmani *et al.*, 1978). The genus *Catharanthus* comprises eight species (Fig. 11.1). Seven species of the genus are known from Madagascar; one is restricted to India and Sri Lanka, while one species is cultivated in China. The plant is believed to be a native of the West Indies. The plant is more commonly known as Madagascar periwinkle.

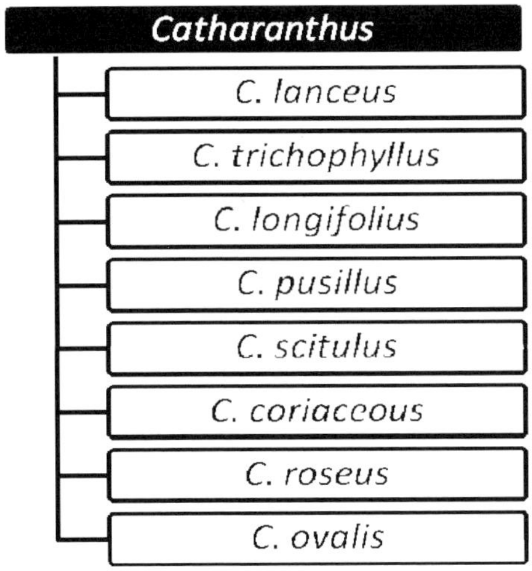

Fig. 11.1. The eight common species of *Catharanthus*.

Catharanthus: A Unique Medicinal Plant

The genus name, *Catharanthus*, refers to 'pure flower'. *Catharanthus* is often considered synonymous to *Vinca*, however, certain authors disagree with the proposition that *Catharanthus* and *Vinca* are one and the same. As of now, *Catharanthus* and *Vinca* are considered two different species.

Catharanthus roseus: A source of medicinally useful terpenoid indole alkaloids (TIAs)

The *Catharanthus* plant has been used in traditional medicine in several parts of the world for a long time for treating a variety of ailments and the existence of local names in most countries is evidence enough regarding the importance of this medicinal plant (Table 11.1). *Catharanthus roseus* comes in various hues and shades, the most common being red and white (Fig. 11.2).

However, in recent years *Catharanthus roseus* has gained popularity as a medicinal plant mainly for its curative and palliative role in cancer therapy (Table 11.2 and Table 11.3). The plant is almost synonymous with cancer therapy and no wonder then that the National Cancer Council of Malaysia (Majlis Kanser National, MAKNA) uses the periwinkle logo as its symbol of hope for cancer patients (Loh, 2008).

The serendipitous discovery of the antineoplastic property of *Catharanthus* alkaloids in the 1950s paved the path for the discovery of a number of plant-derived anticancer drugs. The *Catharanthus* alkaloids are of wide importance in clinical medicine and vincristine (VCR) and vinblastine (VLB) (Fig. 11.3) form an indispensable part of standard chemotherapy regimens. VCR and VLB are integral components of virtually every curative regimen for metastatic malignancy and are extensively used in adjuvant and neoadjuvant drug regimens. Leads obtained from *Catharanthus* have proved that natural products indeed are reliable sources of medicine and can solve several of the health problems being faced by humans (Sharma and Arora, 2006; Arora *et al.*, 2010).

Table. 11.1. A few vernacular names of *Catharanthus roseus*.

Common name(s)	Country
Ainskati, Billaganneru, Nayantara, Sadabahar, Ushamanjairi	India
Atay-biya, Chichirica, Kantotan, Periwinkle, Tsitsirika	Philippines
Boa-noite, Congorca	Brazil
Brown man's fancy, Consumption bush, Old maid, Periwinkle, Pink flower, Ram goat rose, Red rose, Sailor's flower, White tulip	West Indies
Caca poule	Dominica
Chatilla	Guatemala
Chavelita	Peru
Dua can	Vietnam
Liluvha	Venda
Madagascar periwinkle	Madagascar
Maua	Kenya
Mini-mal, Patti-poo	Sri Lanka
Nayantara	Bangladesh
Nichinich-so, Nichinichi-so	Japan
Ninfa	Mexico
Periwinkle	Jamaica, USA
Pervenchede de Madagascar	French Guiana
Phaeng phoi farang, Phang-puai-fa-rang	Thailand
Sadabahar	Pakistan
Saponaire	Rodrigues Islands
Tiare-tupapaku-kimo	Cook Islands

Fig. 11.2. The most common red (top) and white varieties (bottom) of *Catharanthus roseus*.

Table 11.2. *Catharanthus* alkaloids used in the treatment of different neoplasms.

Alkaloid	Neoplasms
Vincristine	Acute lymphocytic leukaemia
	Lymphomas (lymphocytic, mixed cellular, histiocytic, non-differentiated, nodular and diffuse types)
	Hodgkin's disease
	Lymphosarcoma
	Carcinomas (breast, cervix, prostate, choriocarcinoma)
	Primary brain tumours (Astrocytomas)
	Neuroblastoma
Vinblastine	Pediatric tumours
	Embryonal rhabdomyosarcoma
	Wilm's tumour
	Monocytic leukaemia
	Lymphomas
	Hodgkin's disease
	Lymphosarcoma
	Mycosis fungoides
	Reticulum cell sarcoma
Vindesine	Carcinomas
	Methotrexate-resistant choriocarcinoma
	Kaposi's sarcoma
	Metastatic testicular cancer
	Non-small cell lung cancer
	Carcinomas of the breast

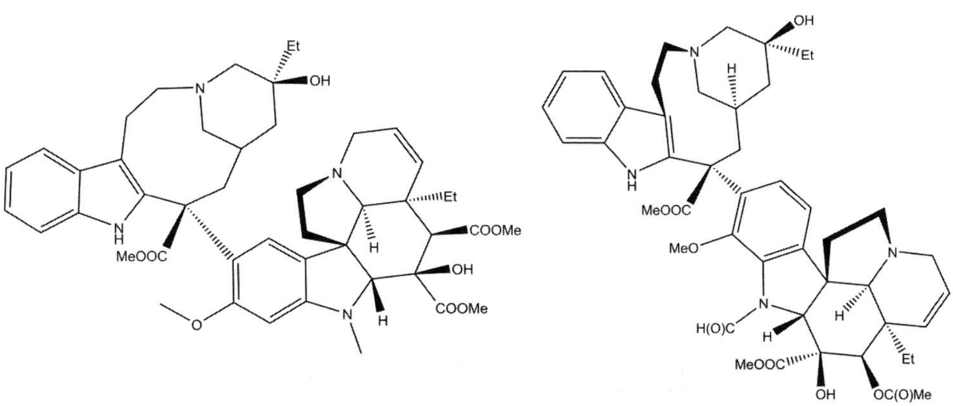

Fig. 11.3. Chemical structures of vincristine (left) and vinblastine (right), the clinically and economically important terpenoid indole alkaloids of *Catharanthus roseus*.

Table 11.3. Curative and palliative regimens containing bisindole *Catharanthus* alkaloids.

Acronym	**Drug Combinations**	**Disease**
ABOS	Doxorubicin, bleomycin, vincristine, streptozocin	
ABV	Doxorubicin, bleomycin, vinblastine	
ABVD	Daunomycin, Bleomycin, vinblastine, dacarbasine	Hodgkin's lymphoma
EVAP	Etoposide,vinblastine, adriamycin, prednisone	
MOPP	Nitrogen mustard, vincristine, procarbazine, prednisone	
CHOP	Cyclophosphamide, daunomycin, vincristine, prednisone	Non-Hodgkin's lymphoma
PVB	Cisplatinum, vinblastine, bleomycin	Testicular cancer
COP	Cyclophosphamide, vincristine, prednisone	Lymphoma
MACOP-B	Methotrexate, daunomycin, cyclophosphamide, vincristine, prednisone, bleomycin	
MVAC	Methotrexate, vinblastine, daunomycin, cyclophosphamide	Bladder cancer
POC	Procarbazine, vincristine, cyclophosphamide	Melanoma, Small-cell lung cancer
VATH	Vinblastine, daunomycin, Thio-TEPA, Halostin	Breast cancer
VP	Vinblastine/vindesine, cisplatinum	Non-small cell lung cancer
-	Doxorubicin, daunorubicin, bleomycin, vincristine and vinblastine	AIDS-Kaposi's sarcoma

Catharanthus roseus contains more than 120 different types of alkaloids, designated as terpenoid indole alkaloids (TIAs). Among the TIAs, vincristine (VCR) and vinblastine (VLB) find extensive use in the treatments of leukaemias, lymphomas, small cell lung cancer and other malignancies (Noble, 1990; Pratt *et al.*, 1994). Vincristine has been an integral component of combination chemotherapy regimens for ALL (acute lymphoblastic leukaemia) and plays an important role in the treatment of Hodgkin's and non-Hodgkin's lymphomas. It has also been used in combination with other agents to treat Wilm's tumour, Ewing's sarcoma, neuroblastoma, and rhabdomyosarcoma in children; and multiple myeloma, breast cancer and small cell lung cancer in adults. Vinblastine is used for the treatment of cell cancers of the testes and advanced Hodgkin's disease, and is commonly used in combination with other agents to treat carcinomas of the breast and bladder, Kaposi's sarcoma and other neoplasms (Rowinsky and Donehower, 1997).

The indole alkaloids of *Catharanthus* represent a market of considerable economic interest and there is a potential in-house demand as well as a world-wide international market. The anti-cancer *Catharanthus* alkaloids have a market value as follows: vincristine ($3,5,00,000 per kg); vinblastine ($1,000,000/ kg) (Veltkamp *et al.*, 1985). Their retail value is more than $20,000/ g (Curtin, 1983). The high prices are mainly due to the costly methods currently available for production of these compounds, and the low yields obtained from the source plants, e.g., the yield of vincristine from C. *roseus* whole plants is of the order of 0.0003–0.0005%, which happens to be the lowest yield of any medicinally useful alkaloid produced on a commercial basis (Taylor and Farnsworth, 1975; Curtin, 1983). Nearly 500 kg of leaves are needed to produce just 1 g of purified vincristine, meaning thereby that about 10,000–15,000 kg leaves are required to produce just 30 g of the drug. Up to 16 weeks may be required for processing. This is one of the reasons behind the high cost of these alkaloids. In the 1970s, the annual world supply of these drugs was of the order of 4.5 kg. The current annual demand for the dimeric/monomeric alkaloids is as follows: VCR: 1 kg per year; VLB: 12 kg per year; ajmalicine: 5000 kg/ year.

India happens to be the third largest manufacturer of vinblastine and vincristine in the world and these alkaloids are exported to Europe and America (Rama Rao and Gurjar, 1990). In 1985–86, Indian exports of *Catharanthus* roots alone were worth US$140,625, and in 1987–88, US$171,875. In 1993–94 *Vinca* alkaloids alone worth > US$210,000 were exported (Chaudhri, 1996). The demand of these alkaloids in the international market has been steadily increasing.

Catharanthus roseus: Model medicinal plant system for biotechnological interventions

World-wide demand of TIAs, and the miniscule amount in which they are produced within the plant, are the factors mainly responsible for the exorbitantly high cost of these compounds. In view of this, tremendous efforts have been directed towards increasing the content of VCR and VLB in the plants by modern biological techniques, including cell line selection, fungal elicitation, biotransformation, and genetic transformation with *Agrobacterium rhizogenes* (van der Heijden *et al.*, 1989, 1998; Saito *et al.*, 1992; Verpoorte *et al.*, 1993; Yeoman and Yeoman, 1997; Arora, 1998; Batra *et al.*, 2004; Arora *et al.*, 2005; Sharma and Arora, 2006; Christie *et al.*, 2009; Junaid *et al.*, 2009; Natali *et al.*, 2009).

The establishment of 5-methyltryptophan (5-MT)-resistant callus lines has been reported (Arora, 1998; Seth and Mathur, 2005). Seth and Mathur (2005) isolated four variant callus lines showing differential tolerance to 5-MT, a tryptophan analogue. The selected lines showed higher alkaloid content in the presence of 5-MT stress. The presence of catharanthine and vindoline were demonstrated in two lines Nir L-60 and C6 L-120, indicating a positive influence of tryptophan accumulation on diverting the metabolic flux towards synthesis of the monomeric indole alkaloids. Shimoda *et al.*, (2007) isolated four disaccharides from the cell suspension of *Catharanthus roseus* after incubation with capsaicin and 8-nordihydrocapsaicin in addition to glucosides and in undifferentiated cultures. Sriram *et al.* (2007) studied flux quantification in central carbon metabolism of C. *roseus* hairy roots by ^{13}C labelling and bondomer balancing. Echevarría-Machado *et al.*, (2007) studied the response of transformed roots of *C. roseus* to very low concentrations of salicylic acid. Pomahačová *et al.* (2009) recently reported improved accumulation of improved accumulation of ajmalicine and tetrahydroalstonine in *Catharanthus* cells expressing an ABC transporter. Jaleel *et al.* (2007) reported on the accumulation of ajmalicine in *Catharanthus roseus* after treatment with giberellic acid.

The molecular events that regulate genes and the activators of terpenoid indole alkaloid metabolism are far from being completely understood and much needs to be unraveled. Several workers have attempted to elucidate the structure and function of key enzymes and the enzymatic pathways and the molecular regulation with a view to augmenting secondary metabolite production (Pasquali *et al.*, 1992; van der Fits and Memelink, 2000; Siberil *et al.*, 2001; van der Heijden *et al.*, 2004; Blondy *et al.*, 2005; Courdavault *et al.*, 2005; Dutta *et al.*, 2005, 2007; Barleben *et al.*, 2007; Bernhardt *et al.*, 2007; Suttipanta *et al.*, 2007; Kutchan *et al.*, 2008; Stöckigt *et al.*, 2008; Pereira *et al.*, 2009). A simplified TIA biogenetic pathway is represented in Fig. 11.4.

Fig. 11.4. The Terpenoid Indole Alkaloid Biogenetic Pathway.

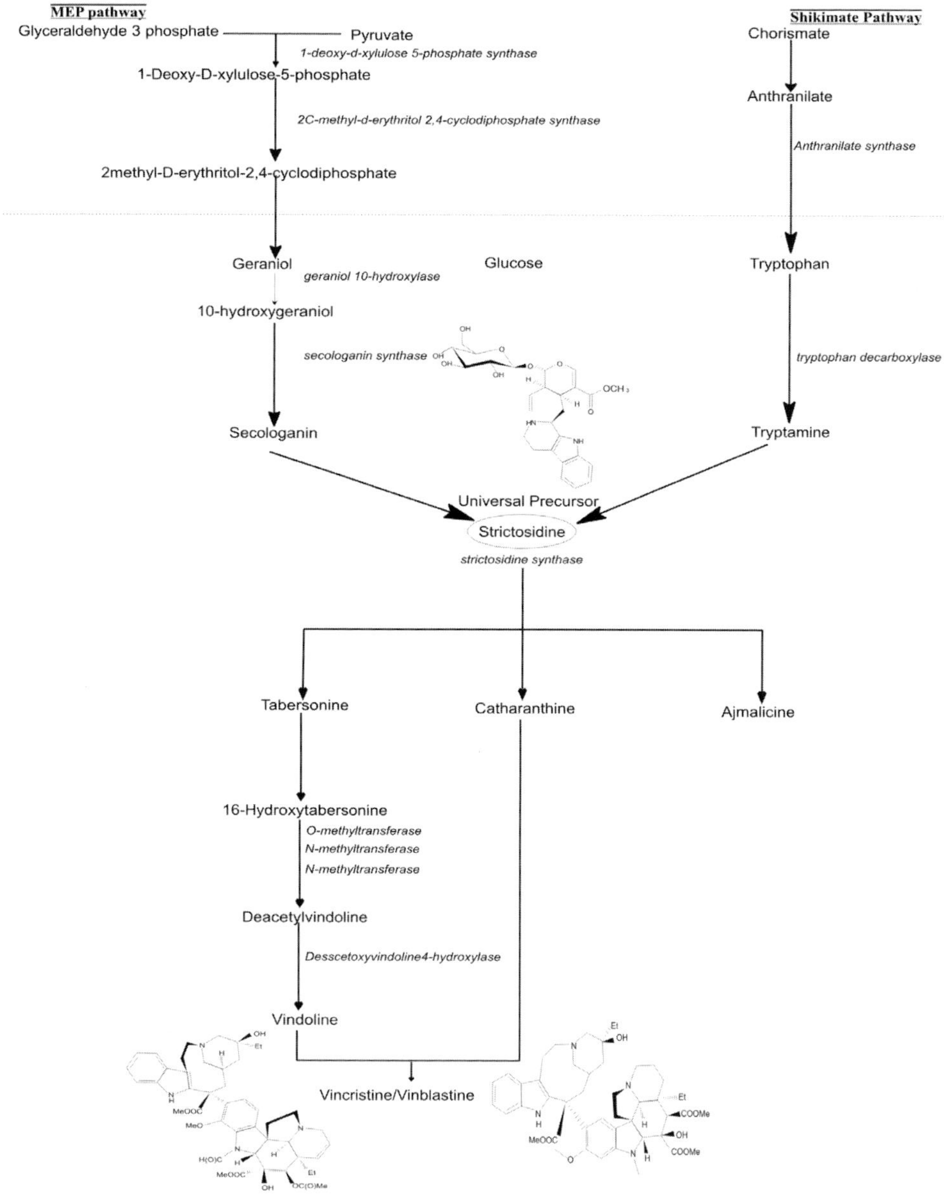

Metabolic engineering approaches have been attempted by several other workers (Rischer *et al.*, 2006; Kutchan *et al.*, 2008; Christie *et al.*, 2009; Pereira *et al.*, 2009).

The Promise of Hairy Roots

Hairy roots, which are produced as a result of transfer of T-DNA from *A. rhizogenes*- a gram-negative bacterium, into the plant genome (Banerjee *et al.*, 1995; Doran, 1997) have been the subject of intense investigation due to their immense potential as an alternative source of TIAs. However, *Catharanthus* has been a fairly recalcitrant crop as far as genetic transformations are concerned, and in particular establishment of fast-growing clones has been rather difficult. Several attempts have been made in recent years to obtain fast-growing clones and increase the levels of TIA accumulation in *in vitro* plant organ systems like hairy roots (Hughes *et al.*, 2004; Arora *et al.*, 2005). Hong *et al.*, (2006) established *Catharanthus roseus* hairy root cultures that were transgenic for the *rol ABC* genes from T-DNA of the agropine-type *Agrobacterium rhizogenes* strain A_4. *The rol ABC* hairy root lines exhibited a wild-type hairy root syndrome in terms of growth and morphology on solid medium. The transgenic hairy roots did not produce detectable levels of mannopine and agropine which, in contrast, are often synthesized abundantly in wild-type hairy root lines. The absence of these opines did not appear to cause the *rol ABC* lines to have higher levels of terpenoid indole alkaloids than wild-type hairy root lines. Unlike wild-type lines, *rol ABC* lines produced very similar levels of total alkaloids despite wide variations in individual alkaloid contents. This work demonstrated that the three genes *rol ABC* are sufficient to induce high-quality hairy roots in *Catharanthus roseus*.

Chung *et al.* (2009) have recently isolated four new flavonoid glucosides viz., 3′,4′-di-O-methylquercetin-7-O-[(4″→13‴)-2‴,6‴,10‴,14‴-tetramethylhexadec-13‴-ol-14‴-enyl]-β-d-glucopyranoside (**1**), 4′-O-methylkaempferol-3-O-[(4″→13‴)-2‴,6‴,10‴,14‴-tetramethylhexadecan-13‴-olyl]-β-d-glucopyranoside (**2**), 3′,4′-di-O-methylbutin-7-O-[(6″→1‴)-3‴,11‴-dimethyl-7‴-methylenedodeca-3‴,10‴-dienyl]-β-d-glucopyranoside (**3**), and 4′-O-methylbutin-7-O-[(6″→1‴)-3‴,11‴-dimethyl-7‴-hydroxymethylenedodecanyl]-β-d-glucopyranoside (**4**), along with the three known compounds, from the methanolic extract of *Catharanthus roseus* hairy roots. The structures of these compounds were elucidated spectroscopically. The new flavonoid glucosides inhibited both MMP-9 activity and TNF-α production in THP-1 cells treated with lipopolysaccharide. These results do show that it is possible to transform *Catharanthus* and produce a variety of chemical compounds. Magnotta *et al.* (2007) have recently reported expression of deacetylvindoline-4-O-acetyltransferase in *Catharanthus roseus* hairy roots.

In this paper we report the successful induction of hairy root cultures in some elite Indian genotypes of *C. roseus* using *Agrobacterium rhizogenes* strains and optimization of culture conditions for some selected fast-growing clones.

Catharanthus roseus: The Medicinal Plant Model System

Five genotypes of *Catharanthus roseus viz.*, Red-flowered (4n=32), White-flowered (4n=32), Nirmal (2n=16), BRT-1 and BRT-2 were tested for their susceptibility towards different *A. rhizogenes* strains. The explants (leaf, petiole, ovule, embryo, petal, microshoots, anthers and sepals), obtained from glasshouse-grown plants, were surface sterilized using 0.1% $HgCl_2$ solution for 2–6 min followed by washing with sterile double distilled water.

Agrobacterium rhizogenes strains: procurement, culture and maintenance

The bacterial strains used for transformation were obtained from a variety of sources, both within India and abroad (Table 11.4.). Transformation efficiency and host-specificity of *Agrobacterium rhizogenes* strains viz., A_4, LBA-9402, NCPPB-1855, SV_2, SV_4, R-1000, ATCC-15834, ATCC-11325 and MAFF-301724 was tested. The *A. rhizogenes* strains were maintained on solid YMB plates (Hooykaas *et al.*, 1977) by streaking with an inoculating loop. The bacteria were then allowed to grow in liquid medium at 25±2°C on a rotary shaker (New Brunswick, USA) at 100 rpm. Bacterial cells in their exponential phase of growth (OD 1.0 at 660 nm) were used for infection.

Table 11.4. List of *A. rhizogenes* strains used for inducing hairy roots in *Catharanthus roseus*.

Strain	Opine	Explants tested	Courtesy
A_4	Mannopine	Leaf Petiole Ovule Embryo Petal Microshoots Anthers Sepals	Dr D. Tepfer, France
A_4	Mannopine		Dr J. Hamill, UK
A_4	Mannopine		Dr D.P.S. Verma, USA
LBA-9402	Agropine		Dr G. Ooms, UK
SV_4	Mannopine		Dr D. Tepfer, France
SV_2	Agropine		Dr D.Tepfer, France
MAFF-301724	Mikimopine		Dr S. Kiyokawa, Japan
NCPPB-1855	Agropine		IARI*, India
R-1000	-		IARI*, India
ATCC-15834	Agropine		ATCC#, USA
ATCC-11325	Mannopine		ATCC#, USA

* IARI: Indian Agricultural Research Institute, New Delhi, India
\# ATCC: American Type Culture Collection, USA

Infection and co-cultivation of explants

Pre-sterilized leaf, petiole, petal, sepal, embryo, ovule, anther and microshoot explants were inoculated with 48 h old cultures of various *A. rhizogenes* strains by wounding with sterile needles. The procedure is represented in Fig. 11.5. Untreated explants were also simultaneously plated to serve as controls. Freshly inoculated explants were kept in darkness during the first 24 h to prevent the bacteria from drying under illumination during the 'window of competence' for transformation of wounded tissue (Binns and Thomashow, 1988). The infected explants were co-cultivated for 3–7days on a semi-solid hormone-free medium, and thereafter transferred onto a maintenance medium containing Kn (1.0 mg/l) + BAP (0.4 mg/l) + NAA (0.2 mg/l). The explants were disinfected by transferring to the maintenance medium supplemented with 250 μg/l cefotaxime (Sigma) or cephalexin (Lupin Laboratories, India) + 100 μg/l ampicillin for three to four passages of 3–5 days each, following which the concentration of cephalexin was reduced to 100 μg/l and the cultures were maintained on this medium for another two to four passages.

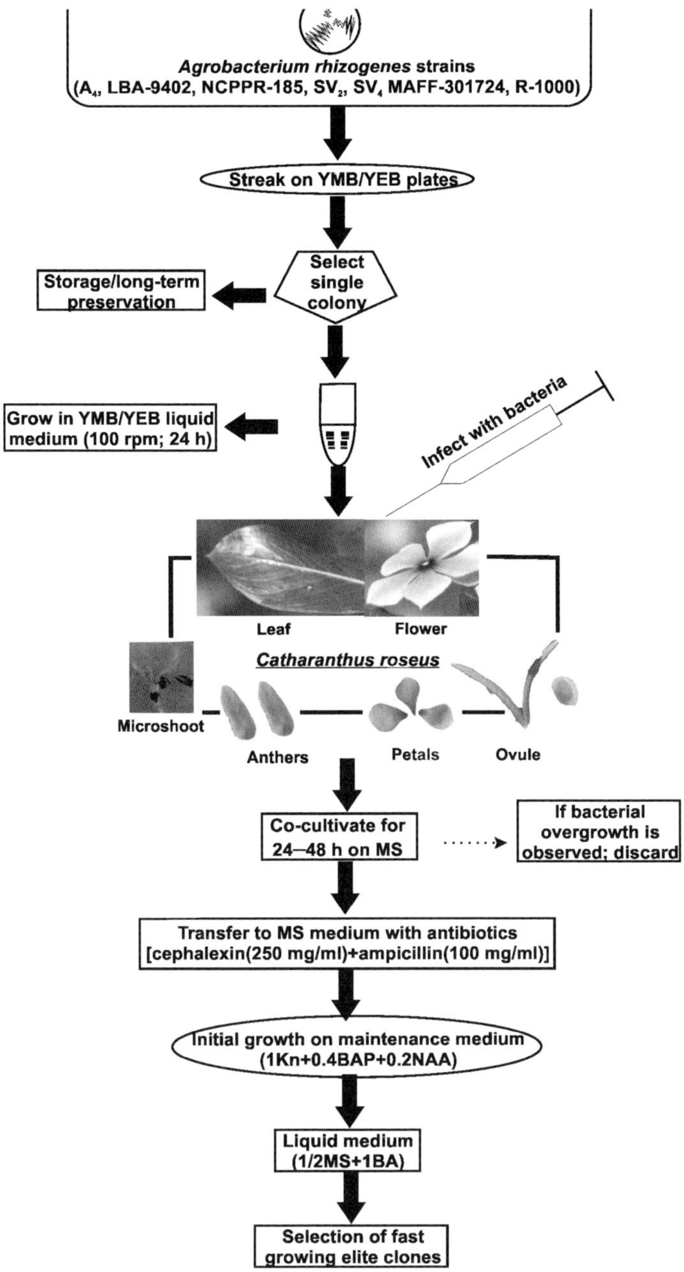

Fig. 11.5. Infection of different explants of *C. roseus* with different strains of *A. rhizogenes*.

Growth and maintenance of axenic hairy root cultures

The frequency of transformation and development of the hairy roots were studied during a 4–8 week culture passage. The frequency of transformation was determined as the number of explants with hairy roots/total number of explants. The hairy roots were excised and transferred to maintenance medium, initially on solid and later on liquid medium. The actively growing, axenic (bacteria-free), hairy roots were then cultured in 60 ml of liquid medium supplemented with IBA (1.0 mg/l). The cultures were kept on the orbital shaker with a constant shaking speed of 80 rpm at 25±2°C under dark conditions. The roots were subcultured at 4 wk intervals, while the medium was changed every 15 days. The individual transformation events were maintained as separate clones.

Confirmation of transformed nature of hairy roots

The transformed hairy roots (50 mg fresh weight) were extracted and analysed by paper electrophoresis and silver staining, according to the method of Petit *et al.* 1983 and Morgan *et al.* 1989. 5–10 μl of the extract (transformed and control) was spotted on Whatman 3MM paper and electrophoresed at 15 V/cm for 30 min in the buffer containing formic acid, acetic acid, distilled water (3:6:91, v/v/v, pH 9.8). After drying overnight, the electrophoretograms were stained with silver nitrate (Trevelyan et al., 1950), fixed with sodium thiosulphate solution (10% w/v in water) and washed with tap water. The spots obtained were compared with authentic samples of mannopine. Molecular characterization of the hairy root lines was also carried out to confirm the transformed nature and the presence of Ri T-DNA in hairy roots. DNA was extracted from non-transformed roots as a control and hairy roots using cetyltrimethyl ammonium bromide (CTAB) method. Primers for detecting *rol A* gene were used. Total isolated DNA from transformed roots was subjected to restriction digestion with EcoRI and was assessed by PCR for confirmation of presence of T-DNA of the bacterial plasmid. The PCR reactions were carried as follows: initial denaturation at 94°C for 1 min, annealing at 50°C for 1 min and extension for 2 min at 72°C for the first cycle, which was followed by 35 successive cycles under identical conditions. For further band amplification the PCR reactions were also carried out at three distinctive annealing temperatures i.e. 55, 60 and 65°C. DNA extracted from putative transformed *Catharanthus* roots demonstrated the amplification product corresponding to *rol A*, when visualized on 1.2% agarose TAE gel stained with ethidium bromide. No amplification products were obtained using non-transformed roots.

Extraction and chemical analysis of TIAs

Alkaloid extraction was done by a modified method of Kutney *et al.*, (1980). Briefly, the method is described as follows: 2.5 g hairy roots (harvested after 5 weeks) were extracted with 50 ml of methanol. The homogenate was incubated at 55°C for 2 h. The methanolic extract was dried and the residue dissolved in 2.5% H_2SO_4 (10 ml) and washed three times with one volume of ethyl acetate. The pH of the aqueous phase was adjusted to 9.5 with NH_4OH (28.4%). This alkaline solution was extracted three times with 25 ml of ethyl acetate. The organic phase was recovered and evaporated to dryness. The residue was finally dissolved in 1 ml of methanol (HPLC grade) and used for HPLC analysis. Control roots from field-grown and *in vitro* non-transformed plants were also extracted in a similar manner. HPLC analysis was done on Water's Modular System consisting of 501 chromatography pump, 484 tunable absorbance detector, 745 B integrator and 46K injector. A μ-Bondapak C18 (3.9 mm, 1.0 x 300 mm) column was used for analysis. The mobile phase was MeOH:CH_3CN:NH_4OAc (0.001M):Triethyl amine (1:2:2.5:0.2%), at a flow rate of 1 ml/min. The absorbance was monitored at λmax 280 nm. The TIAs were identified by retention time data as well as by peak enrichment method.

Optimization of explant vs genotype specificity of the bacterial strains

Five genotypes of *C. roseus* viz., red-flowered, white-flowered, Nirmal, BRT-1 and BRT-2 were tested for standardization of parameters for hairy root induction. The explants utilized were: leaf, petiole, fertilized ovule, embryo, petal, sepal, microshoots and anthers (Table 11.5.). Glasshouse-grown and *in vitro* raised plants proved to be the most ideal sources of explants for *A. rhizogenes* infection. Amongst the different explants tested, leaf and microshoot explants responded best to hairy root induction. Root emergence from non-transformed explants (controls) was not observed.

Table 11.5. Standardization of explant vs *A. rhizogenes* strains for hairy root induction in *Catharanthus roseus*.

Explant	Genotype	*A. rhizogenes* strains					
		A_4	LBA-9402	SV_2	SV_4	NCPPB-1855	MAFF-301724*
Leaf	Red	+	+	+c	+	+	+
	Nirmal	+	+	+c	+	+	-
	White	-	+	+c	+	+	-
	BRT-1	-	-	-	+	-	-
	BRT-2	-	-	-	+	-	-
Petiole	Red	-	-	+	+	-	-
	Nirmal	-	-	-	+	-	-
	White	-	-	-	-	-	-
Embryo	Red	-	-	-	-	-	-
	Nirmal	-	-	+c	+	-	-
	White	-	-	-	-	-	-
Petals	Red	-	-	-	-	-	-
	Nirmal	-	-	-	c	-	-
	White	-	-	-	-	-	-
Microshoot	Red	+	+	+c	+	+	-
	Nirmal	+	+	+c	+	+	-
	White	-	-	+c	+	-	-
Anthers	Red	-	-	-	-	-	-
	Nirmal	-	-	+c	+**	-	-
	White	-	-	-	-	-	-

** Swelling was observed, and white protuberances appeared, but this was accompanied by callusing; + : responded to hairy root induction; - : did not respond; c: callusing

When leaf explants were used for infection with *A. rhizogenes,* hairy roots could be induced in all the genotypes. However, the degree of success depended on strain specificity. While the bacterial strains LBA-9402, SV_2, SV_4 and NCPPB-1855 were able to induce hairy roots in all the genotypes tested. MAFF-301724 was able to induce hairy roots only in red genotype, while A_4 was not able to induce hairy roots in white genotype. The hairy roots produced using MAFF-301724 did not survive due to overgrowth of the bacteria. In general, the efficacy of the different bacterial strains to bring about hairy root transformations in *Catharanthus* was in the order of SV_4>LBA-9402>NCPPB-1855>A_4>SV_2>MAFF-301724. Ri-strain R-1000 could not induce hairy roots in any of the genotypes. Similarly, the capacity of different explants to undergo genetic transformations was in the sequence microshoots>leaf>petiole. The petal, sepal and ovule explants did not respond to most of the strains tested. The susceptibility of the genotypes towards *A. rhizogenes* infection was in the order of nirmal>red>white>BRT-1>BRT-2. BRT-1 and BRT-2 were found to be susceptible only to SV_4 strain. These results are in conformity with the trend observed by Porter (1991) who also reported that genotype is the prime determinant of host specificity towards infection with *A. rhizogenes.* The 'transformation frequency' in red and nirmal genotypes of *C. roseus* ranged from 0.02–0.12, 0.02-0.25 and 0.03–0.06, respectively when leaf and microshoot explants were subjected to co-cultivation with various Ri-strains (Table 11.6; Fig. 11.6.).

Table 11.6 Efficacy of different Ri strains for hairy root induction from leaf and micro-shoot explants of *Catharanthus roseus.*

Bacterial Strain used	Days required for co-cultivation	Transformation frequency*	Days required for root emergence
A_4, SV_4	4-7	0.2–0.25	28–38
SV_2, LBA-9402	5-7	0.02–0.04**	25–30
NCPPB-1855	3-5	0.07–0.11	40–48
MAFF-301724	2-3	0.02***	20–25****
R-1000	Did not respond at all		

* Number of explants with hairy roots/total number of explants; **Roots induced by SV_2 strains exhibited a general tendency towards callusing; ***Bacterial growth after co-cultivation with MAFF-301724 strain was difficult to check; ****Hairy roots were obtained in leaf explants of red genotype only.

The initial transformation events were characterized by the appearance of white protuberances within the next 10–15 days. The time of appearance of hairy roots varied with the strains employed. For A_4 and SV_4 strains, the days required for emergence of hairy roots varied between 28–38, while for SV_2 and LBA-9402 it varied between 25–30 days. NCPPB-1855 strain took maximum time for induction of hairy roots, i.e., 40–48 days, while in case of MAFF-301724 strain, it took 20–25 days. Similarly, the number of days required for co-cultivation to bring about effective transformations varied with the recipient genotype and the bacterial strains employed. The co-cultivation period for A4 and SV_4 strains was 4–7 days, while for SV_2 and LBA-9402 strain it was 5–7 days. In case of NCPPB-1855 and MAFF-301724, fewer number of days were required for co-cultivation i.e., 3–5 and 2–3 days respectively. These strains, when allowed to grow for a longer period, resulted in excessive bacterial growth on and around the explants. The time required for transformations was comparatively longer in contrast to earlier reports (Parr *et al.*, 1988; Bhadra *et al.*, 1993; Bhadra and Shanks, 1995). This might be attributed to the differences in genotype and the *A. rhizogenes* strains employed during the present investigation.

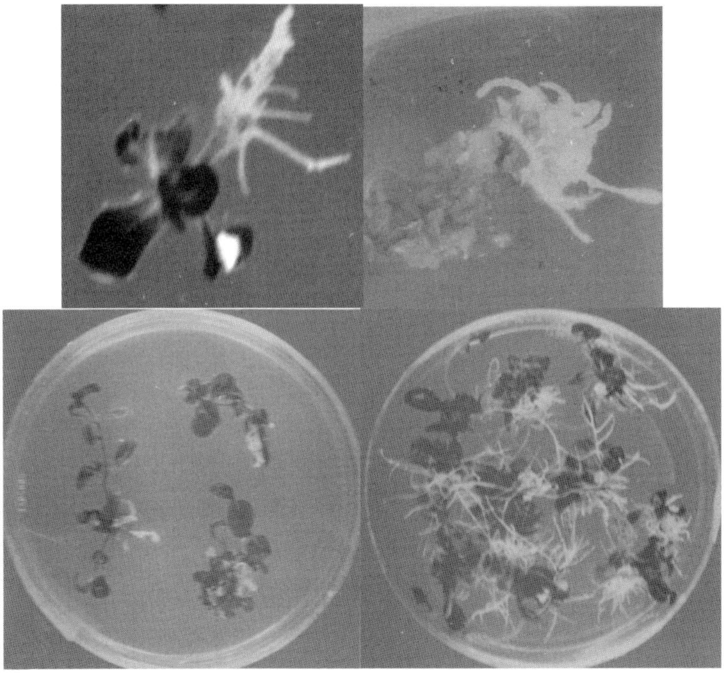

Fig. 11.6. Initiation of transformed roots from microshoot explants of *C. roseus* (top left); Transformed hairy root culture of Nirmal genotype of *C. roseus* showing general tendency towards callusing when SV_2 strain was employed (top middle); Induction of *A. rhizogenes*-mediated hairy root culture from microshoot explants when SV_4 strain was used and White genotype of *C. roseus* was transformed (top right); Rapidly proliferating hairy roots on maintenance medium (bottom left).

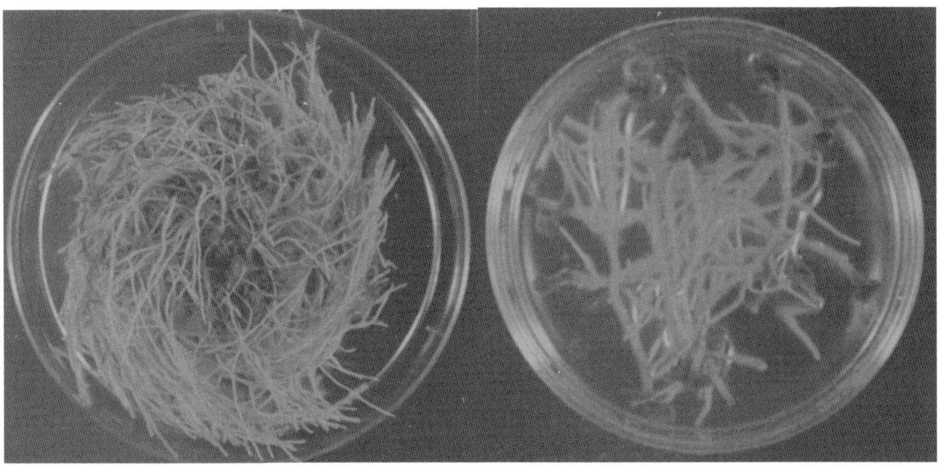

Fig. 11.7. Fast-growing hairy roots of *Catharanthus roseus* after 4 weeks of growth on ½ MS + 1.0 mg/l IBA medium (left). The roots exhibited a tendency to callus when IBA was replaced with NAA (right).

Establishment of axenic root cultures

After the second week of inoculation of explants on plain agar medium with the antibiotics cephalexin (250 mg/l) and ampicillin (100 mg/l), it was found essential to subculture the explants on a maintenance medium containing Kn (1.0 mg/l) + BAP (0.4 mg/l) + NAA (0.2 mg/l) for healthy survival of the co-cultivated explants and subsequent root emergence from them. A high degree of mortality or very inhibited rates of growth of the induced hairy roots was noticed when they were cultured on auxin-free MSO medium.

Among the various accessions, the highest survival and growth of the induced hairy roots was observed in case of variety Nirmal. Hairy root clones of red genotype exhibited a tendency to form callus on the maintenance medium. Experiments were, therefore, devised to optimize culture medium and other incubation conditions for eliciting faster growth and proliferation of the induced hairy root lines. A medium containing half-strength MS salts without NH_4NO_3 and supplemented with IBA (1.0 mg/l) and double dose of organic vitamin supplements was found to be the best for the maintenance and growth of the transformed roots. The hairy root clones obtained as independent transformation events in variety nirmal, and designated as clones N-1–N-5, depicted varying morphologies. Hilton *et al.* (1988) have also reported the existence of variability in terms of morphology and growth in the hairy root cultures of the same plant and also between species. Different morphologies were observed among clones of the same genotype; Nirmal, in the present study, presumably due to random integration of Ri-plasmid T-DNA into the plant genomic DNA. Tepfer (1987) opines that such differences in morphologies may be due to either/all of the following reasons: integration of T-DNA at different sites, differences in copy number, T-DNA structures *per se*, and methylation/demethylation of DNA.

While N-1 and N-5 clones were thin and light brown in appearance, the clones N-2 and N-3 exhibited very limited elongation, giving them a star-shaped appearance. Clones N-4 and N-5 produced regularly spaced branches and shiny root tips with a lot of root hairs. The characteristic features of these clones indicate their suitability for growth in bioreactors. These hairy root clones were then grown in liquid ½ MS (without NH_4NO_3) + IBA (1.0 mg/l) medium for further maintenance and characterization since the best response was observed on this medium (Fig. 11.7; Table 11.7). Shen *et al.* (1988) have also reported that exogenous addition of auxins is sometimes necessary for the growth of hairy roots. In order to establish the transformed nature of the individual root lines, the individual clones were tested for the presence of opines. The transformed hairy roots contained mannopine, which exhibited almost the same mobility in paper electrophoresis as authentic mannopine. Besides, the transformed roots also showed the characteristic features of hairy roots viz., lack of geotropism and a high rate of proliferation of epidermal hairy cells.

Growth analysis of hairy root clones

The growth profile of the different hairy root clones on half-strength liquid MS medium supplemented with IBA (1.0 mg/l) revealed that a very low growth index was achievable, ranging between 1.42 in case of clone N-3 to 3.36 in case of clone N-5 (Table 12.8).

Utilizing a two-stage culture procedure, significant improvement in growth of N-1, N-4 and N-5 clones was noticed.

In the modified procedure, 2-week old hairy root clones were transferred from liquid to agar-gelled medium of similar composition, but with an additional dose of 60 mg/l adenine sulfate. The proliferation of roots on agar medium was still better when the roots were placed on a sterilized filter paper disc placed on top of the medium, and incubation was done in dark. After 1 week of appearance of 3–5 *de novo* root tips, with fine root hairs, the entire root mass was retransferred to liquid medium of similar composition under diffused light (18–22°C). The amount of liquid medium per 250 ml flask (narrow neck) at this stage of culture varied from 60–70 ml. It was found that almost three-fold increase in biomass yield (up to 5 wks of growth) was obtained when 15–20 ml liquid medium was used per flask in comparison to the earlier practice of using 60–70 ml medium per flask. Replenishment of fresh IBA containing medium at the rate of 5 ml every tenth day was required to sustain growth in the modified procedure.

Table 11.7. Different media tested for initial growth and maintenance of hairy root clones of *Catharanthus roseus**.

Media composition*	Hairy Root Clones					Growth Pattern
	N-1	N-2	N-3	N-4	N-5	
MS0	-	-	-	-	-	Root growth could not be sustained
1/2 MS0	-	-	-	-	-	-do-
1/4 MS0	-	-	-	-	-	-do-
MS + IBA (1.0 mg/l)	2	2	2	2	2	-
MS + NAA (1.0 mg/l)	1	1	1	1	1	-
1/2MS + IBA (1.0 mg/l)	3	3	3	3	3	The root tips of hairy roots proliferated very fast, giving out profuse white root hairs
1/2MS + NAA (1.0 mg/l)	3	3	3	3	3	Callusing in all the hairy root clones.
1/4MS + IBA (1.0 mg/l)	2	2	2	2	2	The roots turned brown and collapsed
1/4MS + NAA (1.0 mg/l)	2	2	2	2	2	-

* Data collected on transformed root cultures induced by SV_4 strain of *A. rhizogenes*.
1- Mild branching and elongation
2- Moderate branching and elongation
3- Profuse branching and elongation

Table 11.8. Growth index of hairy root lines of *Catharanthus roseus*.

Clone No.	Growth Index* (initial procedure)	Growth Index* (modified procedure)
N-1	2.94	9.52
N-2	1.64	6.56
N-3	1.42	3.56
N-4	2.02	8.48
N-5	3.36	10.12

* Final fresh weight/initial fresh weight (inoculum) at the end of 5th week of culture.

Conclusion

As has also been previously reported by other workers (Toivonen *et al.*, 1991; Bhadra *et al.*, 1993), the present studies with *Catharanthus* also revealed that the Indian genotypes of *C. roseus* are not a very amenable system for *A. rhizogenes*-mediated genetic transformations. However, the problems were successfully overcome to obtain fast-growing hairy roots as alluded to earlier in the chapter. The reasons for this recalcitrance may include the presence of a large number of alkaloids in the leaf and other explants. These phytochemicals are inhibitory to the bacterial growth and plasmid insertion during the co-cultivation phase of the transformation cycle. In order to overcome the problems of recalcitrance towards *A. rhizogenes*-mediated transformations, in future it would be interesting to explore three possibilities viz., *Agrobacterium* strains resistant to crude alkaloid extracts of *Catharanthus roseus* should be screened, mutant plants with very low alkaloid content be developed and used for bacterial infection, and finally efforts must be initiated to produce haploid hairy roots.

Acknowledgements

The authors are thankful to Dr P.S. Ahuja, Director, Institute of Himalayan Bioresource Technology, Palampur and Prof Sushil Kumar, Former Director, Central Institute of Medicinal and Aromatic Plants, Lucknow for encouragement. The authors are also grateful to Dr Laiq-ur-Rahman, Mr Krishan Gopal and other members of the Division for support in multifarious ways.

References

Arora, R. (1998) Biotechnological Studies in *Catharanthus roseus* (L.) G. Don vis-a-vis Secondary Metabolite Production. PhD thesis. Central Institute of Medicinal and Aromatic Plants, Lucknow and Chaudhary Charan Singh University, Meerut, India, 195 pp. + plates.

Arora, R., Malhotra, P., Chawla, R., Gupta, D., Sharma, R.K. and Baliga, M.S. (2010) Indian Herbal Medicine for Cancer Therapy and Prevention: Bench to Bedside. In: Watson, R.R. and Preedy, V.R. *Bioactive Foods and Extracts: Cancer Treatment and Prevention.* Taylor and Francis Group, LLC, pp. 519–543.

Arora, R., Mathur, A., Mathur, A.K. and Govil, C.M. (2005) *Agrobacterium rhizogenes* mediated genetic transformation in *Catharanthus roseus* (L.) G. Don. National Symposium on Plant Biotechnology, New Frontiers and 27th Annual Meeting of the Plant Tissue Culture Association (India) held at Central Institute of Medicinal and Aromatic Plants (CIMAP), Lucknow, India from 17–19 November, 2005.

Arora, R., Malhotra, P., Mathur, A.K., Mathur, A., Govil, C.M. and Ahuja, P.S. (2010) Anticancer alkaloids of *Catharanthus roseus*: Transition from Traditional to Modern Medicine. In: Arora, R. (ed.) *Herbal Medicine: A Cancer Chemopreventive and Therapeutic Perspective.* Jaypee Brothers Medical Publishers Pvt. Ltd, New Delhi, India, pp. 292–310.

Banerjee, S., Zehra, M., Kukreja, A.K. and Kumar, S. (1995) Hairy roots in medicinal plants. *Current Research on Medicinal and Aromatic Plants* 17, 348–378.

Barleben, L., Panjikar, S., Ruppert, M., Koepke, J. and Stöckigt, J. (2007) Molecular architecture of strictosidine glucosidase, the gateway to the biosynthesis of the monoterpenoid indole alkaloid family. *The Plant Cell* 19, 2886–2897.

Batra, J., Dutta, A., Singh, D., Kumar, S. and Sen, J. (2004) Growth and terpenoid indole alkaloid production in *Catharanthus roseus* hairy root clones in relation to left- and right-termini-linked Ri T-DNA gene integration. *Plant Cell Reports* 23, 148–154.

Bernhardt, P., McCoy, E. and O'Connor, S.E. (2007) Rapid identification of enzyme variants for reengineered alkaloid biosynthesis in periwinkle. *Chemistry and Biology* 14(8), 888–897.

Bhadra, R. and Shanks, J.V. (1995) Statistical design of the effect of inoculum conditions on growth of hairy root cultures of *Catharanthus roseus*. *Biotechnology Letters* 9, 681.

Bhadra, R., Vani, S. and Shanks, J.V. (1993) Production of indole alkaloids by selected hairy root lines of *Catharanthus roseus*. *Biotechnology and Bioengineering* 41, 581–592.

Binns, A.N. and Thomashow, M.F. (1988) Cell biology of Agrobacterium infection and transformation of plants. *Annual Review of Microbiology* 42, 575–605.

Blondy, B. Canto-Canché, Meijer, A.H., Collu, G., Verpoorte, R. and Loyola-Vargas, V.M. (2005) Characterization of a polyclonal antiserum against the monoterpene monooxygenase, geraniol 10-hydroxylase from *Catharanthus roseus*. *Journal of Plant Physiology* 162(4), 393–402.

Chaudhri, R.D. (1996) *Herbal Drugs Industry*. Eastern Publishers, New Delhi, India.

Christie, A.M., Peebles, E.H., Hughes, H., Shanks, J.V. and San, Ka-Yiu (2009) Transcriptional response of the terpenoid indole alkaloid pathway to the overexpression of ORCA3 along with jasmonic acid elicitation of *Catharanthus roseus* hairy roots over time. *Metabolic Engineering* 11(2), 76–86.

Chung, I.-M., Ahmad, A., Ali, M., Lee, O.-K, Kim, M.-Y., Kim, J.-H., Yoon, D.-Y., Peebles, C.A.M. and San, KA. (2009) Flavonoid glucosides from the hairy roots of *Catharanthus roseus*. *Journal of Natural Products* 72 (4), 613–620.

Courdavault, V., Burlat, V., St-Pierre, B. and Giglioli-Guivarc'h, N. (2005) Characterisation of CaaX-prenyltransferases in *Catharanthus roseus*, relationships with the expression of genes involved in the early stages of monoterpenoid biosynthetic pathway. *Plant Science* 168(4), 1097–1107.

Curtin, M.E. (1983) Harvesting profitable products from plant tissue culture. BioTechnology, 1, 649–657.

Doran, P.M. (ed.) (1997) *Hairy Roots*. Harwood Academic Publishers, Amsterdam.

Dutta, A., Batra, J., Pandey-Rai, S., Singh, D., Kumar, S. and Sen, J. (2005) Expression of terpenoid indole alkaloid biosynthetic pathway genes corresponds to accumulation of related alkaloids in *Catharanthus roseus* (L.) G. Don. *Planta* 220, 376–383.

Dutta, A., Singh, D., Kumar, S.and Sen, J. (2007) Transcript profiling of terpenoid indole alkaloid pathway genes and regulators reveals strong expression of repressors in *Catharanthus roseus* cell cultures. *Plant Cell Reports* 26, 907–915.

Echevarría-Machado, I., Escobedo-G.M, R.M., and Larqué-Saavedra, A. (2007) Responses of transformed *Catharanthus roseus* roots to femtomolar concentrations of salicylic acid. *Plant Physiology and Biochemistry* 45(6-7), 501–507.

Hilton, M.G., Wilson, P.D.G., Robins, R.J. and Rhodes, M.J.C. (1988) Transformed root cultures, fermentation aspects. In: Rhodes, M.J.C. (ed.) *Manipulating Secondary Metabolism in Culture*. Cambridge Univerity Press, Cambridge, UK, pp. 239–245.

Hong, S.B., Peebles, C.A.M., Shanks, J.V., San, K.-Y. and Gibson, S.I. (2006) Terpenoid indole alkaloid production by *Catharanthus roseus* hairy roots induced by *Agrobacterium tumefaciens* harboring rol ABC genes. *Biotechnology and Bioengineering* 93(2), 386–390.

Hooykaas, P.J.J., Klapwijk, P.M., Nuti, M.P., Schilperoort, R.A. and Rorsch, A. (1977) Transfer of the *Agrobacterium tumefaciens* Ti-plasmid to avirulent agrobacteria and to rhizobium *ex planta*. *Journal of General Microbiology* 98, 477–482.

Hughes, E.H., Hong, Seung-Beom, Gibson, S.I., Shanks, J.V. and San, Ka-Yiu (2004) Metabolic engineering of the indole pathway in *Catharanthus roseus* hairy roots and increased accumulation of tryptamine and serpentine. *Metabolic Engineering* 6(4), 268–276.

Jaleel, A.C., Gopi, R., Manivannan, P., Sankar, B., Kishorekumar, A. and Panneerselvam, R. (2007) Antioxidant potentials and ajmalicine accumulation in *Catharanthus roseus* after treatment with giberellic acid. *Colloids and Surfaces B: Biointerfaces* 60(2), 195–200.

Junaid, A., Mujib, A., Seikh, A., Nasim and Sharma, M.P. (2009) Screening of vincristine yield in *ex vitro* and *in vitro* somatic embryos derived plantlets of *Catharanthus roseus* L. (G) Don. *Scientia Horticulturae* 119(3), 325–329.

Kutchan, T.M., Frick, S. and Weid, M. (2008) Engineering Plant Alkaloid Biosynthetic Pathways, Progress and Prospects. *Advances in Plant Biochemistry and Molecular Biology* 1, 283–310.

Kutney, J.P., Choi, L.S.L., Kolodziejczyk., Sleigh, S.K., Stuart, K.L., Worth, B.R., Kurz, W.G.W., Chatson, K.B. and Constabel, F. (1980) Alkaloid production in *Catharanthus roseus* cell cultures: isolation and characterization of alkaloids from one cell line. *Phytochemistry* 19, 2589–2595.

Magnotta, M., Murata, J., Chen, J. and Vincenzo De Luca (2007) Expression of deacetylvindoline-4-O-acetyltransferase in *Catharanthus roseus* hairy roots. *Phytochemistry* 68(14), 1922–1931.

Loh, K.Y. (2008) Know the Medicinal Herb: *Catharanthus Roseus (Vinca Rosea). Malaysian Family Physicians* 3(2), 123.

Morgan, A.J., Cox, P.N., Turner, D.A., Peel, E., Davey, M.R., Gartland, K.M.A. and Mulligan, B.J. (1989) Transformation of tomato using an Ri-plasmid vector. *Plant Science* 49, 37–49.

Natali, R., Mustafa, Kim, H.K., Choi, Y.H., Erkelens, C., Alfons., Lefeber, W.M., Spijksma, G., van der Heijden, R. and Verpoorte, R. (2009) Biosynthesis of salicylic acid in fungus elicited *Catharanthus roseus* cells. *Phytochemistry* 70(4), 532–539.

Noble, R.L. (1990) The discovery of the vinca alkaloids-chemotherapeutic agents against cancer. *Biochemistry and Cell Biology* 68, 1344–1351.

Parr, A.J., Peerless, C.L., Hamill, J.D., Walton, N.J., Robins, R.J. and Rhodes, M.J.C. (1988) Alkaloid production by transformed root cultures of *Catharanthus roseus*. *Plant Cell Reports* 7, 309–312.

Pasquali, G., Goddijn, O.J.M., de Wall, A., Verpoorte, R., Schilperoort, R.A., Hoge, J.H.C. and Memelink, J. (1992) Coordinated regulation of two indole alkaloid biosynthetic genes from *Catharanthus roseus* by auxin and elicitors. *Plant Molecular Biology* 18, 1121–1131.

Pereira, D.M., Ferreres, F., Oliveira, J., Valentão, P., Andrade, P.B. and Sottomayor, M. (2009) Targeted metabolite analysis of *Catharanthus roseus* and its biological potential. *Food and Chemical Toxicology*, in press, available online 17 March 2009.

Petit, A., David, C., Dahl, G.A., Ellis, J.G., Gayon, P., Delbart, F.C. and Tempe, J. (1983) Further extension of the opine concept, Plasmids in *Agrobacterium rhizogenes* cooperate for opine degradation. *Molecular and General Genetics* 190, 204–214.

Pomahačová, B., Dušek, J., Dušková, J., Yazaki, K., Roytrakul, S. and Verpoorte, R. (2009) Improved accumulation of ajmalicine and tetrahydroalstonine in *Catharanthus* cells expressing an ABC transporter. *Journal of Plant Physiology*, in press, available online 29 April 2009.

Porter, J.R. (1991) Host range and implications of plant infection by *Agrobacterium rhizogenes*. *Critical Reviews in Plant Sciences* 10, 387–421.

Pratt, W.B., Ruddon, R.W., Ensminger, W.E. and Maybaum, J. (1994) *The Anticancer Drugs*. 2nd edn, Oxford University Press.

Rama Rao, A.V. and Gurjar, M.K. (1990) Drugs from plant sources: an overview. *Pharm. Times* 22, 19–27.

Rischer, H., Oresic, M., Seppanen-Laakso, T., Katajamaa, M., Lammertyn, F., Ardiles-Diaz, W., van Montagu, M.C.E., Inze, D., Oksman-Caldentey, K.-M. and Goossens, A. (2006) Gene-to-metabolite networks for terpenoid indole alkaloid biosynthesis in *Catharanthus roseus* cells. *Proceedings of the National Academy of Sciences USA* 103(14), 5614–5619.

Rowinsky, E.K. and Donehower, R.C. (1997) Antimicrotubule Agents. In: De Vita, V.T. Jr., Hellman, S. and Rosenberg, S.A. (eds) *Cancer Principles and Practice of Oncology*. 5th edn, Lippincott-Raven, Philadelphia, USA, pp. 467–483.

Saito, K., Yamzaki, M. and Murakoshi, I. (1992) Transgenic medicinal plants Agrobacterium-mediated foreign transfer and production of secondary metabolites. *Journal of Natural Products* 55, 149–162.

Seth, R. and Mathur, A.K. (2005) Selection of 5-methyltryptophan-resistant callus laines with improved metabolic flux towards terpenoid indole alkaloid synthesis in *Catharanthus roseus*. *Current Science* 89(3), 544–548.

Sharma, R.K. and Arora, R. (eds) (2006) *Herbal Drugs, A Twenty First Century Perspective*. Jaypee Brothers Medical Publishers Pvt. Ltd, New Delhi, India, pp. 688.

Shen, W.H., Petit, A., Guern, J. and Tempe, J. (1988) Hairy roots are more sensitive to auxin than normal roots. *Proceedings of the National Academy of Sciences of the United States of America* 85, 3417–3421.

Shimoda, K., Kwon, S., Utsuki, A., Ohiwa, S., Katsuragi, H., Yonemoto, N., Hamada, H. and Hamada, H. (2007) Glycosylation of capsaicin and 8-nordihydrocapsaicin by cultured cells of *Catharanthus roseus*. *Phytochemistry* 68(10), 1391–1396.

Siberil, Y., Benhamron, S., Memelink, J., Giglioli-Guivar'h, N., Thiersault, M., Boisson, B., Doireau, P. and Gantet, P. (2001) *Catharanthus roseus* G-box binding factors 1 and 2 act as repressors of strictosidine synthase gene expression in cell cultures. *Plant Molecular Biology* 45, 477–488.

Sriram, G., Fulton, D.B. and Shanks, J.V. (2007) Flux quantification in central carbon metabolism of *Catharanthus roseus* hairy roots by ^{13}C labeling and comprehensive bondomer balancing. *Phytochemistry* 68(16-18), 2243–2257.

Stöckigt, J., Barleben, L., Panjikar, S. and Loris, E.A. (2008) 3D-Structure and function of strictosidine synthase– the key enzyme of monoterpenoid indole alkaloid biosynthesis. *Plant Physiology and Biochemistry* 46(3), 340–355.

Suttipanta, N., Pattanaik, S., Gunjan, S., Xie, C.H., Littleton, J. and Yuan, L. (2007) Promoter analysis of the *Catharanthus roseus* geraniol 10-hydroxylase gene involved in terpenoid indole alkaloid biosynthesis. *Biochimica et Biophysica Acta – Gene Structure and Expression* 1769(2), 139–148.

Taylor, W.I. and Farnsworth, N.R. (1975) *The Catharanthus Alkaloids: Botany, Chemistry, Pharmacology and Clinical Uses*. M. Dekker, New York, USA.

Tepfer, D. (1987) Ri-T-DNA from *Agrobacterium rhizogenes*, a source of genes having applications in rhizosphere biology and plant development, ecology and evolution. In: Kosugue, T. and Nester, E.W. (eds) *Plant Microbe Interactions, Molecular and Genetic Perspectives*. Mc Graw Hill, New York, USA, pp. 294–342.

Toivonen, L., Ojala, M. and Kauppinen, H.V. (1991) Studies on the optimization of growth and whole alkaloid production by hairy root cultures of *Catharanthus roseus*. *Biotechnology and Bioengineering* 37, 673–680.

Trevelyan, W.E., Procter, D.P. and Harrison, J.S. (1950) Detection of sugars on paper chromatograms. *Nature* 166, 444–445.

Van der Fits, L. and Memelink, J. (2000) ORCA3, a jasmonate-responsive transcriptional regulator of plant primary and secondary metabolism. *Science* 289, 295–297.

van der Heijden, R., Verpoorte, R. and ten Hoopen, H.J.G. (1989) Cell and tissue cultures of *Catharanthus roseus* (L.) G. Don, a literature survey. *Plant Cell Tissue and Organ Culture* 18, 231–280.

van der Heijden, R., Jacobs, D.I., Snoeijer, W., Hallard, D. and Verpoorte, V. (2004) The *Catharanthus* alkaloids, pharmacognosy and biotechnology. *Current Medicinal Chemistry* 11, 607–628.

Veltkamp, E., Breteler, H., Huizingh, H. and Bertola, Ma (1985) *Plant Biotechnology in the Netherlands*. Nationale Raad voor Landbouwkumdig Onderzoek, 14g's.

Verpoorte, R., van der Heijden, R., Schripsema, J., Hoge, J.H.C. and ten Hoopen, H.J.G. (1993) Plant biotechnology for the production of alkaloids, present status and future prospects. *Journal of Natural Products* 56, 186–207.

Verpoorte, R., van der Heijden, R., ten Hoopen, H.J.G. and Memelink, J. (1998) Metabolic engineering for the improvement of plant secondary metabolic production. *Plant Tissue Culture and Biotechnology* 4, 3–20.
Yeoman, M.M. and Yeoman, C.L. (1997) Manipulating secondary metabolism in cultured cells. *New Phytologist* 134, 553–561.

Chapter 12

Podophyllum Endophytic Fungi

J.R. Porter

Introduction

Podophyllum spp. are the major sources of podophyllotoxin, precursor to the anticancer drugs etoposide, teniposide and etopophos (Fig. 12.1), as well as a number of compounds currently in clinical trial or development for treatment of cancer, arthritis, inflammation, oxidative or radiative damage, and viral disease (Xu *et al.*, 2009a). Most recently, compounds with insecticidal activity have been developed (Xu *et al.*, 2009b). *P. hexandrum* (syn., *P. emodi, Sinopodophyllum hexandrum*) has been the principal source of podophyllotoxin, but over-collection has led to severe reduction in the wild populations, and the plant has been listed as endangered (Fu and Zhang, 1992; Singh *et al.*, 2001). Studies of the North American *P. peltatum* initially showed that the podophyllotoxin content, particularly of the leaves, was much lower than that of *P. hexandrum*, but the studies of Canel *et al.* (2001) demonstrated that endogenous glycosidases could release substantial podophyllotoxin from leaf tissue.

Efforts are under way to develop *P. peltatum* as a sustainable source of podophyllotoxin and related lignans (Moraes *et al.*, 2002a), but similar earlier attempts were unsuccessful (Bogdanova and Sokolov, 1973). Numerous attempts to synthesize podophyllotoxin completely have not yet led to a commercially viable synthetic route (Wu *et al.*, 2007; Xu *et al.*, 2009a). A sustainable supply of this precursor molecule is required considering the value of the drugs in chemotherapy and the large number of drug leads developed on its scaffold.

Several plant species are known to produce podophyllotoxin. Besides the *Podophyllum* spp., *Linum* spp. (Petersen and Alfermann, 2001), *Anthriscus sylvestris* (Koulman, 2003), *Juniperus virginiana* (Kupchan *et al.*, 1965), *J. chinensis* (Muranaka *et al.*, 1998), *Polygala polygama* (Hokanson, 1978) and at least 14 other species from a wide range of plant families (Gordaliza *et al.*, 2004) have all yielded small but measurable amounts of podophyllotoxin. Closely related lignans are produced by a variety of plants, such as *Forsythia* × *intermedia* (Umezawa *et al.*, 1990), *Hippophae rhamnoides* (Yang *et al.*, 2006), *Bursera* spp. (Koulman, 2003), *Thuja occidentalis* (Chang *et al.*, 2000) and many others. There does not seem to be any phylogenetically significant distribution of the *Podophyllum* lignans, so it is possible that this biosynthetic pathway evolved independently a number of times. As will be seen below, however, an alternate hypothesis for this distribution may be developing.

To complicate the picture still further, the podophyllotoxin content of a plant may be strongly dependent on the plant genetics (*P. peltatum* and *P. hexandrum*) (Moraes *et al.*, 2002a; Sultan *et al.*, 2008), season (*A. sylvestris*) (Koulman *et al.*, 2007), environment (*P. hexandrum*) (Purohit *et al.*, 1999; Alam *et al.*, 2008), or development (*P. hexandrum*) (Purohit *et al.*, 1999).

Plants growing in different environments, with different genotypes, or at different life stages at different times of the year may have widely different podophyllotoxin contents.

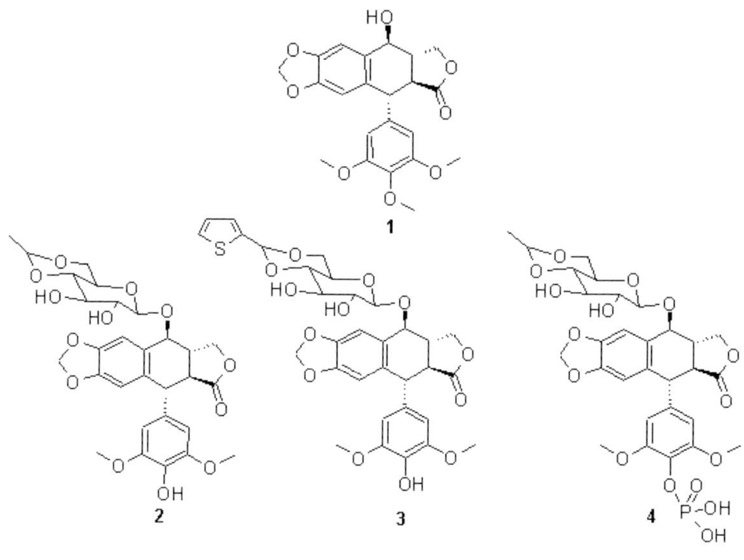

Fig. 12.1. Podophyllotoxin (**1**) and the major anticancer drugs etoposide (**2**), teniposide (**3**) and etopophos (**4**).

Biotechnological approaches to production of podophyllotoxin offer great promise to produce sustainable supplies without biologic, climatic, geographic, or political barriers. Efforts to produce the compound through cell and tissue culture have been ongoing for many years (Kadkade, 1981). Some have focused on biosynthesis by *Linum* spp. cell, tissue and organ cultures (Oostdam *et al.*, 1993; Giri *et al.*, 2001; Baldi *et al.*, 2008b; Berim *et al.*, 2008). Co-culture of *Linum* cell cultures with mycorrhizal fungi (Baldi *et al.*, 2008a) or *Linum* hairy roots with *P. hexandrum* cells (Lin *et al.*, 2003) have led to interesting increases in the desired products.

A different approach was suggested by the studies of Gary Strobel and his group at Montana State University when they discovered that endophytic fungi living in the tissues of a variety of species are capable of producing the anticancer compound taxol (Stierle *et al.*, 1993; Strobel *et al.*, 1996b). The ability of plant endophytic fungi to produce a wide range of interesting biologically-active and medicinal compounds has now been extensively documented; this area has been capably and thoroughly reviewed (Gunatilaka, 2006; Zhang *et al.*, 2006; Porter, 2008). A significant part of this chapter addresses the work that has been done, in this laboratory and others, to isolate, identify and determine the podophyllotoxin production potential of endophytic fungi that live in the tissues of *Podophyllum* spp.

Life History of the Host Plants

As noted, several groups have explored the possibilities of growing the major producing podophyllotoxin producers, *P. peltatum* and *P. hexandrum*, in an agricultural setting (Bogdanova and Sokolov, 1973; Moraes *et al.*, 2002b; Kharkwal *et al.*, 2008), similar to the practice with

Catharanthus rosea for production of the '*Vinca*' alkaloids. Why are the podophyllotoxin producers not more 'croppable'? That is, what are the characteristics of the *Podophyllum* spp. that make them less likely to be grown successfully under agricultural conditions?

The great majority of successful crop plants share a number of characteristics that make them easier to handle and grow successfully in a managed setting: (i) seed dormancy exists and it is relatively easily maintained with simple, inexpensive conditions; (ii) seed dormancy can be maintained for long periods when the seeds are kept dry; (iii) seed dormancy is easily broken, often by simple addition of water or a short cold period; (iv) seed germination occurs at a high percentage, generally well above 50%, because of efficient fertilization, seed development and embryo viability; (v) seedling growth is fairly rapid, and seedling survival is high; (vi) seedling development to the adult, reproductive stage is fairly rapid, generally occurring within 3–60 months; (vii) at least moderate resistance to common pathogens exists at all life stages; (viii) pollination is reliable, and the plants are self-fertile, wind pollinated or pollinated by common insects; (ix) genetic mixing among strains is easy and common and (x) in perennial crops, a well-developed plant dormancy exists for survival in an annual cold or dry season.

Depending on the author, there are between two (Shaw, 2002) and at least 14 (Ying, 1979; Shaw, 2002) species of *Podophyllum.* Those who recognize fewer species often acknowledge at least a few forms and varieties of *P. peltatum* or *P. hexandrum.* Even the number of genera is controversial, with some retaining *P. hexandrum* for the Himalayan species, and an increasing number of authors moving this to *Sinopodophyllum.* In the same publication that proposed this latter genus, Ying (1979) also proposed movement of the majority of *Podophyllum* species other than *P. peltatum* into the genus *Dysosma.* Most recognize *Podophyllum* and *Sinopodophyllum* as members of the Berberidaceae, but some recognize the family Podophyllaceae (Nadeem *et al.*, 2007). Regardless of the taxonomy, only two species have been studied in significant depth with regard to the ecology, biology and podophyllotoxin production.

Virtually all of the species of *Podophyllum*, *Sinopodophyllum*, *Dysosma* and *Diphylleia* studied to date contain detectable or significant quantities of podophyllotoxin (Yin and Chen, 1989; Yu *et al.*, 1991; Ma and Luo, 1992; Canel *et al.*, 2000). *Diphylleia* is recognized as a distinct genus in this group by all authors, although Kim and Jansen (1998) found that the genus is not monophyletic, and this genus may require further study and revision. However, none of the species is distributed over a wide geographic range or, if so, the local populations are not large or abundant. Thus, continued wild collection for podophyllotoxin is highly likely to lead to local or widespread extinction of populations or species.

In addition, all of the species possess the majority of characteristics that would suggest difficulty in converting the species into an agricultural crop. Seed dormancy may be absent or short-lived in most species, particularly if seeds are dried (Kharkwal *et al.*, 2008), or unusual conditions may be required to break seed dormancy (Braun and Brooks, 1987; Shaw, 2002; Simonnet *et al.*, 2008; Sreenivasulu *et al.*, 2009). Dormancy appears to be maintained primarily in rhizomes, rather than in seeds. With the exception of *P. hexandrum*, most of the species appear to be strictly or primarily outcrossing (Swanson and Sohmer, 1976; Liu *et al.*, 2002; Qiu *et al.*, 2006). *P. peltatum* has been shown to possess genetic variation in podophyllotoxin production (Moraes *et al.*, 2002a), and the requirement for outcrossing and significant genetic diversity between populations (Singh *et al.*, 2001; Liu *et al.*, 2002; Alam *et al.*, 2008; Sultan *et al.*, 2008) suggests that many accessions may be necessary to maintain the genetic diversity in response to declining populations. This and other species often exhibit low recruitment of new individuals into a population, low nectar production, low reproductive success and low pollination efficiency (Swanson and Sohmer, 1976; Sohn and Policansky, 1977; Rust and Roth, 1981; Laverty and

Plowright, 1988), although *Dysosma* spp. flowers are noted to give off an odour of faeces or decay and are fly-pollinated. One of the major causes of seedling and juvenile plant mortality is fungal infection within the first five years after germination due to the absence of a protective rhizome tip sheath.

Attempts to cultivate *P. peltatum* and *P. hexandrum* for sustained podophyllotoxin production have met with only limited success (Cushman *et al.*, 2003; Pandey *et al.*, 2007). *In vitro* growth of *Podophyllum*, *Linum* and other plants for production of podophyllotoxin and related lignans similarly is successful only in part (Lin *et al.*, 2003; Wink *et al.*, 2005). There is clearly a need for alternative approaches to assure access to podophyllotoxin for continued drug development.

Fungal Endophytes as Sources of Podophyllotoxin

Recent discoveries show that plant endophytic fungi can produce a wide variety of compounds that traditionally are associated with plant tissues (Porter, 2008). Among the earliest studies in this vein, Strobel's group and other researchers have demonstrated widespread occurrence of endophytic fungi with the ability to produce taxol (Stierle *et al.*, 1993; Strobel *et al.*, 1996a; Wang *et al.*, 2000). Since those studies, an increasing number of researchers have begun study of endophytic fungi in many plants to study the ability of the endophytic fungi to produce a diversity of compounds under controlled culture conditions. The studies of Eyberger (2002) in my laboratory extended the biochemical diversity of plant endophytic fungi to include production of podophyllotoxin (Eyberger *et al.*, 2006). Since that discovery, other laboratories have discovered fungi that similarly produce podophyllotoxin. The endophytes isolated and grown in our laboratory were obtained from the tissues of *P. peltatum*. Podophyllotoxin-producing strains of *Trametes hirsuta* and *Alternaria neesex* were obtained from tissues of *P. hexandrum* (Puri *et al.*, 2006; Li, 2007), of *Fusarium oxysporum* from *Juiperus recurva* (Kour *et al.*, 2007), of *Penicillium implicatum* strain SJ21 in *D. sinensis* (Zeng *et al.*, 2004), and of six endophytic strains from *D. sinensis*, *Dy. veitchii* and *P. hexandrum* identified only to genus (*Penicillium*, *Monilia*, *Alternaria*) or not identified (*mycelia sterilia*) (Yang *et al.*, 2003). To my knowledge, *Linum*, *Anthriscus*, *Polygala* and other genera known to contain podophyllotoxin have yet to be examined for endophytic fungi. It seems to be clear from these studies that the podophyllotoxin-producing endophytic fungi represent a broad taxonomic diversity. The majority of these groups have studied the fungi for the ability to synthesize podophyllotoxin in aseptic culture.

The endophytic fungi isolated in this laboratory have been assigned to the fungal species complex *Phialocephala fortinii* based on the variable D2 region of the large ribosomal subunit rDNA (Eyberger *et al.*, 2006). Additional studies of other genetic regions have not yet caused us to change the identification, but neither has it confirmed the original diagnosis. One of these strains, PPE7 (Genbank accessions DQ485456 and EU382071), is closer both morphologically and genetically to other *P. fortinii* strains. The other strain, PPE5 (Genbank accessions DQ485455 and EU382070), is more distant genetically and has less morphological similarity to other *P. fortinii* strains (Eyberger *et al.*, 2010). Grünig *et al.* (2008) developed techniques for the phylogenetic analysis and placement of *P. fortinii* and *Acephala applanata* strains. We are using these and similar techniques to more precisely define the taxonomic identity of our podophyllotoxin-producing strains.

The opportunity of endophytic fungi that produce valuable medicinal compounds is the same as that of plant tissue and cell culture approaches. If production can be optimized and increased to commercial levels, the compound can be produced without reliance on wild populations, thus saving the genetic diversity of the producing plants, removing geographic, climatic and geopolitical limitations on the production and assuring the supply of vitally-important substances. To this end, we studied the cultural characteristics and podophyllotoxin production in our two producing strains in a variety of fungal growth media and culture conditions. Our two strains grew at varying rates in eight different media, including malt broth (MB), yeast malt broth (YMB), Czapek's minimal medium (CMM), potato dextrose broth (PDB), Sabouraud's dextrose broth (SDB), minimal medium (MM), Murashige & Skoog medium (MS) and Gamborg's B5 medium (GB5) (Mathai *et al.*, 2006). The latter two media are more commonly used for plant cell culture studies, but we were interested in whether these would influence growth and compound production in comparison to the other media. Cultures were shaken to promote aeration or not, and the studies were conducted in 50 ml or 1 l culture volumes. The growth of the fungi was studied over a period of 4 weeks.

The best growth was obtained in GB5 medium for both fungal strains. However, the two strains behaved differently in most growth characteristics. Strain PPE5 showed a volume-specific growth about equal in both culture volumes at 0.1 g/ml; strain PPE7 grew much more slowly in the 1 l culture condition (0.03 g/ml) and more rapidly in the 50 ml culture volume (0.17 g/ml). The overall pattern of growth in the various media was GB5 > MS > PDB = SDB > MB > CMM ≥ YMB > MM for the PPE5 strain. The PPE7 strain showed a pattern of GB5 >> SDB > CMM > MB > MM = YMB > PDB = MS. Aeration or still culture conditions resulted in similar patterns. In seven of the eight culture media, the PPE5 strain grew more quickly in aerated conditions; the growth in SDB was about equal in still or shaken cultures. For strain PPE7, only GB5 had a higher growth in aerated cultures. The growth was about equal by the PPE7 strain in shaken or still cultures for SDB, MS, PDB and YMB media, and the still medium conditions gave higher growth in still conditions in MM, MB and CMM, with the much higher growth in still conditions compared to aerated conditions for the last medium.

The podophyllotoxin production was similarly specific to the strain and culture conditions. In general, 50 ml cultures led to higher production of podophyllotoxin (μg/l) than did 1 l cultures. In fact, the only 1 l cultures that led to a detectable quantity of podophyllotoxin by HPLC-DAD analysis were strain PPE5 cultured in YMB without aeration and strain PPE7 grown in CMM under still conditions. All other 1 l cultures led to no detectable podophyllotoxin. Strain PPE7 generally outperformed strain PPE5 in podophyllotoxin production, and this strain gave the highest level of podophyllotoxin production seen to date, 185 μg/l. This is still well below commercial production levels, but it does show that production is responsive to environmental conditions. PDB gave higher production of podophyllotoxin than all other media tested, including the top level of production, a mean of 16 μg/l for strain PPE7 in 50 ml aerated culture as well as the second highest mean production level of 5 μg/l by the same strain grown in a 50 ml volume under still conditions. Thus, a lower culture volume, PDB medium, and strain PPE7 were associated with the highest production levels. There were no consistent trends in respect to medium aeration – of the 12 conditions that led to detectable podophyllotoxin production, seven were obtained in still culture conditions. The lack of an effect of aeration may be due to the growth habit of the mycelium under the two conditions. The shaken cultures were characterized by separate fungal 'balls' of 5–15 mm diameter suspended in the culture. Most often, still cultures result in a mycelial mat on the top surface with dispersed mycelia adherent to the glass submerged in the medium. Two of the media, MM and MS, led to no detectable podophyllotoxin production. An additional two media, MB and SDB, had only a single condition under which podophyllotoxin production was detected, strain PPE5 in 50 ml aerated cultures in the former medium and strain PPE7 in 50 ml still cultures in the latter. We have observed podophyllotoxin production in a greater number of the tested conditions during subsequent LC-MS analysis of the samples, although at levels below the detection of the HPLC-DAD system.

We have grown the fungi on solid media as well, primarily for morphological studies and maintenance, including Sabouraud's dextrose agar, yeast malt agar and potato dextrose agar. The two strains grew similarly on the solid media and were similar in morphology and cultural characteristics (hyphal size, mycelium colours, hyphal differentiation, presence or lack of conidia) on the various media. As mentioned previously, strain PPE7 is more similar morphologically to *P. fortinii* (*sensu* Wang and Wilcox (1985)), with submerged toroid hyphae, dark septate hyphae and rare hyaline conidia (Eyberger *et al.*, 2006). Strain PPE5 lacks toroid hyphae, and we have never observed sporulation in this strain, even when cultured for up to two years in dark, cold conditions.

Much of our current work is focused on molecular genetic analysis of the podophyllotoxin-producing endophytic fungi. As mentioned, we are analysing a number of loci in order to understand the taxonomic relationships of these strains to other endophytic fungi. A major focus of this work, however, is to determine the genes involved in the biosynthesis of podophyllotoxin. Since the biosynthesis of this product has yet to be completely elucidated, the endophytic fungi, which are usually genetically simpler compared to plant producers, may more readily yield answers for the unknown biosynthetic steps. Work can then focus on enhancement of production through over-expression of rate-limiting gene products, suppression of competing pathways through RNAi and knock-out techniques, and transfer of some or all of the biosynthetic machinery to more tractable hosts.

Among the major questions that arise from these studies is the ultimate origin of the podophyllotoxin biosynthetic pathway. It seems logical that the pathway's existence in both a host plant and its endophytes is due to horizontal gene transfer. However, the direction of that transfer is much in question. Thorough genetic analysis in podophyllotoxin-producing organisms may ultimately answer this question. But, it is interesting to hypothesize that the widespread and phylogenetically disparate occurrence of podophyllotoxin in various plant species may be a consequence of the presence of endophytic fungi, rather than the product of multiple evolutionary events randomly leading to the same end-product. Podophyllotoxin is present at low concentrations in a number of plant species; the endophytes may be the actual source. Even when podophyllotoxin is present at high concentrations, the endophytes could still be the principal source, promoted to greater biosynthesis by the host plant biochemical environment. Regulatory systems similar to the *lae* system of *Aspergillus* (Bok and Keller, 2004) may be discovered in other fungi with major influence on secondary product production. If the biosynthetic machinery is present in the plant genome, which is by no means proven at this point, the origin may be transfer from endophyte to host plant rather than necessarily the other way around. Examination of the genes that have been sequenced to date (the dirigent proteins, secoisolariciresinol dehydrogenase, pinoresinol/lariciresinol reductase), the codon usage bias is equally suggestive of both plant and fungal origins (unpublished data). The presence of podophyllotoxin production in cell cultures of *Linum* spp. and *Podophyllum* spp. would suggest that the biosynthetic machinery is present in the plant genome, but the origin of the pathway remains undetermined. The answers to these questions will be interesting as they unfold, and we will learn much about the details and regulation of podophyllotoxin biosynthesis along the way.

Discovery of the endophytic fungi that can produce important medicinal compounds can help assure the supply of these compounds through biotechnological fermentation systems and molecular manipulations of the fungal strains, while also preserving the genetic diversity of the endangered plants traditionally used for extraction. This scenario is especially pertinent to the availability of podophyllotoxin. The use of this compound for the continued treatment of cancer (etoposide, teniposide, etopophos) and as a platform for the production of antiviral, anti-inflammatory and anti-insect compounds, as well as other applications, make the continued development of an assured, efficient and sustainable source a worthwhile undertaking for research.

Conclusion

Plant endophytic fungi offer a great source of biological and biochemical diversity. The presence of biosynthetic pathways to known medicinal compounds, including podophyllotoxin, offers the promise of a more complete understanding of the biosynthetic mechanisms, biochemical novelty, and development of sustainable biological sources of the compounds. Our studies with podophyllotoxin-producing endophytic fungi from *P. peltatum* have led to a research programme focused on development and exploitation of this resource. Determining the conditions that support enhanced growth and podophyllotoxin production by these fungi suggest additional studies to further the enhancement, which may include separate growth and production media. Molecular analysis of the genetically-simpler fungi (compared to the host plants) may finally yield a complete understanding of biosynthesis as well as offer new opportunities for genetic manipulation and heterologous gene transfer.

Acknowledgements

The work described in this chapter was supported, in part, by grants to the author by the Elsa U. Pardee Foundation for Cancer Research and the National Cancer Institute of the National Institutes of Health (grant no. 1R15CA135589-01A1). I thank Kyung A. Koo and Jason A. Porter for critical review of the manuscript.

References

Alam, A., Naik, P.K., Gulati, P., Gulati, A.K. and Mishra, G.P. (2008) Characterization of genetic structure of *Podophyllum hexandrum* populations, an endangered medicinal herb of Northwestern Himalaya, using ISSR-PCR markers and its relatedness with podophyllotoxin content. *African Journal of Biotechnology* 7, 1028–1040.

Baldi, A., Jain, A., Gupta, N., Srivastava, A. and Bisaria, V. (2008a) Co-culture of arbuscular mycorrhiza-like fungi (*Piriformospora indica* and *Sebacina vermifera*) with plant cells of *Linum album* for enhanced production of podophyllotoxins: a first report. *Biotechnology Letters* 30, 1671–1677.

Baldi, A., Srivastava, A.K. and Bisaria, V.S. (2008b) Improved podophyllotoxin production by transformed cultures of *Linum album*. *Biotechnology Journal* 3, 1256–1263.

Berim, A., Ebel, R., Schneider, B. and Petersen, M. (2008) UDP-glucose:(6-methoxy)podophyllotoxin 7-O-glucosyltransferase from suspension cultures of *Linum nodiflorum*. *Phytochemistry* 69, 374–381.

Bogdanova, V. and Sokolov, V. (1973) Developmental features of *Podophyllum peltatum* L and *Podophyllum emodi* Wall in cultivation conditions in the Lvov Province. *Herba Polonica* 19, 379–384.

Bok, J.W. and Keller, N.P. (2004) LaeA, a regulator of secondary metabolism in *Aspergillus* spp. *Eukaryotic Cell* 3, 527–535.

Braun, J. and Brooks G.R., Jr (1987) Box turtles (*Terrapene carolina*) as potential agents for seed dispersal. *American Midland Naturalist* 117, 312–318.

Canel, C., Moraes, R.M., Dayan, F.E. and Ferreira, D. (2000) Podophyllotoxin. *Phytochemistry* 54, 115–120.

Canel, C., Dayan, F.E., Ganzera, M., Khan, I.A., Rimando, A., Burandt, C.L., Jr and Moraes, R.M. (2001) High yield of podophyllotoxin from leaves of *Podophyllum peltatum* by *in situ* conversion of podophyllotoxin 4-O-β-D-glucopyranoside. *Planta Medica* 67, 97–99.

Chang, L.C., Song L.L., Park E.J., Luyengi, L., Lee, K.J., Farnsworth, N.R., Pezzuto, J.M. and Kinghorn, A.D. (2000) Bioactive constituents of *Thuja occidentalis*. *Journal of Natural Products* 63, 1235–1238.

Cushman, K.E., Maqbool, M., Bedir, E., Khan, I.A. and Moraes, R.M. (2003) Growth and lignan content of field-established mayapple. *Mississippi Agricultural & Forestry Experiment Station Information Bulletin* 398, 363–366.

Eyberger, A. (2002) Investigation of *Podophyllum peltatum* for Podophyllotoxin-Producing Fungal Endophytes. MS Thesis. University of the Sciences, Philadelphia, PA, USA.

Eyberger, A.L., Dondapati, R. and Porter, J.R. (2006) Endophyte fungal isolates from *Podophyllum peltatum* produce podophyllotoxin. *Journal of Natural Products* 69, 1121–1124.

Eyberger, A.L., Newcome, J., Dondapati, R., Martinez, S.D., Wilson, W., Jr, and Porter, J.R. (2010) Fungal endophytes of *Podophyllum peltatum* (Berberidaceae). *Journal of Applied Microbiology*, in submission.

Fu, K.-T. and Zhang, Z.-Y. (1992) *Sinopodophyllum hexandrum* (Royle) Ying. In: *Fu, L.-K.* (ed.) *China Plant Red Data Book– Rare and Endangered Plants*, Vol. 1. Science Press, Beijing, China, pp. 184–185.

Giri, A., Giri, C.C., Dhingra, V. and Narasu M.L. (2001) Enhanced podophyllotoxin production from *Agrobacterium rhizogenes* transformed cultures of *Podophyllum hexandrum*. *Natural Product Letters* 15, 229–235.

Gordaliza, M., Garcia, P.A., Miguel del Corral, J.M., Castro, M.A. and Gomez-Zurita, M.A. (2004) Podophyllotoxin: Distribution, sources, applications and new cytotoxic derivatives. *Toxicon* 44, 441–459.

Grünig, C.R., Duo, A., Sieber, T.N. and Holdenrieder, O. (2008) Assignment of species rank to six reproductively isolated cryptic species of the *Phialocephala fortinii s.l.-Acephala applanata* species complex. *Mycologia* 100, 47–67.

Gunatilaka, A.A.L. (2006) Natural products from plant-associated microorganisms: distribution, structural diversity, bioactivity, and implications of their occurrence. *Journal of Natural Products* 69, 509–526.

Hokanson, G.C. (1978) Podophyllotoxin and 4'-demethylpodophyllotoxin from *Polygala polygama* (Polygalaceae). *Lloydia (Journal of Natural Products)* 41, 497–498.

Kadkade, P.G. (1981) Formation of podophyllotoxins by *Podophyllum peltatum* tissue cultures. *Naturwissenschaften* 68, 481–482.

Kharkwal, A.C., Kushwaha, R., Prakash, O., Ogra, R.K., Bhattacharya, A., Nagar, P.K. and Ahuja, P.S. (2008) An efficient method of propagation of *Podophyllum hexandrum*: An endangered medicinal plant of the Western Himalayas under *ex situ* conditions. *Journal of Natural Medicines* 62, 211–216.

Kim, Y.-D. and Jansen, R.K. (1998) Chloroplast DNA restriction site variation and phylogeny of the Berberidaceae. *American Journal of Botany* 85, 1766–1778.

Koulman, A. (2003) Podophyllotoxin. A Study of the Biosynthesis, Evolution, Function and Use of Podophyllotoxin and Related Lignans. PhD Thesis. Rijksuniversitein Groningen, Groningen, the Netherlands.

Koulman, A., Hendrawati, O., Batterman, S., Van Putten, F., Bos, R. and Kayser, O. (2007) The seasonal variations of lignan profiles in *Anthriscus sylvestris* (L.) Hoffm. *Planta Medica* 73, 858.

Kour, A., Shawl, A.S., Rehman, S., Sultan, P., Qazi, P.H., Suden, P., Khajuria, R.K. and Verma, V. (2007) Isolation and identification of an endophytic strain of *Fusarium oxysporum* producing podophyllotoxin from *Juniperus recurva*. *World Journal of Microbiology & Biotechnology* 24, 1115–1121.

Kupchan, S., Hemingway, J. and Knox, J. (1965) Tumour inhibitors. VII. Podophyllotoxin, the active principle of *Juniperus virginiana*. *Journal of Pharmaceutical Sciences* 54, 659–660.

Laverty, T.M. and Plowright, R.C. (1988) Fruit and seed set in mayapple (*Podophyllum peltatum*): influence of intraspecific factors and local enhancement near *Pedicularis canadensis*. *Canadian Journal of Botany* 66, 173–178.

Li, C. (2007) Fermentation conditions of *Sinopodophyllum hexandrum* endophytic fungus on production of podophyllotoxin. *Food Fermentation Industry* 33, 28–32.

Lin, H.-W., Kwok, K.H. and Doran, P.M. (2003) Production of podophyllotoxin using cross-species coculture of *Linum flavum* hairy roots and *Podophyllum hexandrum* cell suspensions. *Biotechnology Progress* 19, 1417–1426.

Liu, J., Chen, Z. and Lu, A. (2002) Molecular evidence for the sister relationship of the eastern Asia-North American intercontinental species pair in the *Podophyllum* group (Berberidaceae). *Botanical Bulletin Academica Sinica* 43, 147–154.

Ma, C. and Luo, S.R. (1992) Studies on the TLC scanning determination of lignans in *Diphylleia sinensis* Li. *Yao Xue Xue Bao- Acta Pharmaceutica Sinica* 27, 934–938.

Mathai, M., Wilson, W. and Porter, J.R. (2006) Optimization studies of podophyllotoxin production by endophytic fungi. Presented at 47th Annual Meeting of the American Society of Pharmacognosy, Arlington, VA.

Moraes, R.M., Bedir, E., Barrett, H., Burandt, C., Jr, Canel, C. and Khan, I.A. (2002a) Evaluation of *Podophyllum peltatum* accessions for podophyllotoxin production. *Planta Medica* 68, 341–344.

Moraes, R.M., Lata, H., Bedir, E., Maqbool, M. and Cushman, K. (2002b) The American mayapple and its potential for podophyllotoxin production. In: Janickand, J. and Whipkey, A. (eds) *Trends in New Crops and New Uses*. ASHS Press, Alexandria, VA, USA, pp. 527–532.

Muranaka, T., Miyata, M., Ito, K. and Tachibana, S. (1998) Production of podophyllotoxin in *Juniperus chinensis* callus cultures treated with oligosaccharides and a biogenetic precursor – in honour of Professor G. H. Neil Towers' 75th Birthday. *Phytochemistry* 49, 491–496.

Nadeem, M., Palni, L.M., Kumar, A. and Nandi, S.K. (2007) Podophyllotoxin content, above- and belowground biomass in relation to altitude in *Podophyllum hexandrum* populations from Kumaun Region of the Indian Central Himalaya. *Planta Medica* 73, 388–391.

Oostdam, A., Mol, J.N.M. and van der Plas, L.H.W. (1993) Establishment of hairy root cultures of *Linum flavum* producing the lignan 5-methoxypodophyllotoxin. *Plant Cell Reports* 12, 474–477.

Pandey, H., Nandi, S., Kumar, A., Palni, U. and Palni, L. (2007) Podophyllotoxin content in *Podophyllum hexandrum* Royle plants of known age of seed origin and grown at a lower altitude. *Acta Physiologiae Plantarum* 29, 121–126.

Petersen, M. and Alfermann, A.W. (2001) The production of cytotoxic lignans by plant cell cultures. *Applied Microbiology and Biotechnology* 55, 135–142.

Porter, J.R. (2008) Plant fungal endophytes: Interactions, metabolites and biosyntheses. In: Ikan, R. (ed.) *Selected Topics in the Chemistry of Natural Products*. World Scientific Publ., Singapore, pp. 503–580.

Puri, S.C., Nazir, A., Chawla, R., Arora, R., Riyaz-ul-Hasan, S., Amna, T., Ahmed, B., Verma, V., Singh S., Sagar, R., Sharma, A., Kumar, R., Sharma, R.K. and Qazi, G.N. (2006) The endophytic fungus *Trametes hirsuta* as a novel alternative source of podophyllotoxin and related aryl tetralin lignans. *Journal of Biotechnology* 122, 494–510.

Purohit, M.C., Bahuguna, R., Maithani, U.C., Purohit, A.N. and Rawat, M.S.M. (1999) Variation in podophylloresin and podophyllotoxin contents in different populations of *Podophyllum hexandrum*. *Current Science* 77, 1078–1080.

Qiu, Y.-X., Li, J.-H., Liu, H.-L., Chen, Y.-Y. and Fu, C.-X. (2006) Population structure and genetic diversity of *Dysosma versipellis* (Berberidaceae), a rare endemic from China *Biochemical Systematics and Ecology* 34, 745–752.

Rust, R.W. and Roth, R.R. (1981) Seed production and seedling establishment in the mayapple, *Podophyllum peltatum* L. *American Midland Naturalist* 105, 51–60.

Shaw, J.M.H. (2002) The genus *Podophyllum*. In: Stearn, W.T., Shaw, J.M.H., Greenand, P.S. and Mathew, B. (eds) *The genus Epimedium and other herbaceous Berberidaceae*. Timber Press, Portland, Or., pp. 239–314.

Simonnet, X., Hell, B. and Franz, C. (2008) Domestication du podophylle indien (*Podophyllum hexandrum* Royle): Etude sur la germination. *Revue Suisse de Viticulture, Arboriculture, Horticulture* 40, 179–184.

Singh, B.M., Sharma, K.D., Katoch, M., Guleria, S. and Sharma, T.R. (2001) Molecular analysis of variability in *Podophyllum hexandrum* Royle – an endangered medicinal herb of northwestern Himalaya. *Plant Genetic Resources Newsletter*, 57–61.
Sohn, J.J. and Policansky, D. (1977) The costs of reproduction in the mayapple, *Podophyllum peltatum* L. (Berberidaceae). *Ecology* 58, 1366–1374.
Sreenivasulu, Y., Chanda, S.K. and Ahuja, P.S. (2009) Endosperm delays seed germination in *Podophyllum hexandrum* Royle an important medicinal herb. *Seed Science and Technology* 37, 10–16.
Stierle, A., Strobel, G. and Stierle, D. (1993) Taxol and taxane production by *Taxomyces andreanae*, an endophytic fungus of Pacific yew. *Science* 260, 214–216.
Strobel, G., Yang, X., Sears, J., Kramer, R., Sidhu, R.S. and Hess, W.M. (1996a) Taxol from *Pestalotiopsis microspora*, an endophytic fungus of *Taxus wallachiana. Microbiology* 142, 435–440.
Strobel, G.A., Hess, W.M., Ford, E., Sidhu, R.S. and Yang, X. (1996b) Taxol from fungal endophytes and the issue of biodiversity. *Journal of Industrial Microbiology and Biotechnology* 17, 417–423.
Sultan, P., Shawl, A.S., Ramteke, P.W., Kour, A. and Qazi, P.H. (2008) Assessment of diversity in *Podophyllum hexandrum* by genetic and phytochemical markers. *Scientia Horticulturae* 115, 398–408.
Swanson, S.D. and Sohmer, S.H. (1976) The biology of *Podophyllum peltatum* L. (Berberidaceae), the may apple. II. The transfer of pollen and success of sexual reproduction. *Bulletin of the Torrey Botanical Club* 103, 223–226.
Umezawa, T., Davin, L.B., Yamamoto, E., Kingston, D.G.I. and Lewis, N.G. (1990) Lignan biosynthesis in *Forsythia* species. *Journal of the Chemical Society, Chemical Communications*, 1405–1408.
Wang, C. and Wilcox, H. (1985) New species of ectendomycorrhizal and pseudomycorrhizal fungi: *Phialophora finlandia, Chloridium paucisporum,* and *Phialocephala fortinii. Mycologia* 77, 951–958.
Wang, J., Li, G., Lu, H., Zheng, Z., Huang, Y. and Su, W. (2000) Taxol from *Tubercularia* sp. strain TF5, an endophytic fungus of *Taxus mairei. FEMS Microbiology Letters* 193, 249–253.
Wink, M., Alfermann, W.A., Franke, R., Wetterauer, B., Distl, M., Windhovel, J., Krohn, O., Fuss, E., Garden, H., Mohagheghzadeh, A., Wildi, E. and Ripplinger, P. (2005) Sustainable bioproduction of phytochemicals by plant *in vitro* cultures: Anticancer agents. *Plant Genetic Resources* 3, 90–100.
Wu, Y., Zhang, H., Zhao, Y., Zhao, J., Chen, J. and Li, L. (2007) A new and efficient strategy for the synthesis of podophyllotoxin and its analogues. *Organic Letters* 9, 1199–1202.
Xu, H., Lu, M. and Tian, X. (2009a) A review on hemisynthesis, biosynthesis, biological activities, mode of action, and structure-activity relationship of podophyllotoxins: 2003–2007. *Current Medicinal Chemistry* 16, 327–349.
Xu, H., Zhang, L., Su, B.-F., Zhang, X. and Tian, X. (2009b) Natural products-based insecticidal agents 1. Semisynthesis and insecticidal activity of 4ß–benzene–sulfonamide derivatives of podophyllotoxin against *Mythimna separata* Walker. *Heterocycles* 77, 293–300.
Yang, B., Linko, A.-M., Adlercreutz, H. and Kallio, H. (2006) Secoisolariciresinol and matairesinol of Sea Buckthorn (*Hippophaë rhamnoides* L.) berries of different subspecies and harvesting times. *Journal of Agricultural and Food Chemistry* 54, 8065–8070.
Yang, X., Guo, S., Zhang, L. and Shao, H. (2003) Select of producing podophyllotoxin endophytic fungi from podophyllin plant. *Tianran Chanwu Yanjiu Yu Kaifa* 15, 419–422.
Yin, M.L. and Chen, Z.L. (1989) Chemical constituents of *Dysosma aurantiocaulis* (H.-M.) Hu and *Dysosma pleianthum* Woods. *Zhongguo Zhongyao Zazhi* 14, 420–421, 447.
Ying, T.-C. (1979) On *Dysosma* Woodson and *Sinopodophyllum* Ying, *gen. nov.* of the Berberidaceae. *Acta Phytotaxonomica Sinica* 17, 15–23.
Yu, P.-Z., Wang, L.-P. and Chen, Z.-N. (1991) A new podophyllotoxin-type lignan from *Dysosma versipellis* var. *tomentosa*. *Journal of Natural Products* 54, 1422–1424.
Zeng, S., Shao, H. and Zhang, L. (2004) An endophytic fungus producing a substance analogous to podophyllotoxin isolated from *Diphylleia sinensis*. *Weishengwuxue Zazhi* 24, 1–2.
Zhang, H.W., Song, Y.C. and Tan, R.X. (2006) Biology and chemistry of endophytes. *Natural Product Reports* 23, 753–771.

Chapter 13

Biotechnology of Himalayan *Vinca major* and *V. minor*

Rajesh Arora, Ajay K. Mathur, Archana Mathur and C.M. Govil

Introduction

The family Apocynaceae (dogbane family) is an important family of flowering plants that comprises a number of medicinally useful plants, several of which are trees, shrubs, herbs and lianas. The tropical rainforests are home to several species and most occur in the tropics and subtropics and some are found in xeric environments. Chief amongst the medicinal plants belonging to this family include *Catharanthus roseus*, *Carissa carandas*, *Rauvolfia serpentina*, *Tabernaemontana coronaria*, *Vinca minor*, *V. major* etc. (Arora, 1998; Arora *et al.*, 2010). The plants belonging to this family are a source of a variety of alkaloids, several of which find use in the clinic.

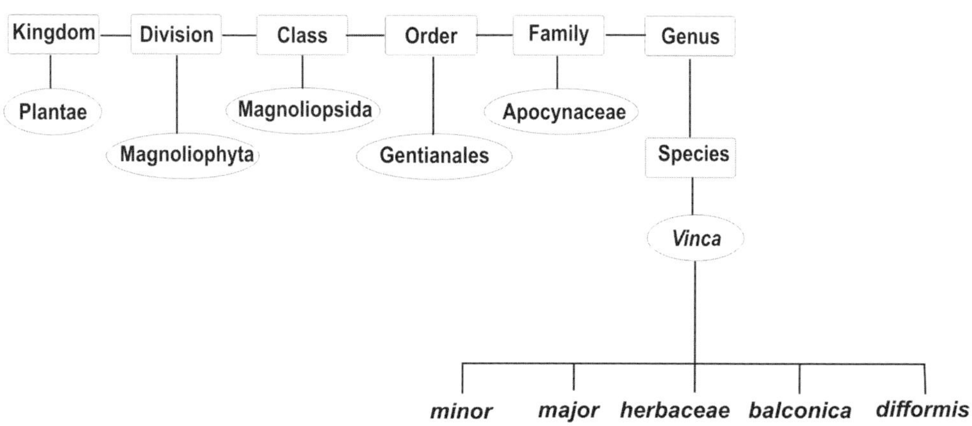

Fig. 13.1. Botanical classification of *Vinca major* and *V. minor.*

Vinca major (Big Leaf Periwinkle)

Fig. 13.2. *Vinca major* flourishing well in the natural environment (note the dark green foliage (left); a leaf (right)).

Fig. 13.3. Plants in flowering stage; a flower of *Vinca major*.

Vinca major is commonly known as large periwinkle, greater periwinkle and blue periwinkle and belongs to the family Apocynaceae. The plant is known as large periwinkle since it possesses bigger leaves in contrast to *V. minor*, the closely related species, which is similar but smaller, with narrower, hairless leaves. *V. major* is a flowering plant native to Southern Europe and is found from Spain and Southern France to the Western Balkans and also in Northeastern Turkey and the Western Caucasus. It is also found in the Himalayan mountainous region of India and Pakistan. *V. major* is an evergreen trailing vine that usually spreads along the ground, rooting along the stems, forming patches of ground cover that can be 2–5 m across and up to 50–70 cm high. The leaves are opposite, 3–9 cm long and 2–6 cm broad, glossy dark green with a leathery texture and an entire but distinctly hairy margin, and a hairy petiole 1–2 cm long (Blamey and Grey-Wilson, 1989; Huxley, 1992). The flowers are produced from early spring to autumn. The flowers are violet-purple, 3–5 cm diameter, and possess a distinct five-lobed corolla.

Vinca major is commonly grown in gardens for its evergreen foliage, spring flowers, and vining habit. A number of cultivars are available, with white to dark violet flowers and variegated foliage. Several botanical reports dating back to the early 1900s exist on the larger periwinkle (Hirsch, 1949; Guseinova, 1986; Huxley, 1992). Aneli *et al.* (1988) have carried out pharmacobotanical studies on *Vinca major* introduced in the Georgian SSR region of the former USSR. De Almeida (2005) reported on the distribution of *Vinca major* from Portugal. Cytostatic properties of the alkaloid fractions isolated from the aerial part of *Vinca major* have been reported by Boratynska *et al.*, (1972).

Phytochemical constituents

Several workers have confirmed the presence of vincamajine, vincamajoreine, vincamine, majvinine, reserpinine, vincawajine, majoridine, methyxyvellosimine and lochvinerine in the plant (Farnsworth, 1960; Farnsworth *et al.*, 1962; Potier *et al.*, 1965; Chatterjee *et al.*, 1970, 1975; Zsadon *et al.*, 1972; Il'Yashenko *et al.*, 1977; Balsevich *et al.*, 1982; Atta-Ur-Rahman *et al.*, 1995). Reserpinine, majdine, akuammicine, strictosidine lactam, pseudoakuammigine, akuammine, and a new alkaloid tentatively named 10-hydroxycathofoline, were isolated from *V. major* cv. variegata leaves (Balsevich *et al.*, 1982). The alkaloid profiles of green and yellowish parts of the leaves, however, did not differ (Balsevich *et al.*, 1982). Ali *et al.*, (1976) reported the isolation of vincarpine and di-hydro vincarpine, two new zwitterionic indole alkaloids, from *Vinca major* var. elegantissima. Banerji and Chakrabarty (1974) reported the isolation of a new alkaloid lochvinerine from *Vinca major*. Ishikura and Minekishi (1978) reported the isolation of new delphinidin glycosides from *Vinca major* flowers. Mukhopadhyay *et al.* (1991) isolated 11 methoxytetrahydroalstonine, a heteroyohimbinoid alkaloid, from *Vinca major*. Detailed studies on alkaloids from *Vinca major*, introduced in the Gerogian SSR (former USSR), have been carried out (Zhukovich, 1987, 1989). Sakushima and Nishibe (1988) carried out mass spectrometry for structural determination of flavonol triglycosides from *Vinca major*. Zhukovich and Vachnadze (1985) isolated vincamaijinine, a new alkaloid, from *Vinca major*, while Zhukovich *et al.* (1986) subsequently quantitatively identified vincamajine in the aboveground part of *Vinca major*.

Medicinal uses of *Vinca major*

In view of the vasoconstricting effect on peripheral blood vessels, *V. major* is considered useful in the treatment of pain and sensitivity in the acute stages of migraine headache. The plant lessens passive haemorrhaging. It is used to stop mild bleeding from haemorrhoids, nose bleeds and urinary tract injury. The plant is useful in treating menorrhea, and mid-cycle bleeding.

Vinca minor

Vinca minor is commonly known as lesser periwinkle, myrtle, running-myrtle, petite pervenche, pervinca, vinca, vincapervinca. *Vinca minor* is an evergreen trailing vine that grows in the wild in the foothills of the Himalayas, particularly in Himachal Pradesh, Kashmir and Uttarakhand region of India. It is a fast grower and spreads quickly after it gets established. *Vinca minor* prefers regular watering, well-drained, fertile and loamy soil. It grows well in shade and can tolerate mild sunlight if regularly watered. It is a low evergreen subshrub or herbaceous perennial. Leaves are opposite, entire, mostly broad at the middle. Flowers are axillary, solitary, ebracteate and pedunculate. Calyx (sepals) are five, 3–5 mm long, fused at base, without internal squamellae or glands. Corolla is blue, infundibuliform, tube gradually expanded with a zone of hairs above the insertion of the stamens in the upper part; petals 5, horizontally spreading, joined at base by a low ridge. There are five stamens, inserted in the middle of the corolla tube; filaments longer than the anthers, flattened, abruptly bent at the base; anthers basifixed, not connivent; connective broad, expanded above the thecae into a hairy flap-like appendage; pollen smooth, with three grooves. Disc consists of two ovoid fleshy nectar-secreting scales (disc glands) alternating with the carpels. There are two carpels, appressed; ovules 4–8 in each loculus; style well-developed, almost clavate, abruptly expanded above into a disc-like clavuncule bearing the conical stigma surmounted by five dense tufts of hair. Fruit consists of two slender follicles (mericarps); seeds cylindric, grooved on one side, without a coma of hairs; embryo straight, embedded in thick endosperm.

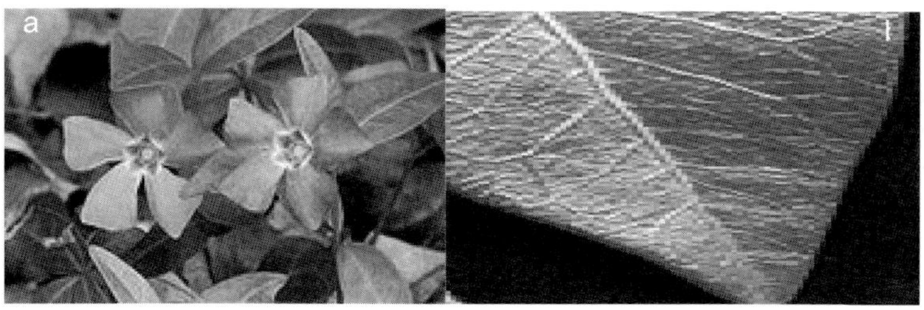

Fig. 13.4 (a) The plant of *Vinca minor* (note the size of the leaves, which are smaller in comparison to *Vinca major*). (b) A leaf of *Vinca minor*.

Phytochemical constituents

Amongst the indole alkaloids (0.15–1.4%), the chief alkaloid present in *Vinca* is vincamine (eburnamine-type, 25–65%), including as well vincine, apovincamine and vincadifformin. The flavonoids present include kempferol-3-O-rhamnoside-7O-galactoside, kempferol-3-O-rhamoglucoside-2-O-glucoside, quercetin-3-O-rhamnoglucoside-7-O-glucoside.

Vincamine (14-hydroxy-14-methoxycarbonyl-l-eburname) and vincamone or l-Eburnamonine (14-oxo-l-eburnane) are two important clinically useful indole alkaloids, used against brain sclerosis and in post-operative states of the CNS, and have been isolated from *Vinca minor*. Vincamine has been proposed for the treatment of drepanocytosis (sicklaemia) (Smeyers *et al*., 1991). Smeyers *et al.* (1991) carried out structural studies on alkaloids of *Vinca minor*. The three-dimensional structures of five indole alkaloids of *Vinca* (vincamine, vincamone, apovincamine, vincaminol, desoxyvincaminol) were determined theoretically and compared with experimental data (Smeyers *et al*., 1991).

Pharmacological properties

The *Vinca minor* plant has a long history of usage as a traditional tonic to treat senile weariness and also as an astringent, for excessive menses, bleeding gums and mouth sores. In traditional systems of medicine in various parts of the world, mainly Asia and Europe, it is used as astringent, bactericide, collyrium. It is used to treat diarrhoea, dysentery, hypertension, lactifuge, menorrhagia, phthisis, piles, scalp, sedative, skin, spasmolytic, tea, tonic, tumour (uvula) catarrh, eczema, haemostat, carminative, diuretic, emetic, hypertension, scurvy, depurative, diuretic, hemostat, lactagogue and tonic. *V. minor* is useful in the following diseases: circulatory disorders, cerebral circulatory impairment and as support for brain metabolism. It is also useful in treating hypertension, memory loss, gastritis and enteritis, diarrhoea. *Vinca minor* is used externally for sore throats, nose bleeds, bruising, abscesses, eczema and to stop bleeding. In homoeopathy, it is used for weeping eczema and bleeding mucous membranes. *Vinca minor* has been reported to be analgesic, amoebicide, antibiotic, antibacterial, cardiotonic, cholagogue, digestive, emmenagogue, febrifuge, hypotensive, laxative, pectoral, stomachic and vermifuge.

Vinca minor has been mentioned in the folklore as useful for treating toothache, snakebite, hypertension and as a carminative, vomitive, haemostatic and astringent. It was the folkloric use of this plant for hypertension that prompted much of the recent phytochemical interest displayed in *V. minor* (Mokry and Compis, 1964; Boulat *et al.*, 1990; Fischoff *et al.*, 1996). The biological effects of *V. minor* extracts have been reported by several workers (Kurmukov, 1967; Zhmurkin, 1968; Moinade and Puvinel, 1969; Federico and Amore, 1984; Fischoff *et al.*, 1996). Aqueous extracts prepared from the whole plant of *V. minor* have been shown to be toxic to American cockroaches when injected into the bloodstream. Extracts of *V. minor* leaves have been found to have an inhibitory effect on *Staphylococcus aureus*, but not against, *Escherichia coli*, *Salmonella typhimurium* and *Mycobacterium tuberculosis* (Taylor and Farnsworth, 1973). Extracts of *V. minor* have been evaluated in a variety of tumor systems, but in contrast to *Catharanthus* they were found to be devoid of anticancer activity (McKenna *et al.*, 1960; Abbott *et al.*, 1966).

Vincamine, the major alkaloid present in *V. minor* is used clinically in certain European countries, especially Hungary, for the treatment of hypertension, angina, migraine headaches and cerebrovascular diseases (Lanfranchi *et al.*, 1984). Vincamine is used for the treatment of memory disorders of the elderly, moderate dementia of degenerative and vascular etiologies (Zhmurkin, 1968; Federrico and Amore, 1984; Kurmukov, 1987; Boulat, 1990; Fischoff *et al.*, 1996).

Vincamine: the neuronal metabolism altering alkaloid

The major component of *Vinca minor*, also reported from *Vinca major*, is vincamine, which is an indole alkaloid (distinctively a tryptamine). Vincamine is found in the aerial parts of *Vinca minor*, primarily in the leaves which possess approximately 25–65% alkaloid. Vincamine is also derived from other plants like *Catharanthus roseus* (Cook and James, 1981), *Vinca major*, *Voaconga* and *Crioceras longiflorus*. Vincamine is hypotensive, negatively chronotrophic, spasmolytic, hypoglycaemic and sympatholytic. Vincamine alters neuronal metabolism by favouring aerobic glycolysis and increases blood flow to the brain and particularly towards ischemic areas. Vincamine also acts to increase cerebral circulation and the utilization of oxygen. Vincamine is known primarily as a vasodilator (peripheral vasodilator) and finds general use as an aid in activities requiring highly focused attention and concentration such as technical writing or computer operation. Vincamine has also been indicated in the treatment of tinnitus or ringing in the ears and for the treatment of poor memory. Vincamine is also commonly used as a nootropic agent to combat the effects of ageing, or in conjunction with other nootropics for a variety of purposes (Cook and James, 1981).

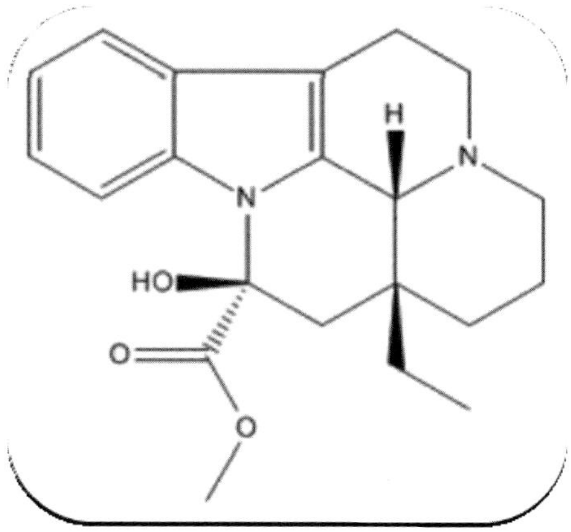

Fig.13.5. Vincamine [(3α,14β,16α)-14,15-dihydro-14-hydroxyeburnamenine-14-carboxylic acid methyl ester] $C_{21}H_{26}N_2O_3$.

Vinpocetine

Vinpocetine ($C_{22}H_{26}N_2O_2$; ethyl apovincaminate) is a semisynthetic derivative alkaloid of vincamine (Lörincz *et al.*, 1976). It is marketed as Cavinton and Intelectol. Vinpocetine possesses cerebral blood-flow enhancing (Szilágyi *et al.*, 2005) and neuroprotective properties (Dézsi *et al.*, 2002). Vinpocetine is used as a drug in Eastern Europe for the treatment of cerebrovascular disorders and age-related memory impairment.

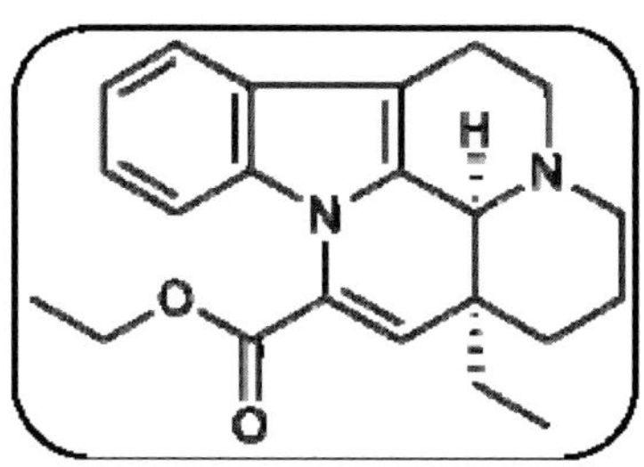

Fig. 13.6. Vinpocetine [(3α,16α)-Eburnamenine-14-carboxylic acid ethyl ester].

Vinpocetine is widely marketed as a supplement for vasodilation and as a nootropic for the improvement of memory. Vinpocetine, like vincamine, helps support brain functions such as enhancing concentration and memory by activating cerebral metabolism. A number of studies on healthy volunteers have demonstrated that vinpocetine may elicit improvement on some aspects of memory. Vinpocetine finds use in the body-building community as a vasodilator. However, no studies have been conducted on the effectiveness of vinpocetine on performance enhancement during exercise. Vinpocetine selectively inhibits voltage-sensitive Na^+ channels, causing a dose-dependent decrease in evoked extracellular Ca^+ ions in striatal nerve endings (Sitges *et al.*, 2005). The Na^+ channel inhibiting properties of vinpocetine are responsible for the neuroprotective effect, which is mediated through blockade of excitotoxicity and attenuation of neuronal damage induced by cerebral ischaemia/reperfusion (Adám-Vizi, 2000).

Vinpocetine is known to produce some side effects like indigestion, nausea, dizziness, anxiety, facial flushing, insomnia, headache, drowsiness and dry mouth and a temporary drop in blood pressure. Vinpocetine has been implicated in one case to induce agranulocytosis, a condition in which granulocytes, an important type of white blood cell, are markedly decreased. The safety of vinpocetine in pregnant women has not been evaluated.

Vinca species: international scenario of biotechnological interventions

Very little tissue culture work has been done on the *Vinca* species. Vincamine is the major alkaloid derived from *Vinca minor*. For the industrial production of this alkaloid, plant cell cultures have been studied as a possible source (Delaby, 1985). Initiation of cell cultures of *V. minor* was reported by Petiard and co-workers in 1972. Small amounts of vincamine were found to be present in these cultures, although these results could not be confirmed during subsequent studies. Although small amounts of spots positive with Draggendorff's reagent were detected on TLC plates, no vincamine was observed in *V. minor* or *V. major* cv. variegata callus cultures. A lignan, lirioresinol B, was found to be the major compound in these cultures. Ostapenko (1986) reported on protoplasm heat resistance in *Vinca major*.

There are very few reports on hairy root production in *Vinca minor* (Tanaka *et al.*, 1994, 1995). Detailed information on growth kinetics in these reports is, however, lacking. Tanaka *et al.*, (1995) investigated the characteristics of regenerated plants obtained from hairy roots (Ri-transformed plants) of *Vinca minor* L. These workers showed that an established Ri-transformed clone, Vm-101, proliferates rapidly *in vitro*, displays a high degree of lateral branching and rapid shoot elongation and has a growth index 2.5 times that of an untransformed plant. The addition of 2.2 μM benzyladenine to the culture medium increased the shoot number but did not decrease the growth index. Vincamine content in the leaves of *in vitro*-cultured Vm-101 was twice that in the cultured untransformed plant suggesting that multiple shoot culture of Ri-transformed plants may be an excellent tool for *in vitro* vincamine production.

Eilert *et al.* (1987) identified strictosidine lactam as the main alkaloid produced in two cell lines of *V. major* cv. variegata. The high yielding cell line produced 0.5 mg/g dry weight. Transferring the cells to an alkaloid production medium resulted in a 6- to 8- fold increase in alkaloid levels combined with greening of the cells. Crespi *et al.* (1990) patented the production of vincamine and epivincamine by means of cell cultures of *V. minor*. Yields of 3.3 and 0.9 g/l, respectively were claimed for the two alkaloids. Two other patents have been reported concerning the production of *Vinca* alkaloids (Synthelabo, 1976, 1982).

Biotechnological Studies on *Vinca major*: Initiatives at CIMAP

The initial expedition to collect *Vinca major* was initiated in the summer season of April 1994. The plants were collected from the slopes of the Himalayan mountains (ca. 2000 m) of the Uttarakhand region in India. The plant was found growing quite well in isolated patches on the mildly shady side of the mountains. The plants were found growing as a climber and were in flowering condition when collected. The plants were collected along with the soil, which had a pH of 5.5–6.0, and brought to the laboratory at Lucknow. Since the weather was quite hot at Lucknow, the plants were initially grown in pots in a growth chamber (22±2°C temp; 60–80% RH and 4000 lux light intensity) for over 3 months for acclimatization and upon acclimatization were transferred to the field. The plant adapted well to the environment of the plains and flourished well in the glasshouse round the year, growing luxuriously in winters. The live plants have been maintained at the CIMAP, Lucknow repository in the field as well as *in vitro* cultures now for over 16 years. The plant is a climber that sends roots at the nodes which serve as propagules (stem runners). The best season for vegetative growth was between November and February. During this season, the plants flowered even in the glasshouse. Blue coloured flowers (3–10 flowers/plant) were produced during this season. Plants transferred to the field grew well till the onset of May, when the growth slowed down appreciably. Propagation was achieved by means of cuttings containing nodal segments, which rooted easily upon transfer to soil. Surprisingly, on a recent expedition (April, 2009), to the same place from where the plants were collected, it was noticed that the plant sparsely grows in the natural habitat at that altitude, possibly due to changes in environment and urban activities.

Development of a simple, efficient, fast and reliable micropropagation protocol

Nodal explants of *V. major* were used for the initiation of aseptic cultures. The explants were washed with savlon and then treated with $HgCl_2$ (0.1% w/v) solution for varying time durations (1–7 min). Optimal sterilization, to the extent of 80%, was achieved upon 5–6 min treatment. Treatment for lesser duration resulted in heavy infection. Nodal segments of *V. major* required much shorter $HgCl_2$ treatment time (2–3 min) to provide 60–70% aseptic cultures.

Successful *in vitro* cultures could be established from juvenile axillary shoots, obtained from glass-house grown plants. In case of *V. major* nodal explants, even a hormone-free MS medium could bring about the bud break and limited axillary shoot elongation in 28–36% of the cultures within 2 weeks of inoculation. The frequency of response was significantly better when 5.0 mg/l Kn was added to the medium. Axillary bud elongation on this medium was observed in about 90% cultures on MS + Kn (5.0 mg/l), which acquired a length of 3–5 cm with 8–10 nodes within 5–6 weeks of culture initiation. A stock of such axillary shoots was multiplied and maintained through repeated subculturing of nodal segments (excised from *in vitro* grown parent cultures) every 4–5 weeks.

With a view to enhancing the rate of *in vitro* multiplication of shoots, efforts were made to devise appropriate medium recipes for the production of multiple shoots. Various media compositions were evaluated to test the ability of cytokinins in improving multiple shoot production *in vitro*. The axillary shoots of *V. major*, however, preferred to remain elongated rather than repeatedly branched on the medium i.e., MS + Kn (5.0 mg/l), referred to as Primary Shoot Elongation Medium (PSEM). For this purpose, nodal explants were excised from 4- wk old axillary shoot cultures and transferred to MS medium fortified with 1.0, 3.0, 5.0, 6.0 and 10.0 mg/l Kn or BAP. The response observed *in vitro* revealed that both cytokinins are appropriate for shoot multiplication. In the next series of experiments, triple combinations of 5.0 mg/l Kn with 1.0–5.0 mg/l BAP and 0.5–1.0 mg/l IBA were tried to see their cumulative influence on the morphogenetic response. The best combination amongst these was found to be MS + Kn (5.0 mg/l) + BAP 2.0 mg/l) + IBA (1.0 mg/l). On this medium, referred to as Primary Multiple Shoot Medium (PMSM), the shoots branched and grew very well *in vitro*.

The influence of number of passages, from initial culture, on multiple shoot formation was also evaluated. The response in terms of time required for bud emergence, elongation and multiple shoot formation was recorded. Multiple shoot formation, even after 24 passages, was fairly constant. The maximum number of shoots were obtained between 18–24 passages, and the average number of shoots were 29.20±1.36 (Table 13.2).

Table 13.2. Influence of culture age on rate of multiple shoot formation on PMSM* in *Vinca major*.

No. of passages from initial culture	Days to Bud Induction	No. of multiple shoots/100 ml flask** (Mean ±SE)
1–6	7–14	28.92± 0.67
6–12	-do-	27.24±0.91
12–18	-do-	26.40±0.97
18–24	-do-	29.20±1.36

* MS medium supplemented with Kn (5.0 mg/l) + BAP (2.0 mg/l) + IBA (1.0 mg/l).
** Data recorded after 4 weeks of culture. Each figure represents an average of 25 replicates.

The multiple shoot cultures of *Vinca major* produced flowers *in vitro* only from 24th culture passage onwards. The number of flowers produced ranged from 1–3 per culture.

Hairy root induction in *Vinca major* using *Agrobacterium rhizogenes*

The following *Agrobacterium rhizogenes* strains were used for transformation studies viz., A_4, LBA-9402, NCPPB-1855, SV_2, SV_4, R-1000, and MAFF-301724. For induction of hairy roots in *V. major*, several different explants viz., leaf, petiole, fertilized ovule, embryo, petals, sepal, microshoots and anthers were tried. The explants were taken from glass-house grown, field-grown and *in-vitro* raised plants. Glass-house-grown and *in vitro* raised plants proved to be the most ideal sources of explants for *A. rhizogenes* infection. Amongst the different explants tested, leaf and microshoot explants responded best to hairy root induction. Root emergence from non-transformed explants (controls) was not observed. When leaf explants were used for infection with various strains of *A. rhizogenes,* hairy roots could be induced by the following strains viz., A_4, LBA-9402, SV_2, SV_4, NCPPB-1855 and ATCC-15834. However, the degree of success depended on strain specificity. Petiole explants of *V. major* responded only to A_4 and SV_4 strains. The transformation frequency in *V. major* ranged from 0.03–0.06 when leaf and microshoot explants were subjected to co-cultivation with various Ri-strains. With microshoots, only SV_2 and SV_4 were able to successfully induce hairy roots, while with anthers as explants only SV_4 gave the results. The efficacy of the different bacterial strains to bring about hairy root transformations in *Vinca major* was in the order of SV_4>SV_2>A_4>LBA-9402>NCPPB-1855>ATCC-15834>MAFF-301724. Ri-strain R-1000 could not induce hairy roots in *Vinca major*. The frequency of transformation was determined as the number of explants with hairy roots/total number of explants. The individual transformation events were grown initially on solid media and thereafter maintained as separate clones in liquid media (Table 13.3; Fig. 13.7). Three distinct clones were isolated viz., BL-1, BL-2 and BL-3. Among the root clones of *V. major*, BL-1 and BL-2 showed a tendency to coil and were very thin. The most healthy looking hairy root line was that of BL-3, which elongated very well and also developed green pigmentation when grown under high illumination. These hairy root clones were then grown in liquid 1/2 MS (without NH_4NO_3) + IBA (1.0 mg/l) medium for further maintenance and characterization. The growth profile of the different hairy root clones on half-strength liquid MS medium supplemented with IBA (1.0 mg/l) revealed that a growth index of 7.04–7.64 was achievable in case of different hairy root clones of *V. major* (Table 13.4).

Table 13.3. Different media tested for initial growth and maintenance of hairy root clones of *Vinca major**.

Media composition	Hairy Root Clones			Growth Pattern
	BL-1	BL-2	BL-3	
MS0	-	-	-	Root growth could not be sustained
½ MS0	-	-	-	-do-
¼ MS0	-	-	-	-do-
MS + IBA (1.0 mg/l)	2	2	1	-
MS + NAA (1.0 mg/l)	1	1	1	-
½ MS + IBA (1.0 mg/l)	3	3	3	The root tips of hairy roots proliferated very fast, giving out profuse white root hairs
½ MS + NAA (1.0 mg/l)	3	3	3	Callusing in all the hairy root clones was observed. BL-2 clone developed green colour.
¼ MS + IBA (1.0 mg/l)	2	2	3	The roots turned brown and collapsed
¼ MS + NAA (1.0 mg/l)	1	1	1	-

* Data collected on transformed root cultures induced by SV_4 strain of *A. rhizogenes*.
1, Mild branching and elongation; 2, Moderate branching and elongation; 3, Profuse branching and elongation

Table 13.4. Growth index of hairy root lines of *Vinca major*.

Clone No.	Growth Index* (initial procedure)
BL-1	7.04**
BL-2	7.22**
BL-3	7.64**

* Final fresh weight/ initial fresh weight (inoculum) at the end of 5th week of culture.
** Mean value of 5 replicates

Vincamine production in selected hairy root clones

HPLC analysis confirmed the enhanced production of vincamine by the hairy root clones of *Vinca major* (BL-1, BL-2 and BL-3 clone). Methanolic extracts of these lines indicated the presence of Dragendorff's reagent positive spots when subjected to TLC initially. For comparisons, leaf and root tissues from glass-house grown plants (6–12 months old) and also the roots of *in vitro* micropropagated plants of *V. major* were used for alkaloid extraction, which served as control. The crude alkaloid content in the three transformed root lines varied form 0.02–0.04% FW after 4 weeks of growth *in vitro*. The corresponding alkaloid levels in the untransformed (controls) varied from 0.04–0.05%. Vincamine was identified by retention time data as well as by peak enrichment method in the three clones of *V. major*.

The transformed root line BL-3 was found to contain the highest level of vincamine (0.40%). HPLC analysis of the transformed roots of *V. minor*, however, did not indicate the presence of other alkaloids in any of these clones, indicating that the repertoire of alkaloids produced *in vitro* is low, which can be augmented by adopting procedures like elicitation, supplementation with specific additives etc.

Induction of photoautotrophy in transformed hairy root cultures

Based on earlier information available in the literature (Husemann, 1984; Tyler *et al.*, 1986; Maldonado-Mendoza and Loyola-Vargas, 1995) on the importance of green hairy roots (possessing functional chloroplast system) to enhance the productivity, experiments were conducted to induce photoautotrophy in the selected hairy root clones of *V. major* (BL-3). These clones were selected because of their inherent high capacity to produce TIAs under photoheterotrophic conditions *in vitro*. Gradual lowering of sucrose in the medium resulted in significant inhibition in the growth of these lines and as such it failed to induce autotrophy. The clones were grown in the CO_2 incubator (New Brunswick, USA). In the initial experiments, the hairy root cultures in liquid medium were exposed to 5% CO_2 level for 1–6 h daily up to 10 days and were then returned back to normal culture room environment after each CO_2 exposure. Only one culture belonging to BL-3 line showed moderate transient greening, which again became non-chlorophyllous when subcultured twice on 2% sucrose containing media. Dual effect of CO_2 enrichment and high illumination was then studied. For this, the root cultures were alternatively exposed to 5% CO_2 and high illumination of 4000 lux for 15, 30, 60 and 120 min and 1, 2, 3 h, respectively. Out of these combined treatments, 1-h exposure to CO_2 followed by 3 h light treatment for 15 continuous days proved highly effective for the induction of photoautotrophy. The green line, thus generated, showed good survival at 2% sucrose level. This was then subjected to further gradual lowering of carbohydrate source in the medium.

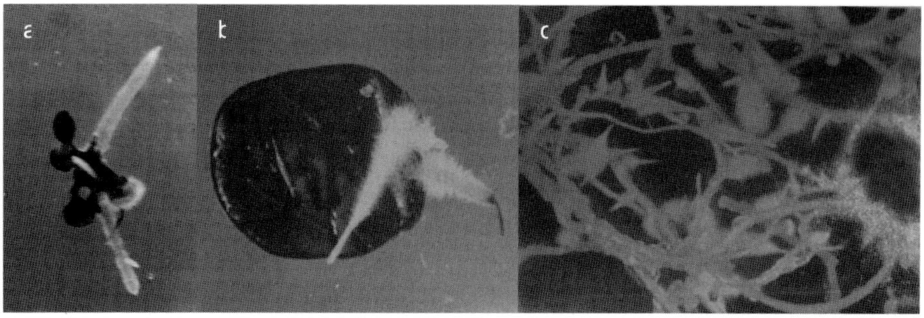

Fig. 13.7. Induction of hairy roots in *Vinca major*. Hairy root emerging from the leaf explants. (a and b); Fast-growing, healthy hairy roots of *Vinca major* (BL-3) (c).

Conclusion

The two *Vinca* species viz., *Vinca major* and *V. minor* are very important medicinal plant species on which scant attention has been paid as far as biotechnological interventions are concerned. The present study reports the efforts made in this direction and the results have been discussed in the light of contemporary literature. The studies in our group have led to development of a simple, efficient, fast and reliable micropropagation protocol for production of true-to-type plants in elite strains of *V. major*. The establishment of such procedures can be effectively utilized for germplasm conservation and mass multiplication of the plant species. The hairy roots established have immense potential for production of the indole alkaloids and there is a need to develop up-scaling processes using bioreactors for large-scale production.

Acknowledgements

The authors are thankful to Dr P.S. Ahuja, Director, Institute of Himalayan Bioresource Technology, Palampur and Prof. Sushil Kumar, Former Director, Central Institute of Medicinal and Aromatic Plants, Lucknow for encouragement. The authors are also grateful to Dr S. Banerjee, Dr AK Kukreja, Dr Laiq-ur-Rahman, Mr Sudeep Tandon, Mr Krishan Gopal and Late Mr M.S. Kidwai of CIMAP, Lucknow for their support in multifarious ways.

References

Abbott, B.J., Leiter, J., Hartwell, J.L., Caldwell, M.E. and Schepartz, S.A. (1966) Screening data from the cancer chemotherapy. National Service Center Screening Laboratories. 33. Plant Extracts. *Cancer Research* 26, 587–743.

Adám-Vizi, V. (2000) Neuroprotective effect of sodium channel blockers in ischemia: the pathomechanism of early ischemic dysfunction. *Orvosi Hetilap* 141(23), 1279–1286.

Ali, E., Giri, V.S. and Pakrashi, S.C. (1976) Vincarpine and di hydro vincarpine 2 new zwitterionic indole alkaloids from *Vinca major* var. elegantissima. *Tetrahedron Letters* 52, 4887–4890.

Aneli, N.A., Zhukovich, E.N., Vachnadze, V.Y. and Aneli, D. (1988) *The pharmacobotanical study of Vinca major introduced in the* Georgian USSR. *Izvestiya Akademii Nauk Gruzinskoi SSR Seriya Biologicheskaya* 14, 227–230.

Arora, R. (1998) Biotechnological Studies in *Catharanthus roseus* (L.) G. Don vis-a-vis Secondary Metabolite Production. Ph.D. Thesis. Central Institute of Medicinal and Aromatic Plants, Lucknow and Chaudhary Charan Singh University, Meerut, India, pp. 195 + plates.

Arora, R., Malhotra, P., Chawla, R., Gupta, D., Sharma, R.K. and Baliga, M.S. (2010) Indian Herbal Medicine for Cancer Therapy and Prevention: Bench to Bedside. In: Watson, R.R. and Preedy, V.R. *Bioactive Foods and Extracts: Cancer Treatment and Prevention.* Taylor and Francis Group, LLC, pp. 519–543.

Atta-Ur-Rahman, A.S., Nighat, F., Bhatti, M.K., Kurucu, S. and Kartal, M. (1995) Alkaloids from *Vinca major. Phytochemistry (Oxford)* 38, 1057–1061.

Balsevich, J., Constabel, F. and Kurz, W.G.W. (1982) Alkaloids of *Vinca major* cultivar variegata. *Planta Medica* 44, 91–93.

Banerji, A. and Chakrabarty, M. (1974) Lochvinerine a new indole alkaloid of *Vinca major. Phytochemistry (Oxford)* 13, 2309–2312.

Blamey, M. and Grey-Wilson, C. (1989) *Flora of Britain and Northern Europe*. Hodder & Stoughton.

Boratynska, W., Kaczmarek, F. and Mrugasiewicz, K. (1972) Cytostatic properties of the alkaloid fractions isolated from the aerial part of *Vinca major. Herba Polonica* 18, 229–241.

Boulat, O., Waldmeier, P. and Maitre, L. (1990) 3,4-dihydroxyphenylacetic acid (DOPAC) as an index of noradrenaline turnover: effects of hydergine and vincamine. *Journal of Neural Transmission Gen.* 82, 181–195.

Chatterjee, A., Biswas, G.K., Mukherjee, B. and Hiremath, S.P. (1970) Indole alkaloids from *Vinca major*– Periwinkle. *Proceedings of the Indian Science Congress Association* 57, 122–123.

Chatterjee, A., Banerji, A. and Chakrabarty, M. (1975) Monoterpenoid alkaloid from *Vinca major. Planta Medica* 28, 109–111.

Cook, P. and James, I. (1981) Cerebral vasodilators. *New England Journal of Medicine* 305(26), 1560–1564.

Crespi, P.N., Garofono, L., Guicciardi, A. and Minghetti, A. (1990) Farmitalia Carlo Erba S.r.l. Ger. Offen. DE 390280 A1, 17-08-1989. *Chemical Abstracts* 112, 196670s.

De Almeida, J.D. (2005) On the distribution of *Vinca major* L. (Apocynaceae) in Portugal continental. *Studia Botanica* 24, 91–92.

Delaby, P.A. (1985) How the production of an alkaloid leads to higher plant tissue cultures-assays of vincamine production. *Ann. Pharm. Franc.* 43, 109.

Dézsi, L., Kis-Varga, I., Nagy, J., Komlódi, Z. and Kárpáti, E. (2002) Neuroprotective effects of vinpocetine *in vivo* and *in vitro*. Apovincaminic acid derivatives as potential therapeutic tools in ischemic stroke. *Acta Pharmaceutica Hungarica* 72(2), 84–91.

Eilert, U., Balsevich, J. and Kurz, W.G.W. (1987) Greening and strictosidine lactam production in two cell lines of *Vinca major* cultivar variegata. *Plant Cell Tissue and Organ Culture* 8, 177–182.

Farnsworth, N.R. (1960) Studies on *Vinca major* L. and its alkaloids. *Dissertation Absts* 20, 3266.

Farnsworth, N.R., Fong, H.H., Blomster, R.N. and Draus, F.J. (1962) Studies on *Vinca major* (Apocynaceae) II. *Jour Pharm Sci* 51, 217–224.

Federico, A. and Amore, I.D. (1984) Vincamine TPS activity on glycoconjugate metabolism of the brain: effect on lysosomal enzymes. *Int. J. Clin. Pharmacol. Res.* 4, 277–279.

Fischoff, P.K., Moslinger, G.R., Herrmann, W.M., Friedmann, A. and Russmann, D.L. (1996) Therapeutic efficacy of vincamine in dementia. *Neuropsychobiology* 34, 29–35.

Guseinova, N.A. (1986) Microsporogenesis morphology and viability of the pollen of *Vinca major* larger periwinkle on apsheron peninsula Azerbaijan SSR USSR. *Izvestiya Akademii Nauk Azerbaidzhanskoi SSR Seriya Biologicheskikh Nauk*, 26–29.

Hirsch, G. (1949) *Vinca major* L. au littoral. *Nat Beiges Bull Mens* 30, 144.

Husemann, W. (1984) Photoautotrophic cell cultures. *Cell Culture and Somatic Cell Genetics Plants* 1, 182–191.

Huxley, A. (ed.) (1992) *New RHS Dictionary of Gardening* 4, 664–665. Macmillan.
Il'Yashenko, L.I., Malikov, V.M., Yagudaev, M.R. and Yunusov, S.Y. (1977) *Vinca major* alkaloids. *Khimiya Prirodnykh Soedinenii* 382–387.
Ishikura, N. and Minekishi, K. (1978) New delphinidin glycosides isolated from *Vinca major* flowers. *Botanical Magazine Tokyo* 91, 181–186.
Kurmukov, A.G. (1967) Electroencephalographic study of the effect of vincamine on the central nervous system in rabbits. *Farmakol. Toksikol.* 30, 286–289.
Lanfranchi, M., Magnani, G., Sinforiani, E. and Casali, R. (1984) Effects induced by vincamine teprosilate on chronic cerebrovascular disease: preliminary results. *International Journal of Clinical Pharmacology Research* 4, 307–312.
Lörincz, C., Szász, K. and Kisfaludy, L. (1976) The synthesis of ethyl apovincaminate. *Arzneimittel-forschung* 26(10a), 1907.
Maldonado-Mendoza, I.E. and Loyola-Vargas, V.M. (1995) Establishment and characterization of photosynthetic hairy root cultures of *Datura stramonium*. *Plant Cell Tissue and Organ Culture* 40, 197–208.
McKenna, G.F., Taylor, A. and Gibson, R.S. (1960) Extracts of plant and cancer chemotherapy. *Texas Reports Biol. Med.* 18, 233.
Moinade, S. and Puvinel, J. (1969) Therapeutic trial of vincamine. *Lyon Med.* 221, 1383–1385.
Mokry, J. and Kompis, I. (1964) On the alkaloids from *Vinca minor*. *Lloydia* 27, 428.
Mukhopadhyay, G., Mukherjee, B., Patra, A., Chatterjee, A., Ghosh, R., Roychowdhury, P. and Kawazura, H. (1991) 11 Methoxytetrahydroalstonine: A heteroyohimbinoid alkaloid from *Vinca major*. *Phytochemistry* (Oxford) 30, 2447–2449.
Ostapenko, A. (1986) Protoplasm heat resistance in *Vinca major* Apocynaceae. *Izvestiya Akademii Nauk Turkmenskoi SSR Seriya Biologicheskikh Nauk*, 24–28.
Potier, P., Beugelmans, R., Men, J.L. and Janot, M.-M. (1965) Alkaloides of roots of *Vinca major* L (Apocynaceae) reserpine, methoxy-10 vellosimine. Alkaloides of Pervenches. *Ann Pharm Franc* 23, 61–64.
Sakushima, A. and Nishibe, S. (1988) Mass spectrometry in the structural determination of flavonol triglycosides from *Vinca major*. *Phytochemistry (Oxford)* 27, 915-920.
Sitges, M., Galván, E. and Nekrassov V. (2005) Vinpocetine blockade of sodium channels inhibits the rise in sodium and calcium induced by 4-aminopyridine in synaptosomes. *Neurochemistry International* 46(7), 533–540.
Smeyers, Y.G., Smeyers, N.J., Randez, J.J., Hernandez-Laguna, A. and Galvez-Ruano, E. (1991) A structural and pharmacological study of alkaloids of *Vinca minor*. *Molecular Engineering* 1(2), 153–160.
Synthelabo 04-02-1975 FR -003347 (05-08-1976) Alkaloid production by culturing *Vinca minor* cells in undifferentiated state in liquid medium. [Patent]
Synthelabo 22-10-1980 FR-022538 (24-04-1982) *In vitro* metabolite production or biotransformation using plant cells is effected in an unstirred liquid medium without any support for fixation of the cellular tissue. [Patent]
Szilágyi, G., Nagy, Z., Balkay, L., Boros, I., Emri, M., Lehel, S., Márián, T., Molnár, T., Szakáll, S., Trón, L., Bereczki, D., Csiba, L., Fekete, I., Kerényi, L., Galuska, L., Varga, J., Bönöczk, P., Vas, A. and Gulyás, B. (2005) Effects of vinpocetine on the redistribution of cerebral blood flow and glucose metabolism in chronic ischemic stroke patients: a PET study. *Journal of Neurological Sciences* 15, 229–230.
Tanaka, N., Takao, M. and Matsumoto, T. (1994) *Agrobacterium rhizogenes*-mediated transformation and regeneration of *Vinca minor* L. *Plant Tissue Culture Letters* 11, 191–198.
Tanaka, N., Takao, M. and Matsumoto, T. (1995) Vincamine production in multiple shoot culture derived from hairy roots of *Vinca minor*. *Plant Cell, Tissue and Organ Culture* 41(1), 61–64.
Taylor, W.I. and Farnsworth, N.R. (1973) *The Vinca Alkaloids*. Marcel Dekker, New York.
Tyler, R.T., Kurz, W.G.W. and Panchuk, B.D. (1986) Photoautotrophic cell suspension cultures of periwinkle (*Catharanthus roseus* (L.) G. Don).: Transition from heterotrophic to photoautotrophic growth. *Plant Cell Reports* 3, 195–198.

Zhmurkin, V.P. (1968) Mechanism of the therapeutic effect of vincamine in disorders of the cerebral circulation. *Zh. Nevropatol. Psikiatr.* 68, 1322–1325.

Zhukovich, E.N. (1987) Alkaloids from the roots of *Vinca major* introduced in the Georgian SSR USSR. *Khimiya Prirodnykh Soedinenii* 611–612.

Zhukovich, E.N. and Vachnadze, V.Y. (1985) Vincamaijinine a new alkaloid from *Vinca major*. *Khimiya Prirodnykh Soedinenii* 720.

Zhukovich, E.N., Bostoganashvili, V.S. and Vachnadze, V.Y. (1986) Method for the quantitative identification of vincamajine in the above ground part of *Vinca major* introduced in the Georgian SSR USSR. *Khimiya Prirodnykh Soedinenii* 653–654.

Zhukovich, E.N., Kikoladze, V.S., Tskitishvili, N.Z., Tsitsishvili, V.G., Vachnadze, V.Y. (1989) Study of alkaloids from *Vinca major* introduced in the Georgian SSR USSR. *Khimiya Prirodnykh Soedinenii* 434–435.

Zsadon, B., Arvay, G. and Tetenyi, P. (1972) Alkaloids from *Vinca major* apocynaceae racemic vincadifformine preliminary communication. *Acta Chimica Academiae Scientiarum Hungaricae* 72, 355–356.

Chapter 14

Analytical Platforms and Databases Ranging from Plant Transcriptomics to Metabolomics

Tetsuya Sakurai, Kenji Akiyama and Kazuki Saito

Introduction

In the post-genomics era, many bioinformatics studies have been aiming to obtain exhaustive information on all cellular components. A comprehensive understanding of the cell based on transcriptomic, proteomic, and metabolomic data is required to understand the complex nature of molecular networks in cells. The study of transcriptomics is a thriving field, and there has been an increase in the number of web-based databases and tools for transcriptome analysis (Galperin, 2008; Galperin and Cochrane, 2009). Although many researchers have noted that metabolomic data are more closely related to biological phenomena than transcriptomic or proteomic data (Dixon *et al.*, 2006; Wang *et al.*, 2006), the availability of metabolomic databases and tools is limited (Kopka *et al.*, 2005; Wishart *et al.*, 2009). Many studies are trying to link '-omics' data (Hirai *et al.*, 2004, 2005 Steinhauser *et al.*, 2004; Sakurai *et al.*, 2005; Tohge *et al.*, 2005; Tokimatsu *et al.*, 2005; Kusano *et al.*, 2007); therefore, there is an increasing need for the construction of metabolomic databases and related web sites.

For instance, chemical resources produced in plants as a result of photosynthetic and metabolic reactions serve as important resources of food, material and energy and constitute the global ecosystem (Fargione *et al.*, 2008). Therefore, many researchers are trying to find a means to increase plant biomass using information from recent biological studies. Although an enormous amount of time and effort has been expended in the empirical selection of better plant varieties because of the classical agricultural approach, much attention has been devoted for elucidating the complex network of interactions among genes and molecular networks that is involved in the production of plant biomass. Further, techniques and databases are rapidly developing in the post-genomic era (Dixon *et al.*, 2006). Therefore, many insights have been provided into both functional and comparative genomics through the use of publicly available databases for the model plant *Arabidopsis thaliana*. Data resources for *Arabidopsis thaliana* are substantially larger and of better quality than those available for other plant species; hence, these resources represent a good starting point for the development of comparable resources for other species.

Here, we describe web databases and tools that support research in transcriptomics, proteomics, and metabolomics. Moreover, we show one of the ways how the transcriptomics and metabolomics information resources are applied.

Transcriptomics

Gene expression and co-expression tools and databases

Genes involved in related biological pathways are often co-expressed, and information on their co-expression helps understand biological systems (Eisen *et al.*, 1998). In fact, co-expression data have been applied to a wide variety of experimental approaches such as gene targeting, regulatory investigations and/or identification of potential partners in protein-protein interactions (PPIs) (Aoki *et al.*, 2007; Shoemaker and Panchenko, 2007; Saito *et al.*, 2008). Large amounts of gene expression data are required to reliably predict relationships between co-expressed genes. Toward this end, DNA microarrays have provided vast amounts of gene expression data that are stored in several public repositories (Craigon *et al.*, 2004; Parkinson *et al.*, 2007; Swarbreck *et al.*, 2008; Barrett *et al.*, 2009). Using these large public data resources, several co-expression databases have been constructed and are widely used, especially in *Arabidopsis* research (Steinhauser *et al.*, 2004; Toufighi *et al.*, 2005; Manfield *et al.*, 2006; Hruz *et al.*, 2008; Obayashi *et al.*, 2009). Tools and databases for gene expression and co-expression are listed in Table 14.1.

Table 14.1. Gene expression and co-expression tools and databases.

AtGenExpress	http://www.weigelworld.org/resources/microarray/AtGenExpress/
AtGenExpress Japan	http://pfg.psc.riken.jp/AtGenExpress/
Genevestigator	https://www.genevestigator.com/gv/
ATTED-II	http://atted.jp/
Arabidopsis thaliana - Tiling Array Express (AT-TAX)	http://www.weigelworld.org/resources/microarray/at-tax/
Bio-Array Resource (BAR)	http://bar.utoronto.ca/
NASCArrays	http://affymetrix.*Arabidopsis*.info/narrays/experimentbrowse.pl
Arabidopsis Co-expression Data Mining Tools (ACT)	http://www.*Arabidopsis*.leeds.ac.uk/act/
AraGenNet	http://aranet.mpimp-golm.mpg.de/aranet
Arabidopsis thaliana co-response databases (AthCoR) at CSB.DB	http://csbdb.mpimp-golm.mpg.de/csbdb/dbcor/ath.html
Correlated Gene Search at PRIMe	http://prime.psc.riken.jp/?action=coexpression_index
CressExpress	http://obiwan.ssg.uab.edu/coexpression/
Gene Co-expression Analysis Toolbox (GeneCAT)	http://genecat.mpg.de/cgi-bin/Ainitiator.py/
A Complete *Arabidopsis* Transcriptome MicroArray (CATMA)	http://www.catma.org/

AtGenExpress
http://www.weigelworld.org/resources/microarray/AtGenExpress/
http://pfg.psc.riken.jp/AtGenExpress/

AtGenExpress is the result of a multinational coordinated effort to uncover the transcriptome of the multicellular model organism *Arabidopsis thaliana*. This large collaborative project combined resources of gene expression patterns observed during the complete plant life cycle; the gene expression data were obtained using the Affymetrix GeneChip.

The Affymetrix ATH1 array- an oligonucleotide-based DNA microarray consisting of 22,746 probe sets, covering approximately 23,700 genes and nearly the entire genome of *Arabidopsis* was designed on the basis of computer-predicted genes (Redman, *et al*., 2004). Using this system, the AtGenExpress consortium, an international undertaking, was established to make a significant contribution to the science community and to add to our knowledge of gene functions in *Arabidopsis*. Previous studies have reported the development of large-scale transcriptome data sets by the AtGenExpress consortium; these data sets provide detailed information on the developmental processes (Schmid *et al*., 2005) and stress responses (Kilian *et al*., 2007) in *Arabidopsis*. Moreover, these web sites present a hormone-response data set, which includes treatment with phytohormones, hormone inhibitors, hormone-related inhibitors, hormone-related mutants, seed germination processes, and sulfate starvation (Goda *et al*., 2008).

Genevestigator
https://www.genevestigator.com/gv/

Genevestigator (Hruz *et al*., 2008) is a reference expression database and meta-analysis system. It allows biologists to study the expression and regulation of genes in a broad variety of contexts by summarizing information from hundreds of microarray experiments into easily interpretable results. A user-friendly interface allows the user to visualize gene expression in many different tissues, at multiple developmental stages, or in response to large sets of stimuli, diseases, drug treatments, or genetic modifications.

ATTED-II
http://atted.jp

ATTED-II (Obayashi *et al*., 2007; Obayashi *et al*., 2009) is a database of co-expressed genes in *Arabidopsis* that can be used to design a wide variety of experiments, including the prioritization of genes for functional identification, or for studies on regulatory relationships. It aims at constructing gene networks by: (i) introducing a new measure of gene co-expression to retrieve functionally related genes more accurately; (ii) providing clickable maps for all gene networks to facilitate step-by-step navigation; (iii) including information on PPIs; (iv) identifying conserved patterns of co-expression; and (v) showing and connecting Kyoto Encyclopedia of Genes and Genomes (KEGG) pathway information (Kanehisa *et al*., 2002) in order to identify functional modules. With these enhanced functions for gene network representation, ATTED-II can help researchers clarify the functional and regulatory gene networks in *Arabidopsis*.

Bio-Array Resource
http://bar.utoronto.ca/

Bio-Array Resource (BAR) (Toufighi *et al.*, 2005) offers various analysis tools such as the electronic Fluorescent Pictograph (eFP) Browser, Expression Angler, and Expression Browser. The *Arabidopsis* expression data was derived from AtGenExpress, BAR and NASCArrays. Expression data in poplar, *Medicago truncatula*, and mouse can be investigated, and gene expression can be visualized by the eFP browser. Promomer is another tool that is useful for identifying over-represented words (n-mer) in the promoters of genes of interest or of co-regulated genes.

cDNA and EST databases

Research in functional genomics demands precise information on gene structure and large volumes of gene catalogues. Expressed sequence tags (ESTs) and full-length cDNAs are essential for the correct annotation and functional analysis of genes and their products. Databases for cDNAs and ESTs are listed in Table 14.2. Below, we have provided two examples of cDNA and EST databases.

Table 14.2. cDNA and EST databases.

Salk Institute Genome Analysis Laboratory (SIGnAL)	http://signal.salk.edu/
RIKEN *Arabidopsis* Full-Length (RAFL) cDNA database at RARGE	http://rarge.psc.riken.jp/cdna/cdna.pl
Arabidopsis thaliana Orphan Transcript Database (AtoRNA DB)	http://atornadb.bio.uni-potsdam.de/
Ceres cDNA database	http://www.tigr.org/tdb/e2k1/ath1/ceres/ceres.shtml
Sequence download at TAIR	ftp://ftp.*Arabidopsis*.org/home/tair/Sequences/

Salk Institute Genome Analysis Laboratory
http://signal.salk.edu/
The Salk Institute Genome Analysis Laboratory (SIGnAL) database displays SALK cDNAs, Salk/Stanford/PGEC (SSP) cDNAs (Yamada *et al.*, 2003), orphan RNAs (AtoRNAs) (Riano-Pachon *et al.*, 2005), etc. on the *Arabidopsis* genome. Moreover, it provides tiling array results, single feature polymorphism information etc.

RIKEN Arabidopsis Genome Encyclopedia
http://rarge.psc.riken.jp/cdna/

The RIKEN *Arabidopsis* Genome Encyclopedia (RARGE) database (Sakurai *et al.*, 2005) contains RIKEN *Arabidopsis* Full-length (RAFL) cDNA (Seki *et al.*, 2002) and RIKEN *Ac/Ds* transposon tagged mutant (Ito *et al.*, 2002a,b; Kuromori *et al.*, 2004, 2006; Ito *et al.*, 2005).

Small RNA databases

High-throughput sequencing has enabled the discovery of hundreds of thousands of unique small RNA sequences in *Arabidopsis* and other plants. These small RNAs include both conserved and non-conserved micro-RNAs (miRNAs). miRNAs are a few dozen nucleotides in length and posttranscriptionally regulate eukaryotic gene expression. Since the discovery of RNA-based silencing systems, there have been some paradigm-shifting changes in understanding gene regulation at the transcriptional and posttranscriptional levels. For example, it was recently shown that miRNAs and small interfering RNAs (siRNAs) inhibit translation (Brodersen *et al.*, 2008), in addition to their more familiar roles in gene silencing and natural antisense interactions. Several research groups have reported data on small RNAs in *Arabidopsis* (Llave *et al.*, 2002; Xie *et al.*, 2005; Axtell *et al.*, 2006; Rajagopalan *et al.*, 2006; Howell *et al.*, 2007). Unlike most animal miRNAs, most plant miRNAs show near-perfect or perfect complementarity to their targets, and slicing is believed to be their predominant, or exclusive, mode of action (Jones-Rhoades *et al.*, 2006). Databases are being generated in order to investigate the function, regulation and evolution of small-RNA-based silencing pathways in plants. Therefore, we list some information resources for small RNA analysis (Table 14.3).

Table 14.3. Small RNA databases.

Arabidopsis thaliana Small RNA Project (ASRP)	http://asrp.cgrb.oregonstate.edu/
Massively Parallel Signature Sequencing (MPSS) Database	http://mpss.udel.edu/
Cereal Small RNAs Database	http://sundarlab.ucdavis.edu/smrnas/
miRNA Precursor Candidates for *Arabidopsis thaliana*	http://sundarlab.ucdavis.edu/mirna/
Plant snoRNA Database	http://bioinf.scri.sari.ac.uk/cgi-bin/plant_snorna/home/
Role of RNA polymerase IV in plant small RNA metabolism	http://epigenomics.mcdb.ucla.edu/smallRNAs/

Arabidopsis Small RNA Project Database
http://asrp.cgrb.oregonstate.edu/

The introduction of high-throughput sequencing technology has driven the development of the *Arabidopsis* Small RNA Project (ASRP) database (Gustafson *et al.*, 2005), which provides information and tools for the analysis of microRNA, endogenous siRNA and other small RNA-related features. In order to accommodate the demands of increased data, numerous improvements and updates have been made to the ASRP database, including new ways to access data, more efficient algorithms for handling data, and increased integration with community-wide resources. New search and visualization tools have also been developed to improve access to small RNA classes and their targets. ASRP contains small RNA profiling data from a series of silencing-defective mutants, developmental stages, and treatments, and it examines small RNA pathways at a genome-wide level. The information can be retrieved as comma-separated value files. Each dataset is hyperlinked to the Gene Expression Omnibus (GEO) (Barrett *et al.*, 2009) accessions; this database is user-friendly.

Massively Parallel Signature Sequencing Database
http://mpss.udel.edu/

Massively Parallel Signature Sequencing (MPSS) is a sequencing-based technology that uses a method to quantify the level of gene expression, and thus generates millions of short sequence tags per library. The website provides a method for the quantification of gene expression in specific plant tissues of *Arabidopsis* (Lu *et al.*, 2005; Nakano *et al.*, 2006), rice (Nobuta *et al.*, 2007) and grape using the MPSS technology. MPSS identifies short sequence signatures produced from a specific position in an mRNA or small RNA molecule, and the relative abundance of these signatures in a given library represents a quantitative estimate of the expression of that gene. Additionally, SBS (sequencing by synthesis) 454 pyrosequencing small RNA data is also available in this database. Detailed information on each signature can be accessed by clicking on the corresponding triangle.

Cereal Small RNA Database
http://sundarlab.ucdavis.edu/smrnas/

The Cereal Small RNA Database (CSRDB) (Johnson *et al.*, 2007) consists of large-scale datasets of maize and rice small modulatory RNA (smRNA) sequences generated by high-throughput pyrosequencing. The smRNA sequences have been mapped to the rice genome and the available maize genome sequence, and these results are presented in two genome browser datasets using the Generic Genome Browser. There are various ways to access the data, including links from the genome browser to the target database. Data linking and integration are the main focus of this interface, and internal as well as external links are present.

Proteomics

Protein localization databases

The location of a protein can be predicted from its protein sequence using bioinformatic means and experimental approaches. In fact, identifying proteins in specific subcellular locations is considered to be very important for understanding cellular functions (Heazlewood *et al.*, 2004). A variety of these bioinformatics programs are available, and several studies have assessed their performances using genome-wide prediction of localization (Richly *et al.*, 2003; Heazlewood *et al.*, 2005).

Many studies have employed mass spectrometry to undertake proteomic surveys of subcellular components in *Arabidopsis*, and have thus yielded localization information on ~2600 proteins. Further information on protein localization may be obtained from other literature on the analysis of locations (~900 proteins) (Gene Ontology Consortium, 2006), location information from Swiss-Prot annotations (~2000 proteins), and location data inferred from gene descriptions (~2700 proteins). Table 14.4 shows the list of the protein localization databases.

Table 14.4. Protein localization databases.

SubCellular Proteomic Database (SUBA)	http://www.plantenergy.uwa.edu.au/applications/suba2/
Plant Proteome DataBase (PPDB)	http://ppdb.tc.cornell.edu/
Plastid Protein Database (plprot)	http://www.plprot.ethz.ch/
Arabidopsis 2010 Peroxisomal Protein Database	http://www.peroxisome.msu.edu/

SUB-cellular Location Database for Arabidopsis Proteins
http://www.plantenergy.uwa.edu.au/suba2/

The SUB-cellular location database for *Arabidopsis* proteins (SUBA) (Heazlewood *et al.*, 2007) houses large-scale proteomic and GFP localization sets from the cellular compartments of *Arabidopsis*. It also contains precompiled bioinformatic predictions for protein subcellular localizations. Currently, SUBA is based on the TAIR8 *Arabidopsis* genome annotation. Presently, the subcellular location data available in SUBA are as follows: 2323 entries based on chimeric fusion protein studies comprising 1538 distinct proteins from 770 publications, and 9793 entries based on subcellular proteomic studies comprising 4642 distinct proteins from 68 publications.

Plant Proteome DataBase
http://ppdb.tc.cornell.edu/

The Plant Proteome DataBase (PPDB) now includes data from maize in addition to data from *Arabidopsis*. This database stores experimental data from in-house proteome and mass spectrometry analysis and curated information on protein function, protein properties and subcellular localization (Friso *et al.*, 2004). The proteins are particularly curated for possible plastid location and plastid function. Protein accessions identified in previously reported *Arabidopsis* and other Brassicaceae proteomics papers are cross-referenced to rapidly determine previous experimental identification by mass spectrometry (MS). Each protein has a protein report page that summarizes all the protein information that is stored in the PPDB. Each accession is hyperlinked to a few other databases to obtain additional information on protein accession. These reports also provide detailed information on similar mass spectrometry data and how these data map to the protein. One can search the PPDB using the accession ID (*Arabidopsis* Genome Initiative (AGI) gene locus or ZmGI accessions), protein name and function, proteome experiments (sample source by species, organ, cell type, or subcellular fraction), biochemical pathway, etc.

PPI databases

High-throughput experiments have resolved genome-scale networks of PPIs (interactomes) in yeast (*Saccharomyces cerevisiae*) (Uetz *et al.*, 2000; Miller *et al.*, 2005), the fruitfly (*Drosophila melanogaster*) (Giot *et al.*, 2003), the nematode worm (*Caenorhabditis elegans*) (Li *et al.*, 2004) and in human (*Homo sapiens*) (Rual *et al.*, 2005; Gandhi *et al.*, 2006). These interactomes have revealed protein interactions in biological processes and the relatedness of interacting partners. Interactomics is rapidly becoming a valuable new area of systems biology through the comprehensive deduction of PPI networks that form the basis for much of the signalling and regulatory control as well as the machinery of cellular function. In the study of plants, several projects have been undertaken to understand protein interaction networks in *Arabidopsis*. It is likely that there will be an increasing need for analytical tools for the visualization of protein interactions to understand the interactome through experimental means. A few examples of databases for PPIs are listed in Table 14.5.

Table 14.5. Protein-protein interaction databases.

Arabidopsis thaliana Protein Interactome Database (AtPID)	http://atpid.biosino.org/
Arabidopsis Interactions Viewer at BAR	http://bar.utoronto.ca/interactions/cgi-bin/*Arabidopsis*_interactions_viewer.cgi
Arabidopsis Membrane Interactome Project	http://www.associomics.org/
Plant Unknown-eome DB (POND)	http://bioinfo.ucr.edu/projects/Unknowns/external/index.html

Arabidopsis thaliana Protein Interactome Database
http://atpid.biosino.org/

Arabidopsis thaliana protein interactome database (AtPID) (Cui *et al.*, 2008a) is a centralized platform for depicting and integrating the information pertaining to PPI networks, domain architecture, orthologue information, and GO annotation in the *Arabidopsis* proteome. The PPI pairs are predicted by integrating several methods with the Naive Bayesian Classifier. All other related information curated in the AtPID is manually extracted from published literatures and other resources by expert biologists. This database includes 28,062 PPI pairs of which 23,396 pairs have been generated by prediction methods. Of the remaining 4666 pairs, 3866 pairs including 1875 proteins were manually curated from the literature, and 800 pairs were obtained from enzyme complexes in the KEGG (Kanehisa *et al.*, 2002). In addition, the subcellular location information of 5562 proteins is available. AtPID was built via an intuitive query interface that provides easy access to the important features of proteins. Orthologue maps, domain attributes, and network displays are developed with crosslinks to other relevant external resources (Hooper and Bork, 2005). In addition to protein interaction data, this database also contains subcellular location annotations to nearly 5000 proteins from the SwissProt and SUBA databases (Boeckmann *et al.*, 2003; Heazlewood *et al.*, 2007) and popular prediction tools, including TargetP (Emanuelsson *et al.*, 2007), Predotar (Small *et al.*, 2004) and MitoProtII (Claros and Vincens, 1996).

Arabidopsis Interactions Viewer at BAR
http://bar.utoronto.ca/interactions/cgi-bin/*Arabidopsis*_interactions_viewer.cgi

The *Arabidopsis* Interactions Viewer (AIV) allows one to view predicted and experimentally determined PPIs in *Arabidopsis*; further, it queries a database of 70,944 predicted and 2779 confirmed interacting proteins in *Arabidopsis* (Geisler-Lee *et al.*, 2007). The protein information in this database is assigned to a subcellular location by using data from SUBA (Heazlewood *et al.*, 2007). Users can search PPI information by using the AGI gene locus ID. The output page of the AIV contains a link to a clickable scalable vector graphics, and the result can be downloaded in Excel and Cytoscape file format.

Metabolomics

Metabolomics databases

Metabolomics is a novel experimental methodology categorized as an 'omics' approach along with genomics, transcriptomics, and proteomics (Fiehn *et al.*, 2001; Fiehn, 2002; Nicholson *et al.*, 2002; Sumner *et al.*, 2003; Saito *et al.*, 2008). Metabolomics is often used in combination with the other 'omics' approaches to gain a deeper understanding of biological processes, especially metabolism. Metabolomics is a powerful tool for the characterization of metabolic phenotypes in highly complex cellular processes. The different metabolites in the plant kingdom are estimated to reach 200,000 (Fiehn *et al.*, 2001; Fiehn, 2002). This fact is due to the great diversity in metabolic pathways that have evolved in each plant species for survival under varying environmental conditions. Detailed metabolite profiling of thousands of plant samples has great potential for directly elucidating plant metabolic processes. Here, we describe a few resources available for metabolomic studies (Table 14.6).

Table 14.6. Metabolomics databases.

DROP Met at PRIMe	http://prime.psc.riken.jp/?action=drop_index
Golm Metabolome Database (GMD) at CSB.DB	http://csbdb.mpimp-golm.mpg.de/csbdb/gmd/gmd.html
KNApSAcK	http://kanaya.naist.jp/KNApSAcK/
NSF2010 Metabolomics	http://lab.bcb.iastate.edu/projects/plantmetabolomics/
MoTo DB	http://appliedbioinformatics.wur.nl/moto/
KOMICS	http://webs2.kazusa.or.jp/komics/
McGill-MD	http://metabolomics.mcgill.ca/

Data Resources of Plant Metabolomics at PRIMe
http://prime.psc.riken.jp/?action=drop_index

Data Resources of Plant Metabolomics (DROP Met) is a repository and distribution site of the dataset obtained from our multiple MS-based metabolomic analyses (Matsuda *et al.*, 2009; Sawada *et al.*, 2009). Various kinds of datasets, ranging from raw fundamental data on analytical conditions to metabolic profiles of plant samples, can be accessed. This site releases 32 raw liquid chromatography-electrospray ionization-mass spectrometry (LC-ESI-MS) chromatogram data derived from four *Arabidopsis* tissues (flower, rosette leaf, cauline leaf, and internode; eight replicates) and recorded in the NetCDF and ASCII format. Detailed information on each data file is described in the meta information (.txt) by following the Metabolomics Standards Initiative (MSI; http://msi-workgroups.sourceforge.net/) recommendations. Users familiar with bioinformatics and metabolomics will be able to develop novel algorithms and methodologies for metabolomics using these datasets.

The Golm Metabolome Database at CSB.DB
http://csbdb.mpimp-golm.mpg.de/csbdb/gmd/gmd.html

The Golm Metabolome Database (GMD) provides public access to custom mass spectra libraries, metabolite profiling experiments, and other necessary information related to the field of metabolomics (Steinhauser *et al.*, 2004). The contents of this database are the 'GMD Page Tree," "GMD Analytics,' 'GMD mass spectral and retention time index (MSRI) libraries,' 'GMD Profiles,' and 'GMD Tools.' The 'GMD Page Tree' refers to the so-called site map. 'GMD Analytics' harbours information related to the analytical technologies, methods and protocols integrated in GMD as well as other necessary information related to the field of metabolomics. 'GMD MSRI' provides public access to custom mass spectra libraries, which are implemented in the MSRI libraries. All libraries can be imported in the National Institute of Standards (NIST) 02 software.

KNApSAcK
http://kanaya.naist.jp/KNApSAcK/

KNApSAcK (Shinbo *et al.*, 2006) includes 20,000 metabolites and the relative species information of these metabolites. Information on natural products has been amassed with special emphasis on the biological origins of these products. The user can retrieve information on metabolites by entering the organism name, the name of the metabolite, molecular weight, or molecular formula. KNApSAcK also has a search engine for the analysis of mass spectrum data.

NSF2010 Metabolomics
http://lab.bcb.iastate.edu/projects/plantmetabolomics/

The NSF2010 Metabolomics project aims at developing metabolomics as the new functional genomics tool for elucidating the functions of *Arabidopsis* genes. Approximately 1800 metabolites have been detected, of which 900 are chemically defined. Users can browse experiment data and download the comma-separated value file and the Microsoft Excel files. With regard to data analysis, ratio plots are available, and users can compare metabolite levels under different experimental conditions by generating dynamic ratio plots of the metabolites. By using this tool, users can obtain the names of all experiments and gain access to the metadata of a chosen experiment. Users can expand each experiment and choose any two experimental factors (for comparison) from two drop-down lists.

Metabolome Tomato Database
http://appliedbioinformatics.wur.nl/moto/

The Metabolome Tomato Database (MoTo DB) (Moco *et al.*, 2006) is a metabolite database dedicated to LC-MS-based metabolomics of the tomato fruit. This database contains all the information that has been obtained thus far using this LC-photodiode array detection (PDA)-quadrupole time-of-flight MS platform, including retention times, calculated accurate masses, PDA spectra, MS/MS fragments and literature references. Unbiased metabolic profiling and comparison of peel and flesh tissues from tomato fruits revealed that all flavonoids and alpha-tomatine were specifically present in the peel, while several other alkaloids and some particular phenylpropanoids were mainly present in the flesh tissue. These findings validated the applicability of the MoTo DB.

Spectral databases

MS and nuclear magnetic resonance (NMR) spectroscopy as well as Fourier-transform infrared (FT-IR) spectroscopy are extensively used in metabolomic studies. Although the comparison of standardized measurements of compounds with data on metabolite mixtures is a key step in metabolome analysis, insufficient standardized experimental spectral data is freely accessible to allow researchers to identify all measurable metabolites (Cui *et al*., 2008a). Furthermore, the use of standardized databases would allow for a more advanced analysis of metabolite mixtures, such as covariance total correlation spectroscopy (Robinette *et al*., 2008). The development of high-throughput measurements of large numbers of metabolites in plants is a result of rapid advances in MS-based methods and computer hardware and software (Last *et al*., 2007). Examples of spectral databases are listed below and in Table 14.7.

Table 14.7. Spectral databases.

National Institute of Standards and Technology (NIST) Scientific and Technical Databases	http://www.nist.gov/srd/analy.htm
MassBank	http://www.massbank.jp/index.html?lang=en
Madison Metabolomics Consortium Database (MMCD)	http://mmcd.nmrfam.wisc.edu/
MS/MS spectral tag (MS2T) libraries at PRIMe	http://prime.psc.riken.jp/lcms/ms2tview/ms2tview.html
NMRShiftDB	http://nmrshiftdb.ice.mpg.de/
Biological Magnetic Resonance Data Bank	http://www.bmrb.wisc.edu/metabolomics/
Standard spectrum search at PRIMe	http://prime.psc.riken.jp/?action=standard_index
Mass Spectral and Retention Time Index (MSRI) libraries at CSB.DB	http://csbdb.mpimp-golm.mpg.de/csbdb/gmd/msri/gmd_msri.html
Spectral Database for Organic Compounds (SDBS)	http://riodb01.ibase.aist.go.jp/sdbs/cgi-bin/direct_frame_top.cgi

NIST Scientific and Technical Database
http://www.nist.gov/data/

For over 40 years, NIST has provided well-documented numeric data to scientists and engineers for technical problem-solving, research and development. These recommended values are based on data that have been extracted from the world's literature, assessed for reliability, and subsequently evaluated to select the preferred values. NIST provides a comprehensive dataset as well as online systems that help the analytical chemist identify unknown materials and obtain physical, chemical and spectroscopic data on known substances. The NIST/EPA/NIH Mass Spectral Library needs the purchase to get available and presently includes 191,436 compounds with spectra. This library is available with the NIST MS Search Program for Windows, which also includes integrated tools for GC/MS deconvolution, mass interpretation, and chemical substructure identification.

MassBank
http://www.massbank.jp/index.html?lang=en

MassBank (Horai *et al.*, 2008) covers high-resolution spectra from a wide range of hybrid MS, ESI-MS and tandem MS analyses, and includes comprehensive data on spectra measured under various experimental conditions (ionization methods, collision methods, etc.). Presently, MassBank contains over 10,000 mass spectra and 1000 compounds. Many compounds were measured by multiple MS methodologies such as quadruple time-of-flight MS (QqTOF-MS), GC/TOF-MS, LC/TOF-MS, LC/triple quadruple tandem-MS/MS (LC/QqQ-MS/MS) and capillary electrophoresis/TOF-MS (CE/TOF-MS). Moreover, MassBank database can be queried by spectral similarity which is based on a vector space search algorithm and is expanded to obtain real value (m/z) data.

Madison Metabolomics Consortium Database
http://mmcd.nmrfam.wisc.edu/

Madison Metabolomics Consortium Database (MMCD) (Cui *et al.*, 2008b) contains data based on NMR spectroscopy and MS measurement. Presently, it serves as a hub for information obtained on small molecules of biological interest from databases and scientific literature. Information on each metabolite entry in the MMCD is available in an average of 50 separate data fields. These data fields provide chemical formulae, names and synonyms, structural data, physical and chemical properties, NMR and MS data of pure compounds under defined conditions where available, NMR chemical shifts determined by empirical and/or theoretical approaches, calculated isotopomer masses, and information on the presence of the metabolite in different biological species. Further, extensive links to images, references and other public databases such as KEGG (Kanehisa *et al.*, 2002) and PubChem (Wang *et al.*, 2009) are provided. The MMCD search engine supports versatile data mining and allows users to make individual or bulk queries on the basis of experimental NMR and/or MS data as well as other criteria. The database mainly contains *Arabidopsis* data; however, it can also be referred to other organisms. In MMCD, the entry for each compound contains two-dimensional and three-dimensional representations of covalent structure, calculated masses for several different monoisotopic compositions, empirically predicted chemical shifts and, where available, experimental NMR chemical shifts and LC-MS data collected under defined conditions.

MS/MS spectral tag libraries at PRIMe
http://prime.psc.riken.jp/lcms/ms2tview/ms2tview.html

The objective of 'non-targeted' metabolic profiling analysis is to understand the mechanisms underlying metabolic events in plants by determining all detectable metabolites. Of the various profiling techniques, non-targeted analysis using LC-MS is a promising tool for investigating the diversity of phytochemicals (Villas-Boas *et al.*, 2005; Dettmer *et al.*, 2007; Bottcher *et al.*, 2008); this tool is known to be as effective as methods employing GC-MS (Tikunov *et al.*, 2005; Moco *et al.*, 2007). MS/MS spectral tag (MS2T) libraries (Matsuda *et al.*, 2009) of metabolites in *Arabidopsis* are created from a set of MS/MS spectra acquired using the automatic data acquisition function of the mass spectrometer. A peak annotation procedure based on the MS2T library was developed for informative non-targeted metabolic profiling analysis involving LC-MS. From this library, users can obtain structural information for peaks detected during metabolic profiling without the need for performing additional MS/MS analysis; structural information can be obtained by searching for the corresponding MS2T accession in the libraries.

Pathway databases

In order to understand the molecular basis of life, it is important to clarify the relationships between cells, organisms and the biosphere. Hence, in the recent years, interactions between and pathways involving genes and molecules are rapidly being organized after which protein activities, gene annotations and metabolites will be integrated into a single database. Some databases for metabolic pathways are listed in Table 14.8.

Table 14.8. Pathway databases.

AraCyc at TAIR	http://www.*Arabidopsis*.org/biocyc/index.jsp
PlantCyc	http://plantcyc.org/
Kyoto Encyclopedia of Genes and Genomes (KEGG) Pathway Database	http://www.genome.jp/kegg/pathway.html
MetaCrop	http://metacrop.ipk-gatersleben.de/
MapMan	http://gabi.rzpd.de/projects/MapMan/
Kazusa Plant Pathway Viewer (KaPPA-View)	http://kpv.kazusa.or.jp/kpv3/guestIndex.jsp
Metabolic Pathway Search at KATANA	http://www.kazusa.or.jp/katana/pathway.html
MetaCyc	http://metacyc.org/
BioCyc	http://biocyc.org/

AraCyc
http://www.*Arabidopsis*.org/biocyc/

AraCyc (Mueller *et al.*, 2003) provides access to manually curated or reviewed information on metabolic pathways in *Arabidopsis*– the model plant. These pathways may be unique to *Arabidopsis* or common to other organisms. Presently, this database features more than 170 pathways, and includes information on compounds, intermediates, cofactors, reactions, genes, proteins, and subcellular locations of proteins. The database uses the Pathway Tools software (Karp *et al.*, 2002), which allows users to visualize all pathways in the database down to the individual chemical structures of the compounds.

PlantCyc of the Plant Metabolic Network
http://plantcyc.org/

PlantCyc contains pathways and the catalytic enzymes and genes involved in pathways as well as compounds from many plant species. Majority of the pathway diagrams in PlantCyc were manually extracted from the plant literature. They are either supported by experimental evidence or are based on expert hypotheses. A small portion of the pathways were computationally predicted. These pathways are known to exist in non-plant organisms but have not been found in plants. However, some hints indicate that plants may utilize the same pathway or similar routes. Both experimentally verified and computationally predicted enzymes and genes are assigned to pathways. Evidence codes are designated to the pathways and genes/enzymes are assigned in order to provide information on data quality. In addition to new pathways curated in-house, the first release of PlantCyc contains all pathways from AraCyc (Mueller *et al.*, 2003), all plant pathways from MetaCyc (Caspi *et al.*, 2008), and a number of curated pathways from other plant pathway databases including RiceCyc (Jaiswal *et al.*, 2006) and MedicCyc (Urbanczyk-Wochniak and Sumner, 2007). Moreover PlantCyc integrates computational assignments of enzymes and genes from other plant pathway databases including AraCyc, RiceCyc, MedicCyc, and SolaCyc.

KEGG PATHWAY
http://www.genome.jp/kegg/pathway.html

The KEGG pathway contains information on metabolites and genes as well as graphical representations of metabolic pathways and complexes derived from various biological processes (Kanehisa *et al.*, 2002). Thus far, 94,739 pathways generated from 327 reference pathways for over 200 organisms, including *Arabidopsis* and rice, have been constructed. Maps of organism-specific metabolic pathways can be generated on the basis of the assignment information in the KEGG/GENES database.

MetaCrop
http://metacrop.ipk-gatersleben.de/

MetaCrop (Grafahrend-Belau *et al.*, 2008) is a database containing manually curated, highly detailed information on metabolic pathways in crop plants; this database includes location information as well as information on transport processes and reaction kinetics. The web interface enables an easy exploration of information from overview pathways to single reactions, and thus helps users understand the metabolism of crop plants. It also supports automatic data export for the creation of detailed models of metabolic pathways in order to support simulation approaches. MetaCrop contains hand-curated information on approximately 40 major metabolic pathways in various crop plants; in particular, the metabolism of agronomically important organs such as seed or tuber is available. In this database, species of both monocotyledons and dicotyledons are represented. Information on involved metabolites, stoichiometry, and the precise location (species, organ, tissue, compartment, and developmental stage) has been incorporated in the reactions. Furthermore, with regard to central metabolism (glycolysis, TCA cycle), kinetic data of the reactions are available. References and relevant PubMed IDs are provided. In order to have a controlled vocabulary for allowing the comparison of data from different sources, ontology terms have been used. In total, the database currently contains information for six major crop plants (approximately 400 enzymatic reactions, 60 transport processes, 6 compartments, and 600 references) and initial information for several more crop plants. All information was manually extracted through an extensive survey of literature and online databases.

MapMan
http://gabi.rzpd.de/projects/MapMan/

MapMan (Thimm *et al.*, 2004; Usadel *et al.*, 2005) is a user-driven tool that displays large genomics datasets onto diagrams of metabolic pathways or other processes. MapMan consists of a Scavenger module that collects and classifies the measured parameters into hierarchical functional categories, and an ImageAnnotator module that uses these classifications to organize and display data on diagrams of the user's choice. Each gene is represented by a discrete signal, allowing individual responses to be identified. Signals for genes that are involved in a particular process are grouped spatially, making it possible to discern general trends that would be less apparent from lists of individual genes. Now, users can browse more than a dozen sample data sets. Moreover the MapMan has been distributed as a JAVA application, therefore, users can perform their analyses by using their own PCs.

Kazusa Plant Pathway Viewer
http://kpv.kazusa.or.jp/kappa-view/

With advances in DNA array technology, vast amounts of transcriptome data have been produced, and mass spectrometers have contributed to the massive production of metabolome data. Interpretation of metabolome data in conjunction with transcriptome data is one of the major concerns associated with research on the identification of metabolism-related gene functions. Kazusa Plant Pathway Viewer (KaPPA-View) (Tokimatsu *et al.*, 2005) facilitates the representation of quantitative data for individual transcripts and/or metabolites on plant metabolic pathway maps, and easy comprehension of the transcripts and metabolites; this in turn facilitates investigation into the hypotheses of gene functions in metabolic pathways.

Assembly of Transcriptomics and Metabolomics

In order to understand the mechanisms underlying the functioning of a living organism by linking a range of information from the genome, transcriptome, proteome and metabolome of the organism, it is essential to have a platform that is capable of integrating a range of 'omics' data so that these data can be analysed together. 'Batch-learning self-organizing map' (BL-SOM) is one of the tools that can support such integrated analyses (Kanaya *et al.*, 2001; Abe *et al.*, 2003). The transcriptome and metabolome datasets of this tool consist of a matrix in which signal intensities are arranged in multiple columns (experimental series) and multiple rows (gene and metabolite IDs). BL-SOM can analyse an integrated matrix of both transcriptome and metabolome data after appropriate normalization of the data and preliminary calculations; thus, this tool allows researchers to visualize the correlations between elements. Genes and metabolites are classified into clusters in a two-dimensional 'feature map' based on the expression and accumulation patterns of genes and metabolites. The high reproducibility and resolution supported by BL-SOM enable the elucidation of gene-to-metabolite, gene-to-gene and metabolite-to-metabolite correlations, which facilitate the identification of a gene's function (Hirai *et al.*, 2004, 2005).

One example stated above is PRIMe (Akiyama *et al.*, 2008). PRIMe is a web site that has been designed and implemented to support research and analysis workflows ranging from metabolome to transcriptome analysis. This site provides access to a growing collection of standardized measurements of metabolites obtained by using NMR, GC-MS, LC-MS, CE-MS, and metabolomics tools that support related analyses (SpinAssign for the identification of metabolites by means of NMR, and KNApSAcK (Shinbo *et al.*, 2006) for searches within metabolite databases). In addition, the transcriptomics tools support Correlated Gene Search, and Cluster Cutting for the analysis of mRNA expression. The use of these tools and databases can contribute to the analysis of biological events at the level of metabolite and gene expression. For *Arabidopsis thaliana*, we describe one example of such an analysis using BL-SOM (Kanaya *et al.*, 2001), which can be accessed from the web site.

Conclusion

Because of space constraints, we were unable to introduce all the existing databases and tools; however, several hundreds of information resources for plant research have been generated and are being updated. This implies that our research environment is also improving. We can reduce the cost of research activities and efficiently focus on further advancements in our studies, which will generate newly analysed results. In other words, new information resources will be more enriched, and will certainly accelerate research in the field of plant sciences. The recent technological advancement in the proteome and metabolome research areas is remarkable. We strongly feel that there have been increasing advances in studies on proteomics and metabolomics. Similarly, there is an increasing need for the study of phenomics. Further, we should consider using new integrative techniques and methods for our future research. If advancements in these are successful, we may obtain further insights into cells and life systems.

References

Abe, T., Kanaya, S., Kinouchi, M., Ichiba, Y., Kozuki, T. and Ikemura, T. (2003) Informatics for unveiling hidden genome signatures. *Genome Research* 13, 693–702.

Akiyama, K., Chikayama, E., Yuasa, H., Shimada, Y., Tohge, T., Shinozaki, K., Hirai, M.Y., Sakurai, T., Kikuchi, J. and Saito, K. (2008) PRIMe: a web site that assembles tools for metabolomics and transcriptomics. *In silico Biology* 8, 339–345.

Aoki, K., Ogata, Y. and Shibata, D. (2007) Approaches for extracting practical information from gene co-expression networks in plant biology. *Plant & Cell Physiology* 48, 381–390.

Axtell, M.J., Jan, C., Rajagopalan, R. and Bartel, D.P. (2006) A two-hit trigger for siRNA biogenesis in plants. *Cell* 127, 565–577.

Barrett, T., Troup, D.B., Wilhite, S.E., Ledoux, P., Rudnev, D., Evangelista, C., Kim, I.F., Soboleva, A., Tomashevsky, M., Marshall, K.A., Phillippy, K.H., Sherman, P.M., Muertter, R.N. and Edgar, R. (2009) NCBI GEO: archive for high-throughput functional genomic data. *Nucleic Acids Research* 37, D885–890.

Boeckmann, B., Bairoch, A., Apweiler, R., Blatter, M.C., Estreicher, A., Gasteiger, E., Martin, M.J., Michoud, K., O'Donovan, C., Phan, I., Pilbout, S. and Schneider, M. (2003) The SWISS–PROT protein knowledgebase and its supplement TrEMBL in 2003. *Nucleic Acids Research* 31, 365–370.

Bottcher, C., von Roepenack-Lahaye, E., Schmidt, J., Schmotz, C., Neumann, S., Scheel, D. and Clemens, S. (2008) Metabolome analysis of biosynthetic mutants reveals a diversity of metabolic changes and allows identification of a large number of new compounds in *Arabidopsis*. *Plant Physiology* 147, 2107–2120.

Brodersen, P., Sakvarelidze-Achard, L., Bruun-Rasmussen, M., Dunoyer, P., Yamamoto, Y.Y., Sieburth, L. and Voinnet, O. (2008) Widespread translational inhibition by plant miRNAs and siRNAs. *Science* 320, 1185–1190.

Caspi, R., Foerster, H., Fulcher, C.A., Kaipa, P., Krummenacker, M., Latendresse, M., Paley, S., Rhee, S.Y., Shearer, A.G., Tissier, C., Walk, T.C., Zhang, P. and Karp, P.D. (2008) The MetaCyc

Database of metabolic pathways and enzymes and the BioCyc collection of Pathway/Genome Databases. *Nucleic Acids Research* 36, D623–631.

Claros, M.G. and Vincens, P. (1996) Computational method to predict mitochondrially imported proteins and their targeting sequences. *European Journal of Biochemistry/FEBS* 241, 779–786.

Craigon, D.J., James, N., Okyere, J., Higgins, J., Jotham, J. and May, S. (2004) NASCArrays: a repository for microarray data generated by NASC's transcriptomics service. *Nucleic Acids Research* 32, D575–577.

Cui, J., Li, P., Li, G., Xu, F., Zhao, C., Li, Y., Yang, Z., Wang, G., Yu, Q., Li, Y. and Shi, T. (2008a) AtPID: *Arabidopsis thaliana* protein interactome database-an integrative platform for plant systems biology. *Nucleic Acids Research* 36, D999–1008.

Cui, Q., Lewis, I.A., Hegeman, A.D., Anderson, M.E., Li, J., Schulte, C.F., Westler, W.M., Eghbalnia, H.R., Sussman, M.R. and Markley, J.L. (2008b) Metabolite identification *via* the Madison Metabolomics Consortium Database. *Nature Biotechnology* 26, 162–164.

Dettmer, K., Aronov, P.A. and Hammock, B.D. (2007) Mass spectrometry-based metabolomics. *Mass Spectrometry Reviews* 26, 51–78.

Dixon, R.A., Gang, D.R., Charlton, A.J., Fiehn, O., Kuiper, H.A., Reynolds, T.L., Tjeerdema, R.S., Jeffery, E.H., German, J.B., Ridley, W.P. and Seiber, J.N. (2006) Applications of metabolomics in agriculture. *Journal of Agricultural and Food Chemistry* 54, 8984–8994.

Eisen, M.B., Spellman, P.T., Brown, P.O. and Botstein, D. (1998) Cluster analysis and display of genome-wide expression patterns. *Proceedings of the National Academy of Sciences of the United States of America* 95, 14863–14868.

Emanuelsson, O., Brunak, S., von Heijne, G. and Nielsen, H. (2007) Locating proteins in the cell using TargetP, SignalP and related tools. *Nature Protocols* 2, 953–971.

Fargione, J., Hill, J., Tilman, D., Polasky, S. and Hawthorne, P. (2008) Land clearing and the biofuel carbon debt. *Science* 319, 1235–1238.

Fiehn, O. (2002) Metabolomics–the link between genotypes and phenotypes. *Plant Molecular Biology* 48, 155–171.

Fiehn, O., Kloska, S. and Altmann, T. (2001) Integrated studies on plant biology using multiparallel techniques. *Current Opinion in Biotechnology* 12, 82–86.

Friso, G., Giacomelli, L., Ytterberg, A.J., Peltier, J.B., Rudella, A., Sun, Q. and Wijk, K.J. (2004) In-depth analysis of the thylakoid membrane proteome of *Arabidopsis* thaliana chloroplasts: new proteins, new functions, and a plastid proteome database. *The Plant Cell* 16, 478–499.

Galperin, M.Y. (2008) The Molecular Biology Database Collection: 2008 update. *Nucleic Acids Research* 36, D2–4.

Galperin, M.Y. and Cochrane, G.R. (2009) Nucleic Acids Research annual Database Issue and the NAR online Molecular Biology Database Collection in 2009. *Nucleic Acids Research* 37, D1–4.

Gandhi, T.K., Zhong, J., Mathivanan, S., Karthick, L., Chandrika, K.N., Mohan, S.S., Sharma, S., Pinkert, S., Nagaraju, S., Periaswamy, B., Mishra, G., Nandakumar, K., Shen, B., Deshpande, N., Nayak, R., Sarker, M., Boeke, J.D., Parmigiani, G., Schultz, J., Bader, J.S. and Pandey, A. (2006) Analysis of the human protein interactome and comparison with yeast, worm and fly interaction datasets. *Nature Genetics* 38, 285–293.

Geisler-Lee, J., O'Toole, N., Ammar, R., Provart, N.J., Millar, A.H. and Geisler, M. (2007) A predicted interactome for *Arabidopsis*. *Plant Physiology* 145, 317–329.

Giot, L., Bader, J.S., Brouwer, C., Chaudhuri, A., Kuang, B., Li, Y., Hao, Y.L., Ooi, C.E., Godwin, B., Vitols, E., Vijayadamodar, G., Pochart, P., Machineni, H., Welsh, M., Kong, Y., Zerhusen, B., Malcolm, R., Varrone, Z., Collis, A., Minto, M., Burgess, S., McDaniel, L., Stimpson, E., Spriggs, F., Williams, J., Neurath, K., Ioime, N., Agee, M., Voss, E., Furtak, K., Renzulli, R., Aanensen, N., Carrolla, S., Bickelhaupt, E., Lazovatsky, Y., DaSilva, A., Zhong, J., Stanyon, C.A., Finley, R.L., Jr., White, K.P., Braverman, M., Jarvie, T., Gold, S., Leach, M., Knight, J., Shimkets, R.A., McKenna, M.P., Chant, J. and Rothberg, J.M. (2003) A protein interaction map of *Drosophila melanogaster*. *Science* 302, 1727–1736.

Goda, H., Sasaki, E., Akiyama, K., Maruyama–Nakashita, A., Nakabayashi, K., Li, W., Ogawa, M., Yamauchi, Y., Preston, J., Aoki, K., Kiba, T., Takatsuto, S., Fujioka, S., Asami, T., Nakano, T.,

Kato, H., Mizuno, T., Sakakibara, H., Yamaguchi, S., Nambara, E., Kamiya, Y., Takahashi, H., Hirai, M.Y., Sakurai, T., Shinozaki, K., Saito, K., Yoshida, S. and Shimada, Y. (2008) The AtGenExpress hormone and chemical treatment data set: experimental design, data evaluation, model data analysis and data access. *Plant Journal* 55, 526–542.

Grafahrend-Belau, E., Weise, S., Koschutzki, D., Scholz, U., Junker, B.H. and Schreiber, F. (2008) MetaCrop: a detailed database of crop plant metabolism. *Nucleic Acids Research* 36, D954–958.

Gustafson, A.M., Allen, E., Givan, S., Smith, D., Carrington, J.C. and Kasschau, K.D. (2005) ASRP: the *Arabidopsis* Small RNA Project Database. *Nucleic Acids Research* 33, D637–640.

Heazlewood, J.L., Tonti-Filippini, J.S., Gout, A.M., Day, D.A., Whelan, J. and Millar, A.H. (2004) Experimental analysis of the *Arabidopsis* mitochondrial proteome highlights signaling and regulatory components, provides assessment of targeting prediction programs, and indicates plant-specific mitochondrial proteins. *The Plant Cell* 16, 241–256.

Heazlewood, J.L., Tonti-Filippini, J., Verboom, R.E. and Millar, A.H. (2005) Combining experimental and predicted datasets for determination of the subcellular location of proteins in *Arabidopsis*. *Plant Physiology* 139, 598–609.

Heazlewood, J.L., Verboom, R.E., Tonti-Filippini, J., Small, I. and Millar, A.H. (2007) SUBA: the *Arabidopsis* Subcellular Database. *Nucleic Acids Research* 35, D213–218.

Hirai, M.Y., Yano, M., Goodenowe, D.B., Kanaya, S., Kimura, T., Awazuhara, M., Arita, M., Fujiwara, T. and Saito, K. (2004) Integration of transcriptomics and metabolomics for understanding of global responses to nutritional stresses in *Arabidopsis thaliana*. *Proceedings of the National Academy of Sciences of the United States of America* 101, 10205–10210.

Hirai, M.Y., Klein, M., Fujikawa, Y., Yano, M., Goodenowe, D.B., Yamazaki, Y., Kanaya, S., Nakamura, Y., Kitayama, M., Suzuki, H., Sakurai, N., Shibata, D., Tokuhisa, J., Reichelt, M., Gershenzon, J., Papenbrock, J. and Saito, K. (2005) Elucidation of gene-to-gene and metabolite-to-gene networks in *Arabidopsis* by integration of metabolomics and transcriptomics. *The Journal of Biological Chemistry* 280, 25590–25595.

Hooper, S.D. and Bork, P. (2005) Medusa: A simple tool for interaction graph analysis. *Bioinformatics (Oxford, England)* 21, 4432–4433.

Horai, H., Arita, M. and Nishioka, T. (2008) Comparison of ESI–MS Spectra in MassBank Database. *Proceedings of the International Conference on BioMedical Engineering and Informatics* 853–857.

Howell, M.D., Fahlgren, N., Chapman, E.J., Cumbie, J.S., Sullivan, C.M., Givan, S.A., Kasschau, K.D. and Carrington, J.C. (2007) Genome-wide analysis of the RNA–DEPENDENT RNA POLYMERASE6/DICER–LIKE4 pathway in *Arabidopsis* reveals dependency on miRNA- and tasiRNA-directed targeting. *The Plant Cell*, 19, 926–942.

Hruz, T., Laule, O., Szabo, G., Wessendorp, F., Bleuler, S., Oertle, L., Widmayer, P., Gruissem, W. and Zimmermann, P. (2008) A Reference Expression Database for the Meta-Analysis of Transcriptomes. *Advances in Bioinformatics* 2008, 420747.

Ito, T., Motohashi, R., Kuromori, T., Mizukado, S., Sakurai, T., Kanahara, H., Seki, M. and Shinozaki, K. (2002a) A new resource of locally transposed Dissociation elements for screening gene–knockout lines *in silico* on the *Arabidopsis* genome. *Plant Physiology* 129, 1695–1699.

Ito, T., Motohashi, R. and Shinozaki, K. (2002b) Preparation of transposon insertion lines and determination of insertion sites in *Arabidopsis* genome. *Methods in Molecular Biology* 182, 209–219.

Ito, T., Motohashi, R., Kuromori, T., Noutoshi, Y., Seki, M., Kamiya, A., Mizukado, S., Sakurai, T. and Shinozaki, K. (2005) A resource of 5,814 dissociation transposon-tagged and sequence-indexed lines of *Arabidopsis* transposed from start loci on chromosome 5. *Plant & Cell Physiology* 46, 1149–1153.

Jaiswal, P., Ni, J., Yap, I., Ware, D., Spooner, W., Youens-Clark, K., Ren, L., Liang, C., Zhao, W., Ratnapu, K., Faga, B., Canaran, P., Fogleman, M., Hebbard, C., Avraham, S., Schmidt, S., Casstevens, T.M., Buckler, E.S., Stein, L. and McCouch, S. (2006) Gramene: a bird's eye view of cereal genomes. *Nucleic Acids Research* 34, D717–723.

Johnson, C., Bowman, L., Adai, A.T., Vance, V. and Sundaresan, V. (2007) CSRDB: a small RNA integrated database and browser resource for cereals. *Nucleic Acids Research* 35, D829–833.

Jones-Rhoades, M.W., Bartel, D.P. and Bartel, B. (2006) MicroRNAS and their regulatory roles in plants. *Annual Review of Plant Biology* 57, 19–53.

Kanaya, S., Kinouchi, M., Abe, T., Kudo, Y., Yamada, Y., Nishi, T., Mori, H. and Ikemura, T. (2001) Analysis of codon usage diversity of bacterial genes with a self-organizing map (SOM): characterization of horizontally transferred genes with emphasis on the *E. coli* O157 genome. *Gene* 276, 89–99.

Kanehisa, M., Goto, S., Kawashima, S. and Nakaya, A. (2002) The KEGG databases at GenomeNet. *Nucleic Acids Research* 30, 42–46.

Karp, P.D., Paley, S. and Romero, P. (2002) The Pathway Tools software. *Bioinformatics (Oxford, England)* 18 Suppl 1, S225–232.

Kilian, J., Whitehead, D., Horak, J., Wanke, D., Weinl, S., Batistic, O., D'Angelo, C., Bornberg-Bauer, E., Kudla, J. and Harter, K. (2007) The AtGenExpress global stress expression data set: protocols, evaluation and model data analysis of UV-B light, drought and cold stress responses. *Plant Journal* 50, 347–363.

Kopka, J., Schauer, N., Krueger, S., Birkemeyer, C., Usadel, B., Bergmuller, E., Dormann, P., Weckwerth, W., Gibon, Y., Stitt, M., Willmitzer, L., Fernie, A.R. and Steinhauser, D. (2005) GMD@CSB.DB: the Golm Metabolome Database. *Bioinformatics (Oxford, England)* 21, 1635–1638.

Kuromori, T., Hirayama, T., Kiyosue, Y., Takabe, H., Mizukado, S., Sakurai, T., Akiyama, K., Kamiya, A., Ito, T. and Shinozaki, K. (2004) A collection of 11,800 single-copy Ds transposon insertion lines in *Arabidopsis*. *Plant Journal* 37, 897–905.

Kuromori, T., Wada, T., Kamiya, A., Yuguchi, M., Yokouchi, T., Imura, Y., Takabe, H., Sakurai, T., Akiyama, K., Hirayama, T., Okada, K. and Shinozaki, K. (2006) A trial of phenome analysis using 4000 Ds-insertional mutants in gene-coding regions of *Arabidopsis*. *Plant Journal* 47, 640–651.

Kusano, M., Fukushima, A., Arita, M., Jonsson, P., Moritz, T., Kobayashi, M., Hayashi, N., Tohge, T. and Saito, K. (2007) Unbiased characterization of genotype-dependent metabolic regulations by metabolomic approach in *Arabidopsis thaliana*. *BMC Systems Biology* 1, 53.

Last, R.L., Jones, A.D. and Shachar-Hill, Y. (2007) Towards the plant metabolome and beyond. *Nature Reviews Molecular Cell Biology* 8, 167–174.

Li, S., Armstrong, C.M., Bertin, N., Ge, H., Milstein, S., Boxem, M., Vidalain, P.O., Han, J.D., Chesneau, A., Hao, T., Goldberg, D.S., Li, N., Martinez, M., Rual, J.F., Lamesch, P., Xu, L., Tewari, M., Wong, S.L., Zhang, L.V., Berriz, G.F., Jacotot, L., Vaglio, P., Reboul, J., Hirozane-Kishikawa, T., Li, Q., Gabel, H.W., Elewa, A., Baumgartner, B., Rose, D.J., Yu, H., Bosak, S., Sequerra, R., Fraser, A., Mango, S.E., Saxton, W.M., Strome, S., Van Den Heuvel, S., Piano, F., Vandenhaute, J., Sardet, C., Gerstein, M., Doucette-Stamm, L., Gunsalus, K.C., Harper, J.W., Cusick, M.E., Roth, F.P., Hill, D.E. and Vidal, M. (2004) A map of the interactome network of the metazoan *C. elegans*. *Science* 303, 540–543.

Llave, C., Kasschau, K.D., Rector, M.A. and Carrington, J.C. (2002) Endogenous and silencing-associated small RNAs in plants. *The Plant Cell* 14, 1605–1619.

Lu, C., Tej, S.S., Luo, S., Haudenschild, C.D., Meyers, B.C. and Green, P.J. (2005) Elucidation of the small RNA component of the transcriptome. *Science* 309, 1567–1569.

Manfield, I.W., Jen, C.H., Pinney, J.W., Michalopoulos, I., Bradford, J.R., Gilmartin, P.M. and Westhead, D.R. (2006) *Arabidopsis* Co-expression Tool (ACT): web server tools for microarray-based gene expression analysis. *Nucleic Acids Research* 34, W504–509.

Matsuda, F., Yonekura-Sakakibara, K., Niida, R., Kuromori, T., Shinozaki, K. and Saito, K. (2009) MS/MS spectral tag-based annotation of non-targeted profile of plant secondary metabolites, *Plant Journal* 57, 555–577.

Miller, J.P., Lo, R.S., Ben-Hur, A., Desmarais, C., Stagljar, I., Noble, W.S. and Fields, S. (2005) Large-scale identification of yeast integral membrane protein interactions. *Proceedings of the National Academy of Sciences of the United States of America* 102, 12123–12128.

Moco, S., Bino, R.J., Vorst, O., Verhoeven, H.A., de Groot, J., van Beek, T.A., Vervoort, J. and de Vos, C.H. (2006) A liquid chromatography-mass spectrometry-based metabolome database for tomato. *Plant Physiology* 141, 1205–1218.

Moco, S., Capanoglu, E., Tikunov, Y., Bino, R.J., Boyacioglu, D., Hall, R.D., Vervoort, J. and De Vos, R.C. (2007) Tissue specialization at the metabolite level is perceived during the development of tomato fruit. *Journal of Experimental Botany* 58, 4131–4146.

Mueller, L.A., Zhang, P. and Rhee, S.Y. (2003) AraCyc: a biochemical pathway database for *Arabidopsis*. *Plant Physiology* 132, 453–460.

Nakano, M., Nobuta, K., Vemaraju, K., Tej, S.S., Skogen, J.W. and Meyers, B.C. (2006) Plant MPSS databases: signature-based transcriptional resources for analyses of mRNA and small RNA, *Nucleic Acids Research* 34, D731–735.

Nicholson, J.K., Connelly, J., Lindon, J.C. and Holmes, E. (2002) Metabonomics: a platform for studying drug toxicity and gene function. *Nature Reviews. Drug Discovery* 1, 153–161.

Nobuta, K., Venu, R.C., Lu, C., Belo, A., Vemaraju, K., Kulkarni, K., Wang, W., Pillay, M., Green, P.J., Wang, G.L. and Meyers, B.C. (2007) An expression atlas of rice mRNAs and small RNAs. *Nature Biotechnology* 25, 473–477.

Obayashi, T., Kinoshita, K., Nakai, K., Shibaoka, M., Hayashi, S., Saeki, M., Shibata, D., Saito, K. and Ohta, H. (2007) ATTED–II: a database of co–expressed genes and cis elements for identifying co-regulated gene groups in *Arabidopsis*. *Nucleic Acids Research* 35, D863–869.

Obayashi, T., Hayashi, S., Saeki, M., Ohta, H. and Kinoshita, K. (2009) ATTED-II provides coexpressed gene networks for *Arabidopsis*. *Nucleic Acids Research* 37, D987–991.

Parkinson, H., Kapushesky, M., Shojatalab, M., Abeygunawardena, N., Coulson, R., Farne, A., Holloway, E., Kolesnykov, N., Lilja, P., Lukk, M., Mani, R., Rayner, T., Sharma, A., William, E., Sarkans, U. and Brazma, A. (2007) ArrayExpress–a public database of microarray experiments and gene expression profiles. *Nucleic Acids Research* 35, D747–750.

Rajagopalan, R., Vaucheret, H., Trejo, J. and Bartel, D.P. (2006) A diverse and evolutionarily fluid set of microRNAs in *Arabidopsis thaliana*. *Genes & Development* 20, 3407–3425.

Redman, J.C., Haas, B.J., Tanimoto, G. and Town, C.D. (2004) Development and evaluation of an *Arabidopsis* whole genome Affymetrix probe array. *Plant Journal* 38, 545–561.

Riano-Pachon, D.M., Dreyer, I. and Mueller–Roeber, B. (2005) Orphan transcripts in *Arabidopsis* thaliana: identification of several hundred previously unrecognized genes. *Plant Journal* 43, 205–212.

Richly, E., Dietzmann, A., Biehl, A., Kurth, J., Laloi, C., Apel, K., Salamini, F. and Leister, D. (2003) Covariations in the nuclear chloroplast transcriptome reveal a regulatory master-switch. *EMBO Reports* 4, 491–498.

Robinette, S.L., Zhang, F., Bruschweiler-Li, L. and Bruschweiler, R. (2008) Web server based complex mixture analysis by NMR. *Analytical Chemistry* 80, 3606–3611.

Rual, J.F., Venkatesan, K., Hao, T., Hirozane-Kishikawa, T., Dricot, A., Li, N., Berriz, G.F., Gibbons, F.D., Dreze, M., Ayivi-Guedehoussou, N., Klitgord, N., Simon, C., Boxem, M., Milstein, S., Rosenberg, J., Goldberg, D.S., Zhang, L.V., Wong, S.L., Franklin, G., Li, S., Albala, J.S., Lim, J., Fraughton, C., Llamosas, E., Cevik, S., Bex, C., Lamesch, P., Sikorski, R.S., Vandenhaute, J., Zoghbi, H.Y., Smolyar, A., Bosak, S., Sequerra, R., Doucette-Stamm, L., Cusick, M.E., Hill, D.E., Roth, F.P. and Vidal, M. (2005) Towards a proteome-scale map of the human protein-protein interaction network. *Nature* 437, 1173–1178.

Saito, K., Hirai, M.Y. and Yonekura-Sakakibara, K. (2008) Decoding genes with coexpression networks and metabolomics – 'majority report by precogs'. *Trends in Plant Science* 13, 36–43.

Sakurai, T., Satou, M., Akiyama, K., Iida, K., Seki, M., Kuromori, T., Ito, T., Konagaya, A., Toyoda, T. and Shinozaki, K. (2005) RARGE: a large-scale database of RIKEN *Arabidopsis* resources ranging from transcriptome to phenome. *Nucleic Acids Research* 33, D647–650.

Sawada, Y., Akiyama, K., Sakata, A., Kuwahara, A., Otsuki, H., Sakurai, T., Saito, K. and Hirai, M.Y. (2009) Widely targeted metabolomics based on large-scale MS/MS data for elucidating metabolite accumulation patterns in plants. *Plant & Cell Physiology* 50, 37–47.

Schmid, M., Davison, T.S., Henz, S.R., Pape, U.J., Demar, M., Vingron, M., Scholkopf, B., Weigel, D. and Lohmann, J.U. (2005) A gene expression map of *Arabidopsis thaliana* development, *Nature Genetics* 37, 501–506.

Seki, M., Narusaka, M., Kamiya, A., Ishida, J., Satou, M., Sakurai, T., Nakajima, M., Enju, A., Akiyama, K., Oono, Y., Muramatsu, M., Hayashizaki, Y., Kawai, J., Carninci, P., Itoh, M., Ishii, Y., Arakawa, T., Shibata, K., Shinagawa, A. and Shinozaki, K. (2002) Functional annotation of a full-length *Arabidopsis* cDNA collection. *Science* 296, 141–145.

Shinbo, Y., Nakamura, Y., Altaf-Ul-Amin, M., Asahi, H., Kurokawa, K., Arita, M., Saito, K., Ohta, D., Shibata, D. and Kanaya, S. (2006) KNApSAcK: A comprehensive species-metabolite relationship database. *Biotechnology in Agriculture and Forestry* 57, 165–184.

Shoemaker, B.A. and Panchenko, A.R. (2007) Deciphering protein-protein interactions. Part I. Experimental techniques and databases. *PLoS Computational Biology* 3, e42.

Small, I., Peeters, N., Legeai, F. and Lurin, C. (2004) Predotar: A tool for rapidly screening proteomes for N-terminal targeting sequences. *Proteomics* 4, 1581–1590.

Steinhauser, D., Usadel, B., Luedemann, A., Thimm, O. and Kopka, J. (2004) CSB.DB: a comprehensive systems-biology database. *Bioinformatics (Oxford, England)* 20, 3647–3651.

Sumner, L.W., Mendes, P. and Dixon, R.A. (2003) Plant metabolomics: large-scale phytochemistry in the functional genomics era. *Phytochemistry* 62, 817–836.

Swarbreck, D., Wilks, C., Lamesch, P., Berardini, T.Z., Garcia-Hernandez, M., Foerster, H., Li, D., Meyer, T., Muller, R., Ploetz, L., Radenbaugh, A., Singh, S., Swing, V., Tissier, C., Zhang, P. and Huala, E. (2008) The *Arabidopsis* Information Resource (TAIR): gene structure and function annotation. *Nucleic Acids Research* 36, D1009–1014.

Thimm, O., Blasing, O., Gibon, Y., Nagel, A., Meyer, S., Kruger, P., Selbig, J., Muller, L.A., Rhee, S.Y. and Stitt, M. (2004) MAPMAN: a user-driven tool to display genomics data sets onto diagrams of metabolic pathways and other biological processes. *Plant Journal* 37, 914–939.

Tikunov, Y., Lommen, A., de Vos, C.H., Verhoeven, H.A., Bino, R.J., Hall, R.D. and Bovy, A.G. (2005) A novel approach for nontargeted data analysis for metabolomics. Large-scale profiling of tomato fruit volatiles. *Plant Physiology* 139, 1125–1137.

Tohge, T., Nishiyama, Y., Hirai, M.Y., Yano, M., Nakajima, J., Awazuhara, M., Inoue, E., Takahashi, H., Goodenowe, D.B., Kitayama, M., Noji, M., Yamazaki, M. and Saito, K. (2005) Functional genomics by integrated analysis of metabolome and transcriptome of *Arabidopsis* plants over-expressing an MYB transcription factor. *Plant Journal* 42, 218–235.

Tokimatsu, T., Sakurai, N., Suzuki, H., Ohta, H., Nishitani, K., Koyama, T., Umezawa, T., Misawa, N., Saito, K. and Shibata, D. (2005) KaPPA-view: a web-based analysis tool for integration of transcript and metabolite data on plant metabolic pathway maps. *Plant Physiology* 138, 1289–1300.

Toufighi, K., Brady, S.M., Austin, R., Ly, E. and Provart, N.J. (2005) The Botany Array Resource: e–Northerns, Expression Angling, and promoter analyses. *Plant Journal* 43, 153–163.

Uetz, P., Giot, L., Cagney, G., Mansfield, T.A., Judson, R.S., Knight, J.R., Lockshon, D., Narayan, V., Srinivasan, M., Pochart, P., Qureshi-Emili, A., Li, Y., Godwin, B., Conover, D., Kalbfleisch, T., Vijayadamodar, G., Yang, M., Johnston, M., Fields, S. and Rothberg, J.M. (2000) A comprehensive analysis of protein–protein interactions in *Saccharomyces cerevisiae. Nature* 403, 623–627.

Urbanczyk-Wochniak, E. and Sumner, L.W. (2007) MedicCyc: a biochemical pathway database for *Medicago truncatula. Bioinformatics (Oxford, England)* 23, 1418–1423.

Usadel, B., Nagel, A., Thimm, O., Redestig, H., Blaesing, O.E., Palacios-Rojas, N., Selbig, J., Hannemann, J., Piques, M.C., Steinhauser, D., Scheible, W.R., Gibon, Y., Morcuende, R., Weicht, D., Meyer, S. and Stitt, M. (2005) Extension of the visualization tool MapMan to allow statistical analysis of arrays, display of coresponding genes, and comparison with known responses. *Plant Physiology* 138, 1195–1204.

Villas-Boas, S.G., Mas, S., Akesson, M., Smedsgaard, J. and Nielsen, J. (2005) Mass spectrometry in metabolome analysis. *Mass Spectrometry Reviews* 24, 613–646.

Wang, Q.Z., Wu, C.Y., Chen, T., Chen, X. and Zhao, X.M. (2006) Integrating metabolomics into a systems biology framework to exploit metabolic complexity: strategies and applications in microorganisms. *Applied Microbiology and Biotechnology* 70, 151–161.

Wang, Y., Xiao, J., Suzek, T.O., Zhang, J., Wang, J. and Bryant, S.H. (2009) PubChem: a public information system for analyzing bioactivities of small molecules. *Nucleic Acids Research* W623–W633.

Wishart, D.S., Knox, C., Guo, A.C., Eisner, R., Young, N., Gautam, B., Hau, D.D., Psychogios, N., Dong, E., Bouatra, S., Mandal, R., Sinelnikov, I., Xia, J., Jia, L., Cruz, J.A., Lim, E., Sobsey, C.A., Shrivastava, S., Huang, P., Liu, P., Fang, L., Peng, J., Fradette, R., Cheng, D., Tzur, D., Clements, M., Lewis, A., De Souza, A., Zuniga, A., Dawe, M., Xiong, Y., Clive, D., Greiner, R., Nazyrova, A., Shaykhutdinov, R., Li, L., Vogel, H.J. and Forsythe, I. (2009) HMDB: a knowledgebase for the human metabolome. *Nucleic Acids Research* 37, D603–610.

Xie, Z., Allen, E., Fahlgren, N., Calamar, A., Givan, S.A. and Carrington, J.C. (2005) Expression of *Arabidopsis* MIRNA genes. *Plant Physiology* 138, 2145–2154.

Yamada, K., Lim, J., Dale, J.M., Chen, H.M., Shinn, P., Palm, C.J., Southwick, A.M., Wu, H.C., Kim, C., Nguyen, M., Pham, P., Cheuk, R., Karlin–Newmann, G., Liu, S.X., Lam, B., Sakano, H., Wu, T., Yu, G.X., Miranda, M., Quach, H.L., Tripp, M., Chang, C.H., Lee, J.M., Toriumi, M., Chan, M.M.H., Tang, C.C., Onodera, C.S., Deng, J.M., Akiyama, K., Ansari, Y., Arakawa, T., Banh, J., Banno, F., Bowser, L., Brooks, S., Carninci, P., Chao, Q.M., Choy, N., Enju, A., Goldsmith, A.D., Gurjal, M., Hansen, N.F., Hayashizaki, Y., Johnson-Hopson, C., Hsuan, V.W., Iida, K., Karnes, M., Khan, S., Koesema, E., Ishida, J., Jiang, P.X., Jones, T., Kawai, J., Kamiya, A., Meyers, C., Nakajima, M., Narusaka, M., Seki, M., Sakurai, T., Satou, M., Tamse, R., Vaysberg, M., Wallender, E.K., Wong, C., Yamamura, Y., Yuan, S.L., Shinozaki, K., Davis, R.W., Theologis, A. and Ecker, J.R. (2003) Empirical analysis of transcriptional activity in the *Arabidopsis* genome. *Science* 302, 842–846.

Chapter 15

Docking-based Virtual Screening of Anticancer Drugs

Feroz Khan, Abha Meena and Ashok Sharma

Introduction

To combat life-threatening diseases such as AIDS, cancer, tuberculosis, malaria etc., the only way to fight is by the development of new, effective drugs, at a fast pace. Drug discovery and development is an intense, lengthy and an interdisciplinary endeavour. Drug discovery is mostly portrayed as a linear, consecutive process that starts with target and lead discovery, followed by lead optimization and pre-clinical *in vitro* and *in vivo* studies to determine if such compounds satisfy a number of pre-set criteria for initiating clinical development. For the pharmaceutical industry, the number of years to bring a drug from discovery to market is approximately 13–15 years and costing up to \$1.3–\$1.5 billion dollars. Traditionally, drugs were discovered by synthesizing compounds in a time-consuming multi-step process against a battery of *in vivo* biological screens and further investigating the promising candidates for their pharmacokinetic properties, metabolism and potential toxicity. Such a development process has resulted in high attrition rates with failures attributed to poor pharmacokinetics, lack of efficacy, animal toxicity, adverse effects in humans and various commercial and miscellaneous factors. In addition to that, toxicity/dosage in the form of IC_{50} (inhibitory concentration) or LD_{50} (lethal dose) of drugs is determined by testing on laboratory animals, unnecessarily killing them. Moreover, if multiple drug resistance is developed against any of these drugs that required so many years to be developed, the whole process would have to be started all over again, where again we are left without a drug against the disease. Along with the current problems, once a biologically active compound has been synthesized in the laboratory not all are suitable to be used as drugs due to toxicity, unfavourable side-reactions, or pharmacokinetics. After the synthesis and aside from the desirable therapeutic properties, the testing and approval of many rigorous tests are required to ascertain whether the compounds are worthy of becoming drugs. An undesirable proportion of compounds with good biological activity fails to progress to later stages of drug development because of inappropriate pharmacokinetic and pharmacodynamic properties. Today, the process of drug discovery has been revolutionized with the advent of genomics, proteomics, bioinformatics and efficient technologies such as combinatorial chemistry, high throughput screening (HTS), virtual screening, *de novo* design, *in vitro*, *in silico* ADMET screening and structure-based drug design. Computational models are useful because they rationalize a large number of experimental observations and therefore allow saving time and money in the drug design process. In addition, they are useful in areas like design of virtual compound collections, computational-chemical optimization of compounds and design of combinatorial libraries with appropriate ADME properties.

An increasing number of protein crystallographic structures are becoming available from high-throughput structural genomics projects, thus prediction of a potential lead and its potential target is a fundamental step in order to investigate the molecular recognition mechanism of a protein (Abagyan *et al.*, 1994; Abagyan and Totrov, 2001). For lead identification and optimization, wet lab and dry lab (*in silico*) methods are effectively applied to speed-up the process of drug discovery (Anderson and Chiplin, 2002). The lead molecules modulate the function of the target proteins and are later optimized to therapeutic drug against a specific disease (Gane and Dean, 2000; Klebe, 2000). Nowadays, to check the binding affinity of the target receptor with the library compounds computational screening method e.g. Virtual Screening (VS) is widely applied and used by many researchers to save the wet lab economy and time (Walters *et al.*, 1998; Subramaniam *et al.*, 2008). Here we attempt to virtually screen the chemical compounds and phytomolecules against known anti-cancer targets based on both Quantitative structure-property/activity relationships (QSPR) and docking studies without any prior knowledge about the nature of the interaction and target's active sites (Shoichet *et al.*, 2002; Veber *et al.*, 2002). We postulate that the structure itself is informative enough, thus consider the whole protein for docking studies. To generate a list of potential drug candidates, drug likeness was established in parallel to high throughput ligand docking (Lipinski, 2000). We have demonstrated that our approach can successfully identify and predict leads and their binding targets in both complexed and apo-structures (Klebe, 2000). For large-scale identification, we used QSPR and docking parameters on diverse compounds e.g. antimalarials, antipruritics, antifungal, antiprotozoal, anti-infective, coccidiostats etc. and are able to generate good results. This fast and accurate method can be used to identify possible binding for orphan receptors and vice versa, or for uncharacterized secondary binding of known receptors. In the studied work, QSPR, Lipinski's rule of five filter and docking studies were performed for the virtual screening of potential leads, suggesting action of compounds on the anticancer targets. A total of 100 compounds were used for library building, mostly belongs to anticarcinogenic pharmacological category. For automated docking strategies, we used 16 known anticancer targets. Results showed that only 68 compounds were found eligible as potential anticancer drug leads. This fast and accurate method can be used to identify possible binding for orphan receptors, or for uncharacterized secondary binding of known receptors. It can also be used to prioritize novel targets.

Methodology

Structure cleaning and preparation

Crystal structures of anticancer receptors obtained from the Brookhaven Protein Databank (Berman *et al.*, 2000) were converted from .pdb to .csf format using standard functionality from the Workspace module of Scigress Explorer 7.7.0.47. This included automatic typing of each ligand atom based on the geometry found in the crystal structure and addition of hydrogens as necessary. Atom typing of protein residues and addition of hydrogens to them was carried out by reference to standard protein libraries keyed to residue type. Terminal rotors such as methyl and hydroxyl groups were then relaxed to avoid distortion of scores by spurious steric clashes with the added hydrogen atoms. To ensure the compound availability or the synthetic feasibility, we used only reported anticarcinogenic compounds for VS. We also analysed and consider tautomeric states of molecules for screening, since these shows interaction with the active site residues of the target (Lipinski *et al.*, 1997).

Collecting uncomplexed structures

To find out the potential active site with minimum docking interaction energy in the uncomplexed protein structure (apo-structures), we consider cleaned monomer unit of target protein as an active site for docking based virtual screening. Note that unliganded pockets may not be as obvious as the liganded pockets due to the ligand induced conformational changes. Side-chains often obstruct a part of the pocket in the absence of the ligand. Scigress Explorer docking runs utilized the default parameters provided with the docking module.

Building of virtual compound/phytomolecules library

We built an *in silico* compounds and phytomolecules library prior to virtual screening based on anticarcinogenic activity. All the compounds and phytomolecules were retrieved through PubChem database at NCBI web server. To made the diversified compound library and also for comparative study, we randomly included compounds and phytomolecules in the library from other pharmacological categories. For comparison we grouped all the compounds in three datasets viz.: (i) known or reference data set, which includes only reported leads e.g. allicin, catechin and camptothecin against studied 16 anticancer receptors; (ii) unknown data set, which includes all the anticarcinogenic compounds and phytomolecules without any information of receptors/targets; and (iii) diverse unknown data set, which includes compounds/phytomolecules other than anticarcinogenic activity e.g. anti-malarial. The basis behind such an analysis is that, similar compounds exhibit similar physicochemical and biological properties. A chemical library enriched with the representatives of dissimilar compounds or diverse chemical structures shall increase the probability of finding different leads with similar biological activities. It also ensures that the library is small but diverse enough to be readily managed.

Descriptors for small molecules

The problem of explicitly representing and storing small molecules as finite-dimensional vectors has a long history in chemoinformatics and a multitude of molecular descriptors have been proposed. These descriptors include in particular physicochemical properties of the molecules, such as its solubility or logP, descriptors derived from the 2D structure of the molecule, such as fragment counts or structural fingerprints, or descriptors extracted from the 3D structure. We evaluated the performance of all combinations of these descriptions on the compound library data, which contains three known and 86 potential anticarcinogenic lead molecules and 11 unknown diverse data. However, in all cases, we obtained significant improvements of the screening accuracy with respect to the individual setting.

Virtual screening of library

In silico virtual screening of receptors is however a daunting task, both for receptor-based approaches (docking) and for ligand-based approaches (QSPR/QSAR). The former relies on the prior knowledge of the 3D structure of the protein, while the latter relies on the prior knowledge of the chemical's structure property and activity relationships. However, their accuracy is fundamentally limited by the amount of known ligands and degrades when few ligands are known. Docking studies includes two goals: to rapidly determine the most likely binding mode of leads and to return some measure of its estimated binding affinity for the target protein. In the studied work, we used a combination of both receptor and ligand based screening methods. Firstly we used cleaned 3D structures of the known and potential anticancer targets to search for potential phytomolecules or compounds that can modulate the receptor function. Each of the library compounds is docked into the known anticancer receptor binding site and the best one is predicted. Secondly we used physicochemical properties of the anticarcinogenic compounds and phytomolecules for screening potential leads based on QSPR and docking studies. For the drug likeness filter we used the 'Rule of Five' screening method. Finally potential leads are ranked on the basis of QSPR and docking scoring function. During molecular docking, we avoided target flexibility parameter and use of homology models since we included the reported anticancer receptors. A brief sketch of methodology has been shown in Fig. 15.1.

Drug-likeness filter

The drug-like compounds represent adsorption, distribution, metabolism and excretion profiles as determined by Lipinski's rule of five. This method is not a QSTR as there is no quantitative structure–toxicity relationship with any experimental data. The number of 'violations' of Lipinski's rules is claimed to be an indication of the absorbability of a molecule. The drug-like compounds resemble the existing drug molecules and show properties such as MW< 500, LogP <5, H bond acceptors <10 and H bond donors <5.

Toxicity filter

To minimize expensive drug failures in the later stage of drug discovery, toxicity related experimental data are essential. Prediction of toxicity before the synthesis of compounds ensures the removal of compounds with potential toxic effects. Considering this fact, we used only those compounds which were experimentally verified in prior toxicity studies. Compound's safety and toxicological data were verified through authentic toxicity databases especially for anticarcinogenic compounds such as CCRIS– carcinogenicity, tumour promotion, tumour inhibition and mutagenicity tests (http://toxnet.nlm.nih.gov), EINECS – European Inventory of Existing Commercial Chemical Substances (http://ecb.jrc.it/esis/), GENETOX– genetic toxicology information (http://toxnet.nlm.nih.gov/), TOXLINE– citations to the toxicological literature (http://toxnet.nlm.nih.gov/) and NTP DBS– toxicological assay results (http://ntp-apps.niehs.nih.gov/ntp_tox/).

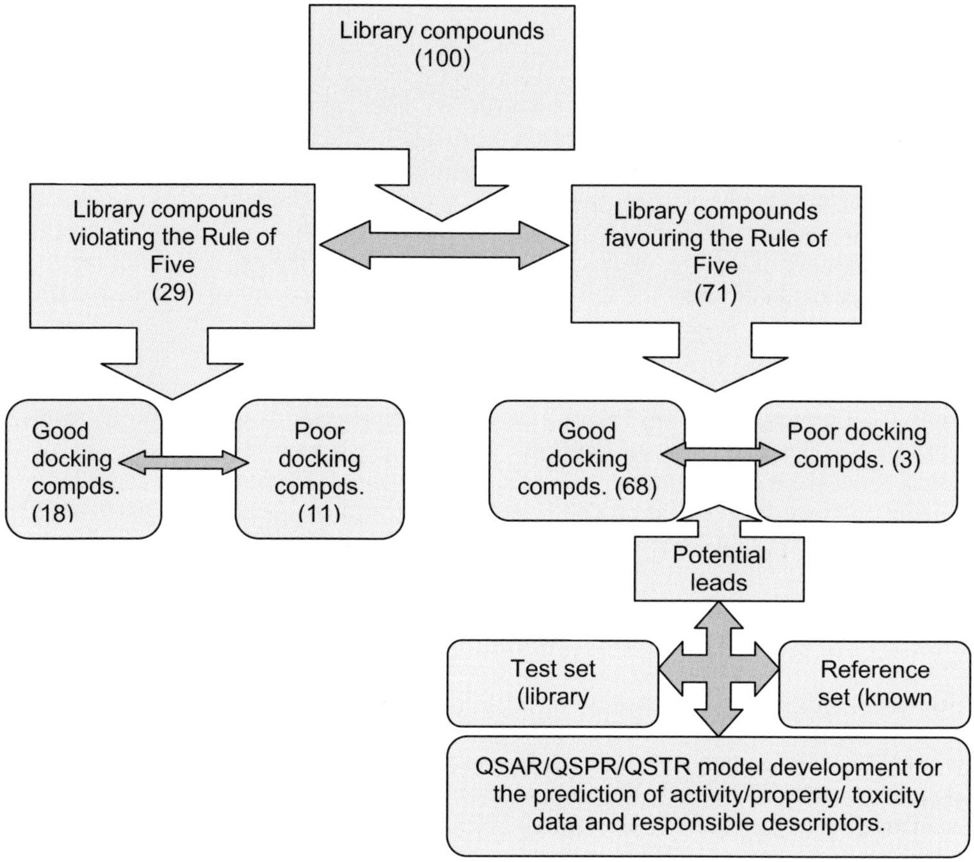

Fig. 15.1. Methodology flowchart for virtual screening of anticancer leads and development of QSAR/QSPR model.

Binding affinity of proteins/ligand

Estimation of ligand–protein affinity/ligand–DNA or drug–target affinity is one of the major and most basic steps in drug discovery, developing QSAR models etc., since only those potential molecules are taken up further for analysis which demonstrate a desirable binding for the targeted receptor. This technique also helps in determining the pharmacological activity of the lead compound and thus develops its QSAR models. Binding affinity is calculated by docking, based on different kinds of scoring functions such as PMF (potential of mean force), PLP, G-score, F-score, LigScore, X-score etc.; the basic assumption here is that a good docking score indicates that the molecules bind to each other.

Quantitative Structure Activity/Property Relationship (QSAR/QSPR)

Structure–function correlation studies aim at broadening the understanding of relationships between molecular intrinsic chemical features and biological properties. While many QSAR analyses involve the interactions of a family of molecules with an enzyme or receptor binding site, QSAR can also be used to study the interactions between the structural domains of proteins. Protein–protein interactions can be quantitatively analysed for structural variations resulted from site-directed mutagenesis. In simple words QSAR help us in understanding how a particular property e.g. high melting point or a pharmacological property such as anti-inflammatory might be attributed to a side chain, chemical group, ring structure etc.

For example, see Table 15.11, summarizing an anti-malarial QSAR model for the prediction of biological activity (IC50) and other dependable chemical structure's descriptors.

QSAR Model Equation:

Predicted log IC_{50} (D) = - 0.071944*F + 0.00188086*H + 0.0844403*I + 0.650836*K + 0.0467254*T + 0.598326*AA - 0.306307*AD + 1.45219

Where, F= Atom Count (all atoms), H= Conformation Minimum Energy (kcal/mol), I= Dipole Moment (debye), K= Dielectric Energy (kcal/mol), T= Polarizability, AA= Ring Count (aromatic 6 membered) and AD= Size of Smallest Ring. With best fit relationship regression line showing regression value of r^2=0.998482 and prediction accuracy of rCV^2=0.994135, where, r^2 = Regression coefficient and rCV^2 = Cross-validated regression coefficient. (Standard error = 0.029061)

Mostly it would be better to consider following classes of chemical structure descriptors, while developing QSAR model:

1. Constitutional descriptors: molecular weight, halogenated atom counts and functional group counts (amine, carbonyl, carboxylic acid, hydroxyl, nitrile, nitro group) etc.
2. Electrostatic descriptors: partial charges, charged surface areas etc.
3. Topological descriptors: Kier and Hall connectivity indices (zero order, first order and second order) and valence connectivity indices (zero order, first order and second order) etc.
4. Geometrical descriptors: solvent accessible surface, isosurface volume, molar volume etc.
5. Quantum-chemical descriptors: dipole moment, quadrupole moment, polarizibility, HOMO and LUMO energies, dielectric energy etc.

For example, **LogP:** The octanol-water partition coefficient. **Solvent accessible surf area (SASA):** The molecular surface area accessible to a solvent molecule. **Polarizability (P):** The molecule's average alpha polarizability. **Smallest ring size (SRS):** Information about rings present in the compound. Rings with more than 12 atoms are ignored. The number of atoms forming the smallest ring in the compound, or 0 if the compound contains no ring of size 12 or less. **Ring count (RC):** The number of rings present in the compound. Rings with more than 12 atoms are ignored. The number of rings with 12 or fewer atoms (All= all aromatic, small, 5-membered, 5-membered aromatic, 6-membered, 6-membered aromatic, large, large aromatic). **Molar refractivity (MR):** The molar refractivity of the compound. **LUMO energy:** The energy gained when an electron is added to the lowest unoccupied molecular orbital (LUMO). **Ionization potential (IP):** The energy required to remove an electron from a molecule in its ground state. **HOMO energy:** The energy required to remove an electron from the highest occupied molecular orbital (HOMO). **Heat of formation (HF):** The energy released or used when a molecule is formed from elements in their standard states. **Conformation minimum energy (CME):** Energy calculated for an optimized conformation of the compound. **Formal Charge (FC):** The net positive or negative charge on the molecule. **Atom count (AC):** The number of atoms. **Bond count (BC):** The number of bonds. Weak and ionic bonds are ignored.

Docking

As mentioned above, for high throughput screening of the lead compounds, whether using the target based or ligand based approaches, we need to dock the ligand into the receptor. Essentially docking is a method which predicts the preferred orientation of one molecule to a second when bound to each other to form a stable complex. Knowledge of the preferred orientation in turn may be used to predict the strength of association or binding affinity between two molecules using scoring functions. Molecular docking can be thought of as a problem of lock-and-key, where one is interested in finding the correct relative orientation of the key which will open up the lock. Here, the protein can be thought of as the 'lock' and the ligand can be thought of as a 'key'. Molecular docking may be defined as an optimization problem, which would describe the 'best-fit' orientation of a ligand that binds to a particular protein of interest.

Results and Discussion

QSPR approach was used to develop a regression equation that describes the correlation between experimental data and certain calculated descriptors based on the molecular structure. The derived equation was then used to predict the properties of new or hypothetical molecules from their structure alone, reducing the need for experimental synthesis and measurement. The purpose of studied work was to find the best correlation coefficient (r^2) and the best cross-validated coefficient (rCV^2), between the experimental data and various descriptors that we calculated, while minimizing the number of descriptors used. Correlation coefficient indicates how well the experimental data fits the regression equation, while cross-validated coefficient indicates the ability of the regression equation to predict unknowns. Results indicate values closer to unity, thus showed better correlation. The 'Lipinski's Rule of Five' predicts that poor adsorption or permeation of a drug is more likely when there are more than 5 H-bond donors, more than 10 H-bond acceptors, the molecular weight is greater than 500, and the calculated LogP is greater than 5. Molecular weight and LogP can be calculated directly. The total number of hydrogen atoms bonded to oxygen (OH) or nitrogen (NH) is considered to be a good index of H-bond-donor count, while the sum of the oxygen atoms and nitrogen atoms is considered to be an acceptable indicator of H bond acceptor count.

Our results demonstrate that VS approaches outperform individual approach, in particular in cases where very limited or no ligand information is available. In the case of well studied targets, more classical ligand-based methods (QSAR) may be better suited to predict new strong binders from a large number of known ligands. Since normally VS approach is a ligand-based approach, it would probably be interesting to use it in combination with docking. Indeed, although prior known ligands can help tuning docking procedures to the receptor under study, it can in principle be used with little or no ligand information. When more experimental 3D structures become available for Targets in the future, this will help building reliable models for a wider range of Targets that would be suitable for docking studies. Joint use of ligand-based VS and docking would certainly improve predictions.

Instead of focusing on each individual target independently from other proteins, we used VS to screen compounds against several targets of the same family simultaneously. This systematic screening of interactions between the compounds and protein targets can be thought of as an attempt to fill a large two dimentional interaction matrix, where rows correspond to compounds and columns to protein targets. Compound entry in the row indicates whether it can bind to target in the column or not. While in general the matrix may contain some description of the strength of the interaction, such as the docking score of the complex, which reflect minimum binding interaction energy. While classical docking or ligand-based virtual screening strategies focus on each single row independently from the others in this matrix, i.e., treat each target independently from each others, the VS approach is motivated by the observation that similar molecules can bind similar proteins, and that information about a known interaction between a ligand and a target could therefore be a useful hint to predict interaction between similar molecules and similar targets. This can be of particular interest when, for example, a particular target has few or no known ligands, but similar proteins have many: in that case it is tempting to use the information about the known ligands of similar proteins for a ligand-based virtual screening of the target of interest. In this context, we can formally define VS as the problem of predicting interactions between a molecule and a ligand (i.e., the unknown entries of the matrix) from the knowledge of all other known interactions or non-interactions (i.e., the known entries of the matrix refer by reference data set).

Example: Case study on anticancer compounds

In the studied work, all the 100 compounds of library were used for QSPR relationship comparison, rule of five filtering and docking studies separately. For the identification of potential leads, we filter out compounds violating the ‘Rule of Five’ for drug likeness. In this case study example, 71 out of 100, library compounds follow the Lipinski’s rule of five (Table 15.1), while rest 29 compounds violate the Lipinski’s rule, thus rejected as potential leads (Table 15.2). These 29 compounds predicted to have poor adsorption or permeation ability because of one or more following features: more than 5 H-bond donors, more than 10 H-bond acceptors, the molecular weight is greater than 500 and the calculated LogP is greater than 5. Out of these 29 compounds, 11 showed poor interaction energy with known 16 anticancer receptors through molecular docking studies (Table 15.3), while 18 showed good interaction energy (Table 15.4) with known 16 anticancer Human and mammalian receptors (Table 15.5), this indicates that apart from rule of five, some compounds exceptionally may act as good leads, but may show low efficacy power in terms of pharmacokinetic studies. On the other hand, 68 out of 71 compounds showed both good docking interaction with known anticancer receptors and also favours the Lipinski's rule of five, thus selected as potential anticancer leads (Table 15.6), while 3 compounds showed poor docking interaction with receptors, thus rejected for further studies. After docking studies, quantitative data prediction analysis have been performed for best selected 68 compounds based on 3D chemical structures (Table 15.7), so that to built the QSAR/QSPR mathematical model for the prediction of biological activity of unknown compounds and also to find out the dependable descriptors responsible for anticancer activity through regression statistical analysis (see example of anti-malarial QSAR model in Table 15.11).

Table 15.1. The library compounds (71) favours the 'Lipinski's Rule of Five' as a test for drug likeness or druggability.

PubChem Compound ID (71 in numbers)	Druggability Test
CID_16741, CID_47318, CID_65036, CID_3712, CID_5280373, SID_12013078, SID_210967, SID_49834379, SID_17396662, CID_24360, CID_16590, CID_11617, CID_9685, CID_5708, CID_5350, CID_5311, CID_5057, CID_5056, CID_65781, CID_3752, CID_4018, CID_10630736, CID_10678455, CID_10923720, CID_11000304, CID_11778794, CID_11821800, CID_23663982, CID_68781, CID_114835, CID_114866, CID_147004, CID_440623, CID_495574, CID_634470, CID_638034, CID_1548910, CID_3001662, CID_3070413, CID_4141487, CID_5281232, CID_5282435, CID_5381733, CID_5385552, CID_5472495, CID_6158261, CID_9064, CID_969516, CID_5281767, CID_442783, CID_10666836, CID_10595440, CID_16666, CID_1254, CID_165675, CID_19244, CID_10607, CID_72, CID_3071, CID_6398761, CID_6712212, CID_9547315, CID_9776851, CID_9802152, CID_9931851, CID_10084664, CID_10251785, CID_10378578, CID_10523508, CID_10534893, CID_10612216	**Lipinski's Rule of Five:** - Mol. weight < 500 - LogP < 5 - H-bond donors < 5 - H-bond acceptors < 10
Inference =	Drug like properties

Table 15.2. The library compounds (29) violating the 'Lipinski's Rule of Five'.

PubChem Compound ID (29 in No.)	Druggability Test
CID_16219940, CID_16197727, CID_122127, CID_65064, CID_446925, CID_5978, CID_5385555, CID_13342, CID_5280531, CID_5319123, CID_5316150, CID_5281235, CID_72495, CID_10794036, CID_5288209, CID_10532024, CID_401405, CID_10428109, CID_23639625, CID_10337961, CID_82146, CID_21773740, CID_9960816, CID_11969852, CID_9847391, CID_9832869, CID_9830646, CID_65076, CID_4259412	**Lipinski's Rule of Five violation:** - Mol. Weight > 500 - LogP > 5 - H-bond donors > 5 - H-bond acceptors > 10
Inference =	Poor permeability/ absorption through cell membrane

Table 15.3. The library compounds violating the Lipinski's Rule of Five and showing poor docking interaction energy (in Kcal/mol) against known anticancer drug targets.

S. No.	PubChemID	Docking score (in Kcal/mol) against known Anticancer drug targets (PDB ID: 1a25, 1a35, 1agw, 1awn, 1cdk, 1cpj, 1crp, 1dvi, 1jsu, 1mnc, 1p38, 1rpa)
1.	CID_11969852	10.2
2.	CID_5385555	4586.9
3.	CID_446925	13605.4
4.	CID_5319123	13686.0
5.	CID_5978	37283.7
6.	CID_5316150	62019.1
7.	CID_5281235	75635.4
8.	CID_13342	76808.1
9.	CID_16219940	979421.3
10.	CID_122127	1118907.0
11.	CID_16197727	12333424.0

Table 15.4. The library compounds violating the Lipinski's Rule of Five and showing good docking interaction energy (in Kcal/mol) against known anticancer drug targets.

S. No.	PubChemID	1a25	1a35	1agw	1awn	1cdk	1mnc
1.	CID_65064	-65.013	-100.77	-63.582	-54.635	**-99.175**	**-119.166**
2.	CID_72495	-42.841	-42.841	-42.841	-42.841	-42.841	-42.841
3.	CID_10794036	-80.007	-80.007	-80.007	-80.007	-80.007	-80.007
4.	CID_5288209	-66.044	-66.044	-66.044	-66.044	-66.044	-66.044
5.	CID_10532024	-31.384	-31.384	-31.384	-31.384	-31.384	-31.384
6.	CID_401405	-47.822	-47.822	-47.822	-47.822	-47.822	-47.822
7.	CID_10428109	-32.073	-32.073	-32.073	-32.073	-32.073	-32.073
8.	CID_23639625	-80.799	-80.799	**-80.799**	**-80.799**	-80.799	-80.799
9.	CID_10337961	-32.061	-32.061	-32.061	-32.061	-32.061	-32.061
10.	CID_82146	-61.877	-61.877	-61.877	-61.877	-61.877	-61.877
11.	CID_21773740	-65.714	-65.714	-65.714	-65.714	-65.714	-65.714
12.	CID_9960816	-63.19	-63.19	-63.19	-63.19	-63.19	-63.19
13.	CID_9847391	-72.317	-72.317	-72.317	-72.317	-72.317	-72.317

Note: Few docking results with modification.

Table 15.5. Known and potential human and mammalian therapeutic targets/receptors. PDB ID is the identification number of a protein in PDB database.

PDB ID	Protein	Phytomolecule	Experimental Finding	Therapeutic Effect	References
1a25	Protein kinase C	Catechin	Inhibitory effect	Anticancer	Komori *et al.* (1993)
	Protein kinase C	Camptothecin	Inhibitor	Induction of apoptosis in tumour	Martelli *et al.* (1999); Nieves-Neira and Pommier (1999)
1a35	Topoisomerase I	Camptothecin	Inhibitor	Anticancer	Sinha *et al.* (1995)
1agw	FGF receptor 1	Catechin	Inhibitor	Anticancer	Liang *et al.* (1997)
1cpj	Cathepsin B	Catechin	Activator	Anticancer	Ahmad *et al.* (2000)
1crp	C-H-ras p21 protein	Catechin	Inhibition of ras-transformed cells	Anticancer	Chung *et al.* (1999)
1dvi	Calpain	Camptothecin	Inhibition of calpain activities	Induces apoptosis in leukemic cells	Eymin *et al.* (1997)
1jsu	Cyclin-dependent kinase-2	Catechin	Inhibitor	Anticancer	Liang *et al.* (1997)
1mnc	Neutrophil collagenase	Catechin	Collagenase inhibitor	Inhibition of metastasis of cancerous cells	Demeule *et al.* (2000)
1p38	MAP kinase p38	Catechin	Inhibition of activation of p38 mitogen-activated protein kinase	Anticancer	Chen *et al.* (1999)

Note: * based on reference Chen and Ung, 2001 with minor modifications.

Table 15.6. The library compounds favour the 'Lipinski's Rule of Five' and showing good docking interaction energy (in Kcal/mol).

PubChemID	1a25	1a35	1agw	1awn	1cdk	1mnc
CID_969516	-93.284	**-110.886**	-60.896	**-79.784**	-62.67	**-92.725**
CID_10595440	-96.161	-84.798	-66.114	-75.358	-61.336	-85.486
CID_24360	-68.579	-88.454	-56.159	-62.99	-73.825	-84.888
CID_442783	-58.505	-95.962	-68.764	-73.19	-58.485	-84.216
CID_9064	-62.434	-78.713	-56.782	-58.508	-78.106	-84.114
CID_5281767	-89.577	-109.977	-64.375	-61.402	-57.156	-82.609
CID_6398761	-78.512	-78.512	**-78.512**	-78.512	**-78.512**	-78.512
CID_10923720	-73.704	-73.704	-73.704	-73.704	-73.704	-73.704
CID_4259412	-72.424	-72.424	-72.424	-72.424	-72.424	-72.424
CID_638034	-71.559	-71.559	-71.559	-71.559	-71.559	-71.559
CID_5057	-71.286	-71.286	-71.286	-71.286	-71.286	-71.286
CID_11000304	-70.201	-70.201	-70.201	-70.201	-70.201	-70.201
CID_65076	-69.511	-69.511	-69.511	-69.511	-69.511	-69.511
CID_9802152	-69.507	-69.507	-69.507	-69.507	-69.507	-69.507
CID_9931851	-69.423	-69.423	-69.423	-69.423	-69.423	-69.423
CID_11821800	-67.853	-67.853	-67.853	-67.853	-67.853	-67.853
CID_5311	-67.649	-67.649	-67.649	-67.649	-67.649	-67.649
CID_9830646	-67.317	-67.317	-67.317	-67.317	-67.317	-67.317
CID_5381733	-67.239	-67.239	-67.239	-67.239	-67.239	-67.239
CID_4018	-66.356	-66.356	-66.356	-66.356	-66.356	-66.356

Note: Docking results of only few leads and receptors are summarized here with modification.

Besides, there are examples of well known drugs, which violate the Lipinski's rule of five but at the same time work well pharmacologically (Table 15.8). A detailed description and pharmacological action of known drugs/bioactive compounds and 68 potential anticancer leads, virtually screened after Rule of five and docking based studies, is given in Table 15.9. Also for comparison, three different set of compounds and their corresponding docking energy (in kcal/mol) against each known anticancer receptor have been summarized in Table 15.10, which revealed that curcumin derivatives showed best interactions with most of the anticancer receptors, compared to other set compounds (Figure 15.2).

Docking and QSPR based virtual screening of anticancer compound

All the targets showed minimum docking of -23.239 kcal/mol with ecabet (CID_3070413). For targets 1a25, bioactive compound curcumin (CID_10666836) showed maximum docking score of -97.054 kcal/mol. For receptors 1a35, 1awn, 1mnc, 1p38, 1rpa, 1ydt, 1yfo, 2gss and 7ice, curcumin (CID_969516) showed maximum docking scores for each. For anticancer target 1a35 and 1awn maximum docking scores were -110.886 and -79.784 respectively, while for targets 1mnc, 1p38, 1rpa, 1ydt, 1yfo, 2gss and 7ice docking score was similar i.e. -92.725 kcal/mol. For receptor 1agw, maxacalcitol (CID_6398761) showed maximum docking score of -78.512 kcal/mol. For receptor 1awn, curcumin (CID_969516) showed maximum docking score of -79.784 kcal/mol. For receptor 1cdk, 1cpj, 1crp, 1dvi and 1jsu compound maxacalcitol (CID_6398761) showed maximum docking score of -78.512 kcal/mol. For receptor 1p38, 1rpa, 1ydt, 1yfo, 2gss and 7ice compound epigallocatechin gallate (CID_65064) showed maximum docking score of -119.166 kcal/mol.

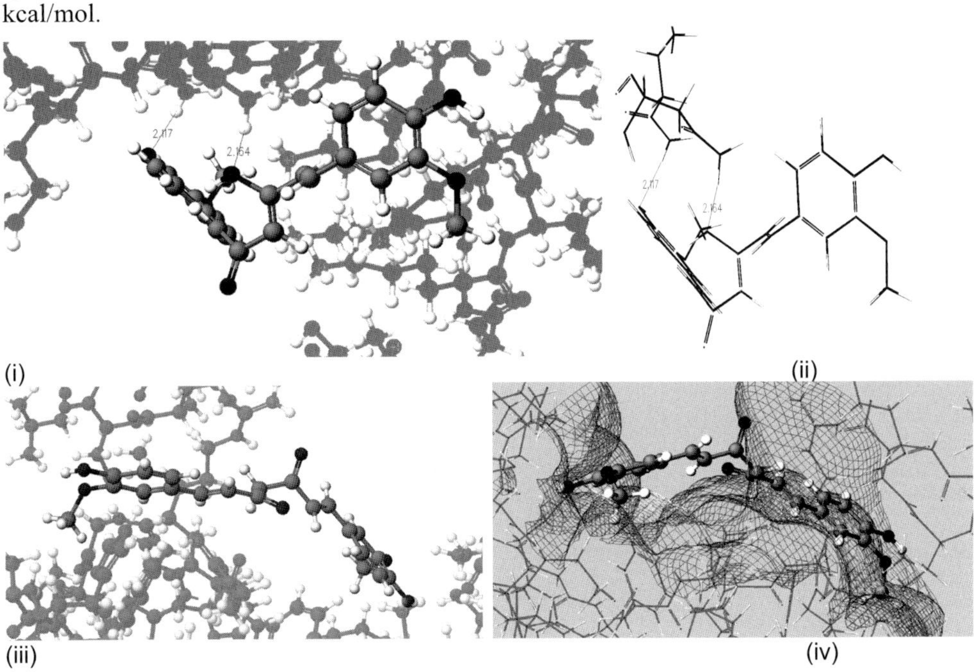

Figure 15.2. Best docked on Curcumin phytomolecule with anticancer known receptors (i, ii) 1a25, (iii) 1a35 and (iv) 1agw, with best docking score -97.054, -110.886 and -68.764 Kcal/mol respectively.

Details of docked compounds are as follows: Eighteen compounds did not show good interaction energy with any anticancer receptors. These are: CID_11969852, CID_1548910, CID_10607, CID_5280373, CID_5385555, CID_446925, CID_5319123, CID_5978, CID_5316150, CID_5281235, CID_13342, CID_16219940, CID_122127, CID_16197727, CID_634470, CID_68781, CID_3001662 and CID_10612216. For receptor 1a25, compound curcumin (CID_10666836, LogP 2.517, solvent accessible surface area 403.32 Å^2, mol. formula $C_{21}H_{20}O_6$, mol.wt. 368.385) showed maximum docking score of -97.054 kcal/mol and minimum -23.239 kcal/mol by ecabet (CID_3070413) with Anticarcinogenic, Anti-Infective, Anti-Ulcer, Protease Inhibitors activity. For receptor 1a35, 1awn, 1mnc, 1p38, 1rpa, 1ydt, 1yfo, 2gss and 7ice compound curcumin (CID_969516 Log P 2.479, solvent accessible surface area 383.405 Å^2, mol. formula $C_{21}H_{20}O_6$ mol.wt. 368.385) showed maximum docking scores.

Table 15.7. Details of quantitative data of best fit compounds*, to be used for QSPR/QSAR model development.

Compound ID	Log P	SASA	Polarity	LRS	SRS	RC	MR	LUMO	IP	HOMO	HF	CME	FC	AC	BC
CID_969516	2.479	383.405	42.632	6	6	2	103.424	-0.853	8.829	-8.829	-156.534	-34.719	0	47	48
CID_10595440	2.517	404.692	44.257	6	6	2	105.359	-1.07	8.804	-8.804	-155.602	-23.394	0	47	48
CID_24360	0.558	334.54	40.783	6	5	5	94.492	-1.586	9.103	-9.103	-89.015	-8.49	0	42	46
CID_442783	2.517	404.092	44.476	6	6	2	105.359	-1.092	8.782	-8.782	-156.042	-23.66	0	47	48
CID_9064	2.113	282.442	29.908	6	6	3	72.852	0.059	8.847	-8.847	-208.638	-12.187	0	35	37
CID_5281767	2.517	402.688	44.28	6	6	2	105.359	-1.085	8.793	-8.793	-155.908	-23.506	0	47	48
CID_6398761	2.984	443.462	50.537	6	5	3	123.828	-0.296	8.688	-8.688	-190.761	44.829	0	72	74
CID_10923720	1.962	272.515	29.338	6	6	3	70.823	-0.43	9.028	-9.028	-134.611	-6.208	0	30	32
CID_4259412	7.089	476.104	51.508	6	5	4	124.565	0.004	9.586	-9.586	-246.985	52.243	0	78	81
CID_638034	4.751	407.713	44.403	6	6	1	107.075	-0.541	8.42	-8.42	-68.454	4.413	0	56	56
CID_5057	4.751	407.212	44.332	6	6	1	107.075	-0.517	8.445	-8.445	-68.775	4.434	0	56	56
CID_11000304	1.962	270.146	29.398	6	6	3	70.823	-0.668	9.17	-9.17	-142.404	-10.93	0	30	32
CID_65076	7.089	474.076	51.404	6	5	4	124.565	0.056	9.532	-9.532	-246.884	52.74	0	78	81
CID_9802152	2.984	453.789	50.665	6	5	3	123.828	-0.218	8.717	-8.717	-191.477	40.306	0	72	74

Note: **LogP:** The octanol-water partition coefficient. **Solvent accessible surf area (SASA):** The molecular surface area accessible to a solvent molecule. **Polarizability (P):** The molecule's average alpha polarizability. **Smallest ring size (SRS):** Information about rings present in the compound. Rings with more than 12 atoms are ignored. The number of atoms forming the smallest ring in the compound, or 0 if the compound contains no ring of size 12 or less. **Ring count (RC):** The number of rings present in the compound. Rings with more than 12 atoms are ignored. The number of rings with 12 or fewer atoms (All= all aromatic, small, 5-membered, 5-membered aromatic, 6-membered, 6-membered aromatic, large, large aromatic). **Molar refractivity (MR):** The molar refractivity of the compound. **LUMO energy:** The energy gained when an electron is added to the lowest unoccupied molecular orbital (LUMO). **Ionization potential (IP):** The energy required to remove an electron from a molecule in its ground state. **HOMO energy:** The energy required to remove an electron from the highest occupied molecular orbital (HOMO). **Heat of formation (HF):** The energy released or used when a molecule is formed from elements in their standard states. **Conformation minimum energy (CME):** Energy calculated for an optimized conformation of the compound. **Formal Charge (FC):** The net positive or negative charge on the molecule. **Atom count (AC):** The number of atoms. **Bond count (BC):** The number of bonds. Weak and ionic bonds are ignored. * Results of only few compounds are summarized in the table.

Table 15.8. Exceptionally some well known bioactive compounds showed violation of Lipinski's rule of five.

Phyto-molecule	**Description**	**PubChemID**	**Mol. Wt.**	**Log P**	**Rule of 5 violation**	**Mol Wt. > 500**	**LogP > 5**	**H-bond donors > 5**	**H-bond acceptors > 10**
Allicin	antibacterial principle of *Allium sativum* (garlic)	CID_65036	162.264	1.415	0	0	0	0	0
Catechin	polyphenolic antioxidant plant metabolites (flavonoids); tea-plant *Camellia sinensis*	CID_9064	290.272	2.113	0	0	0	0	0
Camptothecin	An alkaloid isolated from the stem wood of the Chinese tree, *Camptotheca acuminata*	CID_24360	348.357	0.558	0	0	0	0	0
Indole-3-carbinol	occurs in edible cruciferous vegetables	CID_3712	147.176	0.82	0	0	0	0	0
Epigallocatechin gallate	antimutagen in green tea (*Camellia sinensis*)	CID_65064	458.378	3.017	2	0	0	1	1
Phytoestrogen	biochanin A; Plant extracts and compounds, primarily Isoflavones	CID_5280373	284.268	1.994	0	0	0	0	0
Artemether	quinghaosu derivative (*Artemisia annua*)	SID_12013078 SID_17396662	298.378	3.331	0	0	0	0	0
Artemisinine	sesquiterpene lactone from Chinese medicinal herb *Artemisia annua*	SID_210967 SID_49834379	282.336	3.085	0	0	0	0	0
Curcumin	Phytochemical of	CID_969516	368.38	2.479	0	0	0	0	0

	Curcuma longa	CID_528176 7 CID_442783 CID_106668 36 CID_105954 40	5	2.517	0	0	0	0	0
Menthol	An alcohol produced from mint oils (*Mentha* sp.) or prepared synthetically	CID_16666 CID_1254 CID_165675 CID_19244	156.26 7	2.78	0	0	0	0	0
Vincristine	Antitumour alkaloid isolated from *Vinca rosea*	CID_5978	824.96 9	3.194	2	0.65	0	0	1
Podophyllotoxi n	A lignan found in Podophyllin resin from the roots of *Podophyllum* plants	CID_10607	414.41 1	1.408	0	0	0	0	0
Vinblastine	Antitumour alkaloid isolated from *Vinca rosea*	CID_13342	810.98 6	4.078	2	0.62 2	0	0	1

Table 15.9. Pharmacological action of potential anticancer leads virtually screened after QSPR, Lipinski's Rule and docking studies. PubChem ID in bold characters indicates reference anticancer phytomolecule, while shaded compounds indicate diverse pharmacological action.

PubChem ID	Potential Compound	Description	Pharmacological Action
CID_65036	Allicin	Antibacterial principle of *Allium sativum* (garlic), Free Radical Scavengers	Anticarcinogenic , Anti-Infective, Antilipidaemic, Antioxidants, Hypoglycemic
CID_24360	Camptothecin	An alkaloid isolated from the stem wood of the Chinese tree, *Camptotheca acuminata*. This compound selectively inhibits the nuclear enzyme DNA Topoisomerases Type I.	Anticarcinogenic , Antineoplastic, Phytogenic, Enzyme Inhibitors
CID_9064	Catechin	An antioxidant flavonoid, occurring especially in woody plants as both (+)-catechin and (-)-	Anticarcinogenic , Antioxidant

		epicatechin (cis) forms.	
CID_9685 CID_440623	2-hydroxyestrone	Catechol estrogen which is a major metabolite of estradiol in man and animals	Anticarcinogenic
CID_3071	3,3'-diindolylmethane	Occurs in edible cruciferous vegetables; a cytochrome P-450 antagonist; also downregulates the pro-survival pathway in hormone independent prostate cancer	Anticarcinogenic
CID_11617	Allyl sulfide	Essence of garlic; inhibits CYP2E1	Anticarcinogenic, Antioxidants
CID_10428109	Bexarotene	A retinoid X receptor (RXR) selective agonist	Anticarcinogenic
CID_11778794	Biochainin A	Phytoestrogen	Anticarcinogenic, Phytoestrogens
CID_4259412	Cholesteryl sulfate	Component of human seminal plasma and spermatozoa	Anticarcinogenic, Serine Proteinase Inhibitors
CID_5385552	Crocetin	Increases oxygen diffusion in plasma	Anticarcinogenic, Antioxidants
CID_969516	Curcumin	*Curcuma longa*	Anticarcinogenic
CID_16590	Diallyl disulfide	Major constituent of garlic oil	Anticarcinogenic, Antihypertensive, Antineoplastic, Spermatocidal, Antimutagenic
CID_23663982	Ecabet	Protease Inhibitors	Anticarcinogenic, Anti-Infective, Anti-Ulcer
CID_10923720	Genistein	An isoflavonoid derived from soy products. It inhibits PROTEIN-TYROSINE KINASE and topoisomerase-II (DNA TOPOISOMERASES, TYPE II); activity	Anticarcinogenic, Protein Kinase Inhibitors, Phytoestrogens
CID_3712	Indole-3-carbinol	Occurs in edible cruciferous vegetables	Anticarcinogenic
CID_3752	Irsogladine	Radioprotective	Anticarcinogenic, Anti-Ulcer, Antineoplastic
CID_6398761	Maxacalcitol	Dermatologic	Anticarcinogenic, Antineoplastic
CID_47318	Oltipraz	Reverse Transcriptase Inhibitors	Anticarcinogenic, Antiviral,

			Schistosomicides,
CID_16741	Phenethyl isothiocyanate	A dietary liver aldehyde dehydrogenase inhibitor; also promotes urinary bladder carcinoma	Anticarcinogenic, Carcinogens, Enzyme Inhibitors
CID_72	Protocatechuic acid	-	Anticarcinogenic
CID_5056	Resveratrol	Antineoplastic, Phytogenic, Antioxidants,	Anticarcinogenic, Anti-Inflammatory Non-Steroidal
CID_638034	Retinol acetate	Enzyme Inhibitors, Platelet Aggregation Inhibitors, Antimutagenic	Anticarcinogenic, Adjuvants, Immunologic
CID_114835 CID_147004	Selenomethyl selenocysteine	-	Anticarcinogenic
CID_9830646	Squalamine	From dogfish shark *Squalus acanthias*	Anticarcinogenic, Anti-Bacterial, Angiogenesis Inhibitors
CID_5350	Sulforafan	From *Cardaria draba* L.	Anticarcinogenic
CID_10523508	Sulindac sulfone	Inhibits K-ras-dependent cyclooxygenase-2; sulfated analog of indomethacin	Anticarcinogenic, Antineoplastic, Cyclooxygenase Inhibitors
CID_5311	Vorinostat	Inhibits histone deacetylase I and 3	Anticarcinogenic, Anti-Inflammatory, Non-Steroidal, Antineoplastic, Enzyme Inhibitors
SID_210967	Artemisinin	Sesquiterpene lactone from Chinese medicinal herb *Artemisia annua*	Anti-Infective, Antimalarials
SID_12013078	Artemether	Quinghaosu (*Artemisia annua*) derivative	Antifungal, Antimalarials, Antiprotozoal, Coccidiostats, Schistosomicides
CID_1254	Menthol	An alcohol produced from mint oils or prepared synthetically	Antipruritics

Table 15.10. Docking based virtual screened bioactive library compounds.

Set	Phytomolecule*	PubChemID	Log P	Potential and known Human and Mammalian Anticancer Targets (PDB ID) with Docking PMF Score (kcal/mol)									
				1a25	1a35	1agw	1awn	1cdk	1cpj	1crp	1dvi	1jsu	1mnc
Reference	Allicin	CID_65036	1.415	-43.037	-46.416	-37.286	-48.498	-56.023	-56.023	-56.023	-56.023	-56.023	-48.381
	Catechin	CID_9064	2.113	-62.434	-78.713	-56.782	-58.508	-78.106	-78.106	-78.106	-78.106	-78.106	-84.114
	Camptothecin	CID_24360	0.558	-68.579	-88.454	-56.159	-62.99	-73.825	-73.825	-73.825	-73.825	-73.825	-84.888
Known	Indole-3-carbinol	CID_3712	0.82	-42.206	-55.901	-43.368	-45.251	-58.694	-58.694	-58.694	-58.694	-58.694	-56.348
	Epigallocatechin gallate	CID_65064	3.017	-65.013	-100.77	-63.582	-54.635	-99.175	-99.175	-99.175	-99.175	-99.175	-119.166
	Phytoestrogen	CID_5280373	1.994	**183.896**	**172.614**	**204.157**	**195.431**	**193.097**	**193.097**	**193.097**	**193.097**	**193.097**	**163.819**
	Vincristine	CID_5978	3.194	**37283.7**	**35453.72**	**35519.62**	**35831.84**	**36248.89**	**36248.89**	**36248.89**	**36248.89**	**36248.89**	**48645.82**
	Podophyllotoxin	CID_10607	1.408	**122.043**	**95.506**	**133.323**	**159.248**	**120.838**	**120.838**	**120.838**	**120.838**	**120.838**	**102.247**
	Vinblastine	CID_13342	4.078	**76808.1**	**76664.11**	**76710.02**	**76844.35**	**76681.27**	**76681.27**	**76681.27**	**76681.27**	**76681.27**	**77405.95**
Diverse	Artemether	SID_12013078	3.331	-27.854	-75.02	-34.746	-46.002	-63.802	-63.802	-63.802	-63.802	-63.802	-63.374
		SID_17396662	3.331	-53.178	-76.233	-39.077	-45.211	-62.278	-62.278	-62.278	-62.278	-62.278	-62.944
	Artemisinin	SID_210967	3.085	-48.695	-66.867	-31.04	-36.533	-51.543	-51.543	-51.543	-51.543	-51.543	-65.439
		SID_49834379	3.085	-49.188	-68.085	-31.227	-45.186	-51.642	-51.642	-51.642	-51.642	-51.642	-64.256
	Curcumin	CID_969516	2.479	-93.284	-110.886	-60.896	-79.784	-62.67	-62.67	-62.67	-62.67	-62.67	-92.725
		CID_5281767	2.517	-89.577	-109.977	-64.375	-61.402	-57.156	-57.156	-57.156	-57.156	-57.156	-82.609
		CID_442783	2.517	-58.505	-95.962	-68.764	-73.19	-58.485	-58.485	-58.485	-58.485	-58.485	-84.216
		CID_10666836	2.517	-97.054	-85.034	-67.547	-58.337	-56.308	-56.308	-56.308	-56.308	-56.308	-52.615
		CID_10595440	2.517	-96.161	-84.798	-66.114	-75.358	-61.336	-61.336	-61.336	-61.336	-61.336	-85.486
	Menthol	CID_16666	2.78	-36.111	-46.404	-25.399	-31.926	-49.708	-49.708	-49.708	-49.708	-49.708	-49.456
		CID_1254	2.78	-35.34	-46.768	-29.303	-30.269	-50.201	-50.201	-50.201	-50.201	-50.201	-50.014
		CID_165675	2.78	-35.378	-46.794	-29.295	-30.525	-49.038	-49.038	-49.038	-49.038	-49.038	-49.624
		CID_19244	2.78	-35.988	-46.739	-29.331	-35.357	-49.006	-49.006	-49.006	-49.006	-49.006	-49.497

Note: Epigallocatechin gallate and Curcumin indicates best docking energies, while phytomolecules namely, Phytoestrogen, Vincristine, Podophyllotoxin and Vinblastine indicates poor docking scores against all receptors which suggest that other targets might be responsible for their anticancer activities. * Results of few leads only.

Table 15.11. An anti-malarial QSAR model for the prediction of biological activity and dependable chemical descriptors.

Chemical Sample	Name	Mol. formula	(c)	(D_e)	(D_p)	**AE**	(F)	(H)	(I)	(K)	(T)	(AA)	(AD)
CID_2719	Chloroquine	$C_{18}H_{26}N_3Cl$	3	**0.477**	**0.492**	**0.015**	48	10.775	4.564	-0.34	38.819	2	6
CID_31593	Benznidazol	$C_{12}H_{12}N_4O_3$	4.7	**0.672**	**0.658**	**-0.014**	31	61.863	6.212	-1.09	27.865	1	5
CID_4735	Pentamidine	$C_{19}H_{24}N_4O_2$	1.7	**0.23**	**0.229**	**-0.001**	49	-24.069	6.018	-0.863	40.795	2	6
CID_6437380	Benflumetol	$C_{30}H_{32}NOCl_3$	150	**2.176**	**2.179**	**0.003**	67	-5.975	3.107	-0.282	63.027	3	5
CID_40692	Mefloquine	$C_{17}H_{16}N_2OF_6$	21.84	**1.339**	**1.355**	**0.016**	42	9.36	7.278	-0.532	33.722	2	6
CID_4993	Pyrimethamine	$C_{12}H_{13}N_4Cl$	8.4	**0.924**	**0.99**	**0.066**	30	-8.437	1.985	-0.464	28.923	2	6
CID_8549	Quinine	$C_{20}H_{24}N_2O_2$	100.78	**2.003**	**1.977**	**-0.026**	48	16.247	1.861	-0.367	39.187	2	6
CID_68827	Artemisinine	$C_{15}H_{22}O_5$	10.5	**1.021**	**1**	**-0.021**	42	32.503	6.073	-0.573	29.257	0	6
CID_456408	Artemether	$C_{16}H_{26}O_5$	3.43	**0.535**	**0.551**	**0.016**	47	37.278	1.952	-0.352	31.515	0	6
CID_5464098	Artesunate	$C_{19}H_{28}O_8$	1.17	**0.068**	**0.081**	**0.013**	55	16.644	3.143	-0.757	38.091	0	6
CID_65404	Fansidar	$C_{24}H_{27}N_8O_4SCl$	**63.68**	?	**1.804**		65	-43.088	5.677	-1.491	59.356	4	6
						Standard Error = 0.029061							

Note: (c) = *In vitro* activity (IC_{50}) nM, (D_e) = Experimental log (IC_{50}), = Predicted log (IC_{50}), AE= Absolute error, F= Atom Count (all atoms), H= Conformation Minimum Energy (kcal/mol), I= Dipole Moment (debye), K= Dielectric Energy (kcal/mol), T= Polarizability, AA= Ring Count (aromatic 6 membered), AD= Size of Smallest Ring and ? = unknown activity

For anticancer target 1a35 and 1awn maximum docking scores were -110.886 and -79.784 respectively, while for anticancer targets 1mnc, 1p38, 1rpa, 1ydt, 1yfo, 2gss and 7ice docking score was common i.e. -92.725 kcal/mol. All the targets showed minimum docking score of -23.239 kcal/mol with compound ecabet (CID_3070413) having Anticarcinogenic, Anti-Infective, Anti-Ulcer, Protease Inhibitors activity.

For receptor 1agw, compound maxacalcitol (CID_6398761 Log P 2.984, solvent accessible surface area 443.462 $Å^2$, mol. formula $C_{26}H_{42}O_4$ mol.wt. 418.615) having Anticarcinogenic, Antineoplastic, Dermatologic pharmacological actions, showed maximum docking score of -78.512 kcal/mol and minimum -23.239 kcal/mol by ecabet (CID_3070413) with Anticarcinogenic, Anti-Infective, Anti-Ulcer, Protease Inhibitors activity. For receptor 1awn, compound Curcumin (CID_969516, Log P 2.479, solvent accessible surface area 383.405 $Å^2$, mol. formula $C_{21}H_{20}O_6$ mol.wt. 368.385) having Anticarcinogenic pharmacological actions, showed maximum docking score of -79.784 kcal/mol and minimum -23.239 kcal/mol by ecabet (CID_3070413) with Anticarcinogenic, Anti-Infective, Anti-Ulcer, Protease Inhibitors activity. For receptor 1cdk, 1cpj, 1crp, 1dvi and 1jsu compound maxacalcitol (CID_6398761, Log P 2.984, solvent accessible surface area 443.462 $Å^2$, mol. formula $C_{26}H_{42}O_4$, mol.wt. 418.615) having Anticarcinogenic pharmacological action, showed maximum docking score of -78.512 kcal/mol and minimum -23.239 kcal/mol by ecabet (CID_3070413) with Anticarcinogenic, Anti-Infective, Anti-Ulcer, Protease Inhibitors activity. For receptor 1p38, 1rpa, 1ydt, 1yfo, 2gss and 7ice compound Epigallocatechin gallate (CID_65064, Log P 3.017, solvent accessible surface area 404.253 $Å^2$, mol. formula $C_{22}H_{18}O_{11}$ mol.wt. 458.378) having Anticarcinogenic pharmacological action, showed maximum docking score of -119.166 kcal/mol and minimum -3.606 kcal/mol by retinol palmitate (CID_5280531) having Anticarcinogenic, Antioxidants pharmacological action.

Finally, our results demonstrate that virtual screening approaches outperform the individual approach, in particular in cases where very limited or no ligand information is available. In the case of well studied targets, more classical ligand-based methods (QSAR) may be better suited to predict new strong binders from a large number of known ligands. Since normally VS approach is a ligand-based approach, it would probably be interesting to use it in combination with docking. Although prior known ligands can help in tuning the docking procedures to the receptor under study, it can in principle be used with little or no ligand information. When more experimental 3D structures become available for targets in the future, this will help building reliable models for a wider range of targets that would be suitable for docking studies. Joint use of ligand-based virtual screening and docking would certainly improve predictions.

Conclusion

We showed how sharing information across the targets by considering a VS space of the target-ligand interaction pairs could improve the prediction performances, with respect to the single receptor approach. In addition, we showed that when using such a representation, it was possible to make reasonable predictions even when all known ligands were ignored for a given target, that is, to predict ligands for orphan targets. This VS approach is related to ligand-based approaches. However, sharing information among different targets allows us to perform prediction on orphan targets, which is also possible using target-based methods. Nevertheless, the latter are limited by the number of known receptor structures and the difficulty to apply such methods on homology models. Interesting developments of this method could include application to other important drug target families, like enzymes or ion channels, for which most of the descriptors used for the targets could directly be transposed and other, more specific ones could be designed. VS is a cost-effective and reliable technique that can be applied to identify potential leads and avoid undesirable compounds that would otherwise result in expensive and time consuming experimental methods. However, VS often requires careful preparation of both target and compound library, use of optimal parameters as well as careful analysis of the results. It should be noted that experience and knowledge about the target are very crucial in identifying true positives in such experiments. We have covered some of the essential considerations in designing VS experiment.

References

Abagyan, R. and Totrov, M. (2001) High-throughput docking for lead generation. *Current Opinion in Chemical Biology* 5, 375–382.

Abagyan, R., Totrov, M. and Kuznetsov, D. (1994) ICM: A new method for structure modeling and design: Applications to docking and structure prediction from the distorted native conformation, *Journal of Computational Chemistry* 15, 488–506.

Ahmads N., Guptas S. and Mukhtar, H. (2000) Green tea polyphenol epigallocatechin-3-gallate differentially modulates nuclear factor kappa B in cancer cells versus normal cells. *Archives of Biochemistry and Biophysics* 376, 338-346.

Anderson, S. and Chiplin, J. (2002) Structural genomics: Shaping the future of drug design? *Drug Discovery Today* 7, 105–107.

Berman, H.M., Westbrook, J., Feng, Z., Gilliland, G., Bhat, T.N., Weissig, H., Shindyalov, I.N. and Bourne, P.E. (2000) The protein data bank. *Nucleic Acids Res*earch 28, 235–242.

Chen, X., Ji, Z.L. and Chen, Y.Z. (2002) TTD: Therapeutic Target Database. *Nucleic Acids Research* 30(1), 412–415

Chen, Y.Z. and Ungs, C.Y. (2001) Prediction of potential toxicity and side effect protein targets of a small molecule by a ligand-protein inverse docking approach. *Journal of Molecular Graphics and Modelling* 20(3), 199–218.

Chung, J.Y., Huang, C., Meng, X., Dong, Z. and Yang, C.S. (1999) Inhibition of activator protein 1 activity and cell growth by purified green tea and black tea polyphenols in H-ras-transformed cells : Structure-activity relationship and mechanism involved. *Cancer Research* 59, 4610–4617.

Demeule, M., Buossard, M., Page, M., Ginruas, D. and Beliveau, R. (2000) Matrix metalloproteinase inhibition by green tea catechins. *Biochimica et Biophysica Acta* 1478, 51–60.

Eymin, B., Dubrez, L., Allouche, M. and Solary, E. (1997) Increased Gadd153 messenger RNA level is associated with apoptosis in human leukemic cells treated with etoposide. *Cancer Research* 57, 686–695.

Gane, P.J. and Dean, P.M. (2000) Recent advances in structure-based rational drug design. *Current Opinion in Structural Biology* 10, 401–404.

Klebe, G. (2000). Recent developments in structure-based drug design. *Journal of Molecular Medicine (Berlin)* 78, 269–281.

Komori, A., Yatsunami, J., Okabe, S., Abe, S., Hara, K., Suganuma, M., Kim, S.J. and Fujiki H. (1993). Anticarcinogenic activity of green tea polyphenols. *Japanese Journal of Clinical Oncology* 23(3), 186-90.

Liang, Y.C., Lin-Shiau, S.Y., Chen, C.F. and Lin J.K. (1997) Suppression of extracellular signals and cell proliferation through EGF receptor binding by (-)-epigallocatechin gallate in human A431 epidermoid carcinoma cells. *Journal of Cellular Biochemistry* 67, 55-65.

Lipinski, C.A., Lombardo, F. and Dominy, B.W. (2000) Drug-like properties and the causes of poor solubility and poor permeability. *Journal of Pharmacological and Toxicological Methods* 44, 235–249.

Lipinski, C.A. *et al.* (1997) Experimental and computational approaches to estimate solubility and permeability in drug discovery and development settings. *Advanced Drug Delivery Reviews* 23, 3–25.

Martelli, A.M., Bortul, R., Bareggi, R., Takelloni, G., Grill, V., Baldini, G. and Narducci, P. (1999) The pro-apoptotic drug camptothecin stimulates phospholipase D activity and diacylglycerol production in the nucleus of HL-60 human promyelocytic leukemia cells. *Cancer Research* 59, 3961–3967.

Matsumoto, Y., Gujiwara, T. and Nago, S. (1995) Determinants of drug response in camptothecin-resistant glioma cell lines. *Journal of Neurooncology* 23, 1–8.

Nieves-Neira, W. and Pommier, Y. (1999) Apoptotic response to camptothecin and 7-hydroxy-staurosporine (UCN-01) in the eight human breast cell lines of the NCI anticancer drug screen: Multifactorial relationships with topoisomerase 1, protein kinase C, Bcl-2, p53, MDM-2 and caspase pathway. *International Journal of Cancer* 82, 396–404.

Shoichet, B.K., McGovern, S.L., Wei, B. and Irwin, J.J. (2002) Lead discovery using molecular docking. *Current Opinion in Chemical Biology* 6, 439–446.

Sinha, B.K. (1995) Topoisomerase inhibitors. A review of their therapeutic potential in cancer. *Drugs* 49, 11–19.

Subramaniam, S., Mehrotra, M. and Gupta, D. (2008) Virtual high throughput screening (VHTS) – A perspective. *Bioinformation* 3(1), 14–17.

Veber, D.F., Johnson, S.R., Cheng, H.Y., Smith, B.R., Ward, K.W. and Kopple, K.D. (2002) Molecular properties that influence the oral bioavailability of drug candidates. *Journal of Medicinal Chemistry* 45, 2615–2623.

Walters, W.P., Stahl, M.T. and Murcko, M.A. (1998) Virtual screening– an overview. *Drug Discovery Today* 3, 160–178.

Wang, M.C., Liu, J.H. and Wang, F.F. (1997) Protein tyrosine phosphatase-dependent activation of β-globin and δ-aminolevulenic acid synthase genes in the camptothecin-induced IW 32 erythroleukemia cell differentiation. *Molecular Pharmacology* 51, 558–566.

Chapter 16

Population Structure and Molecular Characterization of *Podophyllum hexandrum* of the Northwestern Himalayas

Pradeep Kumar Naik, Shri Ram and Harvinder Singh

Introduction

Podophyllum hexandrum Royle (Berberidaceae) is an endangered, herbaceous, rhizomatous species of great medicinal importance native to India. In a broad sense the genus *Podophyllum* is represented by two species: *Podophyllum hexandrum*, commonly distributed in the Himalayan regions and popularly called Himalayan mayapple while the other species is *Podophyllum peltatum*, commonly distributed in the Atlantic North America and popularly known as American mayapple. *Podophyllum hexandrum* is believed to have originated in the Himalayan regions. In India the presence of this species has been reported from Zanskar and Suru valleys of Ladakh, Kashmir, Lahaual and Spiti valleys, Kangra, Chamba, Kinnaur of Himachal Pradesh, Sikkim and Arunachal Pradesh. The populations of *P. hexandrum* are found growing on open slopes and alpine pastures in moist humid conditions. These species also grow in forests rich in humus and decayed organic matter as well as in cultivated fields between 2600 and 4500m altitude. The population size of *P. hexandrum* is very low (40–700 plants per location) and is declining every year. Some of the populations in certain pockets have virtually disappeared owing to anthropogenic activities and overexploitation (Bhadula *et al.*, 1996).

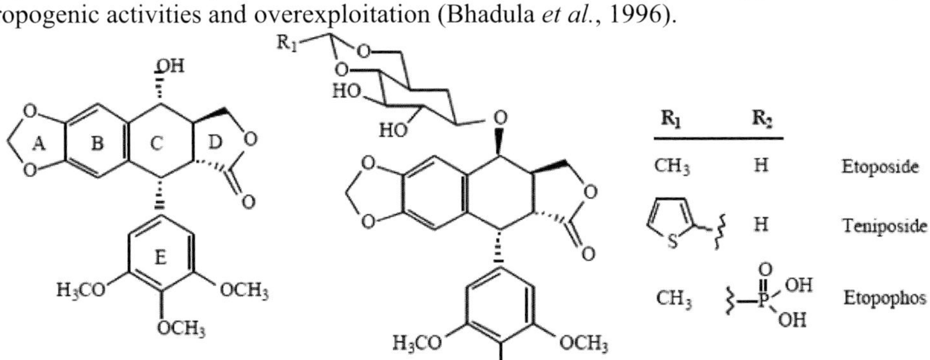

Fig. 16.1. Structure of podophyllotoxin and its derivatives.

Podophyllum hexandrum is recognized for its anticancer properties. It has long been used by the Himalayan natives and the American Indians (Anon, 1970). The rhizome yields a very valuable drug 'podophyllotoxin' which has a great demand in pharmaceutical industries throughout the world (Tyler *et al.*, 1988; Broomhead and Dewick, 1990). In addition, the drug obtained from Himalayan mayapple contains resin of superior quality in comparison to American mayapple (Jackson and Dewick, 1984). These compounds have been used for the treatment of lung and testicular cancers as well as certain types of leukaemia (Stahelin and Wartburg, 1991; Imbert, 1998). Podophyllotoxin is also the precursor to semi-synthetic chemotherapeutic drugs such as etoposide, teniposide and etophos (Stahelin and Wartburg, 1991; Imbert, 1998) (Fig. 16.1).

The populations of *Podophyllum hexandrum* show a range of variation in plant height, number of leaves (1 to 4 leaved), leaf insertion, size and shape of the fruit and the colour of seeds. On the basis of leaf number, four morphotypes viz. single leaved, double leaved, triple leaved and four leaved have been reported by Bhadula *et al.* (1996) from Garhwal Himalaya. The Indian species *P. hexandrum* contains three times more podophyllotoxin than its American counterpart *P. peltatum* (Fay and Ziegler, 1985). So *Podophyllum hexandrum* is being uprooted unscientifically for commercial trade. In the natural habitat, seed germination and seedling establishment are very poor and propagation is mostly through rhizomes (Badhwar and Sharma, 1963). Since the species is already endangered and exploitation of its underground parts continues to exceed the rate of natural regeneration, it needs immediate attention for conservation. A comprehensive understanding of genetic diversity and molecular characterization among and within *P. hexandrum* populations using a set of molecular markers needed for formulating appropriate sampling and management strategies for a *P. hexandrum* repository and for future *P. hexandrum* breeding efforts. A more detailed analysis with large number of genetic markers would provide a better understanding of the genetic diversity that exists in the wild population, which will contribute to useful gene conservation for a long term breeding programme. Among all the molecular markers used for uncharacterized crops, DNA-based molecular-marker techniques such as RAPD (Williams *et al.*, 1990), ISSR (Morgante and Olivieri, 1993) and AFLP (Vos *et al.*, 1995) have been proved powerful in genetic diversity estimations (Lu *et al.*, 1996). These markers are less affected by age, physiological condition of samples and environmental factors. They are not tissue-specific and thus can be detected at any phase of organism development. Only a small amount of sample is sufficient for analysis and the physical form of the sample does not restrict detection. These non-stringent requirements are particularly relevant for study of population structure of *Podophyllum hexandrum* which are in limited supply. In the following, we provide a general account of study of population structure of *P. hexandrum* from the Northwestern Himalayas using different DNA based molecular markers.

Analysis of genetic variations of plants

According to Frankel (1983), an essential prerequisite for a species to survive against environmental pressures is the availability of a pool of genetic diversity and in the absence of that extinction would appear inevitable. Determining how much genetic diversity exists in a species and explaining this diversity in terms of its origin, organization and maintenance are thus of fundamental significance in the application of genetic principles to conservation. Moreover, while assessing genetic diversity it is essential to have a quantitative measure of the hierarchy of organisms as genes, genotypes, populations and species. This is often based on characterization of amount and distribution of genetic diversity in the hierarchy, i.e., the population genetic structure, which is the most fundamental piece of information for species that require genetic management (Brown, 1978).

Variation at DNA level

The ability to investigate DNA sequences directly became available to population biologists only during the late 1970s. Currently, three major DNA-based techniques are widely used for analysing the genetic diversity in natural populations. These include: (i) restriction fragment length polymorphism (RFLP; Botstein *et al.*, 1980); (ii) polymerase chain reaction (PCR; Mullis and Faloona, 1987), and its derivatives, termed random amplification polymorphic DNA (RAPD) (Williams *et al.*, 1990); ISSR (Morgante and Olivieri, 1993); and (iii) a hybrid of both the above techniques named amplification fragment length polymorphism (AFLP; Vos *et al.*, 1995).

RAPD markers

RAPD markers are a modification of PCR contrived in the late 1980s (Wiliams *et al.*, 1990). PCR provides a means by which billions of copies of a particular target DNA fragment can be made from a complex mixture of genomic DNA. Now it is becoming more powerful with the introduction of user-friendly and fully automated techniques (Kreader *et al.*, 2001). The technical ease of RAPD markers and the facility of their application to new species has led their employment in many organisms including forest trees, crop as well as medicinal plants and lower plants for population genetics (e.g. Isabel *et al.*, 1995; Bucci and Menozzi, 1995; Nesbitt *et al.*, 1995; Schierenbeck *et al.*, 1997; Inglis *et al.*, 2001). The technique is one of the best available DNA-based tools for scoring variations between cultivars within species (Lakshmikumaran and Bhatia, 1998). One probable disadvantage, however, is the degree of reproducibility of these markers which can sometimes be low (e.g. Muralidharan and Wakeland, 1993; Skroch and Nienhuis, 1995). This is due to the sensitivity of RAPD banding patterns to reaction conditions, and the difficulty in exactly replicating reaction conditions across laboratories, where different brands of thermocyclers may be used.

RAPD analysis has been widely employed for assessing the genetic diversity and relationships in many plants. In an attempt to study genetic diversity of 28 populations of *Podophyllum hexandrum* collected from 11 forest divisions (Table 16.1) of Northwestern Himalayas of India, Alam *et al.* (2009) successfully used this technique.

Table 16.1. Twenty-eight populations of *Podophyllum hexandrum* collected from different sites at different altitudes covering 11 forest divisions of the NW Himalaya, India.

Forest Division	Sampling site	Altitude (m)
Parvati	R/4 Kasol	1570
	Twin Multivora	1300
	Anganoala	1300
Kullu	Brundhar	1916
	Gulaba	2895
	ChanderKhani	3352
	Kaned Nry	2150
	Sanghar Nry	2100
Dodrakwar	Madhvi Thach	3048
	Kala Pani	2743
Seraj	Sojha Nry	2667
	Jalora-C-3b	2473
Churah	DPF-D-1892-C1	3750
	DPF-D-791-C1	2700
Lahaul	Myar Valley	4300
	Nayan Ghar	4300
Palampur	Bada Bangal	2895
	Chota Bangal	2700
	IHBT	2800
Rampur	Bander Thach	2895
	Saropa Nry	2499
Kinnaur	Nichar Nry	2190
	Rango-NC-8	2710
Pangi	Sach Range	2712
	Killer Range	2850
	Purthi Range	2900
Bharmaur	Ghoei DPF	2080
	Samara RF	2590

Figure 16.2 shows the results of RAPD profiling among the 28 populations of *P. hexandrum*; this enables a direct comparison of genetic differentiation among the populations. The observed high proportion of polymorphic loci suggested that there is a high degree of genetic variation in the *Podophyllum* populations. The analysis of molecular variance revealed 53% of variations among the forest divisions and 47% of variations among populations within a forest division. The total gene diversity (Ht) estimated among populations was 0.338 and within populations (Hs) was 0.104. Shannon's information index was 0.501 and estimated gene flow was found to be 0.110. The RAPD study indicates that *P. hexandrum* populations in the Northwestern Himalayan region are genetically highly diverse.

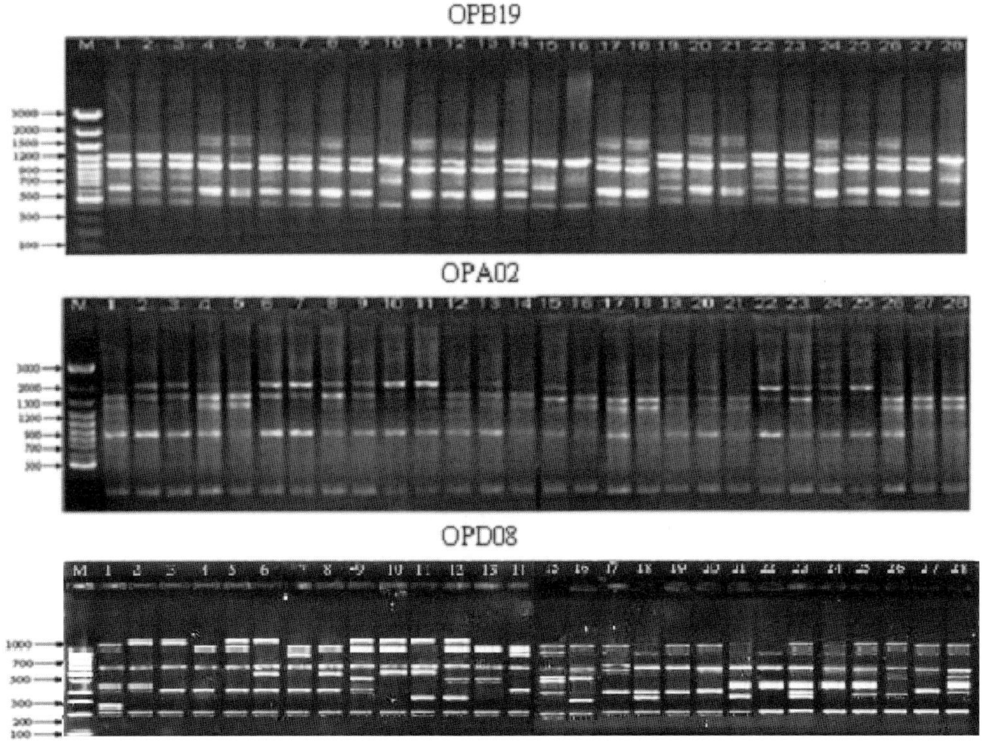

Fig. 16.2. RAPD amplification products obtained from the 28 genotypes of *P. hexandrum* studied. 1. R/4 Kasol 2. Twin Multivora 3. Anganoala 4. Brundhar 5. Gulaba 6. ChanderKhani 7. Kaned Nry 8. Sanghar Nry 9. Madhvi Thach; 10. Kala Pani 11. Sojha Nursery 12. Jalora-C-3b 13. DPF-D-1892-C1 14. DPF-D-791-C1 15. Myar Valley 16. Nayan Ghar 17. Bada Bangal 18. Chota Bangal 19. IHBT 20. Bander Thach 21. Saropa Nry 22. Nichar Nry 23. Rango-NC-8 24. Sach Range 25. Killer Range 26. Purthi Range 27. Ghoei DPF 28. Samara RF. M = the size of molecular markers in base pairs using λ DNA.

ISSR markers

Another powerful derivation of PCR technology is microsatellite markers. In higher plants or animals, inter simple sequence repeat (ISSR) markers are more and more in demand, because they are known to be abundant, very reproducible, highly polymorphic, highly informative, quick to use, are inherited in Mendelian fashion and are scored as dominant markers (Zietkiewicz *et al.*, 1994). ISSR uses the presence of simple sequence repeats (SSR) throughout the genome which are ubiquitous, abundant and highly polymorphic tandem repeat motifs composed of 1 to 6 nucleotides. ISSR analyses offer breeders and geneticists with competent means to link phenotypic and genotypic variations and it is rapidly being used by the research community in various fields of plant improvement (Godwin *et al.*, 1997).

Alam *et al.* (2008) for the first time demonstrated use of ISSR-derived DNA fingerprint patterns in molecular characterization of *Podophyllum hexandrum* populations from the Northwestern region of Himalayas. 28 populations of *P. hexandrum* collected from 11 forest divisions (Table 16.1) were assessed using 11 ISSR primers. Figure 16.3 shows the representative ISSR profiling of *P. hexandrum* populations.

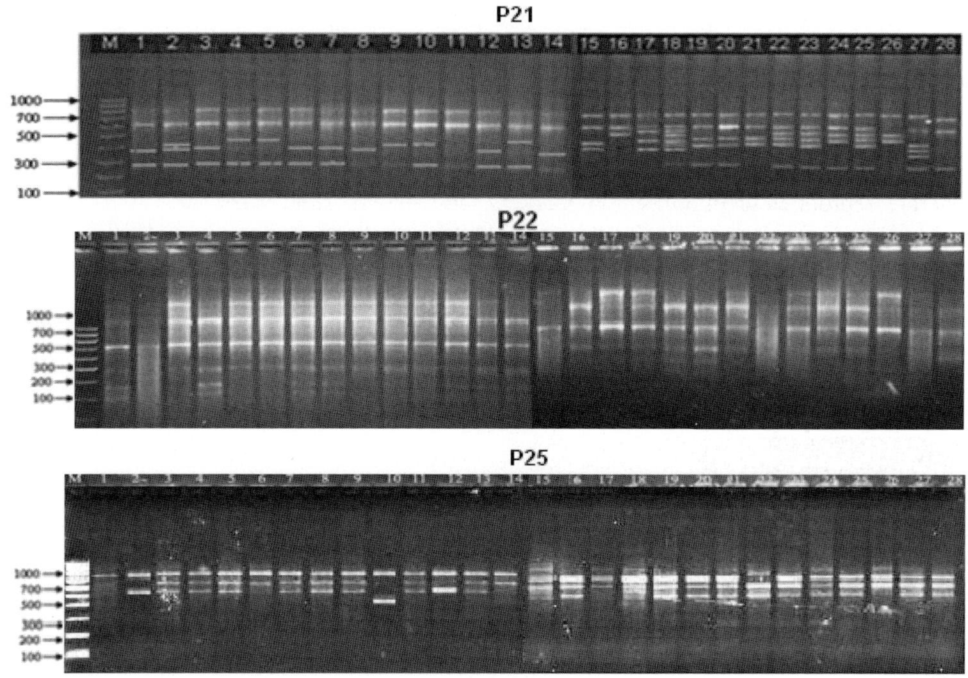

Fig. 16.3. ISSR amplification products obtained from the 28 populations of *Podophyllum hexandrum.*

It revealed high genetic polymorphism among the populations. Genetic polymorphism differed substantially within the discrete groups of plants with an average of 15.05% and was found to be in between 5.88% (Kinnaur forest division) and a maximum of 27.94% (Bahrmaur forest division). The genetic diversity was high (percentage of polymorphic loci, 83.82%; Shannon's information index, I = 0.441) at the populations level. The mean coefficient of gene differentiation (Gst) was 0.630, indicating that 29.44% of the genetic diversity resided within the population. An overall value of mean estimated number of gene flow (Nm = 0.147) indicated that there was limited gene flow among the sampled populations. The high Gst value (0.630) and the low Nm value (0.147) both indicated rapid genetic differentiation among the 28 populations, especially among the regions. The analysis of molecular variance revealed 62% of variations among the 11 forest divisions and 38% of variations among populations within a forest division. The results indicated high genetic diversity in *P. hexandrum* populations from Northwestern Himalayas.

AFLP markers

These fairly genetic markers born after combining positive qualities of both RFLP and PCR techniques are named as amplified fragment length polymorphism (AFLPs – Vos *et al.* 1995). AFLPs have shown high degree of reproducibility in contrast to RAPDs (Akerman *et al.*, 1996; Jones *et al.*, 1997). Due to these advantages, AFLP is now increasingly used in a number of species including many wild plant species (Remington *et al.* 1988; Gaiotto *et al.* 1997). Escaravage *et al.* (1998) have demonstrated the utility of AFLP markers in determining genotypes and clonal diversity in *Rhododendron ferrugineum* L. (Ericaceae). In India, AFLP markers have been used in the assessment of genetic diversity in 37 neem accessions from different eco-geographic regions of India and four exotic lines from Thailand (Singh *et al.*, 1999). The study indicated that neem germplasm within India constitutes a broad genetic base.

It was found suitable for our use with *Podophyllum* populations because of its ability to generate reproducible polymorphic markers. In this investigation all the 28 genotypes (Table 1) were fingerprinted using 13 primer combinations.

The 13 AFLP primer pairs used in this study generated 4616 (mean = 355.08 per primer pair) polymorphic bands across 28 regionally adapted *P. hexandrum* genotypes, giving an estimate of profound (> 84.40%) polymorphism. The AFLP banding pattern of 28 *P. hexandrum* genotypes is shown in the Figure 16.4 obtained with one primer combination.

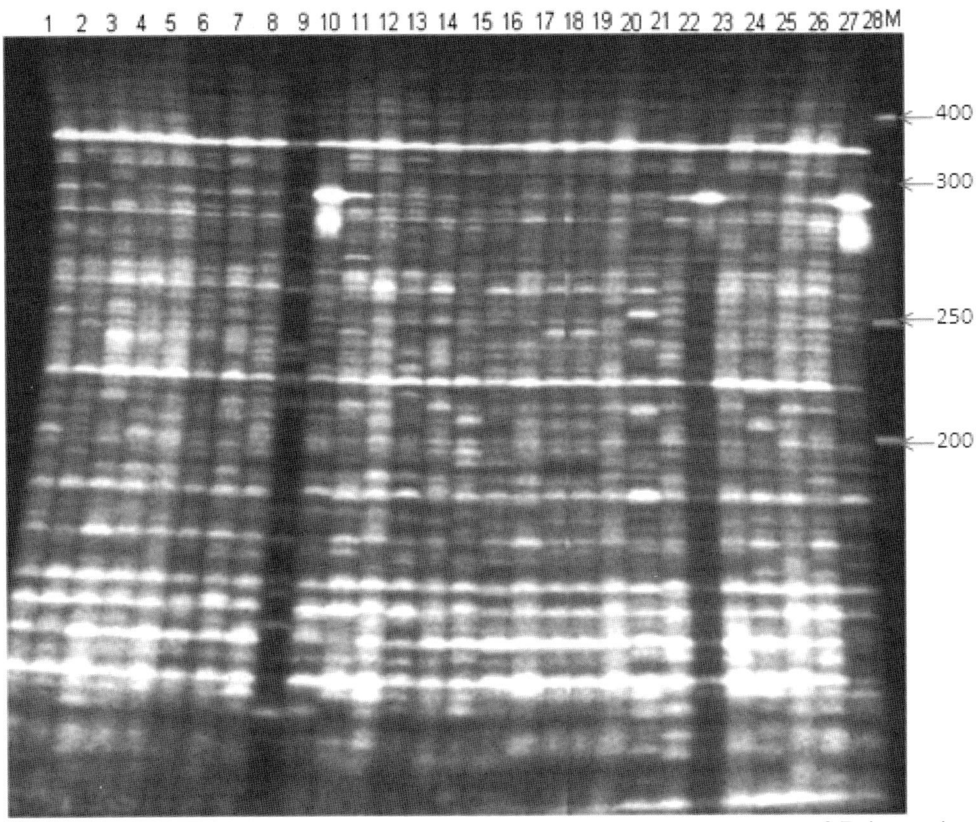

Fig. 16.4. AFLP amplification products obtained from the 28 genotypes of *P. hexandrum* studied using 13 *Eco*RI/*Mse*I primer combinations.

The observed high proportion of polymorphic loci suggests that there is a high degree of genetic variation among the *Podophyllum* genotypes. The analysis of molecular variance revealed 64% of variations among the 11 forest divisions and 36% of variation among populations within a forest division. The gene diversity computed among different groups of populations was recorded in between 0.05 (Bharmaur) and 0.26 (Kullu). Similarly, the total gene diversity (Ht) among populations was 0.26 and within populations (Hs) was 0.13. Shannon's information index was 0.41 and estimated gene flow was found to be 0.13 among the 28 *P. hexandrum* populations. Significant levels of population differentiation were found based on AFLP markers. The study indicates that *P. hexandrum* populations in the Northwestern Himalayan region are genetically highly diverse.

Conclusion

Based on polymorphic feature, genetic diversity, genetic similarity and gene flow among the populations of *Podophyllum* based on RAPD, ISSR and AFLP study, we conclude that all markers proved to be efficient tools in assessing the genetic diversity of *Pododophyllum* genotype and future conservation plans for this species should be specifically designed to include representative populations with the highest genetic variation for both *in situ* conservation and germplasm collection expeditions. These species or at least a large part of its genetic diversity may be lost in the near future owing to its importance and its exploitation as a medicinal plant if appropriate measures for conservation are not adopted. At present, the rate of exploitation of *P. hexandrum* is much more than the rate of propagation. Since no single or even a few plants represent the whole genetic variability of *P. hexandrum*, there appears to be a need to maintain sufficiently large populations in natural habitats to conserve the genetic diversity and avoid genetic erosion. For the conservation aspect, only a few populations may not adequately protect the genetic variations within the species in the Himalayan region. Genetic variations among elite genotypes of *Podophyllum hexandrum* based on RAPD, ISSR and AFLP analysis could be useful in selecting parents to be crossed for generating appropriate populations intended for both genome mapping and breeding purposes.

References

Akerman, S., Tammisola, J., Regina, M., Kauppinen, V. and Lapinjoki, S. (1996) Segregation of AFLP markers in *Betula pendula* (Roth). *Forest Genetics* 3, 117–123.

Alam, A., Naik, P.K. and Mishra, G.P. (2009) Congruence of RAPD and ISSR markers for evaluation of genomic relationship among 28 populations of *Podophyllum hexandrum* Royle from Himachal Pradesh, India. *Turkish Journal of Botany* 33, 1–12.

Alam, A., Naik, P.K., Gulati, P., Gulati, A.K. and Mishra, G.P. (2008) Characterization of genetic structure of *Podophyllum hexandrum* populations, and endangered medicinal herb of Northwestern Himalaya, using ISSR-PCR markers and its relatedness with podophyllotoxin content. *African Journal of Biotechnology* 7, 1028–1040.

Anon (1970) *Podophyllum hexandrum*. The wealth of India: A Dictionary of Indian Raw Material and industrial products. In: *Raw Materials*. Vol. VIII: Ph.Re. Publication and Information Directorate. CSIR, New Delhi, pp. 170–174.

Badhwar, R.L. and Sharma, B.K. (1963) A note on germination of *Podophyllum* seeds. *Indian Forester* 89, 445–447.

Bhadula, S.K., Singh, A., Lata, H., Kunyal, C.P. and Purohit, A.N. (1996) Genetic resources of *Podophyllum hexandrum* Royle, an endangered medicinal species from Garhwal Himalaya, India. *International Plant Genetic Resource Newesletter* 106, 26–29.

Botstein, D., White, R.L., Skolnick, M. and Davis, R.W. (1980) Construction of a genetic linkage map in man using restriction fragment length polymorphisms. *American Journal of Human Genetics* 32, 314–331.

Broomhead, A.J. and Dewick, P.M. (1990) Tumor inhibitory aryltralin lignans in *Podophyllum versipelle, Diphyllcia cymosa* and *Diphyllicia grayi. Phytochemistry* 29, 3831–3837.

Brown, A.H.D. (1978) Isozymes, plant population genetic structure and genetic conservation. *Theoretical Applied Genetics* 52, 145–157.

Bucci, G. and Menozzi, P. (1995) Genetic variation of RAPD markers in a *Picea abies* Karst population. *Heredity* 75, 188–197.

Escaravage, N., Questianu, S., Pornon, A., Doche, B. and Taberlet, P. (1998) Clonal diversity in a *Rhododendron ferrugineum* L. (Ericaceae) population inferred from AFLP markers. *Modern Ecology* 7, 975–982.

Fay, D.A. and Ziegler, H.W. (1985) Botanical source differentiation of *Podophyllum* resin by high performance liquid chromatography. *Journal of Liquid Chromatography* 8, 1501–1506.

Frankel, O.H. (1983) The place of management in conservation. In: Schonewald-Cox, C.M., Chambers, S.M., MacBryde, B. and Thomas, L. (eds) *Genetics and Conservation: A Reference Manual for Managing Wild Animals and Plant Populations*. Benjamin/Cummings, Menlo Park, CA, USA, pp. 1–14.

Gaiotto, F.A., Bramucii, M. and Grattapaglia, D. (1997) Estimation of outcrossing rate in a breeding population of *Eucalyptus urophylla* with dominant RAPD and AFLP markers. *Theoretical and Applied Genetics* 95, 842–849.

Godwin, I.D., Aitken, E.A.B. and Smith, L.W. (1997) Application of inter simple sequence repeat (ISSR) markers to plant genetics. *Electrophoresis* 18, 1524–1528.

Imbert, T.F. (1998) Discovery of podophyllotoxins. *Biochimistry* 80, 207–222.

Inglis, P.W., Teixeira, E.A., Ribeiro, D.M., Valadares-Inglis, M.C., Tigano, M.S. and Mello, S.C. (2001) Molecular markers for the characterization of Brazilian *Cercospora caricis* isolates. *Current Microbiology* 42, 194–198.

Isabel, N., Beaulieu, J. and Bousquet, J. (1995) Complete congruence between gene diversity estimates derived from genotypic data at enzyme and RAPD loci in black spruce. *Proceedings of the National Academy of Sciences USA* 92, 6369–6373.

Jackson, D.E. and Dewick, P.M. (1984) Aryltetralin lignans from *Podophyllum hexandrum* and *Podophyllum peltatum* (isolated from the roots). *Phytochemistry* 23, 1147–1152.

Jones, C.J., Edwards, K.J., Castaglione, S., Winfield, M.O., Sala, F., van de Wiel, C., Bredemeijer, G., Vosman, B., Mathews, M., Daly, A., Brettschneider, R., Bettini, P., Buiatti, M., Maestri, E., Malcevschi, A., Marmiroli, N., Aert, R., Volckaert, G., Rueda, J., Linacero, R., Vazquez, A. and Carp, A. (1997) Reproducibility testing of RAPD, AFLP and SSR markers in plants by network of European laboratories. *Molecular Breeding* 3, 381–390.

Kreader, C., Weber, S. and Song, K. (2001) One-tube preparation and PCR amplifications of DNA from plant leaf tissue. *Life Science* (Newsletter from Sigma, USA) 2:™Extract-N-Amp 16–18.

Lakshmikumaran, M. and Bhatia, S. (1998) DNA fingerprinting of medicinal plants. In: Debry, B. (ed.) *Intellectual Property Rights*. Rajiv Gandhi Institute for Contemporary studies. B.R. Publishing Corporation, Delhi, pp. 293–331.

Lu, J.J., Knox, M.R., Ambrose, M.J., Brown, J.K.M. and Ellis, T.H.N. (1996) Comparative analysis of genetic diversity in pea assessed by RFLP and PCR based methods. *Theoretical Applied Genetics* 93, 1103–1111.

Morgante, M. and Olivieri, A.M. (1993) PCR-amplified microsatelites as markers in plant genetics. *Plant Journal* 3, 175–182.

Mullis, K.B. and Faloona, F.A. (1987) Specific synthesis of DNA *in vitro* via a polymerase-catalyzed chain reaction. *Methods in Enzymology* 155, 335–350.

Muralidharan, K. and Wakeland, E.K. (1993) Concentration of primer and template qualitatively affects products in random amplified polymorphic DNA PCR. *Biotechniques* 14, 362–364.

Nesbitt, K.A., Potts, B.M.,Vaillancourt, R.E., West, A.K. and Reid, J.B. (1995) Partitioning and distribution of RAPD variation in a forest tree species, *Eucalyptus globulus* (Myrtaceae). *Heredity* 74, 628–637.

Remington, D.L., Whetten, R.W. and O'Malley, M. (1988) Genetic map construction with AFLP markers in loblolly pine. Abstract, Plant and Animal Genome VI Conference, San Diego, January 18–22, 1998. (http://probe.nalusda.gov:8300/pag/6/review/640.html).

Schierenbeck, K.A., Skupski, M., Lieberman, D. and Lieberman, M. (1997) Population structure and genetic diversity in four tropical trees in Costa Rica. *Molecular Ecology* 6, 137–144.

Singh, A., Negi, M.S., Rajagopal, J., Bhatia, S., Tomar, U.K., Srivastava, P.S. and Lakshmikumaran, M. (1999) Assessment of genetic diversity in *Azadirachta indica* using AFLP markers. *Theoretical Applied Genetics* 98.

Skroch, P. and Nienhuis, J. (1995) Impact of scoring error and reproducibility of RAPD data on RAPD-based estimates of genetic distance. *Theoretical Applied Genetics* 91, 1086–1091.

Stahelin, H.F. and Wartburg, A.V. (1991) The chemical and biological route from podophyllotoxin glucoside to etoposide: Ninth Cain Memorial Award Lecture. *Cancer Research* 51, 5–15.

Tyler, V.E., Brady, L.R. and Robbers, J.E. (1988) *Pharmacology* 9th ed. Lea and Febiger, Philadelphia, USA.

Vos, P.R., Hogers, M., Bleeker, M., van de Lee Reijans, T., Hornes, M., Fritjers, A., Pot, J., Peleman, J., Kuiper, M. and Zabeau, M. (1995) AFLP: a new concept for DNA fingerprinting. *Nucleic Acids Research* 23, 4407–4414.

Williams, J.G.K., Kubelik, A.R., Livak, K.J., Rafalski, J.A. and Tingey, S.V. (1990) DNA Polymorphisms amplified by arbitrary primers and useful as genetic markers. *Nucleic Acids Research* 18, 6531–6535

Zietkiewicz, E., Rafalski, A. and Labuda, D. (1994) Genome fingerprinting by simple sequence repeat (SSR)- anchored polymerase chain reaction amplification. *Genomics* 20, 176–183.

Chapter 17

Plant Virus Vector Systems for the Production and Delivery of Biopharmaceuticals

Rofina Yasmin Othman and Noor Hasima Nagoor

Introduction

The development of plant virus vectors as *in planta* expression systems for foreign genes is now considered as an attractive alternative biotechnological approach for peptide expression with biopharmaceutical applications. This approach has been exploited for the development of vaccine production systems where small foreign peptides are expressed as a fusion with viral coat proteins to be displayed on the surface of the particle (Marusic *et al.*, 2001; Nichols *et al.*, 2002; Liu *et al.*, 2005; Ooi *et al.*, 2006; Nuzacci *et al.*, 2007; Denis *et al.*, 2008; Teoh *et al.*, 2009). This review looks at two aspects in the use of this system; on factors involved in the construction and on the suitability of the system as immune vehicles for vaccines.

Advantages of Plant Viruses as Vectors

Plant viruses have distinct advantages over other expression or delivery systems for biopharmaceuticals including as vaccine candidates. Their small genomes and inert structures make them ideal structures for genetic modification. The well characterized structure of many viruses including the availability of 3D models of their coat protein subunits and structures means that *in silico* design and predictive tests for stability of new recombinant constructs can be easily carried out and subsequently built. The epitopes of choice can therefore be conveniently displayed on the surface of the virus particle which is then propagated in plants and consequently isolated and used as antigen presenting vehicles. Information on virus propagation in host plants is also widely available and the yield of particles produced for many viruses makes them a potentially economical production system. Although plants themselves have been developed to directly express pharmaceutical proteins the virus system overcomes its major limitation which is the relatively low level of expression that is normally obtained, ranging from 0.01% to 2% of the total soluble protein. In applications requiring purified protein, high-level expression provided by plant virus replication systems would greatly facilitate the purification process from the crude plant extracts and would increase the cost efficiency of the system.

Additionally the use of plant viruses and the appropriate plant hosts provide a relatively safer system for production of human pharmaceuticals compared to the use of mammalian or bacterial cell systems with much less fear of the possibility of contaminating adventitious agents.

Factors in the Construction of Chimeric Plant Vectors

The success of the epitope presentation strategy depends on a detailed knowledge of virus structure at the molecular level. A systematic assessment of the different elements in vector construction has to be taken into account in designing the plant virus vector construct. This includes considerations on the site of insertion, the limitations of the size of inserts (the gene of interest) and the inclusion of modular elements for selection or enhancements. Essentially, an insertion site has to be determined in the virus genome so that the resulting peptide will be displayed on the surface of the virus particle. Modifications that do not interfere with the normal functions of the particular virus are a pre-requisite for this peptide fusion approach. Chen *et al.* (2007) looked at five different insertion sites in testing for stability of expression of proteins in Turnip mosaic virus. Many approaches and modifications in the cloning strategies have been taken. One common strategy suggests that foreign gene segments could to be fused to the terminus of a viral gene in a way that permits the production of both the fusion product and the native viral protein thus avoiding interference with normal gene functions (Sugiyama *et al.*, 1995; Ooi *et al.*, 2006).

Choice of construct

The success of the early experiments opened opportunities for the exploitation of many different viruses with varying degrees of success. Not all plant viruses however are suitable for this strategy. All plant viruses are pathogens and many can cause severe detrimental effects on their host plant. These viruses would not be suitable for use as expression vectors. If the plant were eventually to be grown in open fields it may not be acceptable taking into account the biosafety considerations of the virus if it was not attenuated and could be still transmitted by natural vectors such as insects. Hence viruses which are ideal would preferably be those that naturally are asymptomatic or attenuated unless they were to be grown only in contained facilities. Most of the viruses used as expression vectors are those that infect dicotyledenous plants while only a few such as Wheat streak virus have been developed for monocots (Choi *et al.*, 2000). Among the viruses which have successfully been investigated and exploited to different levels for this approach are Alfalfa mosaic virus (Brodzik *et al.*, 2005), Apple latent spherical virus (Igarashi *et al.*, 2009), Bean pod mottle virus (Zhang and Ghabrial, 2006), Broad bean mottle bromovirus (Sandoval *et al.*, 2008), Citrus tirsteza virus (Folimonov *et al.*, 2007), Clover yellow bean vein virus (Masuta *et al.*, 2000), Cucumber green mottle mosaic virus (Ooi *et al.*, 2006) , Cymbidium ringspot virus, Cucumber mosaic virus (Nazucci *et al.*, 2007), Grapevine virus A, Lettuce infectious yellows virus (Yeh *et al.*, 2001), Papaya mosaic virus (Denis *et al.*, 2007), Potato virus x (Marconi *et al.*, 2006), Soybean mosaic virus (Seo *et al.*, 2009), Tobacco etch virus, Tobacco rattle virus (Ratcliff *et al.*, 2001), Tomato bushy stunt virus, tomato mosaic virus, Turnip mosaic virus (Chen *et al.*, 2007), and zucchini yellow mottle virus (Arazi *et al.*, 2002).

Not all of these viruses were used to express vaccines or biopharmaceuticals. ZYMV for example (Chen *et al.*, 2005) was used to produce proteins from other tosposvirus as antigens for the purpose of raising antibodies for use in diagnostic kits and Apple latent spherical virus was developed specifically as a gene silencing vector in plants (Igarashi *et al.*, 2009).

Factors affecting expression levels

One of the main issues in development of chimeric constructs using plant viruses remains the need to express high levels of the protein. The upper limits reported for expression may among other factors be limited by the natural host and life cycle of the respective virus. Vectors developed for monocot host for instance have generally lower levels of accumulation in their normal hosts. This makes vectors such as Tobacco mosaic virus (TMV) and Tobacco etch virus (TEV) which naturally accumulate to high levels naturally in their host cells ideal when high expression is required. They are also propagated in easy to grow hosts which make them very viable for up-scaling. Brodzik *et al.* (2005) report that high yield and consistency was obtained for the production of recombinant AlMV particles carrying a 15-aa epitope PA-D4s from *Bacillus anthracis* cloned into the +25 aa position from the N-terminus of the virus CP. Recombinant plant virus particles were abundantly present in plant tissues for extended periods of time and were easily harvested and purified.

Additionally modifications can be introduced to further increase the efficiency of these virus vectors. One modification that could be used as an illustration is the introduction of amber stop codons which allow for read through expression of fusion proteins. It has been well established that during nonsense suppression, a nonsense codon (stop codons) can be recognized by certain species of amino acyl-tRNAs (nonsense suppressors) hence allowing the translation of read-through proteins (Beier and Grimm, 2001). Since the discovery of a naturally occurring leaky amber codon within the TMV genome, *in vitro* translation of the TMV genome followed by sequencing of the resulting fusion protein have been conveniently used as a system to screen for naturally occurring amber suppressors. The leaky amber codon of TMV was later used to control the expression of fusion coat protein in chimeric virus system to increase the viability of the resulting chimeric virus (Skuzeski *et al.*, 1991; Sugiyama *et al.*, 1995). However, this type of application was considered impractical since the read-through efficiency was only around 5–10% (Skuzeski *et al.*, 1991; Sugiyama *et al.*, 1995). Ooi *et al.*, (2006) reportedly achieved a significantly higher level of read-through efficiency in a Cucumber green mottle mosaic virus (CGMMV) muskmelon system in which chimeric CGMMV was developed to randomly display the ‘a’ determinant of Hepatitis B surface antigen (HBsAg). In this system the read-through efficiency was significantly higher than that previously reported for tobamoviruses. Mor *et al.* (2002) developed a high-level expression system that utilizes elements of the replication machinery of the single-stranded DNA virus Bean yellow dwarf virus. The replication initiator protein (Rep) mediates release and replication of a replicon from a DNA construct (‘LSL vector’) that contains an expression cassette for a gene of interest flanked by cis-acting elements of the virus. Interestingly it has also been suggested that modifications which cause alleviation of disease symptoms in plants allow better growth and biomass production, which may improve the yields of heterologous proteins. The symptoms caused by the TuMV and ClYVV vectors carrying two heterologous genes were alleviated as compared to the insert-less vector viruses (Masuta *et al.*, 2000; Beauchemin *et al.*, 2005).

Factors affecting stability

Due to their relatively high rate of mutation during replication, RNA viruses are evolving rapidly and this is the basis of their ubiquity and also their adaptability. In the study by Teoh *et al.* (2009) it was shown that the EB4 Dengue virus gene sequence carried by the chimeric CGMMV was systemically removed during the infection process. The order of the removal of the transgene was speculated to be the 5′ to 3′ direction. This report shows the temporal in-host truncation of the transgene from a chimeric virus in a natural host. Other reports have shown truncation occurring in transgenic plants expressing the same or similar transgenes as the chimeric virus (Chung *et al.*, 2007) suggesting targeting by a resistance mechanism or competition with the parental virus as the mechanism involved. The exact mechanism of truncation of the transgene in Teoh's study is less clear as a previous study using the same vector and host with a different transgene did not exhibit the same instability (Ooi *et al.*, 2006). The larger size of the EB4 peptide in comparison to the Hepatitis B epitope, the epitope first used in CGMMV (Ooi *et al.*, 2006) suggests that the truncation mechanism or transgene recognition by the virus was size dependent.

Another approach taken in developing stable constructs is in the use of defective interfering (DI) particles instead of the wild type virus since DI-RNAs are not required for RNAs. Therefore, these molecules can be modified with less restriction than the genomic sequences and are suggested to be more stable. Many experiments have been performed during the last several years to evaluate the use of DI-RNAs as gene expression systems *in planta* (Liao *et al.*, 1995; Yeh *et al.*, 2001). The results have shown a great potential for such applications. The studies showed that BBMV DI-RNAs offer great advantages, such as the wide host range of the helper virus, the ease of mechanical infection, and high levels of virus accumulation in legume hosts.

Teoh *et al.* (2009) also provide information on a rational harvesting timeline for their chimeric virus making this system exploitable for implementation in a plantation scale in the future. It was suggested that once host plants are infected with the chimeric virus carrying the inserted foreign peptides, the optimum harvesting time would be at around 14 days post-inoculation (dpi) or not more than 20 dpi in order to obtain maximum yield of the full-length transgene. Growth of the infected plants for longer periods to obtain higher yields of the chimeric virus may induce unwanted transgene deletions.

Plant Virus Vectors as Vaccine Carrier Candidates

Immune mechanisms relevant to vaccine development

A state of immunity can be induced by passive or active immunization. Each type of immunity is brought about according to a desired situation. Short-term or transient passive immunization is induced by transfer of preformed neutralizing antibodies. Infection or active immunization achieves long-term active immunization with the generation of not only effector cells but long-lived memory cells. Active immunizations with vaccines have been developed for many diseases. There are two basic principles in vaccinology. One, the antigen or specifically the immunogen which can stimulate the adaptive immune system and the second, an adjuvant which is required to induce a certain level of inflammation associated with the innate immune system. Therefore a successful vaccine has to be a combination of immunogen and adjuvant to link adaptive and innate immune defence mechanisms.

A successful vaccine also has to be able to activate both arms of the adaptive immune system, that is, the humoral and cellular-mediated immunity. This is an important strategy as the various components of these two immune systems require collaboration from one another to achieve maximum protection against the vaccinated disease. An ideal vaccine should include both T and B cell epitopes for both intracellular and extracellular pathogens respectively. For a vaccine to have the ability to induce a cellular-mediated or T cell response, it would require recognition by the major histocompatibility complex molecules (MHC). This is because the T cell can only see the antigenic epitope when it is displayed in the groove of an MHC molecule. The antigen is processed and presented by an antigen presenting cell (APC) which expresses the MHC molecules. The most prominent APC is the dendritic cells therefore to be effective; a vaccine must target dendritic cells. However, antigen-specific B cells are the most potent APC for inducing secondary antibody responses. Thus, booster vaccinations must target antigen-specific B cells. Antigen presentation via antigen-processing pathways with class I and class II MHC molecules has been maximized with the use of adjuvants in vaccine production. Adjuvants also have been shown to inhibit the immunosuppressive T regulatory cells. Common examples of adjuvants used in vaccines are TLR (toll-like receptor) ligands which bind to the TLRs on the dendritic cells. In humoral or B cell reponses, the type of immunoglobulin generated is important with respect to the types of pathogens. The route of immunization plays a vital role in generating either an IgG or IgA response. Extracellular and intracellular pathogens would require vaccines that can generate the IgG antibodies whilst mucosal pathogens would require generation of IgA antibodies. The route of immunization for IgG antibodies is normally via subcutaneous injections and for the IgA antibodies is given orally or intranasally.

The current effective types of vaccines are either of the inactivated or live attenuated forms. There are advantages and disadvantages to these forms of vaccines. The main disadvantage of the attenuated vaccines is their ability to revert to virulent forms, and that of the inactivated vaccine is that it can mainly induce the humoral response. The World Health Organisation (WHO) has identified six main properties of an ideal vaccine. They are: affordable worldwide; heat stable; effective after a single dose; applicable to a number of diseases; administered by a mucosal routes; and suitable for administration early in life.

Therefore, active investigations are carried out to develop vaccines to maximize the immune response to selected immunogenic epitopes and to meet the above requirements of an ideal WHO vaccine.

Plant virus vector systems as vaccine vehicles

Plants and plant viruses have recently been proven in animal and preclinical studies as reliable systems for expressing and delivering antigenic proteins or peptides as immunogens to be used for the development of new vaccination strategies. It has become an attractive alternative because when compared to current vaccines such as the conventional live-recombinant, live-attenuated (avirulent), or inactivated (killed) pathogen vaccines, plant viruses are accepted to be nonpathogenic in humans and other animals. Therefore it is regarded as a much safer vaccine vehicle.

Plant virus coat proteins (CP) have been illustrated as carriers to present immunogenic peptides to the immune system. When properly fused at target positions on the capsid proteins, the recombinant viral CP self-assembles and generates chimeric virus particles (CVPs) that display the foreign sequence on their outer surfaces. This 'epitope-displaying' strategy using plant virus CP as carriers for both viral and bacterial antigens has been successfully tested experimentally in animal models. When immunized subcutaneously, the responses showed high titres of specific IgG that displayed the effector function of pathogen opsonization for effective phagocytosis by human neutrophils.

Importantly, the antibodies generated were also able to provide protection when challenged with different immunotype strains. The challenged mice had fewer severe lesions and fewer pathogens compared to mice immunized with wild-type virus (Brennan *et al.*, 1999). There are many evidences on the success of plant virus vector systems as vaccine vehicles. Some of the more common ones are TMV, Potato virus X (PVX), Alfalfa mosaic virus (AMV) and Cowpea mosaic virus (CPMV), each of them with different acceptance ability of peptide lengths. Although the TMV coat proteins can only accept peptides less than 25 amino acids in length, as compared to viruses with flexible coat morphology (spherical, ellipsoid, and bacilliform) such as the AMV which allowed fusions of at least 37 amino acids (Yusibov *et al.*, 1997), it has been successfully developed as a vaccine vehicle for more than 25 diseases with most being in animals (mouse, rat, rabbit, pig) and a few in human. Important examples of serum and mucosal immune responses detected in human with a plant vaccine vehicle are with the hepatitis B surface antigen (HBsAg), enterotoxin B subunit of *E. coli* (LTB), Norwalk virus capsid protein and *Streptococcus mutans* adhesion protein antibody.

Tacket and Mason (1999) performed the first human clinical trial with an oral plant-vaccine thus suggesting that plant based systems were suitable for developing human vaccines. Raw potatoes producing LTB were offered to volunteers at an equivalent dose used for oral immunization against the structural and immunological similar purified cholera B subunit (CTB). Tacket *et al.* (2000) then fed volunteers either 50 or 100 g of potatoes containing ~0.75 mg of LTB per dose at a total of three doses on days 0, 7 and 21. Ten out of the eleven volunteers that ate the transformed tuber tissues showed at least a fourfold increase in serum IgG antibodies against LTB. A four-fold mucosal (IgA) immune response was also seen in six out of the eleven volunteers. Anti-LTB antibodies remained elevated in blood when measured 59 days after ingestion of the first dose. Similar effective doses were given in human trials using Hepatitis B surface antigen in lettuce (200 and 150 μg doses, containing 0.1 to 0.5 μg HBsAg per 100 g raw potato) and Norwalk virus capsid protein (NVCP) in potato (150 g dose, containing 215 to 751 μg NVCP) (Tacket *et al.*, 2000). These results have encouraged more researchers to produce vaccines in plants and subsequently to explore vector carriers such as plant viruses as the expression system in plants thus circumventing the need for developing transgenic plants.

In animal studies TMV was similarly proven as an effective vaccine delivery vehicle for parenteral and mucosal immunization and for protection from challenge with viral infection. Hybrids of TMV were constructed with the fusion to the coat protein peptide of murine hepatitis virus (MHV). The TMV hybrids were propagated in tobacco plants, and the virus particles were purified. Immunogold labelling, with the use of specific monoclonal antibody, showed specific expression of hybrid TMV particles, confirming the display of the MHV epitope on the surface of the TMV. Mice were immunized with purified hybrid viruses for several immunization regimens. Mice that received the TMV hybrids intranasally developed serum IgG and IgA specific for its epitope and for the TMV coat protein. When administered by subcutaneous injections, only high titres of serum IgG that was specific for the epitope and for coat protein was generated, and not specific IgA. Mice that were immunized with hybrid virus by subcutaneous or intranasal routes of administration survived challenge with a lethal dose ($10 \times LD_{50}$) of MHV, whereas control mice administered with wild-type TMV died 10 days post challenge. These results show a positive correlation between the dose of administered immunogen and protection against MHV infection (Koo *et al.*, 1999).

Modelska *et al.* (1998), in their study demonstrated that recombinant plant virus particles containing a chimeric peptide representing two rabies virus epitopes were able to stimulate the synthesis of viral neutralizing antibody in immunized mice. The mice were immunized intraperitoneally or orally (by gastric inturbation or by feeding on virus-infected spinach leaves) with engineered plant virus particles containing rabies antigen was able to mount a local and systemic immune response. After the third dose of antigen, given intraperitoneally, 40% of the mice were protected against challenge infection with a lethal dose of rabies virus, while the oral administration of the antigen stimulated the synthesis of both serum IgG and IgA and showed protection against an intranasal infection with an attenuated rabies virus strain.

Using malaria as a model disease, Turpen *et al.* (1995) utilized the surface of TMV for presentation of selected epitopes to the mammalian immune system. Selected B-cell epitopes were either inserted into the surface loop region of the TMV coat protein or fused to the C terminus using the leaky stop signal derived from the replicase protein reading frame. It was observed that the tobacco plants were systemically infected with each of these constructs and contained high titres of genetically stable recombinant virus. This enabled purification of the chimeric particles in high yields. Symptoms induced in tobacco ranged from a normal mosaic pattern similar to that induced by the parental strain to a unique bright yellow mosaic. When measured by quantitative ELISA against synthetic peptide standards, wild type TMV coat protein and fusion protein synthesized by the leaky stop mechanism co-assembled into virus particles at the predicted ratio of approximately 20:1. This study proved the potential of recombinant plant viruses to meet the requirement for scalable and cost effective production of subunit vaccines that can be easily stored and administered.

Another WHO ideal vaccine requirement is a multivalent subunit vaccine that includes multiple antigen epitopes for protection against different serotypes of a pathogenic species. This ensures better protection from pathogens such as influenza virus, *Salmonella*, *Streptococcus pneumonia* and *Neisseria gonorrhoeae* that regularly undergo subtle genetic changes in their antibody recognition sites to avoid effective immune response during subsequent infections. Denis *et al.* (2008) recently demonstrated the potential of using a Papaya mosaic virus (PapMV) platform carrying the universal M2e influenza epitope (PapMV-CP-M2e) as a candidate flu vaccine. The study showed that PapMV-CP-M2e virus-like particles (VLPs) could induce production in mice of anti-M2e antibodies that can recognize influenza-infected cells. PapMVCP-M2e discs made of 20 coat protein (CP) subunits were shown to be poorly immunogenic compared to PapMV-CP-M2e VLPs composed of several hundred CP subunits. It was also showed that addition of either alum or PapMV-CP VLPs as adjuvant dramatically increased the immunogenicity of the PapMV-CP-M2e-containing vaccine, and led to 100% protection against a challenge of 4LD50 with the WSN/33 strain. These results show, for the first time, the potential of a recombinant plant virus protein to serve as both a peptide delivery system and adjuvant in the crucial field of influenza vaccine development.

There are also efforts to construct a multicomponent vaccine that carries a combination of antigens from several different species of pathogens produced in a single plant to allow the delivery of multiple vaccinations in a single dose. Yu and Langridge (2001) developed potato plants that synthesized cholera, rotavirus and enterotoxigenic *E. coli* antigens as a potential multi-component plant-based vaccine. Their results were most encouraging as oral immunization of mice with five weekly feedings of the transgenic tubers generated significant serum and intestinal antibody formation against all three antigens. Diarrhoea symptoms were reduced in severity and duration by more than 50% following pathogen challenge. Following on with this concept, Kelloniemi *et al.*, (2008) also showed the same possibility in a plant virus vector system. In their study three heterologous proteins were simultaneously expressed from a chimeric potyvirus Potato virus A (PVA) in *Nicotiana benthamiana*. The genes for green fluorescent protein of *Aequoria victoriae*, luciferase of *Renilla reniformis* and glucoronidase of *Escherichia coli* were inserted into the engineered cloning sites at the N-terminus of the P1 domain, the junction of P1 and helper component protein (HC-Pro), and the junction of the viral replicase (NIb) and coat protein (CP), respectively, in an infectious PVA cDNA. The proteins were expressed as part of the viral polyprotein and subsequently released by cleavage at the flanking proteolytic cleavage sites. The engineered viral genome was 39.2% larger than wildtype PVA and infected plants of *N. benthamiana* systemically. pGLUwas stable and expressed all three heterologous proteins, demonstrating the potential of the system for development of multi-component vaccines.

Clearly plant vector virus systems show great potential for the production of both IgG and IgA antibodies as well as for developing protection against recurrent infections. However, there are obstacles to ensure their safety and acceptance, a priority being the formulation and dosage to prevent the induction of tolerance. Other concerns include allergic responses of individuals to certain plants. It might be just a matter of years before all these concerns are addressed and we would be able to benefit from these systems in ensuring protection against dangerous and infectious diseases.

References

Arazi, T., Huang, P.L., Huang, P.L., Li Zhang, Shiboleth, Y.M., Gal-On, A. and Lee-Huang, S. (2002) Production of antiviral and antitumor proteins MAP30 and GAP31 in cucurbits using the plant virus vector ZYMV-AGII. *Biochemical and Biophysical Research Communications* 292, 441–448.

Beauchemin, C., Bougie, V. and Laliberte, J.-F. (2005) Simultaneous production of two foreign proteins from a potyvirus-based vector. *Virus Research* 112, 1–8.

Beier, H. and Grimm, M. (2001) Misreading of termination codons in eukaryotes by natural nonsense suppressor tRNAs. *Nucleic Acids Res* 29, 4767–4782.

Brennan, F.R., Linda, B., Gilleland, J.S., Bendig, M.S., William, D.O., Hamilton and Harry E. Gilleland, Jr (1999) A chimeric plant virus vaccine protects mice against a bacterial infection. *Microbiology* 145, 2061–2067.

Brodzik, R., Bandurska, K., Deka, D., Golovkin, M. and Koprowski, H. (2005). Advances in alfalfa mosaic virus-mediated expression of anthrax antigen *in planta. Biochemical and Biophysical Research Comminication* 338, 717–722.

Chen, C.C., Tsung-Chi Chen, Joseph A.J. Raja, Chin-An Chang, Li-Wen Chen, Shin-Shun lin, Shyi-Dong Yeh (2007) Effectiveness and stability of heterologous proteins expressed in plants by turnip mosaic virus vector at five different insertion sites. *Virus Research* 130, 210–227.

Chen,T.-C., Hsu, H.-T., Jian, R.K., Huang, C.-W., Lin, C.-H., Liu, F.-L. and Yeh, S.D. (2005) Purification and serological analyses of tospoviral nucleocapsid proteins expressed by Zucchini yellow mosaic virus vector in squash. *Journal of Virological Methods* 129, 113–124.

Choi, I-R., Drake, C. Stenger, T., Morris, J. and French, R. (2000) A plant virus vector for systemic expression of foreign genes in cereals. *The Plant Journal* 23(4), 547–565.

Chung, B.-N., Canto, T., and Palukaitis, P. (2007) Stability of recombinant plant viruses containing genes of unrelated plant viruses. *Journal of General Virology* 88(4), 1347–1355.

Denis, J., Majeau, N., Acosta-Ramirez, E., Savard, C., Bedard, M.-C., Simard, S., Lecours, K., Bolduc, M., Pare, C., Willems, B., Shoikry, N., Tessier, P., Lacasse, P., Lamarre, A., Lapointe, R., Lopez Macias, C. and Leclerc, D. (2007) Immunogenicity of Papaya mosaic virus-like particles fused to a hepatitis C virus epitope: evidence for the critical function of multimerization. *Virology* 363, 59–68.

Denis, J., Acosta-Ramirez, E., Zhao, Y., Hamelin, M.-E., Kaukavica, I., Baz, M., Abed, Y., Savard, C., Pare, C. and Lopez Macias, C. (2008) Development of a universal influenza A vaccine based on the M2e peptide fused to the Papaya mosaic virus (PapMV) vaccine platform. *Vaccine* 26, 3395–3403.

Folimonov, A.S., Svetlana, Y., Folimonova, S., Bar-Joseph, M. and Dawson, W.O. (2007) A stable RNA virus-based vector for citrus trees. *Virology* 368, 205–216.

Igarashi, A., Yamagata, K., Sugai, T., Takahashi, Y., Sugawara, E., Tamura, A., Yaegashi, H., Yamagishi, N., Isogai, M., Takahashi, H., and Yoshikawa, N. (2009) Apple latent spherical virus vectors for reliable and effective virus-induced gene silencing among a broad range of plants including tobacco, tomato, *Arabidopsis thaliana*, cucurbits, and legumes. *Virology* 386, 407–416.

Kelloniemi, J., Makinenb, K. and Valkonen, J.P.T (2008) Three heterologous proteins simultaneously expressed from a chimeric potyvirus: Infectivity, stability and the correlation of genome and virion lengths. *Virus Research* 135, 282–291.

Koo, M., Bendahmane, M., Lettieri, G.A., Paoletti, A.D., Lane, T.E., Fitchen, J., Buchmeier, M.J. and Beachy, R. (1999) Protective immunity against murine hepatitis virus (MHV) induced by intranasal or subcutaneous administration of hybrids of tobacco mosaic virus that carries an MHV epitope. *Proceedings of the National Academy of Sciences USA* 96, 7774–7779.

Liao, C.L., Zhang, X. and Lai, M.M.C. (1995) Coronavirus defective interfering RNA as an expression vector the generation of a pseudo-recombinant mouse hepatitis virus expressing hemagglutinin-esterase. *Virology* 208, 319–327.

Liu, Li, M., Canizares, C., Monger, W., Perrin, Y., Tsakiris, E., Porta, C., Shariat, N., Nicholson, L. and Lomonossoff, G.P. (2005) Cowpea mosaic virus-based systems for the production of antigens and antibodies in plants. *Vaccine* 23, 1788–1792.

Marconi, G., Albertini, E., Barone, P., De Marchis, F., Lico, C. Marusic, C., Rutili, M., Veronesi, F. and Porceddu, F. (2006) *In planta* production of two /peptides of the classical swine fever virus (CSFV) E2 glycoprotein fused to the coat protein of potato virus X. *BMC Biotechnology* 6(29), 1–9.

Marusic, C., Rizza, P., Lattanzi, L., Mancini, C., Spada, M., Belardelli, F., Benvenuto, E. and Capone, I. (2001) Chimeric plant virus particles as immunogens for inducing murine and human immune responses against human immunodeficiency virus type 1. *Journal of Virology* 75(18), 8434–8439.

Masuta, C., Yamana, T., Tacahashi, Y., Uyeda, I., Sato, M., Ueda, M., and Matsumura, T. (2000) Development of clover yellow vein virus as an efficient, stable gene-expression system for legume species. *The Plant Journal* 23(4), 539–546.

Modelska, A., Dietzschold, B N., Zhen, F.W., Fu, W., Steplewski, K., Hooper, D.C., Koprowski, H. and Yusibov, V. (1998) Immunization against rabies with plant-derived antigen. *Proceedings of the National Academy of Sciences USA* 95, 2481–2485.

Mor, T.S., Moon,Y.-S., Palmer, K.E. and Mason, H. (2002) Geminivirus vectors for high-level expression of foreign proteins in plant cells. *Biotechnology and Bioengineering* 81(4), 430–437.

Nichols, M.E.K., Stanislaus, T., Keshavarz-Moore, E. and Young, H.A. (2002) Disruption of leaves and initial extraction of wildtype CPMV virus as a basis for producing vaccines from plants. *Journal of Biotechnology* 92, 229–235.

Nuzzaci, M., Piazzolla, G., Vitti, A., Lapelosa, M., Tortorella, C., Stella, I., Natilla, A., Antonaci, S. and Piazzolla, P. (2007) Cucumber mosaic virus as a presentation system for a double hepatitis C virus-derived epitope. *Archives of Virology* 152, 915–928.

Ooi, A.S., Tan, S.H., Mohamed, R., Noorsaadah, A.R. and Othman, R.Y. (2006) The full-length clone of Cucumber green mottle mosaic virus and its application as an expression system for hepatitis B surface antigen. *Journal of Biotechnology* 121, 471–481.

Ratcliff, F., Martin-Hernandez, A.M. and Baulcombe, D. (2001) Tobacco rattle virus as a vector for analysis of gene function by silencing. *The Plant Journal* 25(2), 237–245.

Sandoval, C., Pogany, J., Bujarski, J. and Romero, J. (2008) Use of a defective RNA of Broad bean mottle bromovirus for stable gene expression in legumes. *Archives of Virology* 153, 1755–1758.

Seo, J.-K, Lee, H.-G. and Kin, K.-H. (2009). Systemic gene delivery into soybean by simple rub-inoculation with plasmid DNA of a soybean mosaic virus-based vector. *Archives of Virology* 154, 87–99.

Skuzeski, J.M., Nichols, L.M., Gesteland, R.F. and Atkins, J.F. (1991) The signal for a leaky UAG stopcodon in several plant viruses includes the two downstream codons. *Journal of Molecular Biology* 218, 365–373.

Sugiyama, Y., Hamamoto, Takemoto, S. and Watanabe, Y. (1995) Systemic production of foreign peptides on the particle surface of tobacco mosaic virus. *FEB Letters* 359, 247–250.

Tacket, C.O. and Mason, H.S. (1999) A review of oral vaccination with transgenic vegetables. *Microbes and Infection* 1(10), 777–783.

Tacket, C.O., Mason, H.S., Losonsky, G., Estes, M.K., Levine, M.M. and Arntzen, C.J. (2000) Human immune responses to a novel norwalk virus vaccine delivered in transgenic potatoes. *The Journal of Infectious Diseases* 182, 302–305.

Teoh, P.G., Aik-Seng, Ooi., AbuBakar, S. and Othman, R.Y. (2009) Virus-specific read-through codon preference affects infectivity of chimeric Cucumber green mottle mosaic virus displaying a dengue virus epitope. *Journal of Biomedicine and Biotechnology* 2009, 8.

Turpen, T.H., Reini, S.J., Charoenvit, Y., Hoffman, S.L., Fallarme, V. and Grill, L.K. (1995) Malaria epitopes expressed on the surface of recombinant tobacco mosaic virus. *Nature Biotechnology* 13, 53–57.

Yeh, H.H., Tian, T., Medina, V. and Falk, B.W. (2001) Green fluorescent protein expression from recombinant lettuce infectious yellows virus-defective RNAs originating from RNA2. *Virology* 289, 54–62.

Yu, J. and Langridge, W.H.R. (2001) A plant-based multicomponent vaccine protects mice from enteric diseases, *Nature Biotechnology* 19, 548–552.

Yusibov, V., Modelska, A., Steplewski, K., Agadianyan, M., Weiner, D., Hooper, C., and Koprowski, H. (1997) Antigen produced in plants by infection with chimeric plant viruses immunize against rabies virus and HIV-1. *Proceedings of the National Academy of Sciences United States of America* 94, 5784–5788.

Zhang, C. and Ghabrial, S.A. (2006) Development of Bean pod mottle virus-based vectors for stable protein expression and sequence-specific virus-induced gene silencing in soybean. *Virology* 334, 401–411.

Chapter 18

Novel Medicinal Plants for the Production and Delivery of Vaccines

Simone Reichwein and H. Warzecha

Introduction

Throughout mankind's history, plants and plant products have been used for preventing and curing diseases. Archaeological and historical records document that the use of medicinal plants or plant derived extracts dates back several thousand years and could be found on all continents. For example the use of opium can be traced back to the famous Iliad of the Greek poet Homer, and Indian Ayurveda dates back at least to 400 BC. Although during the past 200 years traditional plant remedies have been replaced by chemical defined substances which were largely chemically synthesized, medicinal plants are still heavily utilized and also provide sources for the extraction of pharmaceutical substances too complicated to synthesize (Newman and Cragg, 2007). Exceptions from traditional medicines were protein therapeutics, which could only be obtained from their natural sources in limited amounts. Insulin is the most prominent example, as it had to be extracted from pig's pancreas (Banting *et al.*, 1991). The pancreas of 50 pigs was used to cover the annual insulin need of one patient, illustrating the huge efforts and the vast amount of resources required to obtain this special remedy. The bottleneck of natural resources could eventually be overcome by a technique first described by Cohen and Boyer in 1973 (Cohen *et al.*, 1973). Recombinant DNA technology, entering the stage more than 30 years ago, enabled a novel class of pharmaceutics to be produced and marketed, the so-called recombinant therapeutics, and human insulin (Goeddel *et al.*, 1979) was the first product entering the market in 1982. Biotechnology has provided a tool through which protein therapeutics for substitution therapy (insulin (Walsh, 2005), glucocerebrosidase (Friedman *et al.*, 1999)), immunotherapy (antibodies (Walsh, 2000)), and even vaccines (hepatitis B (Jilg, 1998), human papilloma virus (Inglis *et al.*, 2006)) can be produced independently from their origin in various host organisms. Today, mainly bacteria (Swartz, 2001), yeast (Gerngross, 2004; Daly and Hearn, 2005) and animal cells (Wurm, 2004) serve as production platforms for recombinant protein therapeutics. However, there has been significant effort and also success in developing new platforms for the cost-effective large-scale production of protein therapeutics. The ratio for novel production systems is manifold: (i) the integrity of the protein product is largely dependent on the production system but also on fermentation conditions, and many protein therapeutics do need special post translational modifications such as glycosylation, proteolytic cleavage and folding;

(ii) to produce large amounts of protein therapeutics at reasonable cost, the price for facilities, fermentation media and supply, as well as subsequent purification of the product need to be significantly reduced; and (iii) contamination of the product with human pathogens should be avoided. Therefore, the use of systems absent of any bacterial cell wall fragments, viral contaminations, as well as oncogenic DNA is preferable.

The emerging field of plant-based biotechnology has led to an enhanced interest in using plants for the production of therapeutic agents. Some of the anticipated advantages of plants over conventional production systems are their inexpensive cultivation, the possibility for rapid scale-up, a variety of post-translational modifications of proteins, their reduced risk of transferring human pathogen contamination, and last but not least the ease of storage and distribution (Table 18.1). The first plant-produced recombinant pharmaceutical protein, initially produced in transgenic tobacco and potato plants, was human serum albumin which is the most abundant protein in the human circulatory system (Sijmons *et al.*, 1990). Since then, a large number of reports have shown that numerous recombinant therapeutics can be efficiently produced in plants. A large variety of crop species has been utilized and several techniques have been developed to optimize structure and yield of the desired proteins. Blood products, growth factors, hormones, antibodies and human as well as veterinary vaccines (Table 18.2) have been successfully obtained from plants and could be tested for efficacy both in animals and humans (Twyman *et al.*, 2003). However, even though more than 130 recombinant peptide and protein pharmaceuticals have been approved by the FDA (US Food and Drug Administration) for human use to date, none of them is plant-derived. There is only one plant-based vaccine approved which is only for use in animals, though, and is not marketed so far (N.N., 2006). This shortfall of plants behind conventional production systems might be at least partly due to a premature technology, but mainly due to general concerns using genetically modified plants for the production of drugs and due to unsettled regulatory processes. However, the approach of plants producing vaccines is coming of age, and the aim of this chapter is to highlight some of the features of the technology, but also to stretch some of the hurdles and drawbacks.

Table 18.1. Advantages of plant-based production systems over conventional systems.

- low cost of raw material
- rapid scale-up
- higher eukaryote protein synthesis pathway: post-translational modifications, similar to animal cells with minor differences in glycosylation pattern
- expression in different intracellular compartments to enhance stability or expression level
- simplified storage of raw material
- reduced need for cold-chain
- reduced concerns over human pathogen contamination
- multiple transgenes may be expressed together/transgene stacking
- enables expression of vaccines in edible plants or plant parts
    - no purification or procession required
    - oral delivery
    - eliminates cost of syringes and needles
    - reduces need for medical assistance in administration
    - eliminate fear of vaccination via injection
    - eliminate concern over transmitting diseases by needle reuse

Vaccines, a Special Medicine

Vaccines are a special medicine regarding their specifications as well as their requirements for safety. As they are usually administered to healthy people, there is very little tolerance regarding side effects or even spreading vaccine transmitted diseases. When Edward Jenner in 1798 inoculated children with pus from pox-infected cows (Jenner, 1798) and Louis Pasteur at the end of the 19th century used an attenuated virus received from the spinal cord of rabbits to vaccinate a child against rabies (Pasteur, 1885), there was little concern about safety. Although physicians as well as patients were aware of these risks (Ballard, 1868), it was considered the less evil compared to the disease. Nowadays, with increasing knowledge about pathogens and production techniques, the risk of acquiring a disease from an incomplete inactivated vaccine is almost zero. Like at the advent of immunization most vaccines used today still comprise live attenuated or killed viruses and bacteria, grown in complex cells or organs (Ulmer *et al.*, 2006). Beside the enormous effort to create a safe formulation, with these techniques only such pathogens could be used for vaccine production which could be cultivated, clearly limiting the array of vaccine-preventable diseases. In the timeline of vaccine development, only very recently, in the 1980s, new strategies for the production of vaccines were developed with the emergence of modern molecular biotechnology, enabling the production of so called subunit vaccines. The active part of these vaccines are not complete pathogens but various proteins or peptides derived from bacteria, viruses and parasites, pathogenic to humans or animals and in general produced not by the pathogens themselves but recombinant using heterologous host organisms. This novel approach enabled the production of antigens irrespective of their origin and regardless of the ability of the pathogen to grow *in vitro*, hence expanding the number of diseases which could be targeted by vaccines.

Today, various expression systems are well established and commercially utilized, which range from prokaryotes such as *Escherichia coli* and *Bacillus subtilis to* eukaryotes such as yeast, fungi, mammalian and insect cell cultures, animals and plants (Yin *et al.*, 2007).

The heterologous expression of genes in bacteria is the most prominent way for obtaining large amounts of recombinant proteins since it is the simplest and most inexpensive method. The gram-negative bacterium *Escherichia coli* is the most widely used host as it is well-characterized, easy to transform and its cultivation is cost-effective (Swartz, 2001). For the production of the vaccine against Lyme disease, LYMErix (Steere *et al.*, 1998), *E. coli* was the expression system of choice. As the active ingredient, the outer surface protein A (OspA), is a bacterial lipoprotein, it could be efficiently produced and purified. However, the major disadvantage in using bacterial systems is that they usually do not perform post-translational modifications (PTM) of proteins which are often required for their function, stability and immunogenicity. The difficulty to express large proteins (> 50 kDa), the absence of glycosylation and proteolytic removal of signal peptides, an incorrect or absent disulfide linkage, and the accumulation of the desired proteins as inclusion bodies are the major problems of using bacterial systems as production platforms.

Another opportunity is the expression of heterologous genes in eukaryotic cells to gain soluble and bioactive proteins. The baker's yeast *Saccharomyces cerevisiae* is a valuable production host due to rapid growth and cost-effective culture conditions (Gerngross, 2004). Since yeasts are eukaryotic organisms, they have the capacity to perform posttranslational modifications, such as glycosylation or phosphorylation. Vaccines produced in yeast are the Hepatitis B surface antigen (*S. cerevisiae*) (Emini *et al.*, 1986) and the human papillomavirus capsid proteins L1 (*S. pombe*) (Inglis *et al.*, 2006).

Table 18.2. Examples of plant-produced antigens and vaccines (modified from Streatfield and Howard, 2003).

Pathogen / host	Antigen	Expression system	Expression level	Efficacy	References
Enterotoxigenic strains of *E. coli* / humans	Heatlabile toxin B subunit	Potato tuber	0.2% TSP[a] / 0.001% FW[b]	Immunogenic and protective by oral delivery to mice and humans	Haq *et al.*, 1995; Mason *et al.*, 1998; Tacket *et al.*, 1998; Lauterslager *et al.*, 2001
		Maize seed	10% TSP/ 0.1% FW	Immunogenic and protective by oral delivery to mice	Streatfield *et al.*, 2001, 2003
Hepatitis B virus / humans	Surface antigen	Tobacco leaf	0.007% TSP/ 0.0008% FW	Immunogenic by intraperitoneal delivery to mice	Mason *et al.*, 1992; Thanavala *et al.*, 1995
Hepatitis C virus / humans	Hypervariable region 1 of envelope protein 2 fused to cholera toxin B subunit	Tobacco leaf (tobacco mosaic virus)	0.04% TSP/ 0.005% FW	Immunogenic by nasal delivery to mice	Nemchinov *et al.*, 2000
Norwalk virus / humans	Capsid protein	Tobacco leaf	0.2% TSP/ 0.03% FW	Immunogenic by oral delivery to mice	Mason *et al.*, 1996
		Potato tuber	0.4% TSP/ 0.003% FW	Immunogenic by oral delivery to mice and humans	Mason *et al.*, 1996; Tacket *et al.*, 2000
Rabies virus / humans and animals	Peptides of glycoprotein and nucleoprotein	Tobacco and spinach leaf (alfalfa mosaic virus and tobacco mosaic virus	10% VPW[c] / 0.0005% FW	Immunogenic and protective by intraperitoneal or oral delivery to mice and immunogenic by oral delivery to humans	Modelska *et al.*, 1998; Yusibov *et al.*, 1997, 2002
***V. cholerae* / humans**	Cholera toxin B subunit	Tobacco leaf	4% TSP/ 0.5% FW	No published data	Daniell *et al.*, 2001
Human immunodeficiency virus type 1 / humans	Peptide of transmembrane protein gp41	Cowpea leaf (cowpea mosaic virus)	2% VPW / 0.002% FW	Immunogenic by subcutaneous, nasal or oral delivery to mice	Durrani *et al.*, 1998; McInerney *et al.*, 1999

Respiratory syncytial virus / humans	Fusion protein	Tomato fruit	0.003% FW	Immunogenic by oral delivery to mice	Sandhu *et al*., 2000
***Plasmodium falciparum* / humans**	Peptides of circumsporozoite protein	Tobacco leaf (tobacco mosaic virus)	0.003% FW	No published data	Turpen *et al*., 1995
Human papillomavirus type 16 / humans	E7 oncoprotein	Tobacco leaf (potato virus X)	0.0004% FW	Immunogenic and protective by subcutaneous delivery to mice	Franconi *et al*., 2002
Foot-and-mouth disease virus / farmed animals	Peptide of structural protein VP1 fused to ß-glucuronidases	Alfalfa leaf	0.004% TSP/ 0.0005% FW	Immunogenic and protective by intraperitoneal delivery to mice	Dus Santos *et al*., 2002
	Structural protein VP1	Tobacco leaf (tobacco mosaic virus)	0.02% FW	Immunogenic and protective by intraperitoneal delivery to mice	Wigdorovitz *et al*., 1999
Transmissible gastroenteritis virus / pigs	Glycoprotein S	Tobacco leaf	0.2% TSP/ 0.03% FW	Immunogenic by intraperitoneal delivery to pigs	Tuboly *et al*., 2000
***Pseudomonas aeruginosa* / humans**	Peptides of outer membrane protein F	Cowpea leaf (cowpea mosaic virus)	4% VPW / 0.005% FW	Immunogenic and protective by subcutaneous delivery to mice	Brennan *et al*., 1999a, b

[a] **TSP**: total soluble protein
[b] **FW**: fresh weight
[c] **VPW**: viral particle weight

However, glycosylation differs from that of mammalian cells and is presumably antigenic in humans (Liu, 1992). Mammalian expression systems, most commonly Chinese hamster ovary (CHO) cells, baby hamster kidney (BHK) cells and myeloma cell lines are the systems of choice to obtain authentically post-translational modified recombinant proteins of human origin (Wurm, 2004).

Many important pharmaceutical proteins are produced by using mammalian cell lines, but the cost of setting up and running mammalian cell cultures can be immense due to their complexity. Moreover, there is an ongoing debate regarding the presence of retroviruses or oncogenic DNA.

However, there are no examples of subunit vaccines produced in mammalian cell cultures, probably due to the sufficient cheaper production systems.

The Rationale for Using Plants

With regard to the immense need of vaccines worldwide and the inadequate immunization particularly of children in developing countries, vaccines are required to be produced and administered in an easy and inexpensive way. Theoretically, the production of vaccine candidates in plants would offer an alternative and cheap production system to any current commercial system. The following considerations build the ratio for favouring plants over conventional production systems: (i) plants have simple requirements regarding their growth and cultivation, needing only sunlight, soil, water, and some fertilizer; (ii) proteins will be built with all necessary post-translational modifications in plants, so plants are an adequate environment to produce correctly processed complex proteins; (iii) plant-derived vaccines reduce the risk of contamination with human pathogens and (iv) eliminate the need for a cold chain since most proteins are highly stable in unprocessed plant material. Last but not least, (v) plants allow the easy oral delivery of the vaccine, if the antigen is produced in edible plants or plant tissues, eliminating the need for expensive processing or purification. Additionally, in case of an edible vaccine, cost of syringes and needles could be eliminated and therefore, the need of medical assistance in administration could be reduced. This ultimately should reduce the overall cost of vaccines and make the concept of edible vaccines especially attractive for developing countries.

Host Plant Options

Other than for microbial, yeast or mammalian cell systems, there is no particular plant species for the production of recombinant proteins which is considered as the standard system. Mainly, plants are utilized which are easy to transform and cultivate, with short regeneration times yielding reasonable amounts of biomass. The great variety of types of plants and types of plant tissues used as factories for the production of recombinant vaccines includes:

- cereals (rice, wheat, maize),
- legumes (pea, soybean, lucerne),
- fruit and root crops (tomato, potato, lettuce, spinach) with tobacco being the most prominent among them,
- aquatic weeds (duckweed, *Lemna* sp.),
- root vegetables (carrot),

- single-cell cultures of the algae *Chlorella* and *Chlamydomonas*,
- suspension cell cultures of a variety of plants.

One of the first crucial points which have to be addressed prior to producing therapeutic proteins and vaccines in plants is the choice of the appropriate plant host species. While tobacco is frequently used as a model system, there may be significant hurdles due to its alkaloid content in terms of developing an edible vaccine. The expression host choice may be influenced by expertise and experience as well as by recombinant protein accumulation levels and production costs, particularly with regard to further processing or purification, and regulatory factors.

In contrast to plants, single-cell cultures thereof offer the advantage of a high degree of containment and the possibility of producing recombinant proteins under Good Manufacturing Practice (GMP) conditions, as it is routinely the case with conventional microbial or mammalian fermentation (Hellwig *et al*., 2004). The advantages offered of plant single-cell culture over mammalian or yeast cell culture are mainly the usually simpler culture media and culture conditions. Potential advantages in terms of accumulation levels or PTMs like glycosylation have to be evaluated for each system. For the enzyme glucocerebrosidase, a substitute for patients with the lysosomal storage disorder Gaucher's disease, the Israel-based Biotech company Protalix has shown a benefit of using plant cells over conventional mammalian cells. When the enzyme is produced in carrot cell cultures, it already adopts the functional form with terminal mannose residues on the glycan structures (Shaaltiel *et al*., 2007). In CHO cells the glycan structure is different and needs to be modified in laborious subsequent production steps (Friedman *et al*., 1999). Hence, this is the first example that a plant cell system could be superior to a traditional production system.

Plant Transformation Methods

Genetic transformation of plants can be accomplished by several techniques, which are the following: (i) stable nuclear transformation of a crop species; (ii) stable plastid transformation of a crop species; (iii) transient transformation of a crop species and (iv) stable transformation of a plant species grown hydroponically with the recombinant protein to be released into the medium. The choice of the adopted method may be influenced by several aspects, including host plant species, its regenerability in tissue culture, and the targeted plant organelle for transgene insertion and expression.

Stable nuclear transformation

Agrobacterium tumefaciens-mediated gene transfer for stable nuclear transformation of plants is the most frequently used method (Fraley *et al*., 1983; Zupan *et al*., 2000). The soil bacterium *A. tumefaciens* is able to infect plants by transferring a circular DNA segment, the so-called Ti-plasmid, into the genome of its host. In its biotechnological use, a modified, disarmed Ti-plasmid harbouring the gene of interest in an expression cassette under control of a strong promoter (Tzfira and Citovsky, 2006) is delivered into the plant cell by the bacterium. The expression cassette is integrated into the plant's nuclear genome and during callus tissue culture, transformed cells are selected under antibiotic treatment and regenerated into transgenic plants. Similar while less efficient results can be obtained by particle bombardment. Plasmid-coated gold particles are biolistically inserted into the nucleus and the expression cassette is integrated into the genome. Detection and regeneration of transformants is again achieved under antibiotic selection. In any case the transgene is integrated into the genome by rather random events in numerous copies. So called position effects make transgenic lines heterogeneous and make screening and crossing costly in terms of labour and time.

Transient transformation via plant viruses

The use of autonomously replicating plant viruses like tobacco mosaic virus (TMV) or cowpea mosaic virus (CPMV) which have the ability to infect plants, is another transformation method resulting in high amounts of recombinant protein (Pogue *et al.*, 2002). A plant virus is modified that the gene of interest replaces the gene for a highly expressed coat protein. Inoculation of the host plant with the chimeric virus and subsequent viral replication initiates transgene expression within days yielding large amounts of recombinant protein.

Transient protein expression systems via plant viral vectors such as recombinant TMV or others are rapidly gaining popularity. The high level of recombinant protein expression (up to 2 g/kg plant tissue) within a very short time makes it an attractive tool for whole-plant based recombinant vaccine production (Liu *et al.*, 2005; Golovkin *et al.*, 2007). However, the number of appropriate vectors is limited, as is the number of plant species which are susceptible to a given virus. The transferred gene is not incorporated into the plant's genome but merely present in the nucleus where it is transcribed, and plants are harvested shortly after the artificial infection. Therefore, no transgenic plants are generated but the use of transgenic viruses generates the need of containment to prevent unintended spread of viral replicons.

Another technique for transient expression is the infiltration of plants with an *Agrobacterium* suspension containing the viral replicon (Gleba *et al.*, 2007). Agroinfiltration of plant tissue is also rather rapid and yields sufficient quantities of recombinant protein for initial analysis or further structural and functional characterization without the need for stably transformed plants. However, the production of a plant virus-derived recombinant protein requires an extra step of manual inoculation of the host plant with virus particles or Agrobacteria containing the viral replicon, hence making this technology not suitable for large scale production.

Plastid transformation

Although nuclear transformation of a crop species is the most common method to date, there are several concerns connected with nuclear genetic transformation of crops. For example, some grains have the potential to cross with native crop species or even food crops. There have been developed technologies to avoid the outcrossing, such as pollen sterility, but these technologies alleviate cost-effectiveness of the particular system due to reduced yields and higher laboratory or genetic efforts.

Expression of recombinant proteins in chloroplasts in several cases eliminates the problems associated with nuclear transformation and some features of chloroplasts make plastid transformants superior to nuclear transformants. The transgene is properly inserted into the plastid genome via homologous recombination and therefore, no position effect takes place. The high number of chloroplasts per cell and genomes per chloroplast (up to 10,000 genome copies per green cell) usually result in high levels of recombinant protein accumulation (Bock, 2007). And, last but not least, the maternal inheritance of organelles in most crop species avoids transgene spread through pollen (Ruf *et al.*, 2007; Svab and Maliga, 2007). The two major downsides of plastid transformants are the extended generation and cultivation periods and that to date only a few plant species can be transformed routinely, with tobacco being the most well-established one (Maliga, 2004). Regarding their protein accumulation, plastid transformants rank almost equal to transient expression levels as they are capable of producing up to 70% of their total soluble protein as recombinant protein (Oey *et al.*, 2009).

Transgene Localization

Plants have been proven as suitable expression system for various recombinant proteins and offer several advantages over conventional expression systems (Table 18.1). In general, integration of the transgene into the nuclear genome enables the production of the desired protein in the cytosol. In addition, as for other higher eukaryotes, proteins can be targeted to various compartments like the endoplasmic reticulum (ER), or can even be secreted to the apoplastic space. The passage of the protein through the ER is generally considered to be mandatory for correct formation of disulfide linkage and it has been shown that plants are capable of introducing the right folding to heterologous mammalian proteins. For example insulin is correctly assembled and functional when expressed in *Arabidopsis thaliana* (Nykiforuk *et al.*, 2006). But plants offer also very special translocation destinations, like vacuoles or plastids, which are unique to this system. This specific targeting of the expressed protein to several particular compartments or organelles, facilitated either by specific signal sequences genetically fused to the protein or, in the case of the plastids, stably integrated into the organellar genome, significantly can change the accumulation levels and post translational modifications. So is the cholera toxin subunit B correctly folded and active as well as present in large amounts when expressed from the chloroplast genome (Daniell *et al.*, 2001).

Beside the subcellular localization there are also several more options on a tissue level when it comes to recombinant protein accumulation. In principle, the expression of recombinant proteins is theoretically feasible in almost every plant part or tissue, including stem tissue, leaf, seed, fruit, and root, or cell cultures (Fischer *et al.*, 2004). This partitioning is highly dependent on the promoters used. The constitutive cauliflower mosaic virus 35S (CaMV35S) promoter facilitates more or less an even expression in all plant tissues (Sanders *et al.*, 1987), but protein yields achieved are rarely exceeding levels of 0.1% of the total soluble protein (TSP). Although considerable from an academic point of view, such levels are by far too low for any commercial application and do not outperform traditional expression hosts.

While constitutive expression of recombinant proteins in whole-plant systems and predominantly in leaves is generally regarded as the easiest way to engineer, there can occur problems. Some proteins could be inhibitory to plant development, for example if they interfere with energy allocation through photosynthesis (Hennig *et al.*, 2007). Leaves and other aerial parts may contain high levels of toxic or irritant secondary metabolites like alkaloids, phenols or terpenoids, which could ultimately contaminate the final product. This exacerbates the purification process or can ultimately prevent the use of the material in an oral applicable form. Moreover, protein stability may change in fresh tissue even through refrigeration.

The use of regulated promoters enables localized protein expression and accumulation in specific tissues, such as the E8-promoter in ripe tomato fruits (Deikman *et al.*, 1992; Sandhu *et al.*, 2000) or the glutelin-1promoter in maize endosperm (Ramessar *et al.*, 2008). The localization has been shown to have a great influence on the stability of the particular protein, due to specific conditions in the localized tissue. Moreover, this technique enables the accumulation in tissue specifically suitable for purification, storage or application purposes, like fruits and grains.

For example, expression and accumulation of recombinant proteins in seeds via seed-specific promoters enables easy protein purification and extremely high protein accumulation levels (Stoger *et al.*, 2005). Up to 36% of TSP could be achieved expressing a recombinant single chain antibody under the control of a seed specific promoter from *Phaseolus vulgaris* (De Jaeger *et al.*, 2002), but this is a rather exceptional result.

In addition, seeds seem to be ideal for protein stability as they protect the recombinant proteins in their mature dry state during long-term storage and transport, even at ambient temperature. A major drawback of seeds as production localization is that protein expression is dependent on the seed set which happens only late in plant development, expanding the cultivation time of the plants. Other than for leaf-produced protein which could be harvested before onset of flowering, growing the transgenic plants to full maturity always bears the risk of spreading transgenes by pollen or even seeds which is contradicting the idea of containment of transgenic plants harbouring biopharmaceuticals. Moreover, most plants yielding sufficient seed material are usually food or feed crops. Therefore, concerns are raised that pharmaceuticals produced in plants could be accidentally mixed with foodstuff, a scenario which should be strictly avoided.

The big variety of options for recombinant protein expression systems demonstrates that there are many critical factors in the development of therapeutic proteins or vaccines in plants, which almost all affect the yield of the antigen produced, including: (i) choice of the host plant species; (ii) expression system; (iii) expression level; and (iv) stability, assembly, antigenicity of the protein expressed.

Similar to other production systems, high expression levels are rather important as they allow to extract and to purify the relevant protein or vaccine candidate not only for characterization but also for subsequent formulation of the medicine. Alternatively, if the plant tissue containing the antigen is edible, a defined antigen dose could be delivered either directly or processed in a small amount of plant material.

Plant Vaccines for Oral Application

The challenging manufacturing process of vaccines and the stringent regulations for production, safety and efficiency have led to the extraordinarily high costs of new drugs and vaccines (Ulmer *et al.*, 2006). Additionally, the administration of most vaccines requires the use of syringes and needles as well as the need of medical assistance, additionally contributing to the total costs and generating an additional burden for the worldwide use of new vaccines, especially in developing countries.

In the past two decades, researchers have sought transgenic plants also as scaffolding for the production of edible vaccines. The desired goal for most plant-derived candidate vaccines is, besides inducing effective immunity, to raise protein expression level to an extent that allows purification and characterization as well as oral administration. Oral vaccination without syringes and needles could have advantages in terms of reduced medical assistance in administration, limited need of plant processing or protein purification, and elimination of blood-borne infections through needle reuse. But an edible vaccine could also eliminate the anxiety against vaccination via injection, thus being less traumatic especially for children (Giudice and Campbell, 2006). In addition, keeping the antigen in a crude form in plant cells may reduce the need for a cold chain as plant tissues provide a sort of 'bioencapsulation' and protection of the antigen (Zhang *et al.*, 2006). Also, oral delivery of antigens is particularly suited to protect against pathogens that infect via mucosal surfaces.

It has been shown that oral administration of antigens has the potential to induce immune responses in the gastrointestinal tract, in serum and at secondary mucosal surfaces, as well as to induce cell-mediated immunity (Brandtzaeg, 2007; Gomez *et al.*, 2008). Due to potential digestive degradation of the antigen delivered orally to the gastrointestinal tract, one option could be the protection of the subunit vaccine against proteolysis by encapsulation in a protective coat. If the antigen is expressed in seeds, the natural endosperm carbohydrate matrix which is rich in protease inhibitors may also protect the antigen. Another option would be the mere drastic increase of the amount administered. It is thought that about 100- to 1000-fold more antigen has to be administered orally compared to the subcutaneous route.

Hence, optimization of the expression system and thus reduction of the amount of plant material to be consumed is an essential step in the development of an edible vaccine.

Today, a number of edible plants or edible plant tissues such as potato, tomato, maize, and soybean have been used for the expression of various candidate oral vaccines against human or animal diseases and already tested in animals as well as humans, including Hepatitis B surface antigen (Kong *et al.*, 2001; Thanavala *et al.*, 1995; 2005, Norwalk virus particles (Mason *et al.*, 1996; Tacket *et al.*, 2000), and heat-labile enterotoxin B subunit (Haq *et al.*, 1995; Tacket *et al.*, 1998). Although the studies mentioned utilized plant material with sub-optimal levels of antigen present in plant tissue, these examples demonstrate that the approach works in principle. However, further improvements in terms of expression levels are required for most antigens to be sufficient for economic products, and to allow the delivery of a protective dose in a manageable amount of plant material.

As food products were consumed in these studies, many concerns arise in terms of human health, such as the potential of induction of tolerance (Weiner, 1994) or the vaccine to enter the food chain.

Conclusions

Since the first report of the use of plants for the production of subunit vaccines in 1992 (Mason *et al.*, 1992), a great variety of antigens has been expressed in transgenic plants, including proteins from bacterial and viral pathogens, allergen peptides and other pathogens that infect human and domestic or wild animals. Research in this issue has been reviewed many times.

However, despite almost 20 years of development and many examples of plant-derived candidate vaccines with proven efficacy or immunogenicity, today there are only two plant-based vaccine-related products that have gone through the major production and regulatory hurdles and reached commercialization (Rybicki, 2009).

There are several issues to be addressed before plant derived vaccines can be considered potential alternatives to the existing and well-established production systems and vaccination programmes, particularly with regard to low yields and inadequate immunogenicity of many products as well as missing grants from the pharmaceutical industry and governments. Considerable effort must go into GMP requirements and regulatory oversight, similar to the situation with conventional production of antigens via fermentation in various cell culture systems.

While technical concerns about plant-based protein production mainly focus on plant molecular biology and immunology, successful development and commercialization of plant-derived proteins and vaccines is also dependent on its social acceptance. Hopefully in the future there will be some successful examples of 'new medicinal plants' adding to the value of herbal medicines and reducing the concerns over plants producing vaccines.

References

Ballard, E. (1868) *On vaccination, its value and alleged dangers*. Longmans, London.

Banting, F.G., Best, C.H., Collip, J.B., Campbell, W.R. and Fletcher, A.A. (1991) Pancreatic extracts in the treatment of diabetes mellitus: preliminary report. 1922. *CMAJ* 145, 1281–1286.

Bock, R. (2007) Plastid biotechnology: prospects for herbicide and insect resistance, metabolic engineering and molecular farming. *Current Opinion in Biotechnology* 18, 100–106.

Brandtzaeg, P. (2007) Induction of secretory immunity and memory at mucosal surfaces. *Vaccine* 25, 5467–5484.

Brennan, F.R., Gilleland, L.B., Staczek, J., Bendig, M.M., Hamilton, W.D. and Gilleland, H.E., Jr (1999a) A chimaeric plant virus vaccine protects mice against a bacterial infection. *Microbiology* 145, 2061–2067.

Brennan, F.R., Jones, T.D., Gilleland, L.B., Bellaby, T., Xu, F., *et al.* (1999b) *Pseudomonas aeruginosa* outer-membrane protein F epitopes are highly immunogenic in mice when expressed on a plant virus. *Microbiology* 145, 211–220.

Cohen, S.N., Chang, A.C., Boyer, H.W. and Helling, R.B. (1973) Construction of biologically functional bacterial plasmids *in vitro*. *Proceedings of the National Academy of Sciences of the United States of America* 70, 3240–3244.

Daly, R. and Hearn, M.T. (2005) Expression of heterologous proteins in *Pichia pastoris*: a useful experimental tool in protein engineering and production. *Journal of Molecular Recognition* 18, 119–138.

Daniell, H., Lee, S.B., Panchal, T. and Wiebe, P.O. (2001) Expression of the native cholera toxin b subunit gene and assembly as functional oligomers in transgenic tobacco chloroplasts. *Journal of Molecular Biology* 311, 1001–1009.

De Jaeger, G., Scheffer, S., Jacobs, A., Zambre, M., Zobell, O., Goossens, A., Depicker, A. and Angenon, G. (2002) Boosting heterologous protein production in transgenic dicotyledonous seeds using *Phaseolus vulgaris* regulatory sequences. *Nature Biotechnology* 20, 1265–1268.

Deikman, J., Kline, R. and Fischer, R.L. (1992) Organization of ripening and ethylene regulatory regions in a fruit-specific promoter from tomato (*Lycopersicon esculentum*). *Plant Physiology* 100, 2013–2017.

Durrani, Z., McInerney, T.L., McLain, L., Jones, T., Bellaby, T., Brennan, F.R. and Dimmock, N.J. (1998) Intranasal immunization with a plant virus expressing a peptide from HIV-1 gp41 stimulates better mucosal and systemic HIV-1-specific IgA and IgG than oral immunization. *Journal of Immunological Methods* 220, 93–103.

Dus Santos, M.J., Wigdorovitz, A., Trono, K., Rios, R.D., Franzone, P.M., Gil, F., Moreno, J., Carrillo, C., Escribano, J.M. and Borca, M.V. (2002) A novel methodology to develop a foot and mouth disease virus (FMDV) peptide-based vaccine in transgenic plants. *Vaccine* 20, 1141–1147.

Emini, E.A., Ellis, R.W., Miller, W.J., McAleer, W.J., Scolnick, E.M. and Gerety, R.J. (1986) Production and immunological analysis of recombinant hepatitis B vaccine. *Journal of Infection* 13 Suppl A, 3–9.

Fischer, R., Stoger, E., Schillberg, S., Christou, P. and Twyman, R.M. (2004) Plant-based production of biopharmaceuticals. *Current Opinion in Plant Biology* 7, 152–158.

Fraley, R.T., Rogers, S.G., Horsch, R.B., Sanders, P.R., Flick, J.S., *et al.* (1983) Expression of bacterial genes in plant cells. *Proceedings of the National Academy of Sciences of the United States of America* 80, 4803–4807.

Franconi, R., Di Bonito, P., Dibello, F., Accardi, L., Muller, A., Cirilli, A., Simeone, P., Dona, M.G., Venuti, A. and Giorgi, C. (2002) Plant-derived human papillomavirus 16 E7 oncoprotein induces immune response and specific tumour protection. *Cancer Research* 62, 3654–3658.

Friedman, B., Vaddi, K., Preston, C., Mahon, E., Cataldo, J.R. and McPherson, J.M. (1999) A comparison of the pharmacological properties of carbohydrate remodeled recombinant and placental-derived beta-glucocerebrosidase: implications for clinical efficacy in treatment of Gaucher disease. *Blood* 93, 2807–2816.

Gerngross, T.U. (2004) Advances in the production of human therapeutic proteins in yeasts and filamentous fungi. *Nature Biotechnology* 22, 1409–1414.

Giudice, E.L. and Campbell, J.D. (2006) Needle-free vaccine delivery. *Advances in Drug Delivery Rev* 58, 68–89.

Gleba, Y., Klimyuk, V. and Marillonnet, S. (2007) Viral vectors for the expression of proteins in plants. *Current Opinion in Biotechnology* 18, 134–141.

Goeddel, D.V., Kleid, D.G., Bolivar, F., Heyneker, H.L., Yansura, D.G., Crea, R., Hirose, T., Kraszewski, A., Itakura, K. and Riggs, A.D. (1979) Expression in *Escherichia coli* of chemically synthesized genes for human insulin. *Proceedings of the National Academy of Sciences of the United States of America* 76, 106–110.

Golovkin, M., Spitsin, S., Andrianov, V., Smirnov, Y., Xiao, Y., Pogrebnyak, N., Markley, K., Brodzik, R., Gleba, Y., Isaacs, S.N. and Koprowski, H. (2007) Smallpox subunit vaccine produced *in planta* confers protection in mice. *Proceedings of the National Academy of Sciences of the United States of America* 104, 6864–6869.

Gomez, E., Chimeno Zoth, S., Carrillo, E., Estela Roux, M. and Berinstein, A. (2008) Mucosal immunity induced by orally administered transgenic plants. *Immunobiology* 213, 671–675.

Haq, T.A., Mason, H.S., Clements, J.D. and Arntzen, C.J. (1995) Oral immunization with a recombinant bacterial antigen produced in transgenic plants. *Science* 268, 714–716.

Hellwig, S., Drossard, J., Twyman, R.M. and Fischer, R. (2004) Plant cell cultures for the production of recombinant proteins. *Nature Biotechnology* 22, 1415–1422.

Hennig, A., Bonfig, K., Roitsch, T. and Warzecha, H. (2007) Expression of the recombinant bacterial outer surface protein A in tobacco chloroplasts leads to thylakoid localization and loss of photosynthesis. *FEBS Journal* 274, 5749–5758.

Inglis, S., Shaw, A. and Koenig, S. (2006) Chapter 11: HPV vaccines: Commercial Research & Development. *Vaccine* 24 Suppl 3, S99–S105.

Jenner, E. (1798) *An inquiry into the causes and effects of the Variolae Vaccinae*. Low, London.

Jilg, W. (1998) Novel hepatitis B vaccines. *Vaccine* 16 Suppl, S65–68.

Kong, Q., Richter, L., Yang, Y.F., Arntzen, C.J., Mason, H.S. and Thanavala, Y. (2001) Oral immunization with hepatitis B surface antigen expressed in transgenic plants. *Proceedings of the National Academy of Sciences of the United States of America* 98, 11539–11544.

Lauterslager, T.G., Florack, D.E., van der Wal, T.J., Molthoff, J.W., Langeveld, J.P., Bosch, D., Boersma, W.J. and Hilgers, L.A. (2001) Oral immunization of naive and primed animals with transgenic potato tubers expressing LT-B. *Vaccine* 19, 2749–2755.

Liu, D.T. (1992) Glycoprotein pharmaceuticals: scientific and regulatory considerations, and the US Orphan Drug Act. *Trends in Biotechnology* 10, 114–120.

Liu, L., Canizares, M.C., Monger, W., Perrin, Y., Tsakiris, E., Porta, C., Shariat, N., Nicholson, L. and Lomonossoff, G.P. (2005) Cowpea mosaic virus-based systems for the production of antigens and antibodies in plants. *Vaccine* 23, 1788–1792.

Maliga, P. (2004) Plastid transformation in higher plants. *Annual Review of Plant Physiology and Plant Molecular Biology* 55, 289–313.
Mason, H.S., Lam, D.M. and Arntzen, C.J. (1992) Expression of hepatitis B surface antigen in transgenic plants. *Proceedings of the National Academy of Sciences of the United States of America* 89, 11745–11749.
Mason, H.S., Ball, J.M., Shi, J.J., Jiang, X., Estes, M.K. and Arntzen, C.J. (1996) Expression of Norwalk virus capsid protein in transgenic tobacco and potato and its oral immunogenicity in mice. *Proceedings of the National Academy of Sciences of the United States of America* 93, 5335–5340.
Mason, H.S., Haq, T.A., Clements, J.D. and Arntzen, C.J. (1998) Edible vaccine protects mice against *Escherichia coli* heat-labile enterotoxin (LT), potatoes expressing a synthetic LT-B gene. *Vaccine* 16, 1336–1343.
McInerney, T.L., Brennan, F.R., Jones, T.D. and Dimmock, N.J. (1999) Analysis of the ability of five adjuvants to enhance immune responses to a chimeric plant virus displaying an HIV-1 peptide. *Vaccine* 17, 1359–1368.
Modelska, A., Dietzschold, B., Sleysh, N., Fu, Z.F., Steplewski, K., Hooper, D.C., Koprowski, H. and Yusibov, V. (1998) Immunization against rabies with plant-derived antigen. *Proceedings of the National Academy of Sciences of the United States of America* 95, 2481–2485.
N.N. (2006) News In Brief. *Nat Biotech* 24, 233–234.
Nemchinov, L.G., Liang, T.J., Rifaat, M.M., Mazyad, H.M., Hadidi, A. and Keith, J.M. (2000) Development of a plant-derived subunit vaccine candidate against hepatitis C virus. *Archives of Virology* 145, 2557–2573.
Newman, D.J. and Cragg, G.M. (2007) Natural products as sources of new drugs over the last 25 years. *Journal of Natural Products* 70, 461–477.
Nykiforuk, C.L., Boothe, J.G., Murray, E.W., Keon, R.G., Goren, H.J., Markley, N.A. and Moloney, M.M. (2006) Transgenic expression and recovery of biologically active recombinant human insulin from *Arabidopsis thaliana* seeds. *Plant Biotechnology Journal* 4, 77–85.
Oey, M., Lohse, M., Kreikemeyer, B. and Bock, R. (2009) Exhaustion of the chloroplast protein synthesis capacity by massive expression of a highly stable protein antibiotic. *Plant Journal* 57, 436–445.
Pasteur, L. (1885) Methode pour prevenir la rage apres morsure. *C.R. Acad. Sci.* 101, 765–772.
Pogue, G.P., Lindbo, J.A., Garger, S.J. and Fitzmaurice, W.P. (2002) Making an ally from an enemy: plant virology and the new agriculture. *Annual Review of Phytopathology* 40, 45–74.
Ramessar, K., Rademacher, T., Sack, M., Stadlmann, J., Platis, D., *et al.* (2008) Cost-effective production of a vaginal protein microbicide to prevent HIV transmission. *Proceedings of the National Academy of Sciences of the United States of America* 105, 3727–3732.
Ruf, S., Karcher, D. and Bock, R. (2007) Determining the transgene containment level provided by chloroplast transformation. *Proceedings of the National Academy of Sciences of the United States of America* 104, 6998–7002.
Rybicki, E.P. (2009) Plant-produced vaccines: promise and reality. *Drug Discovery Today* 14, 16–24.
Sanders, P.R., Winter, J.A., Barnason, A.R., Rogers, S.G. and Fraley, R.T. (1987) Comparison of cauliflower mosaic virus 35S and nopaline synthase promoters in transgenic plants. *Nucleic Acids Research* 15, 1543–1558.
Sandhu, J.S., Krasnyanski, S.F., Domier, L.L., Korban, S.S., Osadjan, M.D. and Buetow, D.E. (2000) Oral immunization of mice with transgenic tomato fruit expressing respiratory syncytial virus-F protein induces a systemic immune response. *Transgenic Research* 9, 127–135.
Shaaltiel, Y., Bartfeld, D., Hashmueli, S., Baum, G., Brill-Almon, E., *et al.* (2007) Production of glucocerebrosidase with terminal mannose glycans for enzyme replacement therapy of Gaucher's disease using a plant cell system. *Plant Biotechnology Journal* 5, 579–590.
Sijmons, P.C., Dekker, B.M., Schrammeijer, B., Verwoerd, T.C., van den Elzen, P.J. and Hoekema, A. (1990) Production of correctly processed human serum albumin in transgenic plants. *Biotechnology (NY)* 8, 217–221.
Steere, A.C., Sikand, V.K., Meurice, F., Parenti, D.L., Fikrig, E., Schoen, R.T., Nowakowski, J., Schmid, C.H., Laukamp, S., Buscarino, C. and Krause, D.S. (1998) Vaccination against Lyme

disease with recombinant *Borrelia burgdorferi* outer-surface lipoprotein A with adjuvant. Lyme Disease Vaccine Study Group. *New England Jounal of Medicine* 339, 209–215.
Stoger, E., Ma, J.K., Fischer, R. and Christou, P. (2005) Sowing the seeds of success: pharmaceutical proteins from plants. *Current Opinion in Biotechnology* 16, 167–173.
Streatfield, S.J. and Howard, J.A. (2003) Plant-based vaccines. *International Journal of Parasitology* 33, 479–493.
Streatfield, S.J., Jilka, J.M., Hood, E.E., Turner, D.D., Bailey, M.R., *et al.* (2001) Plant-based vaccines: unique advantages. *Vaccine* 19, 2742–2748.
Streatfield, S.J., Lane, J.R., Brooks, C.A., Barker, D.K., Poage, M.L., Mayor, J.M., Lamphear, B.J., Drees, C.F., Jilka, J.M., Hood, E.E. and Howard, J.A. (2003) Corn as a production system for human and animal vaccines. *Vaccine* 21, 812–815.
Svab, Z. and Maliga, P. (2007) From the Cover: Exceptional transmission of plastids and mitochondria from the transplastomic pollen parent and its impact on transgene containment. *Proceedings of the National Academy of Sciences of the United States of America* 104, 7003–7008.
Swartz, J.R. (2001) Advances in *Escherichia coli* production of therapeutic proteins. *Current Opinion in Biotechnology* 12, 195–201.
Tacket, C.O., Mason, H.S., Losonsky, G., Clements, J.D., Levine, M.M. and Arntzen, C.J. (1998) Immunogenicity in humans of a recombinant bacterial antigen delivered in a transgenic potato. *Nature Medicine* 4, 607–609.
Tacket, C.O., Mason, H.S., Losonsky, G., Estes, M.K., Levine, M.M. and Arntzen, C.J. (2000) Human immune responses to a novel norwalk virus vaccine delivered in transgenic potatoes. *Journal of Infectious Diseases* 182, 302–305.
Thanavala, Y., Yang, Y.F., Lyons, P., Mason, H.S. and Arntzen, C. (1995) Immunogenicity of transgenic plant-derived hepatitis B surface antigen. *Proceedings of the National Academy of Sciences of the United States of America* 92, 3358–3361.
Thanavala, Y., Mahoney, M., Pal, S., Scott, A., Richter, L., Natarajan, N., Goodwin, P., Arntzen, C.J. and Mason, H.S. (2005) Immunogenicity in humans of an edible vaccine for hepatitis B. *Proceedings of the National Academy of Sciences of the United States of America* 102, 3378–3382.
Tuboly, T., Yu, W., Bailey, A., Degrandis, S., Du, S., Erickson, L. and Nagy, E. (2000) Immunogenicity of porcine transmissible gastroenteritis virus spike protein expressed in plants. *Vaccine* 18, 2023–2028.
Turpen, T.H., Reinl, S.J., Charoenvit, Y., Hoffman, S.L., Fallarme, V. and Grill, L.K. (1995) Malarial epitopes expressed on the surface of recombinant tobacco mosaic virus. *Biotechnology (NY)* 13, 53–57.
Twyman, R.M., Stoger, E., Schillberg, S., Christou, P. and Fischer, R. (2003) Molecular farming in plants: host systems and expression technology. *Trends in Biotechnology* 21, 570–578.
Tzfira, T. and Citovsky, V. (2006) Agrobacterium-mediated genetic transformation of plants: biology and biotechnology. *Current Opinion in Biotechnology* 17, 147–154.
Ulmer, J.B., Valley, U. and Rappuoli, R. (2006) Vaccine manufacturing: challenges and solutions. *Nature Biotechnology* 24, 1377–1383.
Walsh, G. (2000) Biopharmaceutical benchmarks. *Nature Biotechnology* 18, 831–833.
Walsh, G. (2005) Therapeutic insulins and their large-scale manufacture. *Applied Microbiology and Biotechnology* 67, 151–159.
Weiner, H.L. (1994) Oral tolerance. *Proceedings of the National Academy of Sciences of the United States of America* 91, 10762–10765.
Wigdorovitz, A., Perez Filgueira, D.M., Robertson, N., Carrillo, C., Sadir, A.M., Morris, T.J. and Borca, M.V. (1999) Protection of mice against challenge with foot and mouth disease virus (FMDV) by immunization with foliar extracts from plants infected with recombinant tobacco mosaic virus expressing the FMDV structural protein VP1. *Virology* 264, 85–91.
Wurm, F.M. (2004) Production of recombinant protein therapeutics in cultivated mammalian cells. *Nature Biotechnology* 22, 1393–1398.

Yin, J., Li, G., Ren, X. and Herrler, G. (2007) Select what you need: a comparative evaluation of the advantages and limitations of frequently used expression systems for foreign genes. *Journal of Biotechnology* 127, 335–347.

Yusibov, V., Modelska, A., Steplewski, K., Agadjanyan, M., Weiner, D., Hooper, D.C. and Koprowski, H. (1997) Antigens produced in plants by infection with chimeric plant viruses immunize against rabies virus and HIV-1. *Proceedings of the National Academy of Sciences of the United States of America* 94, 5784–5788.

Yusibov, V., Hooper, D., Spitsin, S., Fleysh, N., Kean, R., Mikheeva, T., Deka, D., Karasev, A., Cox, S., Randall, J. and Koprowski, H. (2002) Expression in plants and immunogenicity of plant virus-based experimental rabies vaccine. *Vaccine* 20, 3155.

Zhang, X., Buehner, N.A., Hutson, A.M., Estes, M.K. and Mason, H.S. (2006) Tomato is a highly effective vehicle for expression and oral immunization with Norwalk virus capsid protein. *Plant Biotechnology Journal* 4, 419–432.

Zupan, J., Muth, T.R., Draper, O. and Zambryski, P. (2000) The transfer of DNA from *Agrobacterium tumefaciens* into plants: a feast of fundamental insights. *Plant Journal* 23, 11–28.

Chapter 19

Noscapinoids: a New Class of Anticancer Drugs Demand Biotechnological Intervention

Harish C. Joshi, Asmita Salil, Usha Bughani and P.K. Naik

Introduction

Microtubules (MTs) are cytoskeletal components that play a critical role in many cellular processes such as maintenance of cell shape and polarity, intracellular transport of vesicles and organelles, and cellular motility. During cell division, MTs form a bipolar microtubule (MT) array, called the mitotic spindle responsible for the accurate distribution of chromosomes into two daughter cells (McIntosh, 1985). MTs, assembled from α- and β-tubulin heterodimers, are highly dynamic structures that alternate between periods of growth and shortening (Mitchison and Kirschner, 1984). This dynamic instability property is crucial for MTs to carry out many of their cellular functions (Desai and Mitchison, 1997). Disruption of MT dynamics can lead to the formation of abnormal mitotic spindles, thereby preventing the normal cell proliferation (Jordan and Wilson, 1999). The critical role that MTs play in cell division makes them a suitable target for the development of chemotherapeutic drugs against the rapidly growing cancers. The effectiveness of MT-targeting drugs has been validated by the successful use of several vinca alkaloids and taxanes for the treatment of a wide variety of human cancers. These agents inhibit or promote the polymerization of tubulin, halt mitosis of rapidly dividing cells, induce apoptosis, and are reasonably effective in cancer chemotherapy (Wilson and Jordan, 1995;). As a testament to this, USFDA has approved the use of taxanes such as paclitaxel and taxotere as a first line chemotherapy for ovarian and breast cancers and vincas such as vinblastine, vincristine, and vinorelbine for some haematological pathologies. However, drug resistance is a common problem with repeated and prolonged administration of these agents, possibly owing to the amplification of a membrane glycoprotein involved in efflux of the drug (Stark, 1986). Moreover, these anti-MT agents are frequently toxic to normal tissues and are effective only for certain types of cancer (Rowinsky and Donehower, 1991; Kavanagh and Kudelka, 1993). Another challenge these useful drugs face, is their poor solubility in biocompatible solvents. As a result they are solubilized in unfriendly carriers such as Cremphor EL and ethanol that run risk of anaphylactic shocks and medical emergencies. Hence, there is an urgent need for new and better chemotherapeutic drugs for cancer treatment.

Noscapine: a Plant Derived Alkaloid

Noscapine was originally discovered by French pharmacist and Professor Pierre-Jean Robiquet in 1817. He isolated two natural compounds from opium (*Papaver somniferum*): codeine and noscapine (Warolin, 1999). Noscapine (21%) is one of the more abundant opium alkaloids, the other prominent alkaloids being morphine (42%), codeine (12%), papaverine (18%), thebaine (6.5%) sanguinarine, berberine and tubocurarine. As secondary or specialized metabolites, these alkaloids are not essential for normal growth and development but appear to function in the defence of plants against herbivores and pathogens (Liscombe and Facchini 2008). Realistically, the actual utility of these metabolites in the plants that synthesize them is less well understood than their human use for the medicinal activities.

From a practical point of view, the chemical synthesis of these plant alkaloids can be followed from the time of the appreciation of their medicinal abilities. The pioneering step was achieved by Perkin and Robinson, who obtained noscapine from meconine and cotarnine in the presence of potassium carbonate combined with fractional crystallization (Perkin and Robinson, 1910). Since then, tremendous progress has been made in the development of schemes for total synthesis. The availability of noscapine from natural sources is perhaps more economical than the current synthetic means. However, the elucidation of novel and simpler synthetic methods will continue to provide further synthesis of novel derivatives. Traditionally, a mode of combinatorial chemistry has been of immense value in drug development. Noscapine however faces a bit of a challenge because of the linkage between the two ring system i.e. the iso-quinoline ring and the iso-benzo-furanone ring system are joined by a single rotatable C-C bond involving two chiral centres. This necessarily leads to a racemic mixture of four stereoisomers of noscapine in ordinary chemical reactions. Only one of these stereoisomers, the RS form is active biologically (Karlsson *et al.*, 1990; Ye *et al.,* 1998). Nevertheless, the improvements in the separation techniques of the chiral molecules as well as innovations in the chirally biased synthetic procedures may capitalize on the combinatorial potential of novel synthesis. In addition, many natural compounds such as taxanes can be easily synthesized starting from a natural derived core structure system in a more economical fashion (Nicolaou *et al.*, 1994) similar principles might also apply to noscapine.

Discovery of Noscapine as a New Antimicrotubule Drug

Noscapine is a non-narcotic, phthalide isoquinoline alkaloid derived from the opium poppy *Papaver somniferum*. It is a very safe cough suppressant (antitussive) which has been in use for many decades (Empey *et al.*, 1979; Karlsson *et al.*, 1990). MT assembly inhibitors such as colchicine, podophyllotoxin, MTC [2-methoxy-5-(2′,3′,4′-trimethoxyphenyl)-2,4,6-cycloheptatrien-1-one] and TKB [2,3,4-trimethoxy-4'-acetyl-1,1'-biphenyl] shared a common structural theme with respect to the chemical moieties such as a hydrophobic trimethoxy phenyl group, a variety of other hydrophobic domains such as lactone, tropolone, or other aromatic rings and a small hydrophilic group such as OH and NH_2 (Fig. 19.1). To identify potential new anti-MT drugs, we compared a series of natural compounds including noscapine that show structural similarities with these known agents. Unlike the founding MT-drugs, noscapine arrests mammalian cell cycle with intact bipolar MT-spindles in mitosis even at high concentrations. (Ye *et al.*, 1998; Ke *et al.*, 2000; Zhou *et al.*, 2002).

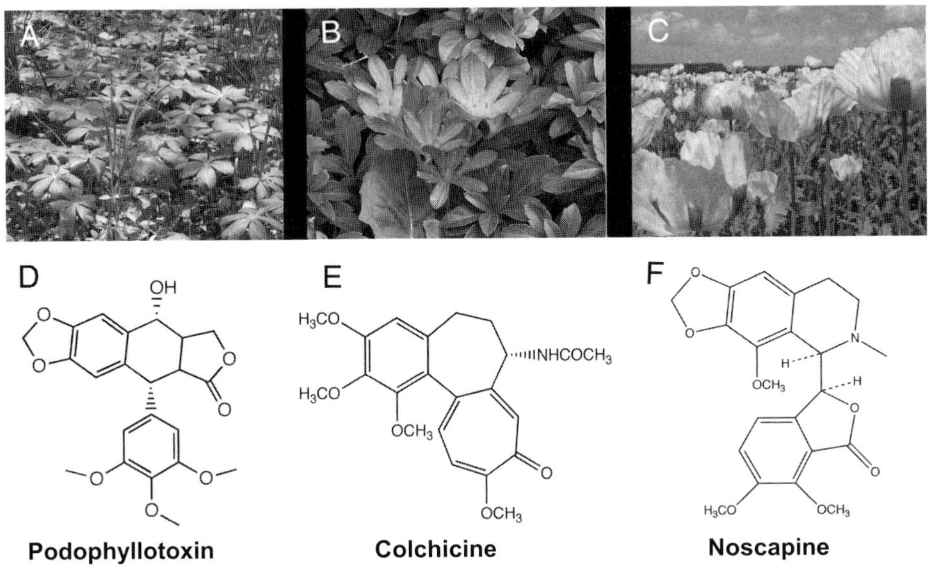

Fig. 19.1. Discovery of anticancer drugs from natural herbs. The foliages of these perennial herbaceous plants, (A) *Podophyllum peltatum* (commonly known as mayapple) and (B) *Colchicum autumnale* (known as meadow saffron), produce the potent drugs podophyllotoxin (D) and colchicine (E) respectively. Both drugs share some structural commonalities such as heavily methoxy-substituted hydrophobic six-membered rings along with some dispersed hydrophilic moieties. (C) *Papaver somniferum* (the poppy plant) produces the alkaloid noscapine (F) which shares these chemical features. Plant photographs adapted from Wikipedia.

Noscapine: Mechanism of Action

Noscapine binds tubulin dimer with a 1:1 stoichiometry (Ye *et al.*, 1998) and alters the auto-fluorescence and circular dichroism spectrum of tubulin suggesting an alteration of the secondary structure of tubulin upon binding and arrests the mammalian cells at mitosis (Ye *et al.*, 1998) (Fig. 19.2). Presence of noscapine does not significantly promote or inhibit MT polymerization *in vitro*. The electron micrographs of the MTs polymerized *in vitro*, in the presence of noscapine were individual polymers, slightly wavy in appearance, but otherwise similar in lengths as in the absence of the drug (Ye *et al.*, 1998). Instead, noscapine alters the steady state dynamics of MT assembly, primarily by increasing the amount of time that MTs spend in an attenuated (pause) state. Therefore, the bulk biochemistry fell short of explaining the observed mitotic arrest.

During the onset of mitosis, tubulin subunits assemble and disassemble vigorously to make the attachment between kinetochores of chromosomes and the plus ends of MTs, a highly dynamic process (Mitchison *et al.*, 1986). Physical tension is generated across kinetochore pairs following attachment to kinetochores and is probably regulated by the combined action of MT dynamics and MT motors within the vicinity of kinetochores (Joshi *et al.*, 1992; Cassimeris *et al.*, 1994; Nicklas, 1997). The careful real time observation of individual polymerizing MTs *in vitro* and tracking the plus end growth over time revealed that noscapine affected MT-dynamics primarily by increasing the amount of time MTs spent in an attenuated pause state rather than engaging into active depolymerization and repolymerization. As a result, noscapine treatment reduced the tension generated across the kinetochore pairs as well as reduced the number of MTs attached to each pair of kinetochores (Zhou *et al.*, 2002). During mitosis, spindle assembly checkpoint (SAC) prevents the onset of anaphase until all chromosomes are correctly attached with MTs and proper tension is applied to the chromosomes (Nicklas, 1997). Owing to its effect on MT dynamics, noscapine reduces tension as well as the number of MTs attached to each pair of kinetochore. It therefore fails to deactivate the spindle assembly checkpoint, and the active checkpoint is responsible for the sustained mitotic arrest (Landen *et al.*, 2002; Zhou *et al.*, 2002).

MT arrays (green) and chromosomes (red) of cells either treated with 25 x 10^{-6} M noscapine dissolved in DMSO (C) and control cells treated with the vehicle solution DMSO alone (D). The control mitotic cells show classical biconical image with MT arrays emanating from two duplicated centrosomes that have moved farthest apart from each other on opposite sides of the centrally aligned chromosomal metaphase plate of the spindle (D). In contrast, the noscapine treated cells arrest with subsets of chromosomes that remain scattered within the otherwise intact looking biconical MT spindle (C). Data are partially adapted from Ye *et al.*, 1998 and Zhou *et al.*, 2003.

Noscapine: Effect on Healthy and Cancerous Cells

Generally, the spindle assembly and cell cycle checkpoints are robust in healthy cells and can sense even minor perturbations of MT dynamics. Many cancer cells, however, have debilitated mitotic checkpoints perhaps due to mutations either directly in the checkpoint genes or indirectly via signal transduction pathways (Stewart *et al.*, 1999). Noscapine treatment leaves the mitotic checkpoint activated and hence arrests mitosis in healthy cells. However, slow MT dynamics in checkpoint-compromised cancer cells fails to arrest them for a long time in mitosis but rather let cells 'slip' into the next G1 phase either with two diploid nuclei, one intact tetraploid nucleus, or one or two major nucleoids surrounded by multiple small micronuclei. Such cells with abnormal ploidy have a further propensity to undergo massive poly- or aneuploidy leading to the activation of apoptotic pathways and death (Fig. 19.3) (Landen *et al.*, 2002).

Because noscapine only affects MT dynamics cellular functions that do not require exquisite control of MT dynamics may not be interrupted. Of particular interest here is the relentless axonal transport required for the proper maintenance of synaptic structures of long peripheral nerves, which is subject to disruption by current anti-MT drugs. Noscapine, therefore has no detectable neurotoxic effect on the histologies of peripheral nerves (Landen *et al.*, 2004). Of particular interest in noscapine therapy is that its levels rise only transiently in plasma (Tiveron *et al.*, 1993), therefore it is conceivable that normal cells resume cell division after the noscapine concentration decreases below the threshold level in a few hours.

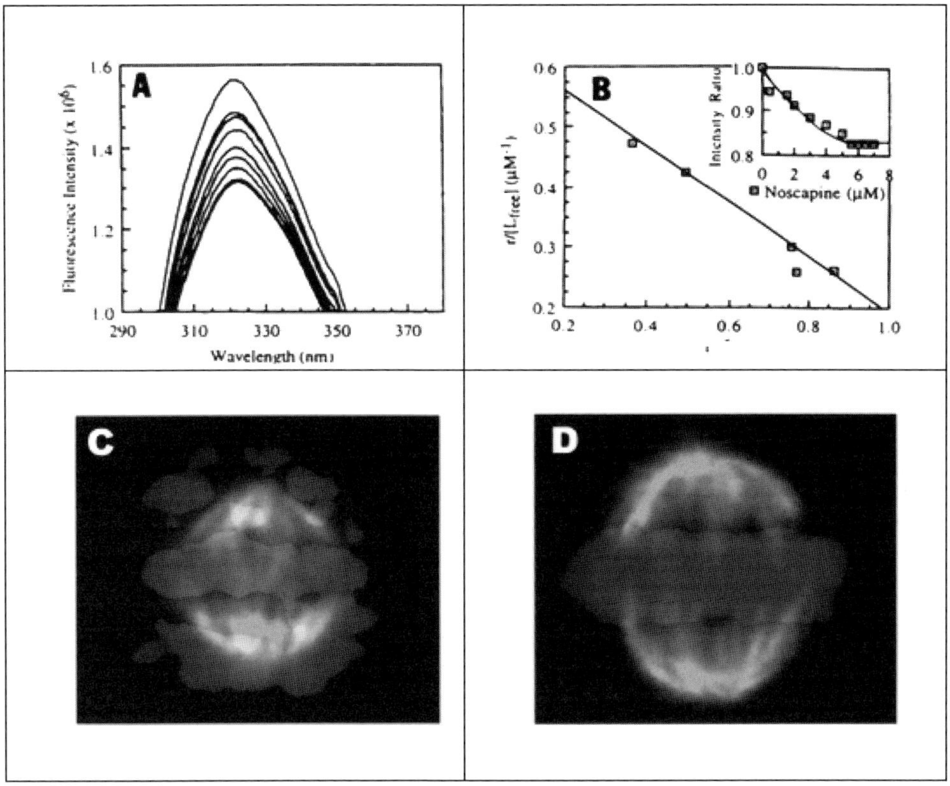

Fig. 19.2. Noscapine binds tubulin and arrests dividing cells in prometaphase with morphologically normal looking mitotic spindles, as would be expected from perturbed-MT dynamics. (A) A concentration dependent but saturable quenching of noscapine dependent fluorescence quenching of tubulin by noscapine. (B) Scatchard plot analysis of quenching data shown in A. These analyses reveal an apparent dissociation constant (K_d) of 1.86 ± 0.34 × 10^{-6} M and a stoichiometry of 0.95 ± 0.02 noscapine molecule per complex of tubulin subunit. (Inset) Saturation of noscapine-induced quenching in tubulin fluorescence intensity (Ye *et al.*, 1998).

Apoptotic cell death with typical features such as DNA fragmentation, upon noscapine treatment, is preceded by activation of c-Jun NH2-terminal kinases (JNK) (Zhou *et al.*, 2002). JNK is known to play an important role in coordinating the cellular response to stress by phosphorylating the transcription factors c-Jun and p53 (Buschmann *et al.*, 2001). Aneja *et al.* showed that noscapine activates growth arrest and apoptosis primarily via a p53-dependent pathway that necessarily involves the function of p21 (Aneja *et al.*, 2007b).

MTs are also important for the integrity of many cellular organelles including mitochondria. Long snake-like morphologies of mitochondria are disrupted by MT depolymerizing drugs. MTs are required for the proper translocation and deployment of mitochondria. Apoptosis induced by noscapinoids is preceded by depolymerization of mitochondria, and this might partially be responsible for apoptosis in noscapinoid treated cells (Aneja *et al.*, 2008). Apoptotic signals are precariously balanced within the cellular milieu by the appropriate ratios of proapoptotic and antiapoptotic proteins and their phosphorylation states. Noscapinoids perturb the level and phosphorylation state of these proteins to tip the balance in favour of apoptosis during noscapinoid treatment (Aneja *et al.*, 2008).

Taken together, our work published so far suggests that besides mitotic failure, many other factors might contribute to varying extents in bringing about apoptosis in noscapine treated cancer cells. These mechanisms may range from genetic stress, cellular homeostasis stress and mitochondrial stress. Whatever the precise contributing mechanistic factors that come into play, it is undeniable that noscapinoids do induce apoptosis in a variety of cancer cells.

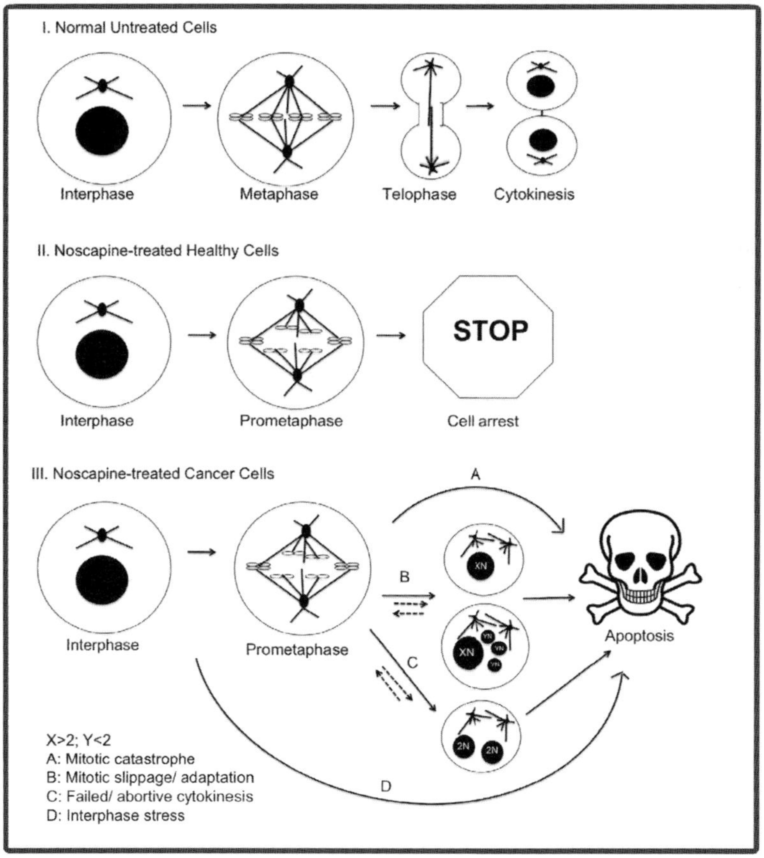

Fig. 19.3. Mechanistic explanations of the cell division cycle responses of normal and cancer cells to noscapine and derivatives.

1. At the onset of mitosis, normal interphase cells (drug untreated) enclose pairs of attached duplicated sister-chromatids within the nucleus, while the nuclear envelope begins to disassemble exposing for the first time, chromosomes in the cytoplasm. At this stage a specialized proteinaceous complex over the centromeric DNA, called the kinetochore, emits a 'wait Anaphase' signal to the cellular surveillance mechanism (called spindle assembly checkpoint (SAC) or Mitotic Checkpoint) that holds the sister kinetochore-glue together, thus inhibiting the onset of subsequent premature anaphase.

As the radial MT interphase arrays disassemble from the duplicated centrosomes, each centrosome begins to organize a much more dynamic MT aster, which slide apart against each other to occupy opposite territories within the cell. In the meanwhile, the dynamic MTs quickly assemble–disassemble–assemble, searching the entire intracellular milieu relentlessly. If however, a growing MT encounters a chromosomal kinetochore, it captures the growing plus end of MT and orients kinetochore of one sister chromatid to one spindle pole (centrosome) thus positioning the other attached sister kinetochore for a better opportunity to grab the growing MT from the opposite centrosome (spindle pole). In this manner a metaphase plate gets assembled that has the entire individual sister chromatid pairs attached to opposite poles of the spindle and perhaps the proper tension is generated across the kinetochores. Thus, robust mitotic checkpoint of normal cells can hold cells in prometaphase until the last one of the sister chromatid pair gets a bi-oriented state (called metaphase), at which point the 'wait anaphase' checkpoint signal is switched off and the glue between the sister chromatids dissolves and sisters are driven apart to their cognate opposite poles (anaphase). Once at their respective poles, a nuclear envelope forms around the each diploid (2n) chromosomal cluster, producing the two daughter nuclei which are destined to produce the nuclei of the daughter G1 cells after mitotic exit and cytokinesis. Under usual circumstances, this all goes very smoothly and happens rather quickly without straining the checkpoint too much.

2. When cells are treated with noscapine, their dynamics are slow due to the prolonged pause or attenuated state in their transition from the growing to the shortening phases. Thus the strong cellular mitotic checkpoints are strained for long periods of time in prometaphase to hold the progression of anaphase (in the absence of metaphase). Fortunately, mammalian somatic cells have very strong checkpoints, primarily because they are constructed from large networks of healthy key proteins and can do the job of arresting cells for up to 20 h. If noscapine is removed after 12 h, arrested cells quickly restore the proper metaphase and undergo anaphase on schedule.
3. Most of the cancer cells have mutational lesions either in checkpoint genes or other signalling pathway components that impinge upon them, hence their checkpoints are not very strong. Hence, when a cancer cell is treated with noscapine it cannot hold the cells in prometaphase for a long time (although a brief arrest of a varying duration always exists), and the cells then follow one of the following fates:
   (i) They go straight from mitosis to cell death, a process called mitotic catastrophe.
   (ii) They slip out into a G1 like phase with a nucleus twice as big and contain twice as many chromosomes (tetraploid/4n) in a process sometimes referred to as adaptation. Such cells can rest for long time, mitotically maintained for some time or die depending upon their next cell cycle checkpoint status (G1/S). In many cases, however, instead of forming one big nucleus, diploid or polyploid cells can also form heterogeneous micronuclei around a predominant aneuploid nucleus.
   (iii) Cells can undergo anaphase and bypass the cytokinesis producing bi-nucleate cells which also follow the same fate as in (ii).
   (iv) A subpopulation of cells dies straight from the interphase prior to attempting mitosis. This is a curious phenomenon and we believe that the dynamic nature of MTs is required for cellular homeostasis and the integrity of cellular organelles including mitochondria, since the health of cellular mitochondria is directly linked to internally triggered apoptosis (cell death) and might contribute to this type of death. Additionally, the cellular stress kinase systems (e.g. jun N-terminal kinases

etc.) do also play a role. More studies are needed to clarify the relative contributions and additional mechanisms.

Noscapine: Non-Toxic Profile in Animals

Noscapine has been shown to inhibit growth of murine and human tumours implanted in mice by inducing apoptosis (Ye *et al.*, 1998). *In vitro* as well as mouse xenograft models have shown that noscapine and its analogues are useful in treatment of tumour cells derived from a variety of cancers, including cancer of colon, non-small cell lung cancer, brain cancer, ovarian cancer, kidney cancer, prostate cancer, leukaemia, breast cancer and bladder cancer (Ye *et al.*, 1998; Ke *et al.*, 2000; Zhou *et al.*, 2003). Noscapine is a non-narcotic derivative of opium that lacks analgesic, sedative and respiratory-depressant properties and it does not produce euphoria or dependence (Martindale, 1977). Most of the MT binding agents such as taxanes and vinca alkaloids are associated with toxicities and adverse side-effects. For example, paclitaxel is associated with toxicities including neurotoxicity (Lipton *et al.*, 1989), cardiotoxicity (Rowinsky *et al.*, 1990), myelosuppression (Wiernik *et al.*, 1987), hypersensitivity reactions (Guchelaar *et al.*, 1994), alopecia (Donehower *et al.*, 1987) and gastrointestinal toxicity (Hruban *et al.*, 1989). In contrast to these MT binding drugs, noscapine is very effective in treatment of a variety of tumours such as murine lymphoma, melanoma, and human breast xenografts. Organ systems such as kidney, heart, liver, bone marrow, spleen, or small intestine in nude mice showed no signs of alterations in their histological profiles, indicative of non-toxicity. Thus, these results suggest that damage to cell cycle is specific to the cancer cell types, while the normal cells arrest till the drug is cleared (Ke *et al.*, 2000). Some toxicity has been reported in experimental animals and in human but only at much higher doses of noscapine (Lasagna *et al.*, 1961; Idänpään-Heikkilä, 1968). Also it has been shown that noscapine does not inhibit primary humoral or cell-mediated immune responses in mice (Ye *et al.*, 1998; Ke *et al.*, 2000). Therefore, noscapine is a promising anticancer drug with minimum toxicity and side effects.

Noscapine and Drug Resistant Cancer Cells

Like in the case of other chemotherapeutics, MT binding drugs such as vinca alkaloids and taxanes also face the challenge of drug resistance. Published data suggest that noscapine treatment may score a few points of victory over this problem. There are potentially three basic mechanisms by which cancer cells acquire resistance toward MT drugs. First, the amplification of multi-drug resistance (MDR) protein that enhances efflux of drugs and thus prevents the accumulation of drugs within cells (Krishna and Mayer, 2000). This MDR protein is a transmembrane pump (P-glycoprotein) and can expel hydrophobic drugs from the cell which is a major cause of chemoresistance to a variety of anti-MT agents (Giannakakou *et al.*, 1997, 2000). Second, mutations might also develop in the genes encoding α- and β-tubulin subunits and thus prevent binding of drugs (Giannakakou *et al.*, 2000). Third, both α- and β-tubulins are expressed as multiple isotypes in varying ratios in different mammalian tissues (Joshi and Cleveland, 1990) and there have been suggestions that differential expression of diverse tubulin isotypes might ameliorate drug binding efficiencies (Ludueña, 1998). Because of this diversity, and the changes in the relative concentrations of various tubulin isotypes, resistance to MT-binding drugs might arise (Kavallaris *et al.*, 2001). Noscapinoids activate JNK, which in turn downregulates the expression of p-glycoprotein drug pumps (Zhou *et al.*, 2002, 2006). In addition, noscapinoids seem to be poor substrates for p-glycoprotein. Indeed, noscapine is found to be highly effective in inhibiting proliferation and inducing apoptosis in human ovarian carcinoma cell lines that are sensitive or resistant to paclitaxel (Zhou *et al.*, 2002).

Noscapine: Pharmacokinetic Profile

Pharmacokinetic evaluation is an important component of the preclinical drug development process. These studies with noscapine showed that upon oral administration in mice, it gets absorbed quickly in mice at all dose levels (75, 150, 300 mg/kg) with a $t_{max} < 2$ h and is distributed rapidly and widely. Noscapine shows a mean bioavailability of ~30–32% across the same dose range (Aneja *et al.*, 2007a). Noscapine, therefore, comes up as a well-tolerated, non-toxic, orally bioavailable drug with desirable pharmacokinetic properties.

Oral administration of noscapine is associated not just with low cost and convenience but also precludes chances of concerning hypersensitivity reactions encountered during infusions of other currently available insoluble chemotherapeutic agents that utilize vehicle agents with several undesirable characteristics. Since biologically inactive doses minimize anti-tumour responses, the therapeutic efficacy of an anticancer drug is directly related to its bioavailable dose. Thus, the oral bioavailability of noscapine offers further support for its clinical advancement as a novel chemotherapeutic agent (Dahlström *et al.*, 1982; Haikala *et al.*, 1986; Karlsson *et al.*, 1990).

Blood–brain barrier remains a major problem for the chemotherapeutic management of brain tumours. Patients diagnosed with glioblastoma have a median survival of 9–12 months despite surgical resection, radiation therapy, and/or chemotherapy (Kleihues *et al.*, 2002). The infiltrative nature of astrocytic tumour growth rarely allows complete surgical resection, and more than 90% of tumours recur within 2 cm of primary tumour site. Post-operative radiotherapy prolongs survival, but the prognosis is still less than 2 years. Intrinsic chemoresistance and poor penetrance of drugs through the blood–brain barrier makes the treatment of gliomas a challenging task (Hofer and Herrmann, 2001). Noscapine can efficiently cross the blood brain barrier better than morphine does and can inhibit the tumour growth of the intracranial rat C6 glioma after noscapine treatment in mice via oral route (Landen *et al.*, 2004). Noscapine showed greater than 78% inhibition of the growth of intracranial glioma in immuno-compromised mice (Landen *et al.*, 2004). Also the doses of noscapine that showed efficacy against brain tumours did not cause any overt pathological changes in the peripheral nervous system. This makes noscapine a promising anticancer agent that provides novel hope for the treatment of malignant gliomas having less than 20% response rate to conventional therapies (Hofer and Herrmann, 2001) and for which existing treatments are associated with debilitating toxic side effects (Cavaletti *et al.*, 1995).

Noscapine and Angiogenesis

It has now been widely appreciated that angiogenesis is a crucial factor in determining tumour growth *in vivo*. Therefore, prevention of neo-vascularization has become a viable strategy for chemoprevention and therapy for tumour growth. To elicit angiogenesis, tumours generally secrete angiogenic factors such as vascular endothelial growth factor

(VEGF) (Semenza, 2003). The transcriptional activation of these factors is under the control of a prominent transcription factor, induced by the intra-tumoural hypoxic conditions, called the hypoxia-inducible factor 1 (HIF-1). Therefore inhibition of HIF-1 plays a direct role in the prevention of angiogenesis. Many MT binding agents such as paclitaxel, show anti-angiogenic properties due to their ability to target endothelial cell proliferation, migration, tubular morphogenesis and processes required for new blood vessel formation required for tumour growth (Sharma *et al.*, 2001). Noscapine, being an MT binding agent and sharing structure function relationship with HIF-1 pathway inhibitors that belong to the benzyl-isoquiniline class of plant metabolites and/or to MT binding agents such as NSC-134754 (Powis and Kirkpatrick, 2004), it became interesting to look at the possible anti-angiogenic role of noscapine. Indeed, Newcomb *et al.* (2006) report that noscapine inhibits HIF-1α expression, VEGF secretion and angiogenesis *in vitro* in glioma cells. HIF-1 is a heterodimeric protein that consists of constitutively expressed HIF-1α and HIF-1β subunits (Belotti *et al.*, 1996), the under-representation of the α subunit results in the inhibition of angiogenesis (Belotti *et al.*, 1996; Chau *et al.*, 2005). Newcomb *et al.* also showed that noscapine inhibits endothelial tubule formation in HUVEC cells (Newcomb *et al.*, 2006). This study suggests that noscapine may possess novel anti-angiogenic activity associated with two broad mechanisms of action; that is by decreasing HIF-1α expression in hypoxic tumour cells and hence VEGF activation is blocked and thereby inhibiting endothelial cells from forming blood vessels in response to VEGF stimulation. These reports together suggest that in addition to cancer cell death by induction of apoptotic mechanisms, noscapine limits the process of neo-vascularization, required for tumour growth. The standardized anti-angiogenesis assays of several noscapine analogues at the NCI now have promising results for some of them tested so far (unpublished data).

Noscapinoids (Noscapine and its Analogues) as Cancer Cell Cytotoxins

The crystal structure of both noscapine and its target molecule tubulin are known (Seetharaman and Rajan, 1995; Löwe *et al.*, 2001). This information allowed the docking studies of various *in silico* noscapine conformer states on to 3.5 Å tubulin coordinates. Noscapine docks onto β-tubulin near the interface with its dimerization partner, α-tubulin (Checchi *et al.*, 2003). This study showed the presence of an empty space around position 9 of noscapine and that can accommodate small chemical moieties such as electronegative halogen atom (e.g. Cl-, Br-). The study suggested that addition of chemical moiety in the empty space might confer additional electrostatic interactions and hence enhanced biological activity (Aneja *et al.*, 2006a). This hypothesis was tested and confirmed with various noscapine analogs described below.

EMO11 is 9-bromo derivative of noscapine and is one of the well characterized analogues so far. It is 10- to 20- fold more toxic to cancer cells than noscapine, and yet it retains the non-toxic nature of noscapine to healthy tissues. It effectively inhibits vinblastine-resistant as well as vinblastine-sensitive human T-cell lymphoid tumour xenografts in nude mice, there by markedly increasing longevity. EMO11 also inhibits cellular proliferation of both hormone-insensitive and hormone sensitive, adriamycin- and tamoxifen-resistant breast cancer phenotypes (Aneja *et al.*, 2007c). Encouraging preclinical data and significant role in treatment of lymphomas and breast cancer (Jaiswal *et al.*, 2009) makes EMO11 a promising drug especially for the therapy of multi-drug resistant (MDR) tumours. It is impressive that noscapine is active against a variety of cell types across the

NCI panel of 60 human tumour libraries organized into sub panels representing leukaemia, non-small cell lung, colon, CNS, melanoma, renal, ovarian, breast and prostate cancers. As compared to noscapine, our rationally designed derivatives (EM001-EM-044) significantly potentiate this anticancer activity (NCI DTP database). *In vivo* tests of each of these cell lines will be necessary to examine if the observed *in vitro* cytotoxic activity is mimicked in animal models of tumour types represented. The mechanistic nature of the tumour reduction in animal models does suggest that rampant apoptosis may be the primary cause (Fig. 19.4).

Unlike vincas and taxanes, EMO11 is orally available and is non-toxic to the duodenum, liver, spleen, kidney, lung, heart, brain and sciatic nerve. Moreover, EMO11 does not alter the cell counts in blood or leave any signatures of haematologic pathologies and is non-immunosuppressive in nature (Aneja *et al.*, 2006c, 2007c, 2010).

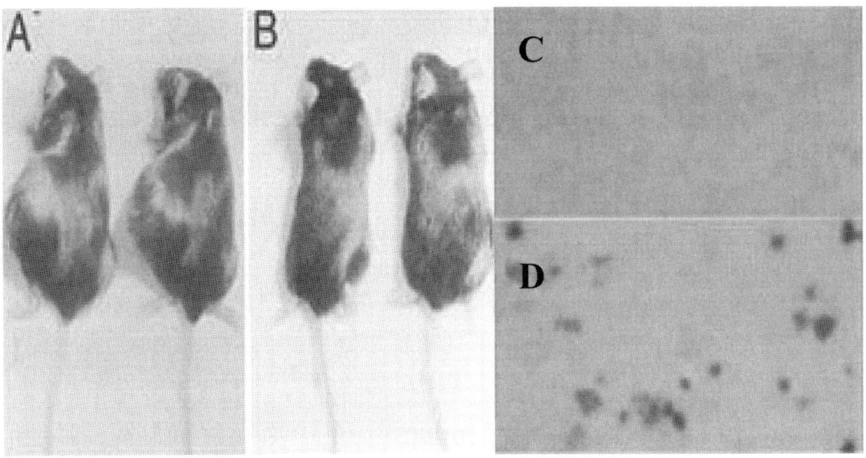

Fig. 19.4. Inhibition of tumour growth by noscapine treatment occurs *via* the induction of tumour cell apoptosis. To determine if noscapine slows tumourogenesis in mice, we engrafted 2 × 10^6 E.G7-OVA thymoma cells subcutaneously in female 8- to 12-week old syngeneic (C57BL/6) mice. After 3 days, one group of the mice was fed intragastrally (gavage) 0.2 ml of vehicle saline solution (pH 4.0, control), and another group received 0.2 ml of noscapine at 15 mg/ml saline pH 4.0 (120 mg/kg body weight). These treatments were performed daily for 3 weeks. At this point, mice were sacrificed and photographed, and tumours were removed and weighed. The tumours were then fixed in 4% formaldehyde, dehydrated, and embedded in paraffin, and sections were cut and processed for histological analysis and for TUNEL staining to detect DNA fragmentation in apoptotic cells. (A) shows the photographs of sham treated control mice bearing large tumours and (B) noscapine treated mice with reduced tumour burden. It is clear from these data that noscapine reduces the tumour size quite dramatically (>75%) (Ye *et al.*, 1998). Additionally, there was no obvious weight loss or any other tissue toxicities detected after noscapine treatment. Furthermore, the residual tumour tissue in the treatment group showed abundant apoptotic cells as revealed by TUNEL stained brown cells in D, compared to rare TUNEL positive cells in control tumour tissue (C).

EMO15 is another analog of noscapine. It is 9-chloro derivative of noscapine that shows a 20-fold lower IC_{50} and induces more potent apoptosis, compared with noscapine, in human breast cancer cells. EMO15 significantly inhibits proliferation of breast cancer cells that are resistant to a variety of anticancer drugs such as tamoxifen, vinblastine, teniposide, adriamycin and paclitaxel (Aneja *et al.*, 2006a). It perturbs cell cycle of breast cancer cells by inducing spindle abnormalities and apoptosis. In addition to its *in vitro* effect, EMO15 effectively inhibits the growth of human breast xenografts and significantly prolongs survival (Aneja *et al.*, 2006a). Like EMO11, this derivative of Noscapine also retains the non-toxic and non-immunosuppressive profile of the parent drug. Therefore, EMO15 emerges as another promising drug for management of breast carcinomas (Aneja *et al.*, 2006a).

9-nitro-noscapine is a novel nitro analog of noscapine, which inhibits cell proliferation of ovarian cancer cells that are sensitive or resistant to paclitaxel and human lymphoblastoid cells that are sensitive or resistant to vinblastine and teniposide. It arrests the cell cycle at G2/M phase and induces apoptosis (Aneja *et al.*, 2006b).

Two brominated noscapine derivatives, 5-bromonoscapine (5-Br-nosc) and reduced 5-bromonoscapine (Rd 5-Br-nosc) are potent MT interfering agents that arrest mitosis and inhibit cell proliferation with much higher efficiency than noscapine. These noscapine derivatives inhibit the proliferation of a series of human breast, cervical, colon, ovarian and prostate cancer cells. Also, these analogs have been shown to inhibit the proliferation of human ovarian cancer cells resistant to paclitaxel and epothilone (Zhou *et al.*, 2003).

Role of Noscapine: Besides Cancer

This chapter has so far been focused upon the discovery of noscapine and its analogs as a unique class of anti-MT, anticancer drugs. As described earlier, the rational basis for this discovery was structural themes that arose from deducing anti-MT drug structure from naturally constructed small library of compounds from Merck Index. Noscapine, being used as an antitussive drug for decades and moreover, its desirable pharmacokinetic and non-toxic profile have been encouraging enough to be taken by human volunteers including pregnant women without any significant signs of toxicities (with couple of subjects reporting nausea). Currently, noscapine is an approved over the counter drug or by prescription in the USA (Link: OTC Ingredient List Updated April 7, 2010).

The excitement in the therapeutic use of noscapine has begun with its unique inclusion as a novel member of anti-MT drugs for cancer chemotherapy. Interphase cells depend upon MTs for the organization of their intracellular organelles and dynamic remodelling requirements of the chemotactic cellular motility. This function of MTs becomes especially important for the sub-population of cells in the animal body that travel long distances in the directed fashion such as the members of the professional phagocytes (macrophages) and other immune cells, to the sites of infection. In fact, colchicine in less-toxic doses had been already in clinical use for gout. Therefore, noscapine being a non-toxic drug, even at high concentration, encouraged some investigators to look at its role in other situations with ischaemia-induced inflammation. Mahmoudian *et al.* (2003) recently reported encouraging results from clinical trials of noscapine as a treatment for stroke. Noscapine antagonizes the effect of bradykinin (Mahmoudian and Mojaverian, 2001) and as expected this group found that noscapine effectively reduces the mortality rate and improves the recovery of the patients suffering from acute ischaemic stroke, without risk of haemorrhage (Mahmoudian *et al.*, 2003).

There are several proliferative non-cancerous conditions such as benign tumours and cysts in various organs that are asymptomatic. However, some of these conditions such as poly-cystic kidney and poly-cystic ovarian syndrome (PCOS) eventually do create serious health problems. New understanding of the mechanism of action of anti-MT drug noscapine and its derivatives leads us to believe that noscapine can be effective in many such diseases. In fact, recently noscapine has been shown to play an important role in subsiding the ovarian cysts and restoring normal menstrual cycle and fertility in female rat model of PCOS induced by anti-progesterone RU486 drug (Priyadarshani *et al.*, 2005).

Clinical Progress

Since noscapinoids tested so far in our laboratory have proven to be safe non-toxic anticancer drugs at concentrations higher then used as antitussive dose in animals, its clinical journey has begun. Table 19.1 shows a time line of its progression so far.

Table 19.1. Journey of noscapine into the drug development pipeline from bench to clinic.

Date	Event
1998	Discovery, CNN coverage, European news telecast
2000	First patent application filed
2002, 2004	Patent application approved (US patent # 6,376,516 and 6,673,814)
2004	Licensed to Cougar Biotechnology
2005	First patient accrued for Phase I trial
2007	Results of phase I trial disclosed to public

In 2004, noscapine was licensed by our university (Emory University, Atlanta, Georgia, USA) to Cougar Biotechnology for clinical development. In 2007, results of phase I trial were disclosed to the public. The trial was an open label dose escalating study where cohorts of subjects with relapsed/refractory non-Hodgkin's lymphoma (NHL) or chronic lymphocytic leukaemia (CLL/SLL) were treated at one of three different dose levels involving total daily doses of 1 g, 2 g and 3 g per day. At each dose level, noscapine was administered orally on a three times a day schedule for 49 days. The interim results are very encouraging showing that noscapine (CB3304) has been well tolerated, with no grade 3 or 4 haematological toxicities (though one grade III neurotoxicity consisting of depressed level of consciousness was experienced at the 3 g dose level). A Phase I/II trial of noscapine in patients with multiple myeloma is currently ongoing at the Center for Lymphoma and Myeloma/Weill Cornell Medical College and Columbia University Medical Center. In addition to this, Cougar is also conducting preclinical development of analogues of noscapine (source: website of cougar biotechnology).

Following the publications of our pre-clinical results, in addition to this official trial, noscapine is also in off-label use by a number of clinicians with patient consent. The cancer patients, who have taken it on their own accord, were desperate due to their failure to respond to standardized chemotherapeutics. These patients range from lymphomas, ovarian, lung, breast, brain and prostate cancers. These experiences however cannot substitute for results from double blind clinical trials which must follow phase I/II results.

Very interestingly, Prostate Cancer Research and Education Foundation (PCREF), in San Diego, CA, USA, has been provided a special permission to use noscapine, based on compassionate use clearance to treat patients with prostate cancer. Dr Israel Barken, founder and director of PCREF, has been very creative in order to be able to glean valuable clinical information by opening 'Smart Chart' a web-based patient tracking graph system (www.pcref.org), which allows data to be collected with ease. This will enable global participation of patients in any trial. This work provides a source of anecdotal information about the personal experiences with noscapine for determined patients who are taking noscapine treatment on their own will, until the results pour into the public domain from official phase II/III trials.

Biotechnological Interventions: *In vitro* Production and Metabolic Engineering

The discovery of noscapinoids has stimulated the biotechnological production of these compounds in plant cell cultures. Yoshimatsu and Shimomura (2001) established three clones of transformed cultures of opium poppy (*Papaver somniferum* L.) by infection with *Agrobacterium rhizogenes* strain MAFF 03-01724. MAFF clone 1 was capable of forming somatic embryos and was selected and its growth and isoquinoline alkaloid production was investigated. Illumination, temperature and nutrient medium composition greatly affected growth, cell morphology and alkaloid accumulation. The MAFF clone 1, cultured in root culture medium, in the dark at 22°C accumulated a high quantity of sanguinarine (652 μg/g dry weight) though the growth was poor (4.4 fold on fresh weight basis after 2 months of culture). The MAFF clone 1 cultured in a quarter macro salt strength Woody Plant medium under 14 h/day light at 22°C developed into plantlets and accumulated significant quantity of codeine (648 μg /g dry wt) together with papaverine, noscapine and sanguinarine. This clone was applied to a rotating drum fermenter (2 l working volume), and ca. 0.3 mg codeine and 0.06 mg sanguinarine were obtained after 4 weeks of culture. One quarter of the codeine produced was found in the culture medium.

Oluk (2006) studied the capacity of alkaloid synthesis in embryogenic callus tissues of Turkish opium poppy, *Papaver somniferum* L. cv. Office-95. The cultures were found to produce noscapine at a level of $2.2 \times 10^{-3}\%$ FW (fresh weight), besides other related compounds like codeine, morphine, papaverine and baine. In addition, efforts have been made to unravel the biogenetic secondary metabolite pathways involved in the synthesis of these and related compounds. The biochemical pathways for their synthesis are far from being completely understood, although scientists have engaged an intense effort for the past several years (Battersby and Hirst, 1965; Facchini and Bird 1998; Frick *et al.*, 2007, Liscombe and Facchini 2008). These efforts may certainly lead to the biotechnological advances in metabolic engineering of these useful metabolites in greater quantities both in homologous or in heterologous systems in coming years.

Future Perspectives

Despite the current cumbersome intravenous slow infusions, requiring multiple hospital visits, uncomfortable side effects of the gastrointestinal and immune systems, neutrocytopenia, and peripheral neuropathies, the first generation MT drugs, vincas, taxanes, epothilones, have proven that MTs are valuable targets for anticancer therapy. Noscapine and its promising derivatives now offer the prospect of a nicer, kinder, and gentler opium derived non-narcotic oral safe class of anticancer drugs that can be taken orally at home. Noscapine has already begun its progress into the pipeline of clinical drug development. New more effective and even targeted derivatives are under development in our group, and we hope many more will follow suite. In addition, biotechnological interventions are underway to obtain improved production of the compound and efforts are on to metabolically engineer the biosynthetic pathways for enhancing yield.

Acknowledgements

We are thankful to all the previous workers of the Joshi lab and collaborators within and outside Emory University, and to the research laboratories that are pursuing noscapinoid research. We are grateful to Dr. Rajesh Arora, the editor, for his valuable assistance during the preparation of this chapter. We thank Ms Vijaya Bhavani Madeti and Ms Krupa Venkatesh for their technical help in referencing the chapter. We apologize to researchers whose work could not be cited in this review for the limitations of size and the focus of this chapter. This research was supported by grants from NIH (NCI R01CA095317).

References

Aneja, R., Lopus, M., Zhou, J., Vangapandu, S.N., Ghaleb, A., Yao, J., Nettles, J.H., Zhou, B., Gupta, M., Panda, D., Chandra, R. and Joshi, H.C. (2006a) Rational design of the microtubule-targeting anti-breast cancer drug EM015. *Cancer Research* 66, 3782–3791.

Aneja, R., Vangapandu, S.N., Lopus, M., Chandra, R., Panda, D. and Joshi, H.C. (2006b) Development of a novel nitro-derivative of noscapine for the potential treatment of drug-resistant ovarian cancer and T-cell lymphoma. *Molecular Pharmacology* 69, 1801–1809.

Aneja, R., Zhou, J., Vangapandu, S.N., Zhou, B., Chandra, R. and Joshi, H.C. (2006c) Drug-resistant T-lymphoid tumours undergo apoptosis selectively in response to an antimicrotubule agent, EM011. *Blood* 107, 2486–2492.

Aneja, R., Dhiman, N., Idnani, J., Awasthi, A., Arora, S.K., Chandra, R. and Joshi, H.C. (2007a) Preclinical pharmacokinetics and bioavailability of noscapine, a tubulin-binding anticancer agent. *Cancer Chemotherapy and Pharmacology* 60, 831–839.

Aneja, R., Ghaleb, A.M., Zhou, J., Yang, V.W. and Joshi, H.C. (2007b) p53 and p21 determine the sensitivity of noscapine-induced apoptosis in colon cancer cells. *Cancer Research* 67, 3862–3870.

Aneja, R., Kalia, V., Ahmed, R. and Joshi, H.C. (2007c) Nonimmunosuppressive chemotherapy: EM011-treated mice mount normal T-cell responses to an acute lymphocytic choriomeningitis virus infection. *Molecular Cancer Therapy* 6, 2891–2899.

Aneja, R., Liu, M., Yates, C., Gao, J., Dong, X., Zhou, B., Vangapandu, S.N., Zhou, J. and Joshi, H.C. (2008) Multidrug resistance-associated protein-overexpressing teniposide-resistant human lymphomas undergo apoptosis by a tubulin-binding agent. *Cancer Research* 68, 1495–1503.

Aneja, R., Asress, S., Dhiman, N., Awasthi, A., Rida, P.C., Arora, S.K., Zhou, J., Glass, J.D. and Joshi, H.C. (2010) Non-toxic melanoma therapy by a novel tubulin-binding agent. *International Journal of Cancer* 126, 256–265.

Battersby, A.R. and Hirst, M. (1965) Concerning the biosynthesis of narcotine. Tetrahedron Letters 11, 669–673.

Belotti, D., Rieppi, M., Nicoletti, M.I., Casazza, A.M., Fojo, T., Taraboletti, G. and Giavazzi R. (1996) Paclitaxel (Taxol®) inhibits motility of paclitaxel resistant human ovarian carcinoma cells. *Clinical Cancer Research* 2, 1725–30.

Buschmann, T., Lin, Y., Aithmitti, N., Fuchs, S.Y., Lu, H., Resnick-Silverman, L., Manfredi, J.J., Ronai, Z. and Wu, X. (2001) Stabilization and activation of p53 by the coactivator protein TAFII31. *The Journal of Biological Chemistry* 276, 13852–13857.

Cassimeris, L., Rieder, C.L. and Salmon, E.D. (1994) Microtubule assembly and kinetochore directional instability in vertebrate monopolar spindles: implications for the mechanism of chromosome congression. *Journal of Cell Science* 107, 285–297.

Cavaletti, G., Tredici, G., Braga, M. and Tazzari, S. (1995) Experimental peripheral neuropathy induced in adult rats by repeated intraperitoneal administration of taxol. *Experimental Neurology* 133, 64–72.

Chau, N.M., Rogers, P., Aherne, W., Carroll, V., Collins, I., McDonald, E., Workman, P. and Ashcroft, M. (2005) Identification of novel small molecule inhibitors of hypoxia-inducible factor-1 that differentially block hypoxia-inducible factor-1 activity and hypoxia-inducible factor-1alpha induction in response to hypoxic stress and growth factors. *Cancer Research* 65, 4918–4928.

Checchi, P.M., Nettles, J.H., Zhou, J., Snyder, J.P. and Joshi, H.C. (2003) Microtubule-interacting drugs for cancer treatment. *Trends in Pharmacological Sciences* 24, 361–365.

Dahlström, B., Mellstrand, T. and Löfdahl, C.G. and Johansson M. (1982) Pharmacokinetic properties of noscapine. *European Journal of Clinical Pharmacology* 22, 535–539.

Desai, A. and Mitchison, T.J. (1997) Microtubule polymerization dynamics. *Annual Review of Cell and Developmental Biology* 13, 83–117.

Donehower, R.C., Rowinsky, E.K., Grochow, L.B., Longnecker, S.M. and Ettinger, D.S. (1987) Phase I trial of taxol in patients with advanced cancer. *Cancer Treatment Reports* 71, 1171–1177.

Empey, D.W., Laitinen, L.A., Young, G.A., Bye, C.E. and Hughes, D.T. (1979) Comparison of the antitussive effects of codeine phosphate 20 mg, dextromethorphan 30 mg and noscapine 30 mg using citric acid-induced cough in normal subjects. *European Journal of Clinical Pharmacology* 16, 393–397.

Facchini, P.J. and Bird, D.A. (1998) Developmental regulation of benzylquinoline alkaloid biosynthesis in opium poppy plants and tissue cultures. *In vitro Cellular and Developmental Biology-Plant* 34, 69–79.

Frick, S., Kramell, R. and Kutchan, T.M. (2007) Metabolic engineering with a morphine biosynthetic P450 in opium poppy surpasses breeding. *Metabolic Engineering* 9, 169–176.

Giannakakou, P., Sackett, D.L., Kang, Y.K., Zhan, Z., Buters, J.T., Fojo, T. and Poruchynsky, M.S. (1997) Paclitaxel-resistant human ovarian cancer cells have mutant beta-tubulins that exhibit impaired paclitaxel-driven polymerization. *The Journal of Biological Chemistry* 272, 17118–25.

Giannakakou, P., Gussio, R., Nogales, E., Downing, K.H., Zaharevitz, D., Bollbuck, B., Poy, G., Sackett, D., Nicolaou, K.C. and Fojo, T. (2000) A common pharmacophore for epothilone and taxanes: molecular basis for drug resistance conferred by tubulin mutations in human cancer cells. *Proceedings of the National Academy of Sciences of the United States of America* 97, 2904–2909.

Guchelaar, H.J., ten Napel, C.H., de Vries, E.G. and Mulder, N.H. (1994) Clinical, toxicological and pharmaceutical aspects of the antineoplastic drug taxol. *Clinical Oncology (Royal College of Radiologists (Great Britain))* 6, 40–48.

Haikala, V., Sothmann, A. and Marvola, M. (1986) Comparative bioavailability and pharmacokinetics of noscapine hydrogen embonate and noscapine hydrochloride. *European journal of clinical pharmacology* 31, 367–369.

Hofer, S. and Herrmann, R. (2001) Chemotherapy for malignant brain tumours of astrocytic and oligodendroglial lineage. *Journal of Cancer Research and Clinical Oncology* 127, 91–105.

Hruban, R.H., Yardley, J.H., Donehower, R.C. and Boitnott, J.K. (1989) Taxol toxicity. Epithelial necrosis in the gastrointestinal tract associated with polymerized microtubule accumulation and mitotic arrest. *Cancer* 63, 1944–1950.

Idänpään-Heikkilä, J.E. (1968) Studies on the fate of 3H-noscapine in mice and rats. *Annales Medicinae Experimentalis et Biologiae Fenniae* 46, 201–216.

Jaiswal, A.S., Aneja, R., Connors, S.K., Joshi, H.C., Multani, A.S., Pathak, S. and Narayan, S. (2009) 9-bromonoscapine-induced mitotic arrest of cigarette smoke condensate transformed breast epithelial cells. *Journal of Cellular Biochemistry* 106, 1146–1156.

Jordan, M.A. and Wilson, L. (1999) The use and action of drugs in analyzing mitosis. *Methods in Cell Biology* 61, 267–295

Joshi, H.C. and Cleveland, D.W. (1990) Diversity among tubulin subunits: toward what functional end? *Cell Motility and the Cytoskeleton* 16, 159–63.

Joshi, H.C., Palacios, M.J., McNamara, L. and Cleveland, D.W. (1992) Gamma-tubulin is a centrosomal protein required for cell cycle-dependent microtubule nucleation. *Nature* 356, 80–83.

Karlsson, M.O., Dahlström, B., Eckernäs, S.A., Johansson, M. and Alm, A.T. (1990) Pharmacokinetics of oral noscapine. *European Journal of Clinical Pharmacology* 39, 275–279.

Kavallaris, M., Tait, A.S., Walsh, B.J., He, L., Horwitz, S.B., Norris, M.D. and Haber, M. (2001) Multiple microtubule alterations are associated with Vinca alkaloid resistance in human leukemia cells. *Cancer Research* 61, 5803–5809.

Kavanagh, J.J. and Kudelka, A.P. (1993) Systemic therapy for gynecologic cancer. *Current Opinion in Oncology* 5, 891–899.

Ke, Y., Ye, K., Grossniklaus, H.E., Archer, D.R., Joshi, H.C. and Kapp, J.A. (2000) Noscapine inhibits tumour growth with little toxicity to normal tissues or inhibition of immune responses. *Cancer Immunology and Immunotherapy* 49, 217–225.

Kleihues, P., Louis, D.N., Scheithauer, B.W., Rorke, L.B., Reifenberger, G., Burger, P.C. and Cavenee, W.K. (2002) The WHO classification of tumours of the nervous system. *Journal of Neuropathology and Experimental Neurology* 61, 215–225.

Krishna, R. and Mayer, L.D. (2000) Multidrug resistance (MDR) in cancer. Mechanisms, reversal using modulators of MDR and the role of MDR modulators in influencing the pharmacokinetics of anticancer drugs. *European Journal of Pharmaceutical Sciences* 11, 265–283.

Landen, J.W., Lang, R., McMahon, S.J., Rusan, N.M., Yvon, A.M., Adams, A.W., Sorcinelli, M.D., Campbell, R., Bonaccorsi, P., Ansel, J.C., Archer, D.R., Wadsworth, P., Armstrong, C.A. and Joshi, H.C. (2002) Noscapine alters microtubule dynamics in living cells and inhibits the progression of melanoma. *Cancer Research* 62, 4109–4114.

Landen, J.W., Hau, V., Wang, M., Davis, T., Ciliax, B., Wainer, B.H., Van Meir, E.G., Glass, J.D., Joshi, H.C. and Archer, D.R. (2004) Noscapine crosses the blood-brain barrier and inhibits glioblastoma growth. *Clinical Cancer Research* 10, 5187–5201.

Lasagna, L., Owens, A.H. Jr, Shnider, B.I. and Gold, G.L. (1961) Toxicity after large doses of noscapine. *Cancer Chemotherapy Reports. Part 1* 15, 33–34.

Lipton, R.B., Apfel, S.C., Dutcher, J.P., Rosenberg, R., Kaplan, J., Berger, A., Einzig, A.I., Wiernik, P. and Schaumburg, H.H. (1989) Taxol produces a predominantly sensory neuropathy. *Neurology* 39, 368–373.

Liscombe, D.K. and Facchini, P.J. (2008) Evolutionary and cellular webs in benzylisoquinoline alkaloid biosynthesis. *Current Opinion in Biotechnology* 19, 173–180.

Löwe, J., Li, H., Downing, K.H. and Nogales, E. (2001) Refined structure of alpha beta-tubulin at 3.5 A resolution. *Journal of Molecular Biology* 313, 1045–57.

Ludueña, R.F. (1998) Multiple forms of tubulin: different gene products and covalent modifications. *International Review of Cytology* 178, 207–275.

Mahmoudian, M. and Mojaverian, N. (2001) Effect of noscapine, the antitussive opioid alkaloid, on bradykinin-induced smooth muscle contraction in the isolated ileum of the guinea-pig. *Acta Physiologica Hungarica* 88, 231–237.

Mahmoudian, M., Mehrpour, M., Benaissa, F. and Siadatpour, Z. (2003) A preliminary report on the application of noscapine in the treatment of stroke. *European Journal of Clinical Pharmacology* 59, 579–581.

Martindale, P. (1977) In: Wade (ed.) *The Extra Pharmacopoeia*. Pharmaceutical Press, London, p.1250.

McIntosh, J.R. (1985) Spindle structure and the mechanisms of chromosome movement. *Basic Life Sciences* 36, 197–229.

Mitchison, T. and Kirschner, M. (1984) Microtubule assembly nucleated by isolated centrosomes. *Nature* 312, 232–237.

Mitchison, T., Evans, L., Schulze, E. and Kirschner, M. (1986) Sites of microtubule assembly and disassembly in the mitotic spindle. *Cell* 45, 515–527.

Newcomb, E.W., Lukyanov, Y., Schnee, T., Ali, M.A., Lan, L. and Zagzag, D. (2006) Noscapine inhibits hypoxia-mediated HIF-1alpha expression and angiogenesis *in vitro*: a novel function for an old drug. *International Journal of Oncology* 28, 1121–1130.

Nicklas, R.B. (1997) How cells get the right chromosomes. *Science* 275, 632–637.

Nicolaou, K.C., Yang, Z., Liu, J.J., Ueno, H., Nantermet, P.G., Guy, R.K., Claiborne, C.F., Renaud, J., Couladouros, E.A. and Paulvannan, K., *et al.* (1994) Total synthesis of taxol. *Nature* 367, 630–634.

Oluk, E.A. (2006) Alkaloid production in tissue cultures of *Papaver somniferum* L. cv. Office-95. *Plant Tissue Culture and Biotechnology* 16(1), 1–4.

Perkin, W.H. and Robinson, R. (1910) Synthesis of dl–narcotine (gnoccopine). Preliminary note. *Proceedings of Chemical Society* 26, 48.

Powis, G. and Kirkpatrick, L. (2004) Hypoxia inducible factor-1alpha as a cancer drug target. *Molecular Cancer Therapeutics* 3, 647–654.

Priyadarshani, A., Katyal, A., Bhatnagar, A. and Chandra, R. (2005) 24th Annual Convention of IACR and International Symposium on 'HPV and Cervical Cancer' Poster presentations.

Rowinsky, E.K. and Donehower, R.C. (1991) The clinical pharmacology and use of antimicrotubule agents in cancer chemotherapeutics. *Pharmacology and Therapeutics* 52, 35–84.

Rowinsky, E.K., Cazenave, L.A. and Donehower, R.C. (1990) Taxol: a novel investigational antimicrotubule agent. *Journal of the National Cancer Institute* 82, 1247–1259.

Seetharaman, J. and Rajan, S.S. (1995) Crystal and molecular structure of noscapine. *Zeftschrm fur Kristamographic* 210, 111–113.

Semenza, G.L. (2003) Targeting HIF-1 for cancer therapy. *Nature reviews. Cancer* 3, 721–732.

Sharma, R.A., Harris, A.L., Dalgleish, A.G., Steward, W.P. and O' Byrne, K.J. (2001) Angiogenesis as a biomarker and target in cancer chemoprevention. *The Lancet Oncology* 2, 726–732.

Stark, G.R. (1986) Cancer chemotherapy. Progress in understanding multidrug resistance. *Nature* 324, 407–408.

Stewart, Z.A., Mays, D. and Pietenpol, J.A. (1999) Defective G1-S cell cycle checkpoint function sensitizes cells to microtubule inhibitor-induced apoptosis. *Cancer Research* 59, 3831–3837.

Tiveron, C., Hartley-Asp, B., Johansson, C.J. and Pacchierotti, F. (1993) Noscapine does not show aneugenic activity in mouse oocytes. *Mutagenesis* 8, 311–315.

Warolin, C. (1999) Pierre-Jean Robiquet. *Revue d'histoire de la pharmacie* 47, 97–110.

Wiernik, P.H., Schwartz, E.L., Strauman, J.J., Dutcher, J.P., Lipton, R.B. and Paietta, E. (1987) Phase I clinical and pharmacokinetic study of taxol. *Cancer Research* 47, 2486–2493.

Wilson, L. and Jordan, M.A. (1995) Microtubule dynamics: taking aim at a moving target. *Chemistry and Biology* 2, 569–573.

Ye, K., Ke, Y., Keshava, N., Shanks, J., Kapp, J.A., Tekmal, R.R., Petros, J. and Joshi, H.C. (1998) Opium alkaloid noscapine is an antitumour agent that arrests metaphase and induces apoptosis in dividing cells. *Proceedings of the National Academy of Sciences of the United States of America* 95, 1601–1606.

Yoshimatsu, K. and Shimomura, K. (2001) Isoquinoline alkaloid production by transformed cultures of *Papaver somniferum. Kokuritsu Iyakuhin Shokuhin Eisei Kenkyusho Hokoku.* 119, 52–56.

Zhou, J., Gupta, K., Yao, J., Ye, K., Panda, D., Giannakakou, P. and Joshi, H.C. (2002) Paclitaxel-resistant human ovarian cancer cells undergo c-Jun NH2-terminal kinase-mediated apoptosis in response to noscapine. *The Journal of Biological Chemistry* 277, 39777–39785.

Zhou, J., Gupta, K., Aggarwal, S., Aneja, R., Chandra, R., Panda, D. and Joshi, H.C. (2003) Brominated derivatives of noscapine are potent microtubule-interfering agents that perturb mitosis and inhibit cell proliferation. *Molecular Pharmacology* 63, 799–807.

Zhou, J., Liu, M., Aneja, R., Chandra, R., Lage, H. and Joshi, H.C. (2006) Reversal of P-glycoprotein-mediated multidrug resistance in cancer cells by the c-Jun NH2-terminal kinase. *Cancer Research* 66, 445–452.

Chapter 20

Recent Developments in Biotechnology and Chemistry of *Podophyllum*

Abha Chaudhary, Bikram Singh and Paramvir Singh Ahuja

Introduction

Plants have provided us with food, fuel and fibre since prehistoric times. They have been an inexhaustible source of a diverse array of chemicals such as flavours, fragrances, natural pigments, pesticides and pharmaceuticals. This seemingly unrelated collection of chemicals can be grouped together under a broad category of plant secondary metabolites. In recent years, the widely occurring natural product class of lignans has become of prominent interest. Enriched in lignans, Podophyllum is the dried roots and rhizomes of species of *Podophyllum*, which was described and its first modern botanical name was given by Linnaeus in 1753, at that time he referred to two of his earlier works, in which he simply called it 'podophyllum'. The first serious chemical investigation was carried out by Podwyssotzki in 1880. Two species of the genus *Podophyllum* (Podophyllaceae), *P. hexandrum* Royle in India and Nepal and *P. peltatum* L. in the USA, are the commercially most exploited sources of podophyllotoxin (Stahelin and Von Wartburg, 1991; Imbert, 1998; Arora *et al.*, 2008). American Podophyllum contains 4–5% podophyllum resin, whereas the Indian sps. contains 6–12%. The variation in percentage of resin is attributed to seasonal differences, different sites of growth and age of the plant. In certain areas as much as 20% resin has also been recorded. The highest percentage of resin is obtained in May–June during the flowering stage. Thus Indian Podophyllum when collected at the proper season contains 2.5 times more resin compared to its American counterpart. Moreover, this resin has double the amount of podophyllotoxin. The currently preferred source of podophyllotoxin is the underground rhizomes and roots of the Indian Podophyllum species, *P. hexandrum* (Bhadula *et al.*, 1996; Nadeem *et al.*, 2000). Podophyllotoxin (Fig. 20.1) is commonly extracted from *P. hexandrum* in which the concentration of podophyllotoxin is around 40% of resin. Other species include *P. aurantiocaule*, *P. delavayi*, *P. pleianthum* and *P. versipelle*. Phenolic quinones, lignans, xanthones, coumarins and other classes exist in considerable numbers. Common lignans present include podophyllotoxin, 4′-demethylpodophyllotoxin, α-peltatin and β-peltatin (Bastos *et al.*, 1996).

 Medicinal Plant Biotechnology (ed. Rajesh Arora)

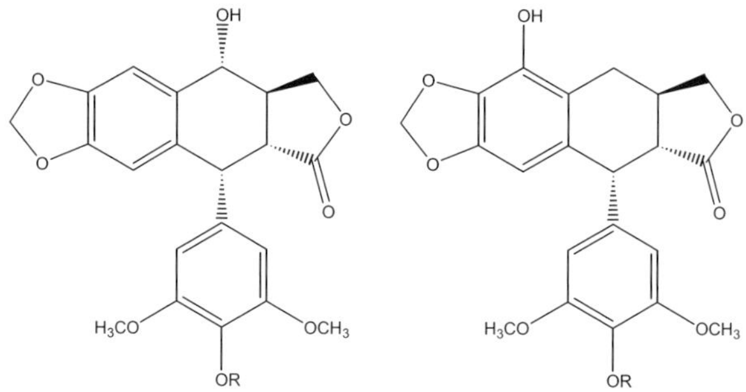

A R=Me
B R=H

C R=Me
D R=H

Fig. 20.1. Podophyllotoxin (A), 4′-demethylpodophyllotoxin (B), α-peltatin (C), β-peltatin (D)

***Podophyllum peltatum* Linn.**

P. peltatum Linn. (Podophyllaceae) is a herbaceous perennial plant originating in North America and distributed from the Hudson Bay to Florida. It is currently referred to as American mandrake, may apple, because of its edible fruit, duck's foot referring to the shape of the leaves, or Devil's apple, hog apple or wild lemon. Native Indians of the American continent used an aqueous extract of the roots as a purgative, antihelmintic or cathartic. At the same time, a traditional use of *Podophyllum* was reported to be against snake bites. *Podophyllum* was used by American pioneers from Maine to Mississippi and Louisiana, as an emetic and cholagogue. It was introduced into the first American pharmacopoeia in 1820 and into the English pharmacopoeia in 1864. Native Americans used the root as a strong laxative, to treat worms and for numerous other things. The root is currently used in cancer medications and may have commercial potential as a cultivated plant.

***Podophyllum hexandrum* Royle**

P. hexandrum Royle syn. *P. emodi* (emodi meaning 'of the Himalayas') is closely related to the American *P. peltatum* and found in the foothills of the Himalayas as well as in the rain forests of Nepal and in Kashmir. *P. emodi*, like *P. peltatum*, has a deep rhizome which is rich in resin. Another species also exists, namely *P. shikkimensis* which is smaller but also contains podophyllotoxins as does *P. peltatum*, native to Taiwan. This has been studied and shown to contain the same substances (podophyllotoxin, deoxypodophyllotoxin and the corresponding 4'-demethylated derivatives).

The lignan derivatives are fairly widespread in the plant world and are not specific to the Podophyllaceae family. Podophyllotoxin and other lignans have been isolated from other plants e.g. deoxypodophyllotoxin from *Thujopsis dolabrata* (L. fil) which belongs to the Cupressaceae family or 4'-demethylpodophyllotoxin, podophyllotoxin and peltatins from *Linum album* which belongs to the Linaceae family. Podophyllotoxin is also found in Polygalaceae spp. This is produced in relatively minute quantities by other plant species, viz., *Linum flavum*, *L. album* and *Juniperus chinensis.* Podophyllotoxin is the major lignan present in the resin and is a dimerized product of the intermediates of the phenylpropanoid pathway. The cytotoxic activity of podophyllotoxin is based on its ability to inhibit the microtubule assembly during cell division. It is used as a starting material for the chemical synthesis of anticancer drugs, etoposide (VP-16-213) and teniposide (VM-26), which have fewer side effects (Canel *et al.*, 2000). The Indian *P. hexandrum* is superior to its American counterpart, *P. peltatum*, in terms of its higher podophyllotoxin content (4% in the dried roots in comparison to only 0.25% for *P. peltatum* (Jackson and Dewick, 1984; Van Uden *et al.*, 1989; Arora *et al.*, 2008).

Podophyllotoxin derivatives belong to the lignan series and are formed by the addition of two cinnamic groups, thereby forming a 2,3-dibenzyl butane framework. On studying the literature concerning podophyllotoxin and related lignans, at least three different numbering systems have been observed in literature. These are shown in Fig 20.2. One of the systems is essentially based on the systematic nomenclature of the IUPAC; it considers naphthalene as the basic system and assigns numbers 9 and 10 to the positions that would correspond to 8a and 4a, respectively. Another numbering system is based on the biogenesis of lignans, where the two phenylpropane residues are numbered from 1 to 9 and 1′ to 9′ for assigning the unprimed locants.

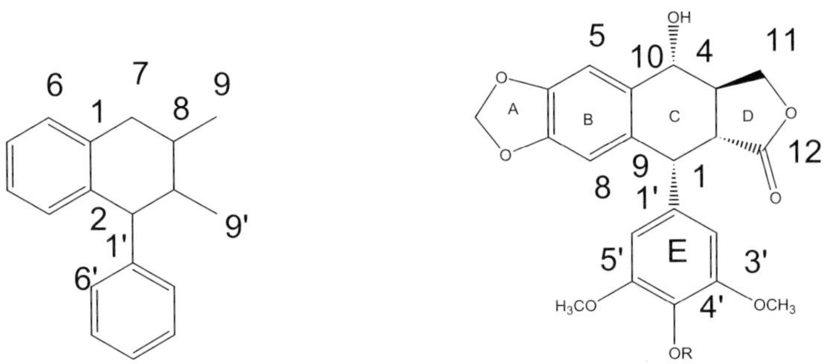

Fig. 20.2. Numbering system for lignans.

The stereochemical convention used indicates that the alpha configuration is below the plane of the molecule and the beta configuration above. The essential characteristics of most podophyllotoxin species are: (i) almost planar tetracyclic group ABCD going from the dioxole ring to the lactone ring; (ii) an aromatic ring E located at position 1 with α-configuration and a bond which has a certain degree of free rotation; (iii) the four adjacent asymmetrical centres; and (iv) special stereo chemical properties at C_4 which determine whether a given class of compounds has an affinity for tubulin or not.

Developments in Chemical Aspects

The first literature report on the extraction of *Podophyllum* was that of King in 1844, who called the resin podophyllin. He also described the effects of podophyllin administered to one of his patients. On the basis of evidence about the traditional therapeutic uses of these extracts, investigators were interested in discovering which of the chemical entities present in podophyllin were responsible for its 'anticancer' activity. In pioneering studies on *P. peltatum* published in 1881, Podwyssotzki managed to achieve the difficult crystallization of the major component of the extract which he named 'podophyllotoxin'. After alkali treatment, he also succeeded in isolating picropodophyllin which he believed to be the active substance in *Podophyllum* but was subsequently shown to be inactive. He also isolated podophyllinic acid and quercetin. The chemical analysis of *P. emodi* began with investigations based on materials from the roots and it was shown that the components of the two different *Podophyllum* species were the same. It was identified along with its epimer picropodophyllotoxin. Various chemical investigations have shown that the resin consisted mainly of podophyllotoxin, deoxypodophyllotoxin, 4'-demethylpodophyllotoxin and α- and β-peltatins. The presence of quercetin, pigments and their glucoside derivatives were also reported.

A microanalytical technique for determination of podophyllotoxin in *P. hexandrum* roots was developed by Mishra *et al.* (2005a) quantitatively through RP–HPLC and RP–HPTLC. Podophyllotoxin in the extract ranged from 4.25 to 5.22% and 4.15 to 5.12% by HPLC and HPTLC respectively. The podophyllotoxin content in resin was estimated as 4.91–5.57% and 4.90–5.58% by HPLC and HPTLC respectively.

A RP–HPTLC method was also developed for the quantification of podophyllotoxin in *P. hexandrum*, collected from different locations of the western Himalaya of India, ranging from 3250 to 4100 m above the mean sea level (Mishra *et al.*, 2005b). Seven pairs of diastereoisomers of podophyllum lignans at the C4 position, including three pairs of spin-labelled compounds, have been separated within 20 min (Slaninaa and Glatz, 2004).

The topical application of podophyllin in *Condyloma acuminatum* produced very satisfactory clinical results and podophyllin caused pronounced cytological changes in normal human and rabbit skin. These reports initiated a renewed medicinal interest in podophyllin from the standpoint of its antimitotic activity. Podophyllin is still the drug of choice in the treatment of *Condyloma acuminatum*. The continued study of podophyllin based upon its recognition as a cytotoxic agent and the growing knowledge of the nature and biological properties of pure constituents involving a multitude of chemical and biological investigations have been conducted over a period of more than a century. Initial expectations regarding the clinical utility of podophyllotoxin were tempered largely due to its unacceptable gastrointestinal toxicity. Using special procedures to inhibit enzymatic degradation, these researchers indeed obtained the podophyllotoxin-β-D-glucopyranoside as the main component and its 4'-demethyl derivative from the Indian *Podophyllum* species. The research efforts were focused on to chemically modify podophyllotoxin, which eventually led to the discovery of clinically important anticancer drugs etoposide and teniposide (Fig 20.3).

During the drug developmental stages, it began to become clear that etoposide induced double stranded breaks by stabilizing a complex formed between topoisomerase II and DNA that is referred to as the cleavable complex. Topoisomerase II was the most likely cellular target in the double stranded breaking activity of epipodophyllotoxin like etoposide and teniposide. The cytotoxicity of analogues of etoposide was associated with inhibiting DNA topoisomerase II by stabilizing the covalent topo II- DNA cleavable complex. Although etoposide is active in the treatment of many cancers and is widely used in the therapy, it presents several limitations, such as moderate potency, poor water solubility, development of drug resistance, metabolic inactivation, and toxic effects. Therefore in order to obtain better therapeutic agents and extensive synthetic efforts have been devoted to overcome these problems.

With increasing information about its structure–activity relationships, wide investigations have generated exciting chemotherapeutic candidates and successful applications of drug development from podophyllotoxin-related lead, such as etopophos, NK-611, GL-331, azatoxin, TOP-53, tafluposide.

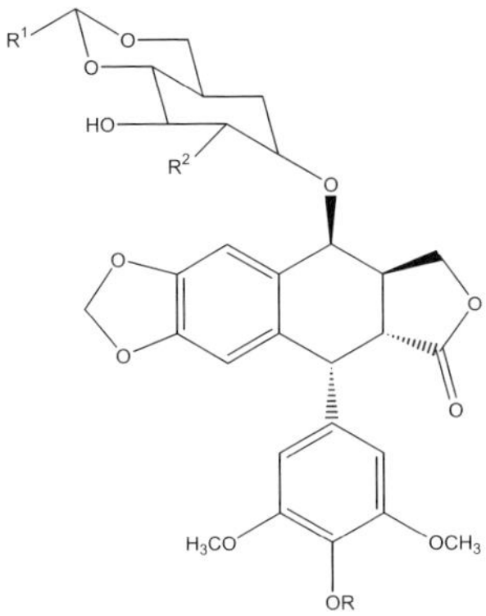

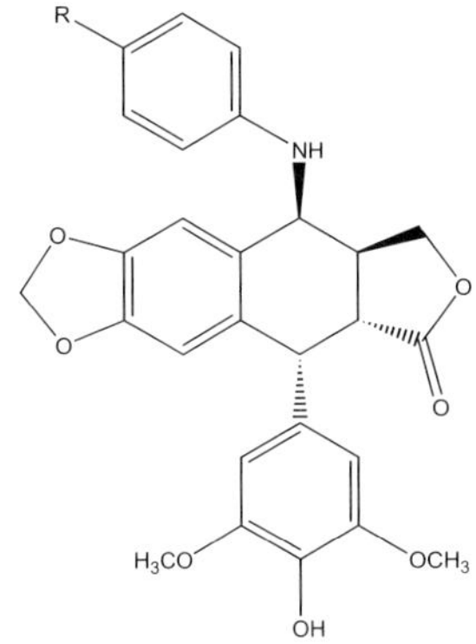

R	R^1	R^2	
H	CH_3	OH	**Etoposide**
H	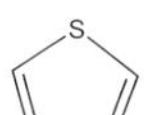	OH	**Teniposide**
$PO(OH)_2$	CH_3	OH	**Etopophos**
H	CH_3	$N(CH_3)_2.HCl.H_2O$	**NK611**

331

R	
NO_2	**GL**
F	**NPF**

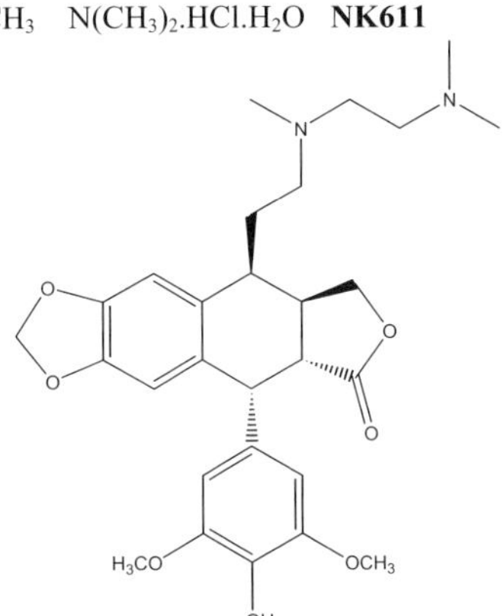

TOP-53

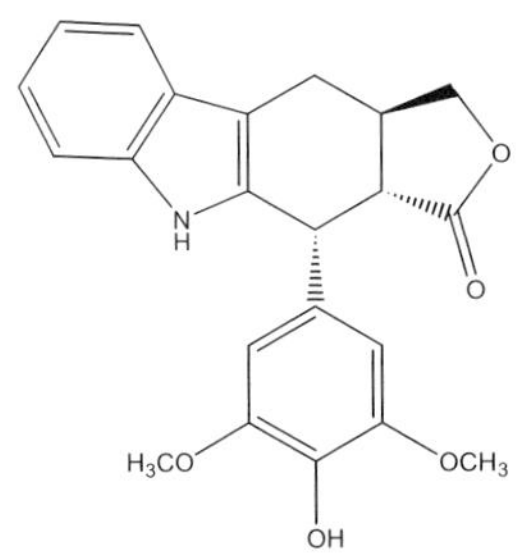

Azatoxin

Fig. 20.3. Chemical structures of podophyllotoxin derivatives.

Etopophos

The problem of the low solubility of etoposide was resolved by introducing a phosphate group at position 4': this led to the selection of etopophos which was better suited for intravenous administration. In fact, the highly soluble compound was shown to be a prodrug for etoposide itself. The phosphate group rapidly hydrolyzed by phosphatases contained in the plasma, thereby generating etoposide in the circulatory system (Schacter *et al.*, 1994; Budman, 1996). This prodrug was developed and many of the side effects of other, more or less toxic analogues of etoposide were eliminated.

NK-611

In NK-611, the alcohol group at position 2" of the ethylidene glucoside species is replaced with a dimethylamino group. Introducing an amine both renders the new compound soluble, because of the generation of its hydrochloride, and improves its bioavailability as a result of the molecule-altered physicochemical properties. The antitumour activity of this compound against human tumour xerograft seemed to be promising and was shown to be similar to that of etoposide; however it proved to be a more potent topoisomersae II inhibitor and was more cytotoxic against a variety of human cancer lines including lung, gastrointestinal, ovarian, testicular, breast, head and neck tumour cells and leukaemia.

GL-331

GL-331 exhibits cytotoxicity, topoisomerase II inhibitory activity and potency against tumours. This new epipodophyllotoxin derivative doesn't have the glucoside group which was believed to be necessary for activity, but rather had a p-nitroanilino group present at 4-β position. Structural studies at this position involving replacement with aniline or an amine (Lee *et al.*, 1989, 1990) showed that compounds containing an aniline group at 4-β position had a higher level of inhibitory activity for human topoisomerase II. Various structural modifications made at this position showed that there was no correlation between human topoisomerase and DNA breaking associated with the protein. GL-331 showed topo II inhibition and caused DNA double-strand breakage and G2 phase arrest, it could induce cell death by stimulating protein tyrosine phosphatase activity and apoptotic DNA formation. GL-331 was also shown to be active in many multidrug-resistant cancer cell lines. It is known to show good stability, biocompatibility and its pharmacokinetic profiles are similar to those of etoposide.

Azatoxin

Azatoxin only differs from podophyllotoxin with respect to: (i) replacement of the benzodioxole group with an indole ring; (ii) the presence of a nitrogen atom at the asymmetrical centre at C and (iii) the presence of a phenol at position 4', essential for the anti-topoisomerase I1 activity. Molecular modelling had shown that only an indole ring at that position gave a tight fit with podophyllotoxin. Nevertheless, the cleavage induced by azatoxin differed from that induced by other topoisomerase II inhibitors like VP-16, amsacrine or mitoxantrone. Although it was almost as potent as VP-16 in inhibiting topoisomerase II, it differed in the fact that it also inhibited the polymerization of tubulin at low concentrations.

TOP-53

Structural studies of podophyllotoxin analogues based on molecular modelling techniques resulted in the rational design of a range of new chemical agents. It was almost as potent as etoposide in inhibiting topoisomerase II. TOP-53 was then selected for evaluation from a series of 4β-alkyl amino derivatives of 4'-O-demethyl-4-desoxypodophyllotoxin synthesized. TOP-53 displayed twice the inhibitory activity of etoposide against topoisomerase II and exhibits *in vivo* superior antitumour activity than etoposide against different types of cancer. In view of high activity and good properties, TOP-53 has been progressed to phase II clinical trials (Terada *et al.*, 1993).

Tafluposide

More recently, 2",3"-bis pentafluorophenoxyacetyl-4',6'-ethylidene-β-D-glucoside of 4'-phosphate-4'-demethylepipodophyllotoxin has been obtained. It is a novel catalytic inhibitor of topoisomerase I and II and has shown outstanding antitumour activity *in vivo* against a panel of experimental human tumour xenografts. Tafluposide showed significant cytotoxicity against cells derived from either haematological or solid tumours, with a marked inter-patient variation (Sargent *et al.*, 2003).

Extensive modifications have been done in the four rings (A, B, C, D) of podophyllotoxin to assess the effects of different structural moieties. Diverse norbornenecarboxylate esters of podophyllotoxin and its epimers and diastereoisomers have been prepared through Diels–Alder cycloaddition by treating the dienophilic acrylates of cyclolignans with cyclopentadiene (Fig. 20.4).

Pharmacological Action

It has been shown that the mechanism of cytostatic action at the cellular level of podophyllin was the same as that of colchicine, i.e. an inhibition of the formation of the mitotic spindle, resulting in an arrest of the cell division process in metaphase and a clumping of the chromosomes (c-mitosis). Later it was shown that, at the subcellular level, this is due to binding of podophyllotoxin to tubulin, preventing these macromolecules to form microtubules, which constitute the fibres of the mitotic spindle. In contrast to podophyllotoxin, neither etoposide nor teniposide has any effect on tubulin assembly. It has been reported that etoposide and teniposide are not inhibitors of microtubule due to the presence of the bulky glucoside moiety, and suggested that they must be inhibiting cell proliferation by some other mechanism. It was concluded that these drugs arrested cells in either the late S or early G2 phase of the cell cycle and they detected single-stranded breaks in DNA by the interaction of drug with DNA in HeLa cells. Further research indicated that etoposide does not interact with DNA *in vitro* to cause cleavage but rather induces single stranded breaks in DNA only *in situ*.

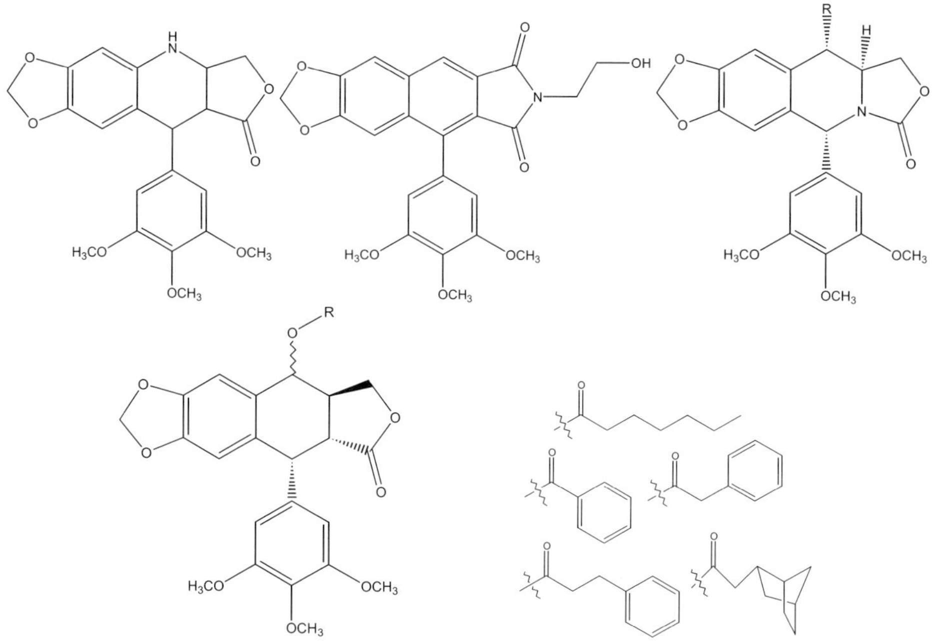

Fig. 20.4. Structures of synthetic analogues of podophyllotoxin.

Apart from antitumour and antiviral activity, insecticidal and phytotoxic activities of podphyllotoxin are also reported. Initially, it was hypothesized that podophyllotoxin exerts its pharmacological effect by producing potent spasm in blood vessels with resultant ischaemia, necrosis and sloughing. The anti-mitotic effects of podophyllotoxin are the result of the drug's ability to act as an inhibitor of microtubule assembly. Podophyllotoxin also appears to attach to cell proteins and acts by increasing the incorporation of amino acids into proteins, inhibition of purine synthesis and inhibition of purine incorporation into RNA. It has also been found to have a direct effect on mitochondria by reducing the activity of cytochrome oxidase and succinoxidase. The mode of their antiviral activity is by tubulin binding, reverse transcriptase inhibition, integrase inhibition and topoisomerase inhibition. Podophyllotoxin is most notable among the tubulin binding lignans. By binding to tubulin, these are able to disrupt the cellular cytoskeleton and interfere with some critical viral processes.

Biosynthesis of Podophyllotoxin and Analogue Compounds

The full biosynthetic route of cyclolignans has not been elucidated yet, but several studies in different species of *Forsythia*, *Linun* and *Podophyllum* led to the proposition of a pathway (Petersen and Alfermann, 2001). In the initial steps, coniferyl alcohol was synthesized from phenylalanine (Fig. 20.5) by phenylpropanoid enzymes (Van Uden *et al.*, 1990). Instead of an alternative route to give the polymeric product lignin, lignans are obtained by dimerization of coniferyl alcohol to yield pinoresinol (Davin *et al.*, 1997). After several steps, this compound is transformed into matairesinol and in multistep reaction the intermediate yatein yields deoxypodophyllotoxin (Bromhead *et al.*, 1991; Xia *et al.*, 2001). Several feeding experiments were carried out in cell cultures to elucidate the biosynthetic relationships among the natural cyclolignans (Van Uden *et al.*, 1995). In *Linum flavum* cell cultures, 6-methoxypodophyllotoxin was obtained as the major cyclolignan and this compound was not obtained when the cell cultures were fed with podophyllotoxin. Otherwise, 6-methoxypodophyllotoxin was obtained when deoxypodophyllotoxin or β-peltatin were added to the cell cultures. In *P. emodi* podophyllotoxin and its β-D-glucoside were the cyclolignans obtained when the cell cultures were fed with deoxypodophyllotoxin. The enzymes deoxypodophyllotoxin-7-hydroxylase and deoxypodophyllotoxin-6-hydroxylase, both isolated from cell cultures of *Linum* spp. (Petersen and Alfermann, 2001; Molog *et al.*, 2001), are the proteins that carry out the hydroxylation on positions 7 or 6 of deoxypodophyllotoxin to give either podophyllotoxin or β-peltatin, respectively.

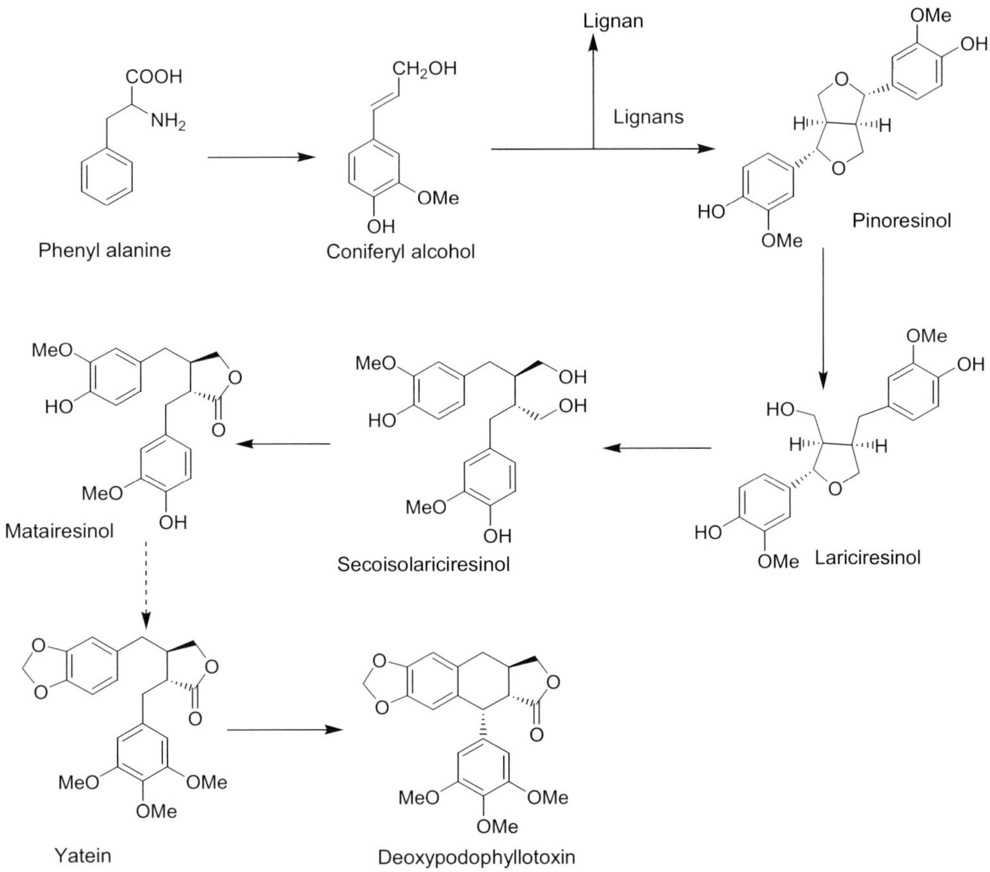

Fig. 20.5. Proposed biosynthesis of cyclolignans related to podophyllotoxin.

Developments in the Biotechnological Aspects

Because of the limited supply of *Podophyllum* rhizomes, due to their intensive collection in the wild, there is considerable interest in both defining and biotechnologically exploiting the podophyllotoxin biosynthetic pathway, thereby making this compound more generally available (Lewis and Davin, 1999).

Cell cultures have been used in the past for the production of natural products *in vitro* as an alternative. The production of podophyllotoxin from callus cultures has been reported for *P. peltatum* (Kadkade, 1981, 1982). It is very difficult to establish callus cultures in *P. hexandrum*. Van Uden *et al.* (1989) reported initiation of cell cultures from the roots of *P. hexandrum* and production of podophyllotoxin in undifferentiated callus and cell suspension cultures. It has also been reported that coniferin was a substrate for podophyllotoxin production (Van Uden *et al.*, 1990). Van Uden *et al.* (1989) tested hundreds of basal medium, hormone combinations and concentrations and found B5 medium supplemented with 2% coconut milk, 4% sucrose, 4 mg/l naphthalene acetic acid to be the best. The cell suspension cultures of *P. hexandrum* do not appear to secrete podophyllotoxin in the culture medium as the amount of podophyllotoxin in the medium was reported to be negligible. Culture conditions and other complex factors are involved in the biosynthesis of podophyllotoxin. Woerdenbag *et al.* (1990) reported increased podophyllotoxin content in *P. hexandrum* cell cultures after feeding coniferyl alcohol complexed with β-cyclodextrin. Cells cultured in media containing 4 mg l^{-1} NAA accumulated 0.001–0.002% podophyllotoxin on dry weight basis. A new route for secondary metabolite production is the use of natural vector system.

Genetically transformed hairy roots produced by infection of plants with *Agrobacterium rhizogenes*, a gram-negative soil bacterium, appear to be a promising tool for secondary metabolite production. A variety of fungal products can elicit inducible defensive plant responses in both host and nonhost plants and that such responses can also be triggered by plant products released during cell-wall degradation. Physical damage to the leaves of *P. hexandrum* resulted in the accumulation of podophyllotoxin glucosides and aglucone at the wound site (Oliva *et al.*, 2002).

Since podophyllotoxin has been recognized as a starting compound for the chemical synthesis of etoposide (VP-16-213) and teniposide (VM-26), its ever increasing demand has led to the investigations on the biotechnological production of podophyllotoxin (Petersen and Alfermann, 2001; Puri *et al.*, 2006). Submerged cultivation of *P. hexandrum* cells for podophyllotoxin production was studied in a 3l stirred tank bioreactor fitted with a low-shear Setric impeller (Chattopadhyay *et al.*, 2002). Batch cultivation of *P. hexandrum* was conducted in a 3l stirred-tank bioreactor using statistically optimized process parameters. The plants exhibit delayed emergence of functional leaves or hypocotyl dormancy. On GA_3 treatment the functional leaves (Khushwaha *et al.*, 2007) were found to emerge at a favourable temperature of 25°C in a higher percentage of seedlings and in a shorter time. Functional leaves emerged even at 10°C, a temperature when hypocotyl dormancy generally prevails. For further growth and development in all the seedling parts there is a considerable increase in the biochemical parameters related to carbon and nitrogen metabolism, respiration and total dehydrogenase activity also indicated an enhancement of metabolic processes as influenced by GA_3.

Puri *et al.*, (2006) isolated an endophytic fungus *Trametes hirsuta*, which has been shown to synthesize podophyllotoxin and other aryltetralin compounds *in vitro*. Culture conditions for growth and productivity were also standardized by these workers. The availability of a biotechnological route for production of aryltetralin lignans is a step forward in the direction of gaining independence from the endangered plant species.

Kharkwal *et al.* (2002) have developed a method for inducing improved seed germination in a high altitude *P. hexandrum* species through hot water treatment at 40–100°C for 30–120 seconds. Plants of *P. hexandrum*, collected from lower, mid, and upper distribution limits in alpine Himalaya were studied under greenhouse conditions to evaluate the photosynthetic response (Vats and Kumar, 2006). The maximum net photosynthetic rates (P_N) and stomatal conductance (g_s) were measured in the considered population during the 3–6th week of development.

Etoposide is derived chemically from podophyllotoxin by two step conversion. It has been demonstrated that a cell line of *P. peltatum* active in the biosynthesis of podophyllotoxin was able to maintain repeated biotransformation by oxidative coupling of the butanolide to the podophyllotoxin analogue. It has been carried out in a bioreactor for a total of more than 15 cycles each of 24 h duration giving a yield of around 50%. This result indicates the feasibility of a biotransformation route to podophyllotoxin using a chemically synthesized precursor (Franssen and Walton, 1999).

Conclusion

Lignans are believed to act as defense compounds in the producing plants. For mankind, the lignans play a role as health-promoting agents in dietary plants on the one hand and in cancer treatment on the other. Although *P. hexandrum* has been considered a rare and threatened species, large scale removal (generally during the middle of growing season and well before seed set in the wild) of its underground parts still continues at rates well over natural regeneration. Therefore, special attention needs to be given for its propagation and conservation. It must be emphasized that greater awareness needs to be created and the local inhabitants should be motivated to cultivate medicinal plants as alternative (high value) crops, including trials at lower altitudes.

Acknowledgements

Authors are grateful to the Director, IHBT (CSIR), Palampur for providing required facilities. Financial assistance to Abha Chaudhary during the course of the project (NWP-0037) from CSIR, New Delhi, India is gratefully acknowledged. This is IHBT Communication No. 1002.

References

Arora, R., Singh, S., Puri, S.C. and Sharma, R.K. (2008) Himalayan Mayapple (*Podophyllum hexandrum* Royle): Traditional Uses, Clinical Indications and Future Prospects. In: Watson, R. and Preedy, V.R. (eds) *Botanical Medicine in Clinical Practice*. CABI Publishing, Wallingford, Oxon, United Kingdom, pp. 71–84.

Bastos, J.K., Burandt Jr, C.L., Nanayakkara, N.P.D., Bryant, and McChesney, J.D. (1996) Quantitation of aryltetralin lignans in plant parts and among different populations of *Podophyllum peltatum* by reversed-phase high-performance liquid chromatography. *Journal of Natural Products* 59, 406–408.

Bhadula, S.K., Singh, A., Lata, H., Kunyal, C.P. and Purohit, A.N. (1996) Genetic resources of *Podophyllum hexandrum* Royle, an endangered medicinal species from Garhwal Himalaya, India. *Plant Genetic Resources Newsletter* 106, 26–29.

Bromhead, A.J., Rahman, M.M., Dewick, P.M., Jackson, D.E. and Lucas, J.A. (1991) Matairesinol as precursor of *Podophyllum* lignans. *Phytochemistry* 30, 1489–1492.

Budman, D.R. (1996) Early studies of etoposide phosphate, a water-soluble prodrug. *Seminars* in *Oncology* 23, 8–14.

Canel, C., Moraes, R.M., Dayan, F.E. and Ferreira, D. (2000) Molecules of interest podophyllotoxin. *Phytochemistry* 54, 115–120.

Chattopadhyay, S., Srivastava, A.K., Bhojwani, S.S. and Bisaria, V.S. (2002) Production of podophyllotoxin by plant cell cultures of *Podophyllum hexandrum* in bioreactor. *Journal of Fermentation* and *Bioengineering* 93, 215–220.

Davin, L.B., Wang, H.B., Crowell, A.L., Bedgar, D.L., Martin, D.M., Sarkanen, S. and Lewis, N.G. (1997) Stereoselective bimolecular phenoxy radical coupling by an auxiliary (dirigent) protein without an active center. *Science* 275, 362–366.

Franssen, M.C.R. and Walton, M.J. (1999) Chemicals from plants, perspectives on plant secondary products. In: Walton, M.J. and Brown, D.E. (eds), *Biotransformations*. Imperial College Press, London, pp. 277–325.

Imbert, T.F. (1998) Discovery of podophyllotoxins. *Biochimie* 80, 207–222.

Jackson, D.E. and Dewick, P.M. (1984) Aryltetralin lignans from *Podophyllum hexandrum* and *Podophyllum peltatum*. *Phytochemistry* 23, 1147–1152.

Kadkade, P.G. (1981) Formation of podophyllotoxins by *Podophyllum peltatum* tissue cultures. *Naturwissenschaften* 68, 481–482.

Kadkade, P.G. (1982) Growth and podophyllotoxin production in callus tissues of *Podophyllum peltatum*. *Plant Science Letters* 25, 107–115.

Kharkwal, A.C., Prakash, O., Bhattacharya, A., Nagar, P.K. and Ahuja, P.S. (2002) Method for inducing improved seed germination in *Podophyllum hexandrum* Royle, US 6449899 B1.

Kushwaha, R., Pandey, S., Chanda, S., Bhattacharya, A. and Ahuja, P.S. (2007) GA3 induced changes in slow growing endangered Himalayan plant *Podophyllum hexandrum* and hastening of vegetative growth. *Plant Growth Regulation* 51, 207–215.

Lee, K.H., Imakura, Y., Haruna, M., Beers, S.A., Thurston, L.S., Dai, H.J., Chen, C.H., Liu, S.Y. and Cheng, Y.C. (1989) Antitumour Agents. 107. New cytotoxic 4-alkylamino analogues of 4'-demethyl-epipodophyllotoxin as inhibitors of human DNA topoisomerase II. *Journal of Natural Products* 52, 606–613.

Lee, K.H., Beers, S.A., Mori, M., Wang, Z.Q., Kuo, Y.H., Li, L., Liu, S.Y., Chang, J.Y., Han, F.S. and Cheng, Y.C. (1990) Antitumour agents. 111. New 4-hydroxylated and 4-halogenated anilino derivatives of 4'-demethylepipodophyllotoxin as potent inhibitors of human DNA topoisomerase II. *Journal of Medicinal Chemistry* 33, 1364–1368.

Lewis, N.G. and Davin, L.B. (1999) Lignans: biosynthesis and function. In: Barton, Sir D.H.R., Nakanishi, K. and Meth-Cohn, O. (eds) *Comprehensive Natural Products Chemistry*, Vol. 1. Elsevier, London, pp. 639–712.

Mishra, N., Acharya, R., Gupta, A.P., Singh, B., Kaul, V.K. and Ahuja, P.S. (2005a) A simple microanalytical technique for determination of podophyllotoxin in *Podophyllum hexandrum* roots by quantitative RP-HPLC and RP-HPTLC. *Current Science* 88, 1372–1373.

Mishra, N., Gupta, A.P., Singh, B., Kaul, V.K. and Ahuja, P.S. (2005b) A rapid determination of podophyllotoxin in *Podophyllum hexandrum* by reverse phase high performance thin layer chromatography. *Journal of Liquid Chromatography & Related Technologies* 28, 677–691.

Molog, G.A., Empt, U., Kuhlmann, S., Van Uden, W., Pras, N., Alfermann, A.W. and Petersen, M. (2001) Deoxypodophyllotoxin 6-hydroxylase, a cytochrome P450 monooxygenase from cell cultures of *Linum flavum* involved in the biosynthesis of cytotoxic lignans. *Planta* 214, 288–294.

Nadeem, M., Palni, L.M.S., Purohit, A.N., Pandey, H. and Nandi, S.K. (2000) Propagation and conservation of *Podophyllum hexandrum* Royle: an important medicinal herb. *Biological Conservation* 92, 121–129.

Oliva, A., Moraes, R.M., Watson, S.B., Duke, S.O. and Dayan, F.E. (2002) Aryltetralin lignans inhibit plant growth by affecting the formation of mitotic microtubular organizing centers. *Pesticide Biochemistry* and *Physiology* 72, 45–54.

Petersen, M. and Alfermann, A.W. (2001) The production of cytotoxic lignans by plant cell cultures. *Applied Microbiology and Biotechnology* 55, 135–142.

Puri, S.C., Nazir, A., Chawla, R., Arora, R., Riyaz-ul-Haq, S., Amna, T., Ahmed, B., Verma, V., Singh, S., Sagar, R.K., Sharma, A.K., Kumar, R., Sharma, R.K. and Qazi, G.N. (2006) The endophytic fungus *Trametes hirsuta* as a novel alternative source of podophyllotoxin and related aryl tetralin lignans. *Journal of Biotechnology* 122, 494–510.

Sargent, J.M., Elgie, W., Williamson, J.C. and Hill, B.T. (2003) *Ex vivo* effects of the dual topoisomerase inhibitor tafluposide (F 11782) on cells isolated from fresh tumour samples taken from patients with cancer. *Anti-cancer Drugs* 14, 467–473.

Schacter, L.P., Igwemezie, L.N., Seyedsadr, M., Morgenthien, E., Randolph, J., Albert, E. and Santabfirbara, P. (1994) Clinical and pharmacokinetic overview of parenteral etoposide phosphate. *Cancer Chemotherapy* and *Pharmacology*. 34, 58–63.

Slaninaa, J. and Glatz, Z. (2004) Separation procedures applicable to lignan analysis. *Journal of Chromatography* B 812, 215–229.

Stahelin, H.F. and Wartburg, Von, A.V. (1991) The chemical and biological route from podophyllotoxin glucoside to etoposide. *Cancer Research* 51, 5–15.

Terada, T., Fujimoto, K., Nomura, M., Yamashita, J., Wierzba, K., Yamazaki, R., Shibata, J., Sugimoto, Y. and Yamada, Y. (1993) Antitumour agents. 3. Synthesis and biological activity of 4.beta.-alkyl derivatives containing hydroxy, amino, and amido groups of 4'-O-demethyl-4-desoxypodophyllotoxin as antitumour agents. *Journal of Medicinal Chemistry* 36, 1689–1699.

Van Uden, W., Pras, N., Visser, J.F. and Malingre, T.M. (1989) Detection and identification of podophyllotoxin produced by cell cultures derived from *Podophyllum hexandrum* Royle. *Plant Cell Reports* 8, 165–168.

Van Uden, W., Pras, N. and Malingre, T.M. (1990) On the improvement of the podophyllotoxin production by phenylpropanoid precursor feeding to cell cultures of *Podophyllum hexandrum* Royle. *Plant Cell, Tissue and Organ Culture* 23, 217–224.

Van Uden, W., Bouma, A.S., Bracht Waker, J.F., Middel, O., Wichers, H.J., De Waard, P., Woerdenbag, H.J., Kellogg, R.M. and Pras, N. (1995) The production of podophyllotoxin and its 5- methoxy derivative through bioconversion of cyclodextrin complexed deoxypodophyllotoxin by plant cell cultures. *Plant Cell, Tissue and Organ Culture* 42, 73–79.

Vats, S.K. and Kumar, S. (2006) Photosynthetic response of *Podophyllum hexandrum* Royle from different altitudes in Himalayan ranges. *Photosynthetica* 44, 136–139.

Woerdenbag, H.J., Van Uden, W., Frijlink, H.W., Lerk, C.F., Pras, N. and Malingre, T.M. (1990) Increased podophyllotoxin production in *Podophyllum hexandrum* cell suspension cultures after feeding coniferyl alcohol as a β-cyclodextrin complex. *Plant Cell Reports* 9, 97–100.

Xia, Z.Q., Costa, M.A., Pelissier, H.C., Davin, L.B. and Lewis, N.G. (2001) Secoisolariciresinol dehydrogenase purification, cloning and functional expression-implications for human health protection. *Journal of Biological Chemistry* 276, 12614–12623.

Chapter 21

Plant-expressed Griffithsin: A Protein with Potent Broad Spectrum Inhibitory Effects against Enveloped Viruses

J. Calvin Kouokam and Kenneth E. Palmer

Introduction

Entry of enveloped viruses is the first event of the several steps that are critical for the infection of their human host. The mechanisms surrounding the viral entry have been studied for many families and genera of enveloped viruses. Although viral entry might slightly differ from one group of viruses to another, this step is usually mediated by envelope glycoproteins which span the viral membrane bilayer and project from the virion surfaces as spikes. Envelope glycoproteins are essential for two principal tasks including binding of virions to the host cell surface (attachment) and mediation of membrane fusion (Marsh, 1984). The envelope proteins are targets for infection-neutralizing antibodies; some viruses have evolved immune evasion strategies that involve masking critical domains of the viral envelope with a shield of N-linked glycans (NLG), and targeting the glycan shield has recently been proposed as an antiviral strategy relevant to several important human pathogens such as HIV-1, hepatitis C virus and influenza (Balzarini, 2007).

In general, antiviral drugs target specific steps in the viral life cycle. Many antiviral drugs are nucleoside analogues that interfere with viral nucleic acid metabolism, for example Ribovirin (active against HCV and several other viruses) and the antiretroviral reverse transcriptase inhibitor Zidovudine (ZDV). However, despite the recent advances in understanding of viral cycles as well as the development of new antiviral therapies and other preventive methods, enveloped viruses continue to pose a serious health threat. HIV continues to spread worldwide with about 15,000 new infections every day, Hepatitis C virus is becoming a serious clinical challenge in many countries and seasonal and pandemic influenza remains a major public health problem. Some of the limitations of existing antiviral drugs include the resistance displayed by some viruses towards them which has led drug designers to opt for complementary combinations with high manufacturing costs, which makes delivery in resource-poor areas (where the disease burden is often highest) unfeasible. In order to better combat viral diseases, new, more efficient, inexpensive, broad spectrum antiviral drugs are needed. Carbohydrate-binding products like lectin proteins have the ability to tightly and specifically bind to NLG that are present on the surface of enveloped viruses. Since these glycoproteins often act as ligands that target receptors on the host cell surface, lectins could inhibit initial viral entry events and therefore act as prophylactics to block initial infection or as therapeutics in preventing disease progression. Griffithsin (GRFT) is a relatively newly discovered lectin that was identified and purified using an anti-HIV bioassay guided fractionation from an aqueous extract of the red alga *Griffithsia* sp. collected from the waters off New Zealand (Mori *et al.*, 2005). Of note, the discovery of GRFT supports the superiority of bioactivity guided fractionation of natural products over the isolation of individual substances or fractions followed by screening for active substances. In fact in the case of the marine alga *Griffithsia* sp., although there were many studies done on this plant, no report had been published on its antiviral activity, probably because the organic fractions and/or compounds were the exclusive targets (Mori *et al.*, 2005). GRFT belongs to the jacalin related lectin family and has' been shown to display potent inhibitory effect against enveloped viruses, including HIV and SARS (Mori *et al.*, 2005; Ziolkowska *et al.*, 2006, 2007a; O'Keefe *et al.*, 2009;). Therefore, GRFT is considered a candidate vaginal microbicide for prevention of HIV transmission; it may also have application in the context of prevention and treatment of acute and chronic infections with viruses bearing dense clusters of high mannose NLG such as SARS CoV, influenza and hepatitis C, amongst others. Further studies are underway in our laboratory and others in order to completely define pharmacological and toxicological properties of this lectin which could become an effective weapon against many enveloped viruses including HIV. In this review, we will discuss the GRFT structure and sugar binding properties as well as described antiviral activities of this lectin and the suggested mechanisms of antiviral activity. For use as a vaginal microbicide product in particular, a low cost source of bulk griffithsin product is needed, and we discuss our recent efforts to use the economies of scale of agriculture to produce this lectin derived from a marine alga in land plants.

GRFT Structure and Binding Properties

A thorough characterization of the molecular structure and mode of action of any drug candidate is an important prerequisite for clinical development. The structure and function of griffithsin has been reported in several recent publications (Mori *et al.*, 2005; Ziolkowska *et al.*, 2006, 2007a,b; Emau *et al.*, 2007; O'Keefe *et al.*, 2009). Preliminary studies of the anti-HIV active fraction of an aqueous extract of the red alga *Griffithsia* by Mori *et al.* (2005) suggested the active component to be a protein that binds the viral envelope glycoprotein gp120; further purification was done utilizing ammonium sulfate precipitation and various chromatography techniques. Peptide sequencing of the purified GRFT polypeptide after cleavage with cyanogen bromide (CNBr) showed a 121 amino acid protein with no significant homology with any known protein. Amino acid residue 31 did not match any of the 20 standard amino acids, and was substituted with alanine in synthetic cDNAs used to express recombinant GRFT in *E. coli* (Mori *et al.*, 2005; Giomarelli *et al.*, 2006;) and plants (Ziolkowska *et al.*, 2006; O'Keefe *et al.*, 2009). Another remarkable feature in GRFT primary structure is the presence of three separate glycine rich regions (GGSGG) found at positions 9–12, 40–44 and 86–90, respectively. SDS-PAGE and ESI-MS studies of the purified lectin determined a molecular mass of around 13 kDa. Further structural studies using antibodies raised against GRFT suggested a possible existence of a dimeric form of this protein (Mori *et al.*, 2005). GRFT protein sequence similarities search did not show any significant homologies of greater than eight continuous amino acids and no known protein was more than 30% similar in the amino acid composition. These findings confirmed GRFT as a unique protein but gave no indication of what its secondary or tertiary structures could be. In 2006, GRFT became the first recombinant protein derived from a plant to undergo detailed structural studies (Ziolkowska *et al.*, 2006). It is currently proposed that GRFT belongs to the family of jacalin related lectins (JRL), a group of proteins that is widespread in the plant kingdom which form a β-prism fold I structure consisting of three repeats of an antiparallel four-stranded β sheet that form a triangular prism (see Ziolkowska *et al.*, 2006 and public protein structure databases for graphic depictions of GRFT structure). Jacalin is the prototype of this diverse group of lectins and is derived from seeds of *Artocarpus integrifolia* (jack fruit). While the structure of each GRFT monomer is similar to that of the JRLs, its dimeric structure is unique (Ziolkowska *et al.*, 2006). Advanced structural studies revealed that GRFT forms an intimate dimer in which the first two β strands of one monomer are associated with ten strands of the other chain. This unique feature has led to describe GRFT as a domain swapped dimer and it has been suggested that the GRFT minimum biological unit is the dimeric form of this lectin, since domain swapping enhances the inter subunit interaction (Chandra, 2006; Ziolkowska *et al.*, 2006). Indeed, whether the monomeric form of GRFT actually exists is still an unanswered question (Ziolkowska *et al.*, 2006). It has been shown that GRFT dimer displays six principal mannose binding sites, where each monomer bears a group of three (Man1, Man2 and Man3 for the first GRFT monomer and Man4, Man5 and Man6 for the second) and that the interactions between the three mannose molecules with the first monomer were virtually identical to those of the other mannose ligands of the other subunit (Ziolkowska *et al.*, 2006). The binding site of Man1 (homologous to Man4) corresponds to a common carbohydrate binding site found in all the jacalin related family of lectins. Asp109, Tyr110 and Asp112 are actively involved in the sugar binding. However, in the case of GRFT, this site includes amino acids from both monomers due to domain swapping. Site 2 (Man2 and Man5, respectively) has been found so far only in banana lectin which belongs to the same family. Here the hydrogen bonds are established between the mannose molecule and the amides of Ser27, Tyr28 and Gly44. The third site (Man3 and Man6, respectively) is unique to GRFT since Asp70, with which these mannose molecules seal their primary interaction, is not found in any other β prism-I lectin. Amino acids 67, 68 and 90 are involved in the formation of hydrogen bonds (Ziolkowska *et al.*, 2006). In the case of GRFT, the three carbohydrate binding sites form an almost perfect equilateral triangle on the edge of the lectin, a phenomenon never observed for the JRLs, although seen in lectins with β-prism-II fold (Ziolkowska *et al.*, 2006). A careful observation revealed that all three carbohydrate binding sites have some common features in their sequences. A GGSGG segment where the main chain amides of C-terminal Gly residues (Gly12, 44 and 90 to site 1, 2 and 3, respectively) provides the final interactions with the saccharides. In addition, a tyrosine precedes the aspartic acid by two positions, and a glycine by four in these conserved sequences (Mori *et al.*, 2005; Ziolkowska *et al.*, 2006). In further work, Ziolkowska and collaborators studied the crystallization of GRFT using different disaccharides including three forms of mannobiose and maltose (Ziolkowska *et al.*, 2007a): addition of 1→2α-mannobiose and 1→3α-mannobiose lead to precipitation of the GRFT solution and co-crystallization with 1→6α-mannobiose as well as maltose was successful. In addition, many techniques were used to show that mannose and 1→6α-mannobiose binding constants to GRFT were in the same range (binding constants of 102 and 83.3 for mannose and mannobiose, respectively). Maltose showed a fourfold lower affinity to the lectin in comparison with the latter sugars although the interactions at the first sugar unit of 1→6α-mannobiose or maltose were very similar to interactions present in the complex with mannose (Ziolkowska *et al.*, 2007a). Previous data indicated that 100 mM of mannose, glucose and N-acetylglucosamine prevented GRFT binding to gp120, which was not the case for some other saccharides tested including fucose, xylose, galactose, N-acetylgalactose, and the α-acid glycoprotein (Mori *et al.*, 2005). In a recent publication, it was shown that all six monosaccharide-binding sites of GRFT were occupied in the complexes formed by GRFT and glucose or N-acetylglucosamine and that the conformation of the sugar molecules is similar to that of mannose. However, calorimetry experiments revealed that binding between mannose and GRFT was fully saturable whereas saturation of binding was not achieved for the other monosaccharides, suggesting that GRFT has a thermodynamic preference for mannose and 1→ 6α-mannobiose over glucose and maltose (Ziolkowska *et al.*, 2007a) although structural data indicate very little difference in the ability of GRFT to bind mannose, glucose or N-acetylglucosamine (Ziolkowska *et al.*, 2007b). Further biophysical studies are needed in order to better understand the basis of this tight affinity of GRFT to mannose. However, the data reviewed above describe GRFT as a mixed specificity lectin that has the ability to bind glycan rich glycoproteins found on the viral envelope in a monosaccharide-dependent manner (Mori *et al.*, 2005). Modelling studies were carried out using one monomer of GRFT. Here, $Man_9GlcNAc_2$, a high mannose oligosaccharide found on the surface of many viral glycoproteins, was modelled to interact in a tridentate fashion to the three close binding sites, in all three possible configurations irrespective to which terminal mannose or which binding site was involved (Ziolkowska *et al.*, 2007a). These studies also showed that none of the smaller oligomannose sugars could bind in a tridentate way since the loss of one or two terminal mannose residue ($Man_8GlcNAc_2$ and $Man_7GlcNAc_2$, respectively) leads to a prediction of a bidentate complex instead. It was therefore suggested that the high affinity of $Man_9GlcNAc_2$ to GRFT is a result of its interaction with up to three binding sites on the lectin. Interestingly, another lectin, the snowdrop agglutinin, which also have three carbohydrate binding sites arranged in a form of equilateral triangle, did not allow the tridentate modelling even with the high mannose oligosaccharide $Man_9GlcNAc_2$, probably because of the large separation of the mannose binding sites on this protein, more than 20Å compared to approximately 15Å for GRFT (Ziolkowska *et al.*, 2007a).

Antiviral Activity and Available Safety Data of GRFT

As mentioned above, GRFT was discovered as part of a vast programme at the National Cancer Institute (NCI) aiming at screening natural product extracts for their inhibitory effects against HIV (Mori *et al.*, 2005; Ziolkowska *et al.*, 2007a). In a bioactivity guided fractionation, preliminary findings allowed Mori *et al.* to hypothesize that the anti-HIV compound of *Griffithsia* sp. extracts was likely a protein that bound soluble gp120 (Mori *et al.*, 2005). Indeed, purified GRFT was shown in ELISA experiments to have a direct interaction with the HIV envelope glycoproteins gp120, gp160 and gp41 and the binding was concentration and glycosylation-dependent in the case of gp120 (Mori *et al.*, 2005). It has been observed that GRFT rapidly precipitates upon addition of multivalent carbohydrates (Ziolkowska *et al.*, 2006, 2007a). It has been suggested that the precipitation should not be considered to negatively affect the antiviral activity of the lectin since cross-linking of the glycoproteins on the viral envelope might be as efficient in preventing fusion as the multivalent interactions with a single molecule (Ziolkowska *et al.*, 2007a). In fact, it is generally accepted that the enhanced binding potential of GRFT for the high-mannose sugars present on the HIV envelope gp120 is due to its structure which offers six binding sites for mannose in a compact domain-swapped dimer (Ziolkowska *et al.*, 2006). Cyanovirin (CV-N), another lectin originally isolated from blue green algae *Nostoc ellipsosporum*, has attracted much attention these past years in the field of microbicide development due to its potent anti-HIV inhibitory effects. To date, CV-N is the most potent anti-HIV plant lectin described apart from GRFT with an EC50 of 0.1 nM which is at least 2.5 higher than that of GRFT (Table 21.1) and structural studies show that CV-N displays only four carbohydrate binding sites in its dimeric form, which might explain this significant difference in the antiviral activity of both proteins (Ziolkowska *et al.*, 2006; Ziolkowska and Wlodawer, 2006). Indeed, using HIV-1_{RF} (a T-topic strain) in CEM-SS cells, it was shown that GRFT exhibit potent cytoprotective effects along with decreased concentrations of viral replication markers (including the reverse transcriptase and the viral core antigen P24) in the cell culture supernatants and this antiviral activity of GRFT was shown to be in the mid-to-high picomolar range against both T-tropic and M-tropic viruses, laboratory adapted or primary isolates (Mori *et al.*, 2005).

Table 21.1. Selected algal and plant lectins with potent anti-HIV activity.

Lectin	Origin	EC_{50} or IC_{50} in nM (HIV)
GRFT	*Griffithsia* sp.	0.043
CV-N	*Nostoc ellipsosporum*	0.1
SVN	*Scytonema varium*	0.3
MVL	*Microcystis viridis*	30
Psophocarpus tetragonolobus lectin	*P. tetragonolobus*	52

Recently, it was reported that GRFT exhibited a high potency in blocking R5- and X4-viruses at concentrations below 1 nM (Emau *et al.*, 2007). In addition it has been shown that GRFT is able to inhibit syncytium formation between uninfected and chronically infected cells in a concentration dependent manner, with an eventual complete abolishment at high concentrations (Mori *et al.*, 2005). These findings suggest that GRFT has the ability to act using a dual mode which consists of prevention of initial infection, as well as cell-to-cell transmission–both modes are probably used in HIV infection. Binding of oligosaccharides to multiple sites on a single molecule of GRFT provide the basis of its unique antiviral properties (Ziolkowska *et al.*, 2007b) and it has been hypothesized that the lectin would inactivate other enveloped viruses, especially those that have highly glycosylated proteins on their surface. In fact, CV-N which inactivates HIV by binding to the viral envelope glycoprotein gp120 with high affinity (similar to GRFT) has been shown to possess a rather broad range of inhibitory activities against many other enveloped viruses from diverse genera. For example, CV-N binds to the influenza haemagglutinin surface glycoprotein, blocks viral infection and this lectin has recently be used to provide treatment of influenza infections in mice and ferrets (Smee *et al.*, 2008). Previous studies have reported the antiviral effects of CV-N against human pathogens namely herpes simplex virus, ebola virus, measles virus as well as hepatitis C virus (Ziolkowska and Wlodawer, 2006), and even some viruses belonging to the order of Nidovirales which include toroviruses, arteriviruses, roniviruses and the coronaviruses (van der Meer *et al.*,

2007). These inhibitory effects have been attributed to the binding of CV-N to surface envelope glycoproteins of the respective viruses. Interestingly, the antiviral properties of GRFT were evaluated against the coronavirus which causes SARS (Severe Acute Respiratory Syndrome) and the algal lectin was able to inhibit at nanomolar concentrations both viral replication and the cytopathicity induced by the virus, most likely by binding to the SARS associated coronavirus (CoV) surface 'spike' protein that is heavily glycosylated. Despite the structural properties of GRFT which predict potent inhibitory effects against a broad range of enveloped viruses, this protein is yet to be fully characterized as for its antiviral activities. More studies are expected in the near future to examine the inhibitory effects of this unique carbohydrate binding protein against other enveloped viruses beside HIV and SARS-CoV, especially those that are transmitted via mucosal routes, for example human influenza viruses A and B, ebola virus and diverse coronaviruses. It would be particularly significant if griffithsin is shown to have *in vivo* activity against herpes simplex viruses (HSV) -1 and 2 since these viruses have epidemiological synergy with HIV/AIDS, i.e. play a major role in increasing the risk for acquisition and transmission of HIV. If GRFT is found to be useful as a therapeutic, it could also have valuable utility for treatment of HIV and HCV co-infections, which are presently difficult to treat effectively.

Since GRFT is a relatively newly discovered carbohydrate binding agent (CBA), there is still a realm of questions which deserve full attention, especially about its safety profile in the case it should be developed as a topical microbicide. There are many reports which indicate that some CBAs are mitogenic and/or able to agglutinate human red blood cells, to induce inflammation and cellular toxicity. Until now, there is little documentation about the eventual toxic effects of GRFT. In their pioneering work, Mori *et al.* (2005) tested the cytotoxicity of the algal lectin at concentrations up to 783 nM (1,000 to more than 18,000 times the EC_{50} values in the same study system) and concluded that there was no significant cellular toxicity of GRFT to any tested host cell types. More studies should be conducted to evaluate the effects of the lectin on a variety of cervico-vaginal epithelial cell lines since the latter are likely the first cells to be exposed to GRFT formulated as a vaginal microbicide. To this end, a recent study from our laboratory showed that GRFT at 2.0 μM induced no significant alterations in levels of an extensive panel of cytokines and chemokines in human cervical explants (O'Keefe *et al.*, 2009). In addition, no mitogenic activity of GRFT was detected after treatment of human peripheral blood mononuclear cells (hPBMCs) by the CBA and rabbit vaginal irritation (RVI) assays showed good safety profile for 0.01%, 0.05% and 0.1% GRFT with no gross anatomical pathology observed at any dose (O'Keefe *et al.*, 2009). It would be interesting to further evaluate the effect of GRFT on a more comprehensive set of cytokines and chemokines as well as other molecules that mediate the innate immune response especially those that are implicated in boosting the migration of CD4+ cells. Other animal studies are needed as well to further establish the safety profile of the drug where the effects of long term, repeated vaginal/rectal administration of GRFT would be evaluated. The animal models should include the rabbit 'Gold Standard' model recommended by the US Food and Drug Administration for safety evaluation of vaginal products as well as rodents like mice and/or rats which have the advantage of the availability of a greater number of reagents with relatively low costs for preclinical studies. Ongoing projects in our laboratory evaluating the epithelial toxicity *in vitro* and *in vivo* show highly encouraging results (Kouokam *et al.*, unpublished data).

Large scale production of GRFT

It is widely acknowledged in the field of microbicide research and product development that low cost of production is one of the key features of a topical antiviral since the end product would be affordable to the populations in poor countries as well as less wealthy people in the developed west. In fact, in the case of HIV/AIDS, current drugs are very expensive and barely accessible to the poorest countries where the disease prevalence is even higher. Many strategies have been used in order to solve the problem of cost/availability of the algal lectin GRFT. The first attempt used the *Escherichia coli* bacterial system to overexpress GRFT which yielded a product that displayed a similar antiviral effect against HIV as the natural, algal derived lectin (Mori *et al.*, 2005). However it is still debatable whether the existing bacterial expression systems could allow a production large enough to supply all the current need, taking into consideration the large population of AIDS patients and the eventual use of GRFT in the treatment and/or prevention of other diseases caused by enveloped viruses. One of the drawbacks of the bacterial systems is that recombinant GRFT accumulates in both soluble and non–soluble fractions of *E. coli* which complicates the purification process as well as reducing the yield of purified protein (Giomarelli *et al.*, 2006). Since a topical microbicide needs to be produced at extremely high levels and low cost to have an impact on global health, bacteria or any other cell based system such as yeasts, which require sterile environment, are unlikely to provide a satisfying solution for the mass production of GRFT. In the case of CV-N, *Lactococcus lactis* and *L. plantarum* were engineered to secrete CV-N that inhibited HIV-1 infection *in vitro* and *L. jensenii*, found in the normal vaginal flora, was bioengineered to deliver this lectin directly to the mucosa (Liu *et al.*, 2006). This concept of 'live microbicide' appeared very attractive since it would allow development of inexpensive yet durable protein-based topical antiviral agents. However, many questions remain unanswered as for the safety of such procedure, for instance it is difficult to predict eventual mutations that will appear in the inserted lectin/protein gene during the course of time or to be certain that the transformed strain will still optimally colonize the vaginal mucosa without creating any disastrous imbalance in the existing complex flora.

Transgenic plants have been used to express a variety of molecules including lectins. There are many advantages in using plant systems to produce protein microbicides: transgenic plant production can be easily scaled up in order to obtain large supplies and the creation of agricultural infrastructure is usually of low cost enough to allow inexpensive end products. For example, when transformed with a gene encoding CV-N, *Nicotiana tabacum* (tobacco) transgenic plants were reported to yield impressive amounts of recoverable protein at levels of 130 ng/mg of fresh leaf tissue corresponding to at least 0.85% of total soluble plant protein (Sexton *et al.*, 2006). In the same work published by Sexton *et al.*, it was shown that the plant expressed recombinant CV-N binds to soluble gp120 in a concentration dependent manner with a similar affinity to the natural counterpart and therefore exerts an antiviral activity against HIV. These studies have demonstrated the superiority of transgenic plant production versus bacterial over-expression of lectins, CV-N in this case. Similar conclusions have been recently made after studies on GRFT production. In fact, using a close relative of tobacco, *Nicotiana benthamiana*, it was possible to produce a non–His-tagged version of GRFT whose crystals diffracted significantly better than the bacterial recombinant protein (Ziolkowska *et al.*, 2006). It is therefore believed that the plant GRFT (pGRFT) bears a closer resemblance to the natural, algal lectin. Recently, we reported a breakthrough in manufacturing of pGRFT where *N. benthamiana* seedlings were transduced with a tobacco mosaic virus expressing

GRFT (O'Keefe *et al.*, 2009). A greenhouse of *N. benthamiana* plant expressing GRFT is shown in Fig. 21.1. We found that pGRFT was the most abundant protein in a pH 4.5 extract with the viral coat protein being the major contaminant; the plant produced lectin was relatively easy to purify compared to the bacterial counterpart; purified pGRFT could be recovered at around 45-50% of the *in planta* level of 1g/kg fresh plant biomass and was able to bind HIV envelope glycoprotein gp120 and therefore exert antiviral activity similar to the natural or bacterial forms of GRFT. The final purified pGRFT product was sterile with no residual TMV and has extremely low endotoxin contamination of 0.0896 EU/ml which is much lower than is tolerated by the FDA for an injectable drug (O'Keefe *et al.*, 2009). Even though the yield described above is high, further optimization of the purification process, for example reducing the number of steps and/or increasing the yield is needed in order to minimize as much as possible the production costs of pGRFT.

Fig. 21.1. A greenhouse of *N. benthamiana* plants expressing GRFT.

GRFT has the potential to become a topical microbicide against a broad range of enveloped viruses

Enveloped viruses are still a huge health challenge despite lots of efforts and use of modern techniques in the development of new antivirals. Indeed, there has been enormous progress in the development of antiretrovirals against HIV, vaccine products against influenza, hepatitis A and B, etc. However, the manufacturing cost constitutes a major obstacle to these products to achieve global distribution for prophylaxis, especially to those people in the poorest regions of the world who are most heavily affected by infectious diseases. It is well established that an antiretroviral combination therapy can significantly improve life quality and life expectancy of AIDS patients but again this treatment option comes at high cost, out of reach for most people living with HIV/AIDS. Antiretroviral treatment has been proposed as an appropriate 'pre-exposure prophylaxis', either orally or in vaginal microbicide products. Aside from the cost, there is concern that prophylactic use of antiretrovirals already used in therapeutic regimens could lead to undesirable antiretroviral resistance becoming common in the circulating virus population. Moreover, many antiretroviral drugs have significant side-effects. However, using malaria pre-exposure prophylaxis as a model, the same strategy may be useful against HIV. An injectable protein product is unlikely to find widespread use as first-line antiretroviral therapy–it is more likely to be useful as salvage therapy in people for whom most other antiretrovirals have become ineffective due to viral resistance. GRFT therefore has some advantages as a prophylaxis product versus many small molecule antiretrovirals. Since GRFT has been proven manufacturable in large amounts in plants (O'Keefe *et al.*, 2009), the final cost of the end-product may be affordable and the production could be easily scaled up to meet the global need. For most vaccines, a cold chain is required for activity preservation and this adds to the already high costs of the current products not to mention the lack of suitable refrigerating equipments in many reclusive areas of the world. In contrast, topical microbicides can be formulated in gels and the end products are therefore able to stand extreme physical storage conditions including high tropical temperatures, without significant loss of activity. Colourlessness, tastelessness and odourlessness constitute additional important properties of GRFT (and lectins in general) which favour its development as a topical microbicide. In some enveloped viruses exemplified by HIV, the constant modification of the glycan shield of the envelope renders even more difficult to design vaccines or drugs capable of dealing with the mutant viruses and to control infection.

The binding and activity of lectins including GRFT to envelope glycoproteins is not influenced by these viral mutations since the anti-HIV CBAs described so far generally inhibit the growth of all the HIV clades tested. Thus, CBAs can be an invaluable tool to fight the emerging drug-resistant viruses. This justifies in part the rising attention given to lectins for their ability to bind glycoproteins that are present on the surface of the viral envelope. As mentioned above, GRFT has been shown to exert potent inhibitory effects on enveloped viruses including HIV and SARS and is expected, like other lectins with the same mode of action (CV-N for instance), to show antiviral effects against a broad range of enveloped viruses. The potent inhibition of HIV entry and syncytium formation suggests that the lectin could be used both to prevent infection and to stop or reduce disease progression, in the case of AIDS. Bioavailability is a key feature for any product which undergoes drug development. The mode of action of GRFT and other CBAs primarily suggests a use at mucosal surfaces to inactivate enveloped viruses and protect the host cells from infection. As mentioned above, CV-N has been successfully used in the treatment of flu in mice and ferrets by intra-nasal administration. To be qualified for a vaginal administration, for treatments of AIDS or genital herpes for example, a microbicide should be able to maintain stability and therefore appreciable antiviral activity in the acidic (pH 4–6) cervico-vaginal environment which include the vaginal microflora as well as various macromolecules such as secreted proteins that could affect the diffusion of the drug. In an elegant study, it was shown that GRFT retained similar antiviral efficacy after a week of incubation in culture media at pH 4–8 and incubation in macaque cervico-vaginal lavages or bovine serum albumin at pH ranging from 4 to 8 for 24 h did not alter the anti-HIV-1 inhibitory effects (Emau *et al.*, 2007). These and many previous findings let the authors suggest that GRFT should be considered as an excellent candidate for anti-HIV microbicide. This lectin could easily become the 'next generation' anti-HIV drug and be used in the fight against other enveloped viruses providing that the protein passes all the safety tests and that a suitable delivery system of the product is developed. Many vaginal products including contraceptives are formulated in gels and GRFT might easily follow this path. However, it is not clear as whether a gel formulated GRFT for frequent use as a vaginal microbicide will be appreciated. In addition, it is not yet known as whether GRFT will keep its stability in gels and optimally diffuse out of this milieu to show appreciable antiviral efficacy in the vagina and/or cervix. It is therefore useful to explore the efficacy of other formulation methods including thin films and solid lipid particulate systems such as solid lipid nanoparticles, lipid microparticles and lipospheres which have been recently developed for vaginal formulations.

Conclusion

The algal GRFT is a relatively newly discovered lectin with proven potent antiviral activity against HIV and SARS-CoV. GRFT is the most potent anti-HIV lectin described so far and thanks to this, the CBA from *Griffithsia* sp. became the first algal lectin to undergo detailed structural studies although the structure of the amino acid at position 31 in the natural, algal derived lectin is still unknown. However, the recombinant GRFT (containing an alanine in lieu of the unknown amino acid) expressed both in bacteria and plants has been shown to possess similar biological properties, especially for its binding features. Therefore no special effort has been deployed in order to disclose the nature and properties of this unusual amino acid. Structural studies have suggested that the dimeric form of GRFT should be considered as the minimum biological unit due to the domain-swapped structure of this protein. With 6 sugar binding sites on a dimer, GRFT exerts its antiviral inhibitory effects by binding oligosaccharides of the viral envelope glycoprotein with high affinity. Such mode of action predisposes GRFT to inhibit a wide range of enveloped viruses and studies are expected in the near future to unveil the potential inhibitory effects of this lectin against many other glycan shielded viruses, especially those that are pathogenic to human. Although GRFT as a lectin is not expected to carry any substantial risk of systemic side effects and is unlikely to be absorbed efficiently (O'Keefe *et al.*, 2009) as components of microbicides, the safety profile of this compound should be thoroughly studied since it would be destined for a chronic use especially if it would be developed to serve as a prophylactic means as well. Some studies have been done to this end with encouraging results reviewed here. Future works using cell lines or primary cells and animal models should include a more comprehensive evaluation of the lectin on the mediators of the innate immune system, in particular those that trigger the CD4+ T helper cells which are needed by the virus to initiate infection and disease in the case of HIV/AIDS. One of the biggest problems that the drug development of a macromolecule such as a protein encounters is the availability. In fact it is unlikely that any cell based system in the current settings, be it prokaryotic or eukaryotic would allow the production of sufficient amounts of GRFT to be used in a huge global demand. In addition, cell based systems which require sterile conditions are very expensive to establish and maintain and therefore could not be used to manufacture the lectin at a low cost that would be affordable to the poor(est) populations dealing with viral infectious diseases. Fortunately a plant based system where large amounts of GRFT were produced in an agricultural setting has been recently developed and this carries the promise to solve both issues including cost and availability.

In the prevention of AIDS and sexual transmitted diseases, such a product would have a very low cost (in the same price range as the male condom) with the advantage that women would have a word to say in their own protection. In fact male condom encounters a lot of reticence in some cultures and/or religions. In addition, some women (especially in developing countries or sex workers in the western world) do not always consent to have sex and when they do, it is not obvious to convince their partners to use the latex. As a result, the proportion of women living with HIV has steadily increased. Discrete use of a topical microbicide would bring a solution to this and dramatically decrease the rate of heterosexual transmission of HIV which is currently the dominant mode.

Optimal formulation of GRFT would yield a product that can undergo preclinical and clinical trial using various animal models according to the target enveloped virus with hope that this lectin will help tackle viral diseases caused by a large number of enveloped viruses. The discovery of GRFT supports the bioactivity guided fractionation as an efficient means to identify new active pharmaceutical principles from natural products.

Acknowledgements

This work was supported by a NIH grant AI076169 to KEP. We acknowledge ongoing collaborations with many scientists, but particularly Dr Barry O'Keefe and colleagues at the National Cancer Institute who have made many key contributions to the development of griffithsin and other antiviral protein products as broad spectrum antivirals.

References

Balzarini, J. (2007) Targeting the glycans of glycoproteins: a novel paradigm for antiviral therapy. *Nature Reviews Microbiology* 5(8), 583–597.

Chandra, N. (2006) Common scaffolds, diverse recognition profiles. *Structure* 14(7), 1093–1094.

Emau, P., Tian, B., O'Keefe, B.R., Mori, T., McMahon, J.B., Palmer, K.E., Jiang, Y., Bekele, G., and Tsai, C.C. (2007) Griffithsin, a potent HIV entry inhibitor, is an excellent candidate for anti-HIV microbicide. *Journal of Medical Primatology* 36(4–5), 244–253.

Giomarelli, B., Schumacher, K. M., Taylor, T. E., Sowder, R. C., 2nd, Hartley, J. L., McMahon, J. B., and Mori, T. (2006) Recombinant production of anti-HIV protein, griffithsin, by auto-induction in a fermentor culture. *Protein Expression and Purification* 47(1), 194–202.

Liu, X., Lagenaur, L.A., Simpson, D.A., Essenmacher, K.P., Frazier-Parker, C.L., Liu, Y., Tsai, D., Rao, S.S., Hamer, D.H., Parks, T.P., Lee, P.P., and Xu, Q. (2006) Engineered vaginal

lactobacillus strain for mucosal delivery of the human immunodeficiency virus inhibitor cyanovirin-N. *Antimicrobial Agents and Chemotherapy* 50(10), 3250–3259.
Marsh, M. (1984) The entry of enveloped viruses into cells by endocytosis. *Biochemical Journal* 218(1), 1–10.
Mori, T., O'Keefe, B.R., Sowder, R.C., 2nd, Bringans, S., Gardella, R., Berg, S., Cochran, P., Turpin, J.A., Buckheit, R.W., Jr, McMahon, J.B., and Boyd, M.R. (2005) Isolation and characterization of griffithsin, a novel HIV-inactivating protein, from the red alga *Griffithsia* sp. *Journal of Biological Chemistry* 280(10), 9345–9353.
O'Keefe, B.R., Vojdani, F., Buffa, V., Shattock, R. J., Montefiori, D.C., Bakke, J., Mirsalis, J., d'Andrea, A.L., Hume, S.D., Bratcher, B., Saucedo, C.J., McMahon, J.B., Pogue, G.P. and Palmer, K.E. (2009) Scalable manufacture of HIV-1 entry inhibitor griffithsin and validation of its safety and efficacy as a topical microbicide component. *Proceedings of the National Academy of Sciences of the United States of America* 106(15), 6099–6104.
Sexton, A., Drake, P.M., Mahmood, N., Harman, S.J., Shattock, R.J., and Ma, J.K. (2006) Transgenic plant production of Cyanovirin-N, an HIV microbicide. *FASEB Journal* 20(2), 356–358.
Smee, D.F., Bailey, K.W., Wong, M.H., O'Keefe, B.R., Gustafson, K.R., Mishin, V.P., and Gubareva, L.V. (2008) Treatment of influenza A (H1N1) virus infections in mice and ferrets with cyanovirin-N. *Antiviral Research* 80(3), 266–271.
van der Meer, F. J., de Haan, C. A., Schuurman, N. M., Haijema, B. J., Peumans, W. J., Van Damme, E. J., Delputte, P. L., Balzarini, J., and Egberink, H. F. (2007) Antiviral activity of carbohydrate–binding agents against Nidovirales in cell culture. *Antiviral Research* 76(1), 21–29.
Ziolkowska, N. E., and Wlodawer, A. (2006) Structural studies of algal lectins with anti-HIV activity. *Acta Biochimica Polonica* 53(4), 617–626.
Ziolkowska, N.E., O'Keefe, B.R., Mori, T., Zhu, C., Giomarelli, B., Vojdani, F., Palmer, K.E., McMahon, J.B. and Wlodawer, A. (2006) Domain-swapped structure of the potent antiviral protein griffithsin and its mode of carbohydrate binding. *Structure* 14(7), 1127–1135.
Ziolkowska, N.E., Shenoy, S.R., O'Keefe, B.R., McMahon, J.B., Palmer, K.E., Dwek, R.A., Wormald, M.R. and Wlodawer, A. (2007a) Crystallographic, thermodynamic, and molecular modeling studies of the mode of binding of oligosaccharides to the potent antiviral protein griffithsin. *Proteins* 67(3), 661–670.
Ziolkowska, N.E., Shenoy, S.R., O'Keefe, B.R., and Wlodawer, A. (2007b) Crystallographic studies of the complexes of antiviral protein griffithsin with glucose and N-acetylglucosamine. *Protein Science* 16(7), 1485–1489.

Chapter 22

Camptothecins: SAR, QSAR and Biotechnology

Rajeshwar P. Verma

Introduction

Camptothecin (CPT, **I**), a naturally occurring pentacyclic quinoline alkaloid firstly isolated from a Chinese tree *Camptotheca acuminata* in 1966 (Wall *et al.*, 1966), is one of the prominent lead compounds in the field of anticancer drug design and development. Although CPT has shown remarkable promise as an anticancer agent, its success has been tempered due to its poor water solubility, severe toxicity, and hydrolyzation to the biologically inactive carboxylate, which binds to serum albumin. Many efforts had been spent to improve the pharmacological features of CPT with respect to increase their water solubility and lactone stability by the design of several semi-synthetic derivatives. Two most promising CPT derivatives of such types are 9-[(dimethylamino)methyl]-10-hydroxycamptothecin (topotecan, **II**) and 7-ethyl-10-[4-(1-piperidino)-1-piperidino]carbonyloxycamptothecin (irinotecan, **III**), which exhibit a much better water solubility and antitumour activity than that of camptothecin (Soepenberg *et al.*, 2003). These two CPT analogues have already been approved by the FDA and currently prescribed in clinics (Oberlies and Kroll, 2004; Li *et al.*, 2006). Topotecan (**II**) is used for the treatment of the ovarian and small-cell lung cancers (Kingsbury *et al.*, 1991; Ormrod and Spencer, 1999; Markman, 2005), whereas irinotecan (**III**) is used for the treatment of metastatic colorectal cancers (Pavillard *et al.*, 2004; Fuchs *et al.*, 2006). Irinotecan (**III**) is a prodrug, which is converted into the active metabolic form i.e. 10-hydroxy-7-ethylcamptothecin (SN-38, **IV**) (Fig. 22.1).

Fig. 1. 20(*S*)-Camptothecin (CPT, **I**), Topotecan (**II**), Irinotecan (**III**) and SN-38 (**IV**)

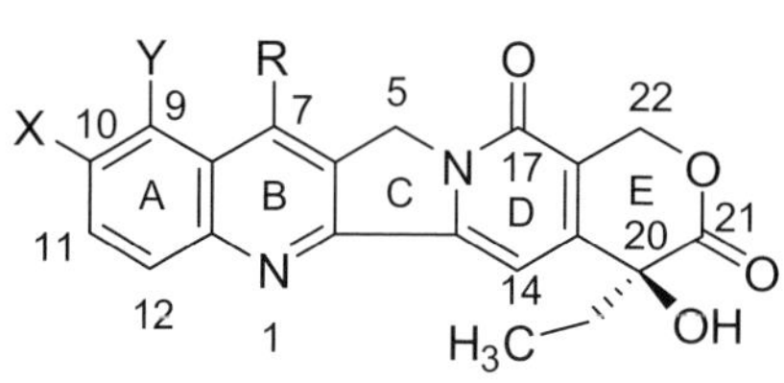

I-IV

20(*S*)-Camptothecin (CPT, **I**); X = Y = R = H
Topotecan (**II**); X = OH, Y = $CH_2N(CH_3)_2$, R = H
Irinotecan (**III**); X = 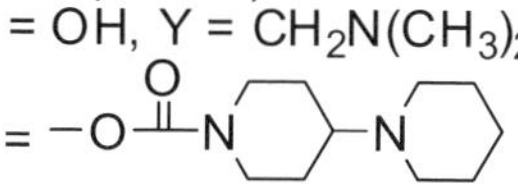

Y = H, R = C_2H_5
SN-38 (**IV**); X = OH, Y = H, R = C_2H_5

Topoisomerase I (topo I), a vital DNA-manipulating enzyme, is the only known target for the anticancer compounds of CPT family, which binds to the interface of the covalent protein-DNA complex and enhances apoptosis through blocking the advancement of replication forks. Once the CPT analogue is intercalated into topo I-DNA cleavable complex, the collision between the complex and the replication fork during S-phase is thought to be responsible for the DNA double strand breaks that ultimately lead to the cell death (Hsiang *et al.*, 1989; Liu *et al.*, 2000; Legarza and Yang, 2006). The crystal structure of human topo I in ternary complexes with DNA and topotecan (Chrencik *et al.*, 2004) may provide valuable information that might improve efficacy of the CPT analogues. Scientific efforts in this field are continued to develop new CPT derivatives in order to improve their pharmacokinetics, drug resistance, and toxicity profiles. Currently, over a dozen new CPT derivatives are already in various stages of clinical trials (Legarza and Yang, 2005). Interestingly, plants are not the only source of CPT. Recent advancement in the cloning techniques and characterization of biosynthetic enzymes involved in CPT biosynthetic pathway may be helpful in the development of genetically engineered CPT-producing plants. Thus, biotechnology may be proved to be an alternative route for rapid and large scale production of CPT. In the present chapter we describe SAR, QSAR and biotechnology of CPT derivatives.

Structure–Activity Relationships (SAR)

Structure–activity relationship (SAR) is qualitative in nature. It displays structural alerts that include molecular substructures or fragment counts related to the presence or absence of the biological activity (Eriksson *et al.*, 2003; Verma and Hansch, 2007). The essential structural features of CPT analogues for their activity are the following: α-OH lactone ring, the pyridone moiety of the D-ring, the lactone moiety of the E-ring, conformation at C-20, and the planarity of the five-membered ring system (Sirikantaramas *et al.*, 2007). Although the details about the SAR of CPT derivatives with respect to their pentacyclic ring system have already been discussed in several publications (Li *et al.*, 2006; Verma and Hansch, 2009), a brief discussion has been given below.

Modification of A ring (C-9, C-10, C-11and C-12 positions)

(i) DNA topo I inhibition generally increases if CPT is substituted at C-9 and C-10 positions by halides and other electron-rich groups (Carrigan *et al.*, 1997). (ii) DNA topo I inhibition activity also increases by the modification at C-11 position e.g. 11-azacamptothecin has approximately twice the activity in comparison to CPT (Uehling *et al.*, 1995). (iii) Monosubstitution by NH_2 or OH group at position C-12 reduces greatly the antitumour activity, whereas substitution at C-9, C-10, or C-11 positions increases the activity (Wani *et al.*, 1986). (iv) C-10 and C-11 substitutions are generally unfavourable to the biological activity (Wani *et al.*, 1987a), but it is not true for C-10,11-methylenedioxy and C-10,11-ethylenedioxy functional group (Wall *et al.*, 1993; Carrigan *et al.*, 1997).

Modification of B ring (C-7 position)

C-7 substitution generally increases both the potency and water solubility of CPT analogues depending upon the nature of their C-7 substituent (Wall *et al.*, 1993). Examples: (i) 7-N-isopropylaminoethyl-CPT showed a marked increase in cytotoxicity and lower toxicity in comparison to CPT and topotecan (Jew *et al.*, 1999); (ii) 7-X-CPT (where, X = alkyl, alkenyl, cyano, carbethoxy, $CH{=}NC_6H_5$, $CH{=}NC_6H_{11}$, $CH{=}N(CH_2)_4OH$, $CH{=}NCH_2C_6H_5$, $CH{=}NC_6H_4(4\text{-}NO_2)$, and $CH{=}N(CH_2)_2N(CH_3)_2$) showed potent cytotoxicity against human non-small-cell lung carcinoma H460 cell-line, most of them exhibiting IC_{50} values in the 0.05–1μM range, and were more active than topotecan (IC_{50} = 1.38μM) (Dallavalle *et al.*, 2000, 2001).

Modification of C ring (C-5 position)

5-X-CPT (where, X = acetoxy, alkoxy, amino, or hydroxyl groups) generally exhibited diminished antitumour activity (Subrahmanyam *et al.*, 1999a; Thomas *et al.*, 2004;), whereas two CPT analogues (**V**; X = $CH_2COOCH_2CH_2F$) and (**VI**; X = [$=CHCH_3$]) showed a good or maintained the same potency equivalent to CPT (Sugimori *et al.*, 1994; Subrahmanyam *et al.*, 1999b).

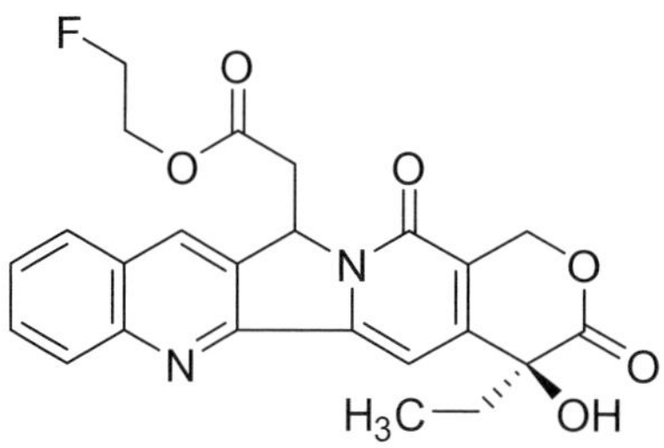

V

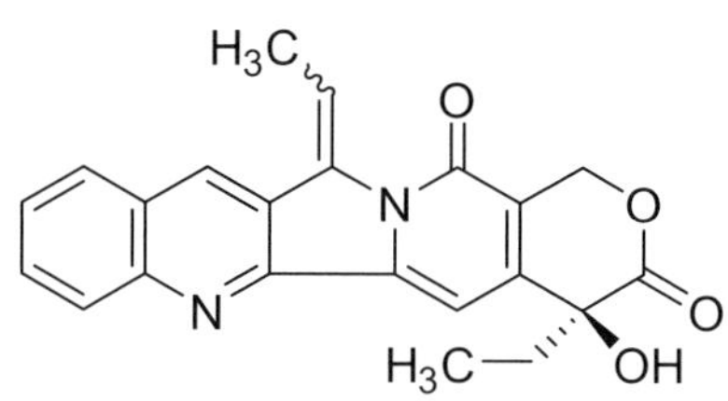

VI

Modification of D ring (C-14 position)

Steric factor at C-14 is very crucial for the activity e.g. methyl ester group, as C-14 substituent, reduces the antitumour activity (Crow and Crothers, 1992). On the other hand, 14-aza CPT (**VII**) inhibits the relaxation of super-coiled plasmid DNA more potently than CPT (Hecht, 2005).

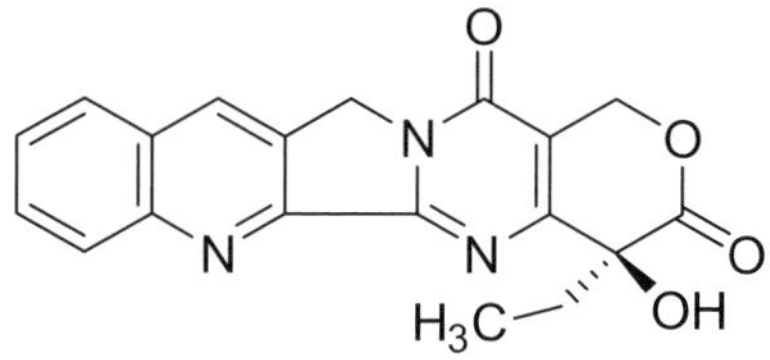

VII

Modification of E ring (C-20 and C-21 positions)

(i) The lactone (E-ring) with 20-*S* configuration is vital for both the topo I inhibition and *in vivo* activity (Wani *et al.*, 1987b). Any changes in this ring e.g. replacement of lactone by lactum, reduction of lactone, removal of carbonyl oxygen, removal of 20-OH or replacement of 20-OH by other groups, inactivate the CPT analogues (Bailly, 2003; Hertzberg *et al.*, 1989; Thomas *et al.*, 2004; Li *et al.*, 2006). (ii) The presence of α-OH (20-*S* configuration) results in an equilibrium under physiological conditions that favours the (inactive) open carboxylate form over the (active) ring-closed lactone (Burke and Mi, 1993; Hatefi and Amsden, 2002; Thomas *et al.*, 2004; Li *et al.*, 2006). (iii) The importance of α-OH (20-*S* configuration) for topo I inhibition is mainly due to the two possible reasons: (a) hydrogen bond formation between the hydroxyl group and the enzyme-DNA complex, and (b) an intra-molecular hydrogen bond formation between the hydroxyl group and the lactone carbonyl (C-21 position). Both interactions may facilitate the E-ring opening (Hertzberg, 1989; Crow and Crothers, 1992; Li *et al.*, 2006). (iv) Esterification of 20-OH group can either eliminate the formation of intra-molecular hydrogen bonding or increase the steric hindrance of the carbonyl group of E-ring, which results in the stability of lactone ring of CPT derivatives *in vivo* (Cao *et al.*, 2000; Li *et al.*, 2006). (v) Homocamptothecin (homoCPT, **VIII**), obtained by the replacement of α-hydroxylactone with β-hydroxylactone, exerted potent topo I inhibition, enhanced cytotoxicity, and increased stability of the lactone after 24 h at physiological pH (Lavergne *et al.*, 1997).

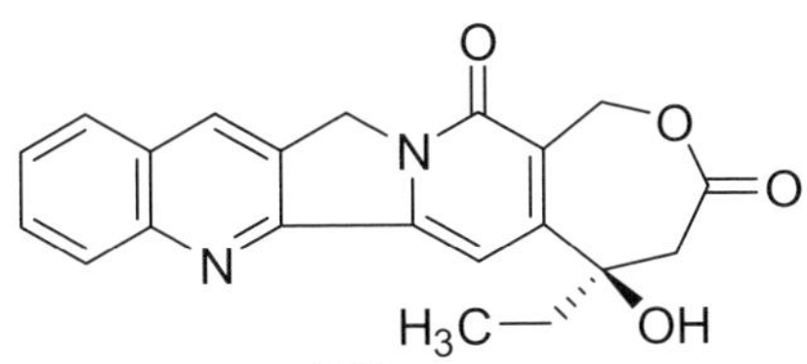

VIII

Modification of A/B ring

CPT analogues with modifications at A/B ring are of great interest because only two CPT derivatives (topotecan and irinotecan) approved by the FDA for clinical use are the derivatives of CPT with substitutions within the quinoline ring. The SAR of the quinoline (A/B) ring suggests the following: (i) 7-C_2H_5-CPT substituted at C-11 position either by -F or -CN group showed enhanced cytotoxicity against KB and L1210 cells as well as strong DNA topo I inhibition (Yaegashi *et al.*, 1994); (ii) Stability comparison among CPT-, SN-38- and 10-OH-CPT-induced cleavable complexes has demonstrated that the -OH group at C-10 enhances the stability of cleavable complexes. Further, the stability difference between SN-38- and 10-OH-CPT-induced cleavable complexes indicated that the ethyl group at C-7 is an important factor for stabilizing the interaction between CPT derivatives and topo I-DNA complex (Tanizawa *et al.*, 1995); (iii) 7-alkyl-10-OH-CPT has shown improved drug stability in the presence of human albumin. SN-38 is one of the best examples with this substitution pattern (Burke and Mi, 1993); (iv) CPT derivatives with an additional ring combined with positions C-10 and C-11, C-9 and C-10, or C-7 and C-9 were predicted to have superior antitumour activities in comparison to the original CPT. The best examples of these types are lurtotecan and extatecan etc. (Li *et al.*, 2006); (v) C-7 and C-9 substitutions do not affect the inhibition of DNA topo I, suggesting the absence of tight interaction between the receptor site and the regions around C-7 and C-9 positions (Kim and Lee, 2002).

Modification of A/C ring

Substitutions of alkoxy or other groups at C-5 (C ring) are reasonably tolerated when either hydroxyl or nitro group is present to the A-ring. One of the best examples of this kind is 9-OH-5-OC_2H_5-CPT (Subrahmanyam *et al.*, 2000; Sriram *et al.*, 2005).

Modification of A/E ring

Homocamptothecin (homoCPT, **VIII**) shows enhanced plasma stability and a decreased binding to serum albumin. The activity of homoCPT can further be enhanced by mono- or bis-fluorination of the A ring (Sriram *et al.*, 2005). The most important homoCPT derivative is 10,11-difluoro-homoCPT (**IX**) which exhibits enhanced antiproliferative activity against various cell lines (Lavergne *et al.*, 1998).

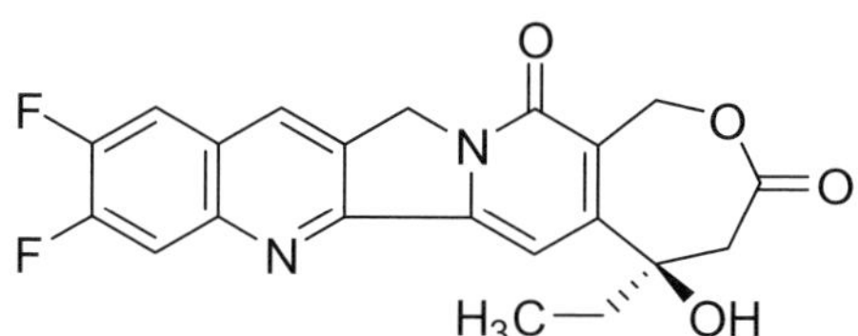

IX

Quantitative Structure–Activity Relationships (QSAR)

Quantitative structure–activity relationship (QSAR) is one of the well-developed areas in computational research with established methodology and successful history (Hansch *et al.*, 1962). It has long been considered to play a vital role in drug discovery and development. This technique uses extra-thermodynamically derived and computational-based descriptors to correlate biological activity in isolated receptors, in cellular systems, and *in vivo*. The use of meaningful descriptors, which can explain the molecular features responsible for the relevant activity, is an important aspect for this paradigm. This method can shed light on mechanisms of action of physiologically active substances to increase their effectiveness and reduce the cost of development for the final medicinal/drug product. Although a detailed QSAR study of CPT analogues has already been discussed (Verma and Hansch, 2009), a few examples are given below.

QSAR for the inhibition of DNA topo I

*QSAR for the inhibition of DNA topo I by 7-X-9-Y-10-Z-camptothecins (**X**) (Hansch and Verma, 2007)*

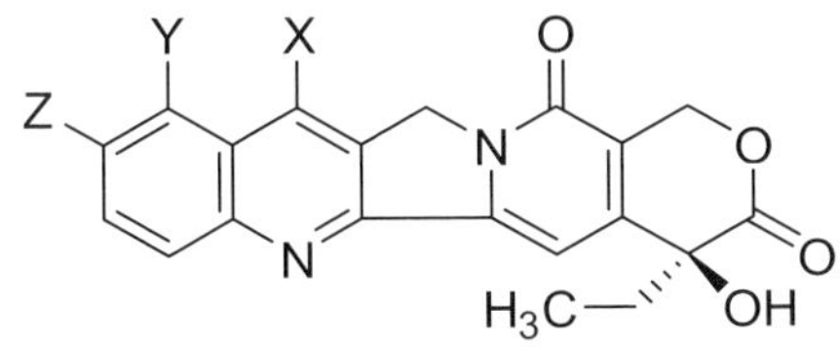

$$\log 1/C = 0.84\ \mathrm{Clog}\ P - 0.62\ B5_X - 0.59\ MR_Y - 0.45\ MR_Z + 0.88\ I_1 - 0.52\ I_2 + 6.37 \quad (1)$$
$n = 30$, $r^2 = 0.854$, $s = 0.258$, $q^2 = 0.738$

where, C is the molar concentration of CPT analogues (X) that cause 50% inhibition of topo I cleavable complex formation, n is the number of data points, r^2 is the square of the correlation coefficient, q^2 is the cross-validated r^2, and s is the standard deviation.

Clog P is the hydrophobic parameter for the whole molecule and its positive coefficient suggests that the highly hydrophobic molecules will be more active. $B5_X$ is the Verloop's sterimol descriptor for the maximum width of X-substituents. MR_Y and MR_Z are, respectively, the calculated molar refractivities of Y and Z substituents. A negative coefficient associated with $B5_X$, MR_Y and MR_Z brings out steric effects for X, Y and Z substituents, respectively. I_1 and I_2 are the indicator variables for the unusual activities associated with Z = OH and X = *n*-alkyl groups, respectively. The presence of Z = OH increases the activity as evidenced by the positive coefficient of I_1. On the other hand, the negative coefficient of the indicator variable (I_2) suggests that the presence of branched alkyl groups at C-7 position is preferred over *n*-alkyl groups.

QSAR for the inhibition of DNA topo I by 7-X-10-Y-camptothecins ***(XII)*** *(Vladu* et al.*, 2000; Verma and Hansch, 2009)*

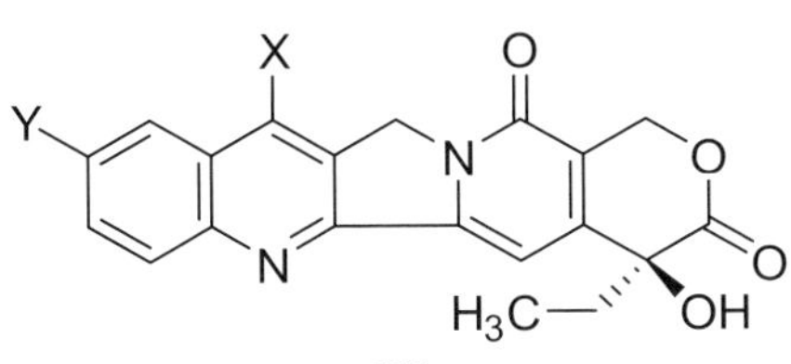

XI

$$\log 1/EC_{50} = -0.30\ \pi_X - 1.37\ \pi_Y + 5.87 \quad (2)$$
$n = 14$, $r^2 = 0.859$, $s = 0.227$, $q^2 = 0.750$

where, EC_{50} is the molar concentration of CPT analogues (**XI**), which are required for producing cleavable complexes with purified topo I in 50% of the plasmid DNA. The hydrophobic parameters π_X and π_Y are for the substituents at C-7 and C-10 positions, respectively. Its negative coefficients suggest that the presence of more hydrophilic substituents at both the positions will enhance the activity.

QSAR for the inhibition of various cancer cells

QSAR for the cytotoxicity of 7-X-camptothecins (***XII***) *to H460 human NSCLC cells (Dallavalle et al., 2001; Verma and Hansch, 2009)*

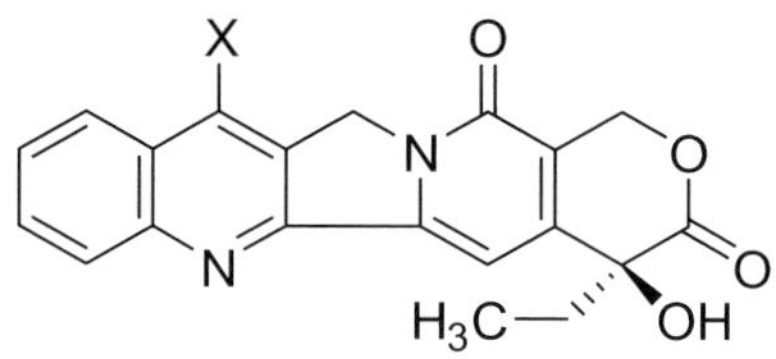

XII

$$\log 1/IC_{50} = 0.86 \text{ Clog } P - 0.27 \text{ Clog } P^2 + 5.97 \quad (3)$$

$n = 10$, $r^2 = 0.924$, $s = 0.180$, $q^2 = 0.864$
optimum Clog $P = 1.62(1.33\text{-}2.30)$

where, IC_{50} is the molar concentration of CPT analogues (**XII**) that cause 50% growth inhibition of H460 human NSCLC cells. The parabolic correlation (Eq. **3**) in terms of Clog *P* suggests that the cytotoxicity of CPT derivatives (**XII**) against H460 human NSCLC cells first increases with an increase in their hydrophobicity up to an optimum Clog *P* of 1.62 and then decreases.

QSAR for the cytotoxicity (IC_{50}; mol/l) of 20-OCOCH2CH2X-camptothecins (XIII) to A2780 (human ovarian cancer) cells (Wang et al., 2004; Verma and Hansch, 2009)

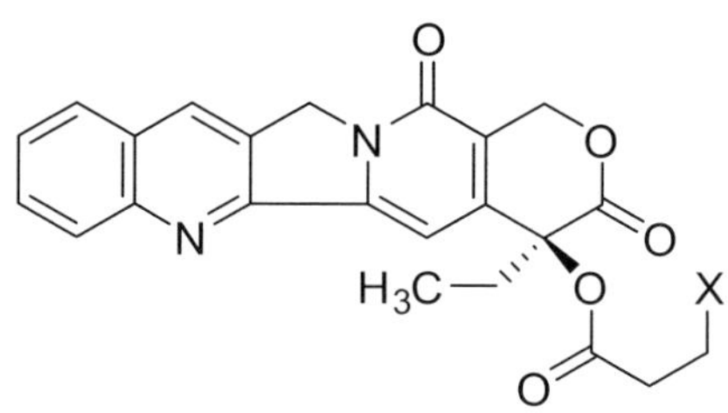

XIII

$$\log 1/IC_{50} = 2.05\ \pi_X - 0.42\ \pi_X^2 + 4.70 \quad (4)$$

$n = 9$, $r^2 = 0.908$, $s = 0.076$, $q^2 = 0.602$
optimum $\pi_X = 2.46(2.22\text{-}2.59)$

QSAR for the cytotoxicity (IC_{50}; mol/l) of 20-$OCOCH_2CH_2X$-camptothecins (***XIII***) *to Bel7402 (human liver cancer) cells (Wang et al., 2004; Verma and Hansch, 2009)*

$$\log 1/IC_{50} = 12.81\ MR_X - 1.25\ MR_X^2 - 24.84 \quad (5)$$

$n = 12$, $r^2 = 0.892$, $s = 0.258$, $q^2 = 0.841$
optimum $MR_X = 5.14(5.03\text{-}5.26)$; π_X vs. MR_X; $r = 0.654$

π_X and MR_X are the calculated hydrophobic and molar refractivity parameters of X substituents, respectively. Equations **4** and **5** are parabolic correlations, respectively, in terms of π_X and MR_X. This suggests that CPT analogues (**XIII**) may act by different mechanisms or interaction with different cellular targets in addition to the DNA topo I in these two human cancer (A2780 and Bel7402) cell lines.

Biotechnology

At present, CPT, supplied for pharmaceutical use, is still exclusively extracted from the plants, *Camptotheca acuminata* and *Nothapodytes foetida*. Recently, it has been discovered that CPT is also present in other plant species but in very low concentration i.e., *Ophiorrhiza pumila* and *Ophiorrhiza mungos* (Efferth *et al.*, 2007). Although the chemical synthesis of CPT analogues has been studied extensively, a large-scale synthesis has not yet been established due to the shortage of plants and the environmental issues. Thus, it is essential to explore the other resources to fulfill the great demand gap of CPT in the pharmaceutical market. The recent advancement in the cloning techniques and characterization of biosynthetic enzymes involved in the CPT biosynthetic pathway may provide valuable and rich information for the development of genetically engineered CPT-producing plants. Thus, biotechnology may be helpful in the rapid and large production of CPT. The details about the biotechnology of CPT have been well described in recent reviews (Efferth *et al.*, 2007; Srikantaramas *et al.*, 2007). Some important biotechnological approaches for the production of CPT alternative to field cultivation are described, in brief, below.

Cell suspension culture

By altering nitrogen source to NH_4^+/NO_3^- (5:1 molar ratio; 40 mM) in cell suspension culture of *Camptotheca acuminata*, the CPT yield has been increased by 340%, as compared to the control (Pan *et al.*, 2004). This method may be helpful in commercial production of CPT.

Hairy root culture

A hairy root culture of *Ophiorrhiza pumila*, using a modified 3 l bioreactor, gave CPT with concentration 0.0085% of fresh weight tissue from 8-week culture and the total CPT production reached 22 mg, for which about 16.5% (3.6 mg) was excreted into the culture medium. This method with a large-scale bioreactor can be applied for the commercial production of CPT (Sudo *et al.*, 2002).

Genetically modified plant

Recently, a method has been developed for the regeneration of *Ophiorrhiza pumila* plants from hairy roots. The regeneration frequency was over 83% and the regenerated plants accumulated CPT in amounts of 66–111% as compared to that of wild-type plants. The regeneration of transformed plant may contribute, in the future, to the establishment of genetically modified plants feasibly producing CPT (Watase *et al.*, 2004).
Endophytes

It is interesting to note that the isolation of CPT-producing organisms in the future, may provide an easily accessible source for the production of CPT. In a recent study, an endophytic fungal producing CPT has been isolated from the inner bark of the plant *Nothapodytes foetida* from the western coast of India (Puri *et al.*, 2005) and its immunomodulatory, antioxidant activity (Amna *et al.*, 2007; Arora *et al.*, 2010) and radioprotective efficacy were established (Arora *et al.*, unpublished work).

The expression of biosynthetic pathways

The genes of some key enzymes such as strictosidine synthase, tryptophan decarboxylase, and NADPH: cytochrome P450 reductase are responsible for the biosynthesis of CPT that have been cloned from hairy roots of *Ophiorrhiza pumila.* Heterologous expression of the cDNAs coding for these enzymes expressed in *Escherichia coli* cells yielded functional recombinant proteins. This may prove to be an important step towards the genetic engineering for recombinant CPT biosynthesis in large-scale production (Yamazaki *et al.*, 2003).

Conclusion

Isolation and structural elucidation of camptothecin (CPT; a natural product) enabled pharmacological and molecular biological investigations to establish CPT as one of the lead compounds for bioactivity-based generation of semi-synthetic derivatives e.g. topotecan and irinotecan, the FDA approved anticancer drugs. SAR and QSAR studies on CPT analogues may provide key information in further design and development of new CPT derivatives in order to improve pharmacokinetics, drug resistance and toxicity profiles. At present, over a dozen new CPT derivatives are in various stages of clinical trials. The recent advancement in biotechnology has provided a great hope for the alternative route of CPT production to field cultivation. We hope the biotechnology may fulfil the demand gap of CPT in the pharmaceutical market in the future.

References

Amna, T., Puri, S.C., Gupta, A., Khajuria, A., Arora, R., Spiteller, M. and Qazi, G.N. (2007) Immunomodulatory activity of an extract of the novel endophyte *Entrophosphora infrequens* isolated from *Nothapodytes foetida* (Wight) Sleumer. *Acta Microbiologica et Immunologica Hungarica* 54(3), 237–260.

Arora, R., Chawla, R., Dhaker, A.S., Adhikari, M., Sharma, J., Jaiswal, S., Gupta, D., Amna, T., Puri, S.C., Kumar, R., Sharma, A., Sharma, R.K. and Tripathi, R.P. (2010) Pro-antioxidant activities of fractions of a novel camptothecin-producing endophyte (*Entrophosphora infrequens*). *Trakia Journal of Sciences* 8(2), 1–15.

Bailly, C. (2003) Homocamptothecins: potent topoisomerase I inhibitors and promising anticancer drugs. *Critical Reviews in Oncology/Hematology* 45, 91–108.

Burke, T.G. and Mi, Z. (1993) Ethyl substitution at the 7 position extends the half-life of 10-hydroxycamptothecin in the presence of human serum albumin. *Journal of Medicinal Chemistry* 36, 2580–2582.

Cao, Z., Pantazis, P., Mendoza, J., Early, J., Kozielski, A., Harris, N., Vardeman, D., Liehr, J., Stehlin, J.S. and Giovanella, B. (2000) Structure-activity relationship of alkyl camptothecin esters. *Annals of the New York Academy of Sciences* 922, 122–135.

Carrigan, S.W., Fox, P.C., Wall, M.E., Wani, M. and Bowen, J. (1997) Comparative molecular field analysis and molecular modeling studies of 20-(S)-camptothecin analogs as inhibitors of DNA topoisomerase I and anticancer/antitumour agents. *Journal of Computer-Aided Molecular Design* 11, 71–78.

Chrencik, J.E., Staker, B.L., Burgin, A.B., Pourquier, P., Pommier, Y., Stewart, L. and Redinbo, M.R. (2004) Mechanisms of camptothecin resistance by human topoisomerase I mutations. *Journal of Molecular Biology* 339, 773–784.

Crow, R.T. and Crothers, D.M. (1992) Structural modifications of camptothecin and effects on topoisomerase I inhibition. *Journal of Medicinal Chemistry* 35, 4160–4164.

Dallavalle, S., Delsoldato, T., Ferrari, A., Merlini, L., Penco, S., Carenini, N., Perego, P., De Cesare, M., Pratesi, G. and Zunino, F. (2000) Novel 7-substituted camptothecins with potent antitumour activity. *Journal of Medicinal Chemistry* 43, 3963–3969.

Dallavalle, S., Ferrari, A., Merlini, L., Penco, S., Carenini, N., De Cesare, M., Perego, P., Pratesi, G. and Zunino, F. (2001) Novel cytotoxic 7-iminomethyl and 7-aminomethyl derivatives of camptothecin. *Bioorganic & Medicinal Chemistry Letters* 11, 291–294.

Efferth, T., Fu, Y.-J., Zu, Y.-G., Schwarz, G., Konkimalla, V.S.B. and Wink, M. (2007) Molecular target-guided tumor therapy with natural products derived from traditional Chinese medicine. *Current Medicinal Chemistry* 14, 2024–2032.

Eriksson, L., Jaworska, J., Worth, A.P., Cronin, M.T.D., McDowell, R.M. and Gramatica, P. (2003) Methods for reliability and uncertainty assessment and for applicability evaluations of classification- and regression-based QSARs. *Environmental Health Perspectives* 111, 1361–1375.

Fuchs, C., Mitchell, E.P. and Hoff, P.M. (2006) Irinotecan in the treatment of colorectal cancer. *Cancer Treatment Reviews* 32, 491–503.

Hansch, C. and Verma, R.P. (2007) 20-(S)-camptothecin analogs as DNA topoisomerase I inhibitors: a QSAR study. *ChemMedChem* 2, 1807–1813.

Hansch, C., Maloney, P.P., Fujita, T. and Muir, R.M. (1962) Correlation of biological activity of phenoxyacetic acids with Hammett substituent constants and partition coefficients. *Nature* 194, 178–180.

Hatefi, A. and Amsden, B. (2002) Camptothecin delivery methods. *Pharmaceutical Research* 19, 1389–1399.

Hecht, S.M. (2005) Camptothecin: Role of the D and E rings in binding to the topoisomerase I-DNA covalent binary complex. *Current Medicinal Chemistry – Anti-Cancer Agents* 5, 353–362.

Hertzberg, R.P., Caranfa, M.J., Holden, K.G., Jakas, D.R., Gallagher, G., Mattern, M.R., Mong, S.M., Bartus, J.O., Johnson, R.K. and Kingsbury, W.D. (1989) Modification of the hydroxylactone ring of camptothecin: inhibition of mammalian topoisomerase I and biological activity. *Journal of Medicinal Chemistry* 32, 715–720.

Hsiang, Y.H., Lihou, M.G. and Liu, L.F. (1989) Arrest of replication forks by drug-stabilized topoisomerase I-DNA cleavable complexes as a mechanism of cell killing by camptothecin. *Cancer Research* 49, 5077–5082.

Jew, S.S., Kim, H.J., Kim, M.G., Roh, E.Y., Hong, C.I., Kim, J.K., Lee, J.H., Lee, H. and Park, H.G. (1999) Synthesis and in vitro cytotoxicity of hexacyclic camptothecin analogs. *Bioorganic & Medicinal Chemistry Letters* 9, 3203–3206.

Kim, D.K. and Lee, N. (2002) Recent advances in topoisomerase I-targeting agents, camptothecin analogues. *Mini-Reviews in Medicinal Chemistry* 2, 611–619.

Kingsbury, W.D., Boehm, J.C., Jakas, D.R., Holden, K.G., Hecht, S.M., Gallagher, G., Caranfa, M.J., McCabe, F.L., Faucette, L.F., Johnson, R.K. and Hertzberg, R.P. (1991) Synthesis of water-soluble (aminoalkyl)camptothecin analogs: inhibition of topoisomerase I and antitumour activity. *Journal of Medicinal Chemistry* 34, 98–107.

Lavergne, O., Lesueur-Ginot, L., Rodas, F.P. and Bigg, D.C.H. (1997) BN 80245: an E-ring modified camptothecin with potent antiproliferative and topoisomerase I inhibitory activities. *Bioorganic & Medicinal Chemistry Letters* 7, 2235–2238.

Lavergne, O., Lesueur-Ginot, L., Rodas, F.P., Kasprzyk, P.G., Pommier, J., Demarquay, D., Prevost, G., Ulibarri, G., Rolland, A., Schiano-Liberatore, A.M., Harnett, J., Pons, D., Camara, J. and Bigg, D.C.H. (1998) Homocamptothecins: Synthesis and antitumour activity of novel E-ring-modified camptothecin analogs. *Journal of Medicinal Chemistry* 41, 5410–5419.

Legarza, K. and Yang, L.X. (2005) Novel camptothecin derivatives. *In Vivo* 19, 283–292.

Legarza, K. and Yang, L.X. (2006) New molecular mechanisms of action of camptothecin-type drugs. *Anticancer Research* 26, 3301–3305.

Li, Q.Y., Zu, Y.G., Shi, R.Z. and Yao, L.P. (2006) Review camptothecin: Current perspectives. *Current Medicinal Chemistry* 13, 2021–2039.

Liu, L.F., Desai, S.D., Li, T.K., Mao, Y., Sun, M. and Sim, S.P. (2000) Mechanism of action of camptothecin. *Annals of the New York Academy of Sciences* 922, 1–10.

Markman, M. (2005) Topotecan as second-line therapy for ovarian cancer: Dosage versus toxicity. *Oncologist* 10, 695–697.

Oberlies, N.H. and Kroll, D.J. (2004) Camptothecin and taxol: Historic achievements in natural products research. *Journal of Natural Products* 67, 129–135.

Ormrod, D. and Spencer, C.M. (1999) Topotecan: a review of its efficacy in small-cell lung cancer. *Drugs* 58, 533–551.

Pan, X.-W., Xu, H.-H., Xin, L., Gao, X. and Lu, Y.-T. (2004) Improvement of growth and camptothecin yield by altering nitrogen source supply in cell suspension cultures of *Camptotheca acuminata. Biotechnology Letters* 26, 1745–1748.

Pavillard, V., Charasson, V., Laroche-Clary, A., Soubeyran, I. and Robert, J. (2004) Cellular parameters predictive of the clinical response of colorectal cancers to irinotecan. A preliminary study. *Anticancer Research* 24, 579–585.

Puri, S.C., Verma, V., Amna, T., Qazi, G.N. and Spiteller, M. (2005) An endophytic fungus from *Nothapodytes foetida* that produces camptothecin. *Journal of Natural Products* 68, 1717–1719.

Soepenberg, O., Sparreboom, A. and Verweij, J. (2003) Clinical studies of camptothecin and derivatives. *The Alkaloids: Chemistry and Biology* 60, 1–50.

Srikantaramas, S., Asano, T., Sudo, H., Yamazaki, M., Saito, K. (2007) Camptothecin: Therapeutic potential and biotechnology. *Current Pharmaceutical Biotechnology* 8, 196–202.

Sriram, D., Yogeeswari, P., Thirumurugan, R. and Bal, T.R. (2005) Camptothecin and its analogues: a review on their chemotherapeutic potential. *Natural Product Research* 19, 393–341.

Subrahmanyam, D., Sarma, V.M., Venkateswarlu, A., Sastry, T.V.R.S., Kulakarni, A.P., Srinivasa, R.D. and Krishna Reddy, K.V.S.R. (1999a) In vitro cytotoxicity of 5-aminosubstituted 20(S)-camptothecins. Part 1. *Bioorganic & Medicinal Chemistry* 7, 2013–2020.

Subrahmanyam, D., Venkateswarlu, A., Rao, K.V., Sastry, T.V.R.S., Vandana, G. and Kumar, S.A. (1999b) Novel C-ring analogues of 20(S)-camptothecin. Part-2: synthesis and in vitro cytotoxicity of 5-C-substituted 20(S)-camptothecin analogues. *Bioorganic & Medicinal Chemistry Letters* 9, 1633–1638.

Subrahmanyam, D., Sarma, V.M., Venkateswarlu, A., Sastry, T.V., Srinivas, A.S., Krishna, C.V., Deevi, D.S., Kumar, S.A., Babu, M.J. and Damodaran, N.K. (2000) Novel C-ring analogues of 20(S)- camptothecin. Part 3: synthesis and their in vitro cytotoxicity of A-, B- and C-ring analogues. *Bioorganic & Medicinal Chemistry Letters* 10, 369–371.

Sudo, H., Yamakawa, T., Yamazaki, M., Aimi, N., Saito, K. (2002) Bioreactor production of camptothecin by hairy root cultures of *Ophiorrhiza pumila. Biotechnology Letters* 24, 359–363.

Sugimori, M., Ejima, A., Ohsuki, S., Matsumoto, K., Kawato, Y., Yasuoka, M., Tagawa, H. and Terasawa, H. (1994) Antitumour agents. VI. Synthesis and antitumour activity of ring A-, ring B-, and ring C-modified derivatives of camptothecin. *Heterocycles* 38, 81–94.

Tanizawa, A., Kohn, K.W., Kohlhagen, G., Leteurtre, F. and Pommier, Y. (1995) Differential stabilization of eukaryotic DNA topoisomerase I cleavable complexes by camptothecin derivatives. *Biochemistry* 34, 7200–7206.

Thomas, C.J., Rahier, N.J. and Hecht, S.M. (2004) Camptothecin: current perspectives. *Bioorganic & Medicinal Chemistry* 12, 1585–1604.

Uehling, D.E., Nanthakumar, S.S., Croom, D., Emerson, D.L., Leitner, P.P., Luzzio, M.J., McIntyre, G., Morton, B., Profeta, S., Sisco, J., Sternbach, D.D., Tong, W.Q., Vuong, A. and Besterman, J.M. (1995) Synthesis, topoisomerase I inhibitory activity, and *in vivo* evaluation of 11-azacamptothecin analogs. *Journal of Medicinal Chemistry* 38, 1106–1118.

Verma, R.P. and Hansch, C. (2007) Matrix metalloproteinases (MMPs): Chemical-biological functions and QSARs. *Bioorganic & Medicinal Chemistry* 15, 2223–2268.

Verma, R.P. and Hansch, C. (2009) Camptothecins: A SAR / QSAR Study. *Chemical Reviews* 109, 213–235.

Vladu, B., Woynarowski, J.M., Manikumar, G., Wani, M.C., Wall, M.E., von Hoff, D.D. and Wadkins, R.M. (2000) 7- And 10-substituted camptothecins: dependence of topoisomerase I-DNA cleavable complex formation and stability on the 7- and 10-substituents. *Molecular Pharmacology* 57, 243–251.

Wall, M.E., Wani, M.C., Cook, C.E., Palmer, K.H., McPhail, A.T. and Sim, G.A. (1966) Plant antitumour agents. I. The isolation and structure of camptothecin, a novel alkaloidal leukemia and tumor inhibitor from *Camptotheca acuminata. Journal of the American Chemical Society* 88, 3888–3890.

Wall, M.E., Wani, M.C., Nicholas, A.W., Manikumar, G., Tele, C., Moore, L., Truesdale, A., Leitner, P. and Besterman, J.M. (1993) Plant antitumour agents. 30. Synthesis and structure activity of novel camptothecin analogs. *Journal of Medicinal Chemistry* 36, 2689–2700.

Wang, C.-Y., Pan, X.-D., Liu, H.-Y., Fu, Z.-D., Wei, X.-Y. and Yang, L.-X. (2004) Synthesis and antitumour activity of 20-O-linked nitrogen-based camptothecin ester derivatives. *Bioorganic & Medicinal Chemistry* 12, 3657–3662.

Wani, M.C., Nicholas, A.W. and Wall, M.E. (1986) Plant antitumour agents. 23. Synthesis and antileukemic activity of camptothecin analogs. *Journal of Medicinal Chemistry* 29, 2358–2363.

Wani, M.C., Nicholas, A.W. and Wall, M.E. (1987b) Plant antitumour agents. 28. Resolution of a key tricyclic synthon, 5'(RS)-1,5-dioxo-5'-ethyl-5'-hydroxy-2'H,5'H,6'H-6'-oxopyrano[3',4'-f]Δ6,8-tetrahydroindolizine: total synthesis and antitumour activity of 20(S)- and 20(R)-camptothecin. *Journal of Medicinal Chemistry* 30, 2317–2319.

Wani, M.C., Nicholas, A.W., Manikumar, G. and Wall, M.E. (1987a) Plant antitumour agents. 25. Total synthesis and antileukemic activity of ring A substituted camptothecin analogues. Structure-activity correlations. *Journal of Medicinal Chemistry* 30, 1774–1779.

Watase, I., Sudo, H., Yamazaki, M. and Saito, K. (2004) Regeneration of transformed *Ophiorrhiza pumila* plants producing camptothecin. *Plant Biotechnology* 21, 337–342.

Yaegashi, T., Sawada, S., Nagata, H., Furuta, T., Yokokura, T. and Miyasaka, T. (1994) Synthesis and antitumour activity of 20(S)-camptothecin derivatives. A-ring-substituted 7-ethylcamptothecins and their E-ring-modified water-soluble derivatives. *Chemical & Pharmaceutical Bulletin* 42, 2518–2525.

Yamazaki, Y., Sudo, H., Yamazaki, M., Aimi, N. and Saito, K. (2003) Camptothecin biosynthetic genes in hairy roots of *Ophiorrhiza pumila*: Cloning, characterization and differential expression in tissues and by stress compounds. *Plant and Cell Physiology* 44, 395–403.

INDEX